Unit Root, Cointegration, Granger-Causality, Threshold Regression and Other Econometric Modeling with Economic and Financial Data

Bwo-Nung Huang and Chin-Wei Yang

美商EHGBooks微出版公司
www.EHGBooks.com

EHG Books Micro-Publishing Company, Inc. USA

Visit Amazon.com for ordering

Registered for US Copyright © 2018

First edition: 2018 by EHG Books

Manufactured in United States

Contact: info@EHGBooks.com

ISBN-13 : 978-1-62503-467-0

Table of Contents

Preface

Both deductive and inductive methods in scientific research have undergone significant changes since the beginning of the 20th century as sciences advance rapidly. Deductive method reached its pinnacle when Russel's paradox became popular in the field of mathematical logics. The famous barber's paradox illustrates the inevitable logical dilemma: the only barber in an isolated village does the following: (1) he cuts hair for those who do not cut their own and (2) does not cut hair for those who cut their own. Suppose 90 members in the village do not cut their hair (so barber cuts their hair) and 9 cut their own hair (thus barber does not cut their hair). The question is who cuts the barber's hair? The intrinsic contradiction arrives in either way. If barber cuts his own hair, then it contradicts the condition the barber cuts hair for those who do not cut their own hair: the barber cuts his own hair if he (the barber) does not cut his hair. On the other hand, if the barber does not cut his hair, then he (the barber) will cut his own hair, another contradiction. In either case, we seem to arrive at an inescapable contradiction: such a barber cannot possibly exist in the logical world.

Godel took a step further to show contradictions are intrinsically inevitable in his famous Incomplete Theorems. Let us start it by trying to prove the statement that "ghost exists" via valid arithmetic rules and true axioms. Suppose at halfway, we arrive at "that ghost does not exist is provable (which is quite acceptable to some of us)" with all correct logical steps and well-known and time-tested axioms. Assuming for one moment we reject the hypothesis that ghost exists and hence conclude ghost does not exists. The conclusion that "ghost does not exist "can clearly be translated into "that ghost does not exist is provable". However given the hypothesis is false, its proposition logically derived halfway (that ghost does not exist is provable) cannot be true because all the arithmetic rules and axioms are valid. As a result we reject the proposition derived halfway that "ghost does not exist is provable "so that we have "ghost does not exist is not provable or ghost exist is provable" because in a complete system we have only two possible outcomes: either ghost exists or does not exist or it is provable or not provable. In a nutshell, we have arrived at both that "ghost does not exist is provable" and "ghost exists is provable". Reader can find out when the hypothesis is supported, we have two contradictory propositions as well: that "ghost exists is provable" and that "ghost does not exist is provable". There is an intrinsic dilemma between consistency and completeness in the formal deductive logic.

On the other hand, as is commonly understood in inductive methods, one counter-example does not kill a theory. It is the p value that counts: the smaller the p value is, the more likely a null hypothesis is rejected. So those on the inductive track can enjoy fruit born in statistics. Rapid development in time series and panel data econometrics in recent years has rendered modeling in Economics, Finance and other Social Sciences much easier and results more reliable. However, like Godel's Incomplete Theorem in deductive methodology, there is a nearly insurmountable hurdle to overcome: most statistical results had limited interpretations. As a proverbial example, an ordinary regression that fits beautifully between numbers of new-born babies and numbers of storks arrived at a locality does not provide causality. No a priori causal direction can be given so we arrive at a dead end: do arrivals of storks cause new-born babies or vice versa? It tells a close association only, which does not really yield much useful information. One must wait for a new methodology to break through and this is exactly where Granger causality models come into play.

In an important paper in *Econometrica*, Professor C.W.J. Granger came up with a valid framework within which causality can be statistically tested (Granger, 1969). Regression analysis with variables in level or differenced forms must first be studied lest one may encounter spurious relationship (Granger and Newbold, 1974). As a consequence, concepts of unit root, cointegration vector and error correction term must be examined before the regression analysis (Engle and Granger, 1987) can be appropriately formulated. It has become a standard procedure nowadays. Our sincere thanks go to Professor C.W.J. Granger and other prominent statisticians, who have advanced the inductive methodology significantly where their contributions can never be over-emphasized.

The aim of the book is to provide hands-on modeling techniques which are virtually indispensable in Economics and Finance and other social sciences. The availability of statistical or econometric software (RATS, Eviews, Stata, Gauss, SAS, SPSS, R)has rendered empirical testing much more convenient. However many caveats are to be avoided which researchers frequently encounter. Numerous real world examples are provided especially in Economics and Finance. Most data are time series or panel data; one is cross-sectional. Note that all papers in this book were published in peer-reviewed journals This book is suitable to graduate or first year doctoral students who have some the familiarity with an econometric text such as Greene (2003).

We are grateful to late Professor C.W.J. Granger, the originator of the Granger Causality in the domain of inductive methodology. His encouragement motivated us to pursue this academic endeavor. His model has made nearly all the economics and finance theories readily testable. The first author would like to thank his late father- Mr. Hai-Nan Huang (黃海南) and mother Mrs. May-Chu Chang (張美珠) for their supports, his wife Wen-Chien Tsai (蔡雯倩), daughter Chun-Jean Huang (黃春靜) and son Brian Jason Huang (黃春浩) for their thoughtful consideration and for their loving and care. The second author would take this opportunity to thank his late parents- Mr. Chien Yang (楊健)and Mrs. Hanying Chang Yang （楊張涵英）- for their unconditional sacrifices, his wife Alice Shichia Tang Yang （唐錫嘉）and daughter Felicia Yang (楊天恩） Gorman for their unlimited patience and his granddaughter Fern Lindsey Gorman (楊長霏) for being so adorable. Without them, this book could not have been made possible.

I. Granger Causality Models in Mean and Variance

Granger causality has become perhaps the most important household names in econometrics and financial statistics due to its rigor and popularity. The workable concept of causality instead of association is the milestone that represents a major methodological breakthrough in the field of econometrics. It has fundamentally changed the way we model volatile financial and macroeconomic data. We no longer have to decide which variable is cause and which is effect. As such a clear causal relationship can be established statistically especially with time series data; that is, past records can be used to predict future outcome, but not vice versa. However, causal variables need to be in the same "wavelength", otherwise the relationship can be spurious as was demonstrated by Granger and Newbold in their famous 1974 paper in *Journal of Econometrics*.

Needless to say, the Granger causality model has a standard protocol. First, one needs to examine if the variables are stationary or I (0) via one of famed unit root tests before performing the cointegration test. If a variable is not stationary : I(1) or higher order I(k) , variables in first or kth difference is recommended . Perhaps the most widely used unit root test is the Dickey-Fuller or Augmented Dickey-Fuller (Dickey, Fuller 1979, 1981) or ADF test followed by its variant, the Phillips-Perron (Phillips and Perron, 1988) or PP test. According to Perron and Ng (*Review of Economic Studies*, Vol. 63, pp 435-465, 1996), PP test performs less well than the ADF test in finite samples. However, this applies to modeling economic or financial data. In general, the ADF and the PP tests seem to be comparable and complementary in Economics and Finance. Can the same be said about mathematical data?

As we examine the decimals of the famous pie (ratio of circumference/ diameter of a circle) to see if these never-ending and non-repeating decimals (two decimals as a unit for 200 units or 400 decimal places after the decimal point) have a unit root, it is interesting to find out that the ADF test do not reject the Ho of a unit root (p value of 0.2348) whereas the PP test resoundingly reject the Ho of a unit root (p value of 0.000). The contradictory results suggest that ADF and PP tests may very well complement each other in economic or financial data. But it may produce a very different conclusion in mathematical data. For this reason, we limit our discussion to economic and financial data.

Second, after the unit root test, one would like to examine if a cointegration vector exists between the dependent and independent variables. If a co-integration relationship exists, one needs to add a lagged error correction term - which reflects the speed of adjustment toward the equilibrium- into the regression model. Granger's causality approach has not only provided the causal directions, but also has enriched the estimation greatly when compared with comparative static models before. The estimated causal directions, if the relationship is deemed statistically significant, tells us either Y Granger-causes (leads) X or X Granger-causes (leads) Y or there is a two-way feedback relationship. It is little wonder that that Granger richly deserved the top honor in Economics: 2003 Nobel Prize winner along with Engel. The causality can be modeled in terms of means or variances. In this section we apply the Granger causality to different models to illustrate both the power and elegance of the causality model.

In the paper of "A Bivariate Causality between Stock Prices and Exchange Rate: Evidence from Recent Asian Flu " (Granger, Huang and Yang, 2000), we had the honor and opportunity to work with Professor Granger. The idea is to see how much information a Granger causality model can actually reveal during a tumultuous scenario: the Asian Flu when large investment funds raided and wreaked havoc on Asian financial markets. This paper is important in testing if the Granger causality model can truly describe and predict outcome satisfactorily in a time of major changes.

Sample period spans from January 3, 1986 to June 16, 1998, a total of 3247 observations of exchange rates and stock prices from Hong Kong, Indonesia, Japan, South Korea, Malaysia, the Philippines, Singapore, Thailand and Taiwan. We first take logarithmic transformation on stock prices and exchange rates to dampen the volatility. To take into consideration the structural break, we opted for the procedure by Zivot and Andrew (1992) for the unit root test. After taking co-integration relationships into consideration, we obtain the following results during the Asian financial debacle. For South Korea, change in the exchange rate was found to Granger-cause the stock price. For the Philippines, change in stock price was found to Granger-cause the exchange rate. For Hong Kong, Malaysia, Singapore, Thailand, and Taiwan, there existed feedback relationships. For both Indonesia and Japan, no recognizable patterns could be found.

The causality direction from the exchange to stock market in South Korea can readily be explained by the unique business structure of chaebols (huge business conglomerates) : their unusually high debt/equity ratio (6.4), which over-leveraged itself in capital investment denominated in foreign currencies. With a sizable current account deficit and overcapacity of ambitious chaebols, Moody downgraded its credit rating from A1 to A3 and then to B2. As a result, when baht of Thailand started its rapid devaluation, South Korea Central Bank did not have enough currency reserve to defend her won: thus it ushered in the beginning of the financial avalanche. The South Korean government begged her citizens to turn in gold items to be melted down in order to boost the reserve. As was well known in Korean, the propaganda was : by gathering dust, you can create a mountain. Korean government took in a total of 15 tons of gold and raised $2 billion. The patriotic movement barely made a dent as compared to the amount owed to foreigner investors.

On the other hand, the Philippines has the smallest debt/equity ratio (1.9) due to her unsatisfactory economic growth rate. The relatively poor ability to borrow or attract foreign capital into the Philippines might be viewed as a blessing in disguise. Nonetheless, the Philippines markets could not steer clear of the crisis. The causality direction was from the stock to the exchange market.

For Hong Kong, Malaysia, Singapore, Thailand and Taiwan, the feedback relationships began when the central bank of Thailand on May 4, 1997 defended baht unsuccessfully. Massive selling of local currencies caused a great deal of panic, which was transmitted to their economies. Bleak economic uncertainty led to significant layoffs of labor. To prevent local currency from depreciating further, central banks had to raise interest rates, which was in turn detrimental to business and stock markets as well. Hedge funds investors, at these perilous moments, took the opportunity to short local currencies to line their pockets and worsen the situation. The feedback loop continued until the IMF extended loans with conditions attached. The flow chart started from massive selling of Asian local currencies (exchange rate market) to raised interest rates, which would surely damage economies and depressed stock prices (stock market). Not having enough reserve account to defend domestic currency leads to the shorting of the local currency which in turn leads to even higher interest rates to prevent domestic currencies from depreciation further. Seemingly endless and vicious loop goes on and on.

The president of Indonesia, Suharto, refused the IMF rescue package until May 21, 1998 when he finally resigned. Our sample ends on June 16 and thus it failed to capture the causality direction: from the exchange rate (raids on rupiah) to stock market (severe recession). Japan had suffered from her stagnation for 23 years on June 12, 1997 at the beginning of the Asian Flu. Both US and Japan bought and boosted Yen to prevent it from further sliding. With the largest reserve, Japan did not experience the currency attack as other economies did. As a consequence, our paper did not capture a causal pattern for Japan.

The Granger causality model which takes unit root, cointegration into consideration faithfully explains the causal directions for the nine Asian economies. For South Korea, it is the exchange rate market that started the downfall: over-investment coupled with current account deficit made serving foreign debts nearly impossible. In the case of the Philippines, foreign debt is relatively minuscule. However, the poor performance of the economy was the culprit as was correctly predicted by the Granger causality model.

In both cases of Indonesia and Japan, our sample period ended too soon. As mentioned before, Suharto refused the loan from IMF until May 21, 1998 when he was forced to resign, 25 days before our sample ends. As a result, its economy was in constant shambles with riots and protests, and unfortunately they lasted for a long time. Both economies suffered from long recession and our limited sample period fails to capture such effect.

The rest economies exhibited the two-way causality direction indicating a loop : massive speculative selling of local currencies clearly caused panic because central banks did not have enough major reserve currencies to back up local money supply at fixed rates. As such they had to raise interest rate to defend their currencies, only to find out higher interest rate was detrimental to stock price. In the meantime, output reduction and hence lay-offs of labor force, more often than not, becomes necessary to accommodate to the rising uncertainty. Unfortunately low output and high unemployment provide a fertile ground for hedge fund investors to short local currencies, thus the vicious cycle goes on until the IMF stepped in with large loans and conditions attached to the debtor countries.

The paper " An Analysis of Factors Affect Price Volatility of the US Oil Market " utilized Generalized Autoregressive Conditional Heteroskedasticity(GARCH) model to illustrate that shocks to the conditional variance are indeed persistent : sum of alpha and beta exceeds one indicating that excessive price volatility gives rise to the uncertainty for both oil exporting and importing countries . The volatility is seeded in the intrinsically unstable demand for oil developed by Greenhut, Otha and Hwang (1974). Our results echo their theory which pinpoints the problematic section of a nearly vertical demand function.

In order to estimate the demand function for oil, we first employ the ADF test on logarithmic variables but fail to reject the unit root hypothesis. In addition there exists a long-term cointegration relationship and as such we included an error correction term to better estimate the demand. The price elasticity values were found to be between -0.272 and - 0.194. Demand for oil is indeed highly inelastic again indicating a great deal of future price volatility, hence the ensuing uncertainty in the world oil market.

The paper " Causality and Cointegration of Stock Market among the US, Japan, and the South China Growth Triangle " investigates the causality direction of the US, Japan, South China Growth Triangle (SCGT). The unit root by Zivot and Andrews (1992) that takes structural break into consideration is employed. No cointegration among the markets was found except for the Shanghai and Shenzhen markets. After taking non-synchronized trading problems into consideration, the paper finds the following interesting causality results.

First, change in stock price in the US Granger-caused that of South China Growth Triangle more than it did on Japan. Second, change in stock price in the US Granger-caused that of Taiwan and Hong Kong. Third, change in stock price in Hong Kong Granger-caused or led that of Taiwan by one day. Fourth, Shanghai and Shenzhen stock markets exhibited feedback relationship: two-way Granger causality perhaps due to geographical proximity in Southeast Asia. This paper illustrates the power of the Granger causality model even during the presence of volatile Asian stock markets.

The paper "Long-run Purchasing Power Parity Revisited : A Monte Carlo Simulation " examines a long-disputed economic phenomenon: does the long run purchasing power or PPP really exist ? Here we have an example where the economic theory and statistics dovetail nicely. The existence of a long-run PPP implies the existence of a cointegration vector of nominal exchange rates, domestic price and foreign prices. The question is reduced to testing empirically whether such a cointegration vector did exist.

We find that the Engle-Granger method has a tendency to reject the long-run PPP hypothesis. On the other hand, the Johansen model (1988) has a tendency to support the long run PPP hypothesis. Via Monte Carlo simulations, this paper finds the Johansen model has a bias toward supporting long-run PPP especially when the assumption of normally or/and independently and identically distributed disturbance terms is violated. For instance, first difference of exchange rates and consumer prices of the US, Switzerland and the UK are not normally distributed and they exhibit the ARCH phenomenon. The utility of using Monte Carlo simulations was clearly and specifically discussed.

The paper "Oil Price Movements and Stock Markets Revisited: A Case of Sector Stock Price Indexes in the G-7 Countries " used a Sup F procedure by Bai et al. (1998) to locate and test multiple structural breaks for the G-7 countries. It is found that oil price shock does not impact composite index in each of the G-7 countries. In addition stock price changes in the US, Germany and the UK leads oil price changes. Furthermore, 4 of 7 negative causal relationships were observed in Germany (1), the US(2) and France (1) respectively. Finally, oil price changes have direct impact on sector prices of information technology, consumer staples, financial, utilities and transportation industries.

The paper " Volatility of Changes in G-5 Exchange Rates and Its Market

Transmission Mechanism " applies two-stage causality in variance by Cheng and Ng (1996) to three major exchange rate markets, namely London, New York and Tokyo. The London and New York markets transmitted its volatility to or Granger-caused in terms of variance the Tokyo Market with the New York market led slightly the London market.

During the Euro Monetary System crisis, the frequency of both the volatility and the mutual feedback effects intensified greatly. The volatility spillover effect in London was found to lead the Tokyo market after the Asian financial debacle. It is to be pointed out that our causality-in-variance model predicted G-5 exchange rate better than the traditional ARMA model.

In the paper " Stock Market Integration —An Application of the Stochastic Permanent Breaks Model " Huang and Fok (2001) employed the stochastic permanent breaks model developed by Engel and Smith (1998) to investigate the concept of temporary cointegration between the US, Japan and 8 European markets. By temporarily co-Integrated markets, it means markets may not follow random walk and hence market inefficiency may be taken advantage of to make a profit. The Engle-Smith technique can identify otherwise unidentified cointegration relationships. The results indicated the US market was temporarily cointegrated with the markets of Japan, Germany, Netherland and Switzerland, whereas the standard Johansen cointegration model (1988) found the cointegration relationship only between the US and the Netherlands stock markets.

In their paper "State Dependent Correlation and Lead-Lag Relation when Volatility of Markets is Large: Evidence from the US and Asian Emerging Markets ", Huang et al. (1999) calculated correlation coefficients for both high and low volatility states. Then a Granger causality model was employed to detect the interaction between the US and the Asian emerging markets.

The switching ARCH or SWARCH model developed by Hamilton and Susmel (1994) was employed to calculate the filtered probability, then the asymmetrical state-dependent correlation coefficients were computed between the US and the Asian emerging markets. The result showed that there existed much greater uncertainty in the state of high volatility than that in the state of low volatility between the Asian and the US markets.

The last paper in this section combines both theoretical and empirical aspects of the unstable demand structure discovered by Greenhut, Hwang and Ohta (1974) more than 40 years ago. By unstable demand structure, we mean that price elasticity is expected to remain inelastic (absolute value less than one) and hence there is always a pressure for the price of the commodity to increase. The first part of the paper deals with the generalized theoretical framework in which supply is affected by the market price. As is well known, petroleum is price inelastic for it is essential in production process. As a necessary consequence, it is always to the seller's advantage to collude and raise the price.

To test the hypothesis of the unstable petroleum demand, the second part of the paper provides empirical estimation of the demand function. To do that, we first need to examine the stationarity of the variables: petroleum consumption, its price, income, price of coal and natural gas for the time period of 1949 to 1998. All variables are found to be non-stationary and as such authors took first difference in the ensuing estimation. To explore the co-integration relationship, the two-stage technique by Engle and Granger was used. The lagged error correction term is included in the estimation with R square of 0.71. The long-term price elasticity for the entire sample period is found to be -0.198 = -0.052/(1-0.7462), a value far less than unity. Short-term price elasticity before and after the energy crisis were found to be -0.1778 and - 0.0625 respectively: a 65% decrease after the 1975 energy crisis.

It is little wonder that demand for oil becomes more inelastic after the energy crisis. The inelastic demand by definition means price volatility is much greater than quantity volatility. The GARCH model indicates there was a persistency in oil price volatility in Japan as the sum of alpha and beta exceeds one. In the US, the volatility of high oil price leads negatively the stock price.

The QUARTERLY REVIEW
of ECONOMICS
And FINANCE

NORTH-HOLLAND   The Quarterly Review of Economics and Finance 40 (2000) 337–354

A bivariate causality between stock prices and exchange rates: evidence from recent Asian flu☆

Clive W.J. Granger[a], Bwo-Nung Huang[b],*, Chin-Wei Yang[c]

[a]Department of Economics, University of California, San Diego, La Jolla, CA 92093-0508, USA
[b]Department of Economics, National Chung-Cheng University, Chia-Yi, Taiwan 621
[c]Department of Economics, Clarion University of Pennsylvania, Clarion, PA 16214-1232, USA

Abstract

This paper applies recently developed unit root and cointegration models to determine the appropriate Granger relations between stock prices and exchange rates using recent Asian flu data. Via impulse response functions, it is found that data from South Korea are in agreement with the traditional approach. That is, exchange rates lead stock prices. On the other hand, data of the Philippines suggest the result expected under the portfolio approach: stock prices lead exchange rates with negative correlation. Data from Hong Kong, Malaysia, Singapore, Thailand, and Taiwan indicate strong feedback relations, whereas that of Indonesia and Japan fail to reveal any recognizable pattern. © 2000 Bureau of Economic and Business Research, University of Illinois. All rights reserved.

JEL classification: F300; G150

Keywords: Asian Flu; Bivariate Causality between Stock Prices and Exchange Rates

1. Introduction

The financial crisis sparked in Thailand in July 1997 has sent shock waves throughout southeast Asia, South Korea, and Japan. On October 27, the short-run interest rate in Hong Kong took a huge jump to maintain its exchange rate to the US dollar. As a result, the Hang Seng Index plummeted 1438 points, setting off the crash in the US with the Dow Jones

☆ Financial assistance from the National Science Council of Taiwan (NSC88-2415-H-194-002) and Fulbright Scholarship Program is gratefully acknowledged. This paper is written during my tenure as a visiting scholar at UCSD. We are grateful to two anonymous referees for valuable suggestions. However, we wish to absorb all the culpability.
  * Corresponding author. *E-mail address:* ecdbnh@ccunix.ccu.edu.tw (B.-N. Huang).

Industrial Averages down by 554.26 points. On November 24, Yamaichi—the fourth largest financial corporation—filed for bankruptcy, which gave rise to a 854-point drop in Nikkei Index. In mid-December, the Korean Won depreciated drastically from 888 wons (per dollar) on July 1 to more than 2000 wons. The currency crisis in South Korea set off a financial avalanche in its stock markets, which witnessed a 50.3% freefall. Similar debacles also occurred in other Asian markets. After the devaluation of the Thai Baht (July 1997), other southeast Asian currencies over the next 2 months abandoned their close links to the US dollar and began to depreciate. The most severe pressures in foreign exchange markets, in the third quarter of 1997, were experienced by Thailand, Malaysia, the Philippines, and Indonesia, but the currencies of Singapore and a number of other Asian countries also weakened. As the pressure spread to Hong Kong and Korea in late October after the depreciation of the New Taiwan dollar, the scale of the crisis worsened significantly. The financial tsunami continued to exert its devastating force and did not slow down until the first quarter of 1998. This is known as the Asian flu in which stock prices and exchange rates plunged in tandem. Turmoils in both currency and stock markets are of paramount interest: does currency depreciation lead to stock market downfall or vice versa? Such a study provides an opportunity for analyzing dynamics between stock prices and exchange rates. If a statistical lead-lag relation can be ascertained, practitioners can profit from the arbitrage especially during a severe financial crisis.

The bridge between the exchange rate and the asset market equilibrium is the interest parity condition according to Krugman and Obstfeld, (1997, Chapt. 16): $R = R^* + (E^e - E)/E$, where R and R^* are interest rates of domestic and foreign currencies and E and E^e denote exchange rate and expected future exchange rate respectively. For asset markets to remain in equilibrium, ceteris paribus, a decrease in domestic output (hence lower R due to reduced demand for money) must be accompanied by a currency depreciation (a greater value for E). Aggarwal (1981) has argued that a change in exchange rates could change stock prices of multinational firms directly and those of the domestic firms indirectly. In the case of a multinational firm, a change in exchange rate will change the value of that firm's foreign operation, which will be reflected on its balance sheet as a profit or a loss. Consequently it contributes to current account imbalance. Once the profit or a loss is announced, that firm's stock price will change. This argument shows that devaluation could either raise or lower a firm's stock price depending on whether that firm is an exporting firm or it is a heavy user of imported inputs. If it involves in both activities, its stock price could move in either direction. This is true especially when most stock prices are aggregated to investigate the effects of devaluation on stock markets. From this viewpoint, exchange rate change is expected to give rise to stock price change. Such a causal relation is known as the traditional approach.

However, as capital market become more and more integrated, changes in stock prices and exchange rates may reflect more of capital movement than current account imbalance. The central point of such a portfolio approach lies in the following logical deductions: A decrease in stock prices causes a reduction in the wealth of domestic investors, which in turn leads to a lower demand for money with ensuing lower interest rates. The lower interest rates encourage capital outflows ceteris paribus, which in turn is the cause of currency depreciation.[1] Under the assumption of the portfolio approach, stock price is expected to lead exchange rate with a negative correlation. If a market is subject to the influences of both approaches simultaneously, a feedback loop will prevail with an arbitrary sign of correlation between the two variables.

In the case of the US markets, empirical examinations of relation between stock returns and exchange rate changes have largely provided only weak evidence on the following relationship: exchange rate changes are significantly related to firm or portfolio stock returns (Jorion, 1990; Bahmani–Oskooee and Sohrabian, 1992; Amihud, 1993; Bartov and Badnar, 1994). Similar tests are performed by Bondnar and Gentry (1993) for Japan and Canada. In the case of emerging markets, only skimpy literature is available. For instance, the result of Abdalla and Murinde (1997) supports the traditional approach: exchange rate changes lead stock returns. However, their study includes only India, South Korea, Pakistan, and the Philippines. The recent Asian flu raises a rather puzzling question: in the wake of depressing stock and currency markets, which one is the culprit? The majority of previous studies support the traditional approach: exchange rate change leads stock return. Only a few indicate that opposite direction holds true or the feedback phenomenon exists. But they do not discuss the sign between exchange rate and stock price. Well-known in the literature, using monthly data may not be adequate in describing the effect of capital movement, which is intrinsically a short-run occurrence. In this paper, we employ daily data of the emerging markets in the hope that it will capture such effects. To further the analysis, we employ impulse response functions to explore the relation (signs) and its dynamics. In addition, we apply more advanced unit root and cointegration techniques, which account for structural break (for example, the Asian flu), to avoid the potential estimation bias. The organization of the paper is as follows: Section 2 describes empirical models; Section 3 introduces the unit root and cointegration models with structural break; Section 4 discusses the empirical result; Section 5 investigates some related problem of the model; and Section 6 provides a conclusion.

2. Discussion of empirical models

Since (Nelson and Plosser's (1982)), well-known paper, the unit-root property of macroeconomic variables has been widely accepted. As such, a unit-root test is often necessary before empirical studies. Based on the result by Dickey and Fuller (1979), the Augmented Dickey and Fuller (ADF) test is generally employed as shown below:

$$\Delta y_t = \alpha + \beta t + (\rho - 1)y_{t-1} + \sum_{i=1}^{k-1} \theta_i \Delta y_{t-i} + a_t \tag{1}$$

where $\Delta = 1 - L$; y_t is a macroeconomic variable such as exchange rate or stock price; t is a trend variable; and a_t is a white noise term. The null hypothesis is H_0: $\rho = 1$ and y_t is said to possess the unit root property if one fails to reject H_0.

Nevertheless, the ADF test is suspect when the sample period includes some major events (for example, great depression, oil shocks). Failure to consider it properly can lead to erroneous conclusions in the case when the null is not rejected. To circumvent this problem,

Perron and Vogelsang (1992) introduce dummy variable into Eq. (1) and recalculate the new set of critical values. However, as pointed out by Zivot and Andrew (1992, p. 251), a skeptic of Perron's approach would argue that his choices of breakpoints are based on prior observation of the data, and hence problems associated with 'pre-testing' are applicable to his method.[2] Consequently, they introduce an alternative formulation to overcome the pre-testing problems.[3]

$$\Delta y_t = \alpha + \beta t + (\rho - 1)y_{t-1} + \gamma DU_t(\lambda) + \sum_{i=1}^{k-1} \theta_i \Delta y_{t-i} + a_t \tag{2}$$

where $DU_t(\lambda) = 1$ for $t > T\lambda$, otherwise $DU_t(\lambda) = 0$; $\lambda = T_B/T$ represents the location where the structural break lies; T is sample size; and T_B is the date when the structural break occurred. Evident from Eq. (2), the estimation result hinges critically on the value λ as well.

One of better ways to test the existence of a unit root is to choose the breakpoint that gives the least favorable result for the null of H_0: $\rho = 1$ using the test statistics $t_{\hat\rho}$ (λ). That is, λ is chosen to minimize the one-sided t statistic for testing $\rho = 1$, when small values of the statistic lead to the rejection of the null. Let $\hat\lambda_{\inf}$ denote such a minimizing value for the model. Then, by definition, $t_{\hat\rho}[\hat\lambda_{\inf}] = \inf_{\lambda \in \Lambda} (\lambda)$ where Λ is a specified closed subset of (0, 1) (Zivot and Andrew, 1992, p. 254). As is shown in this paper, a set of simulated critical values is employed for hypothesis testing during the period of the Asian flu. To investigate the stationarity assumption of several I(1) variables, majority of academicians still rely on the widely-accepted and easy-to-apply model proposed by Engle and Granger (1987) despite its normalization problem. Just as the ADF model fails to consider problems associated with structural breaks, the Engle–Granger formulation bypasses the same difficulty. Applying the similar approach by Zivot and Andrews (1992), Gregory and Hansen (1996) revise the Engle and Granger (1987) model to consider the regime shift via residual-based cointegration technique. The Gregory and Hansen model is a two-stage estimation process of which the first step is to estimate the following multiple regression:

$$y_{1t} = \alpha + \beta t + \gamma DU_t(\lambda) + \theta_1 y_{2t} + e_t \tag{3}$$

in which y_{1t} and y_{2t} are of I(1) and y_{2t} is a variable or a set of variables; and $DU_t(\lambda)$ has the same definition as that in Eq. (2). The second step is to test if e_t in Eq. (3) is of I(0) or I(1) via the ADF or Phillips–Perron technique. If e_t is found to be consistent with I(0), one may claim that cointegration exist between y_{1t} and y_{2t}. Once the statistical property of e_t is established, one may adopt the bivariate VAR model to test the Granger causality. If the cointegration does not exist, the following formulation is needed in testing the hypotheses:

$$\Delta y_{1t} = \alpha_0 + \sum_{i=1}^{k} \alpha_{1i} \Delta y_{1t-i} + \sum_{i=1}^{k} \alpha_{2i} \Delta y_{2t-i} + \varepsilon_{1t}$$

$$\tag{4}$$

$$\Delta y_{2t} = \beta_0 + \sum_{i=1}^{k} \beta_{1i} \Delta y_{1t-i} + \sum_{i=1}^{k} \beta_{2i} \Delta y_{2t-i} + \varepsilon_{2t}$$

in which y_{1t} and y_{2t} represent stock prices and exchange rates. Failing to reject the H_0: $\alpha_{21} = \alpha_{22} = \cdots = \alpha_{2k} = 0$ implies that exchange rates do not Granger cause stock prices. Likewise, failing to reject the H_0: $\beta_{11} = \beta_{12} = \cdots = \beta_{1k} = 0$ suggests that stock prices do not Granger cause exchange rates. If cointegration exists between y_{1t} and y_{2t}, an error correction term is required in testing Granger causality as shown below:

$$\Delta y_{1t} = \alpha_0 + \delta_1(y_{1t-1} - \gamma y_{2t-1}) + \sum_{i=1}^{k} \alpha_{1i}\Delta y_{1t-i} + \sum_{i=1}^{k} \alpha_{2i}\Delta y_{2t-i} + \varepsilon_{1t}$$

(5)

$$\Delta y_{2t} = \beta_0 + \delta_2(y_{1t-1} - \gamma y_{2t-1}) + \sum_{i=1}^{k} \beta_{1i}\Delta y_{1t-i} + \sum_{i=1}^{k} \beta_{2i}\Delta y_{2t-i} + \varepsilon_{2t}$$

in which δ_1 and δ_2 denote speeds of adjustment. According to Engle and Granger (1987), the existence of the cointegration implies a causality among the set of variables as manifested by $|\delta_1| + |\delta_2| > 0$. Failing to reject the H_0: $\alpha_{21} = \alpha_{22} = \cdots \alpha_{2k} = 0$ *and* $\delta_1 = 0$ implies that exchange rates do not Granger cause stock prices while failing to reject H_0: $\beta_{11} = \beta_{12} = \cdots = \beta_{1k} = 0$ *and* $\delta_2 = 0$ indicates stock prices do not Granger cause exchange rates.

3. Data, time series trends, unit root, and cointegration results

Included in this sample study are exchange rates and stock prices from Hong Kong (HKN), Indonesia (IND), Japan (JPN), South Korea (KOA), Malaysia (MAL), the Philippines (PHI), Singapore (SIG), Thailand (THA), and Taiwan (TWN). The sample period starts from January 3, 1986 to June 16, 1998. Daily data (5 days a week) in total of 3247 observations are from Datastream.[4] Three subperiods are used to better dissect the relations between exchanges and stock prices. Period 1 (1987-Crash) covered from January 3, 1986 to November 30, 1987; Period 2 (After Crash) started on December 1, 1987 and ended on May 31, 1997; Period 3 (the Asian-Flu Period) continued from June 1, 1997 through June 16, 1998.[5] All data points are transformed into logarithmic scale and are shown in time series plots (see Fig. 1).

Evident from Fig. 1, all the nine economies exhibit pronounced structural breaks. That is, barring the Hong Kong dollar that maintained its peg to the US dollar, all other eight currencies suffered noticeable depreciations since July, 1997. During the period of July 1 ~ June 16, 1998, the Rupiah of Indonesia experienced the greatest slide in its value (144.83%), followed by Thai Bhat (55.50%), Won of South Korea (46.73%), Ringgit of Malaysia (45.82%), and Peso of the Philippines (45.52%). The rest of currencies witnessed between 18% and ~23% depreciation (Table 1). Similar freefalls in stock prices were witnessed ranging from 19.8% of the Taiwan market to 84.9% of the Malaysian market. As the Asian flu worsened, the performance of Pacific funds outside Japan fell 35.5% on average. Because the existence of the structural break is very clear from both Table 1 and Fig. 1, a revised model is necessary to explore the subtle relations between exchange rates and stock prices.

To account for the structural change, we employ the Zivot and Andrew (1992) model to test the unit root property for the nine economies, and the result is reported in Table 2.

15

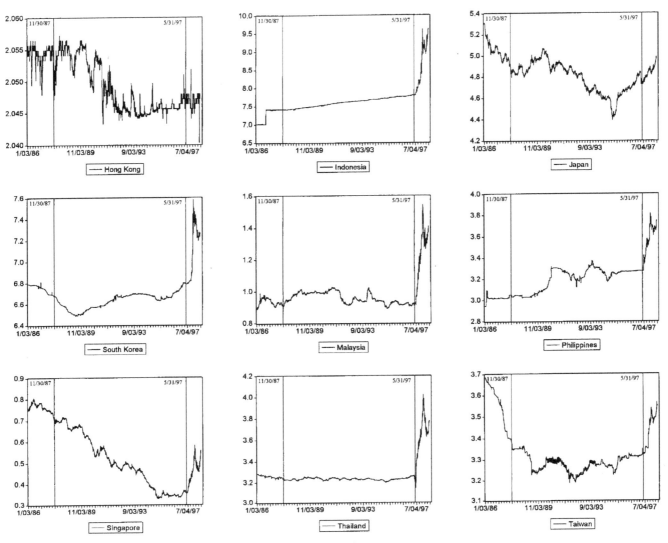

Fig. 1a. Time Series of the Nine Asian Exchange Rates (in Logrithm)

Table 1
A comparison of daily stock prices and exchange rates between July 1 and June 16, 1998

Exchange rate	HKN	IND	JPN	KOA	MAL	PHI	SIG	THA	TWN
07/01/97	7.75	2431.32	114.90	888.10	2.52	26.37	1.43	24.45	27.81
06/16/98	7.75	15197.57	143.33	1429.59	4.02	41.91	1.73	43.23	34.82
% Change	−0.05%	144.83%	22.02%	46.73%	45.82%	45.52%	18.82%	55.50%	22.38%
Stock index	HKN	IND	JPN	KOA	MAL	PHI	SIG	THA	TWN
07/01/97	15196.79	731.61	20175.52	758.03	1078.90	2815.54	494.00	527.28	9030.28
06/16/98	7526.45	406.50	14720.38	380.91	435.84	1726.55	289.04	263.39	7404.27
% Change	−67.51%	−57.13%	−31.27%	−66.22%	−84.91%	−47.95%	−52.35%	−66.75%	−19.79%

HKN, Hong Kong; IND, Indonesia; JPN, Japan; KOA, Korea; MAL, Malaysia; PHI, Philippines; SIG, Singapore; THA, Thailand; TWN, Taiwan. All prices are based on daily market close; and all exchange rates are expressed as number of local currencies per US dollar.

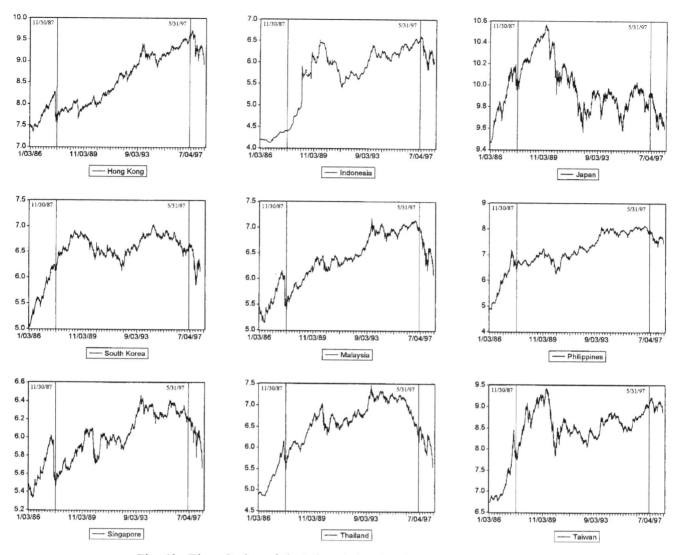

Fig. 1b. Time Series of the Nine Asian Stock Indexes (in Logrithm)

Table 2
Zivot and Andrew unit root results

	HKN	IND	JPN	KOA	MAL	PHI	SIG	THA	TWN
Exchange	−7.28*	−6.77	−2.47	−4.98*	−6.38*	−4.21**	−3.10	−11.0*	−2.95
rate	[0.43]	[0.98]	[0.78]	[0.97]	[0.95]	[0.94]	[0.94]	[0.94]	[0.97]
Stock	−3.25	−3.17	−3.65	−3.18	−3.08	−3.47	−2.40	−1.82	−3.36
index	[0.96]	[0.17]	[0.36]	[0.95]	[0.95]	[0.92]	[0.95]	[0.88]	[0.06]

Note: The estimation result is based on $\Delta y_t = \alpha + \beta t + (\rho - 1)y_{t-1} + \gamma DU_t(\lambda) + \sum_{i=1}^{k-1} \theta_i \Delta y_{t-i} + a_t$, where $DU_t(\lambda) = 1$ for $t > T\lambda$; otherwise, $DU_t(\lambda) = 0$. $\lambda = T_B/T$ denotes the location of the structural break, T = sample size, T_B = date of structural break; * = 1% significance level; critical value for various λ are from Table 3 provided by Zivot and Andrew (1992). Value of λ are in brackets. For λ = 0.1, 0.2, 0.3, 0.4, 0.5, 0.6, 0.7, 0.8, 0.9, the 5% critical values are −3.65, −3.80, −3.87, −3.94, −3.96, −3.95, +3.95, −3.82, −3.68, respectively.

Table 3
Results of Gregory and Hansen cointegration test

	JPN	SIG	TWN
y_{1t} on y_{2t}	−3.08	−4.65	−3.00
y_{2t} on y_{1t}	−3.59	−3.11	−3.17

Note: The first-stage estimation is based on $y_{1t} = \alpha + \beta t + \gamma DU_t(\lambda) + \theta_1 y_{2t} + e_t$. The second-stage estimation is to apply traditional ADF or Phillips–Perron approach to test if e_t is of I(0) or I(1). The y_{1t} on the first row represents exchange rate and y_{2t} denotes stock price. The reverse is true for the second row. The critical values are taken from Table 1 in Gregory and Hansen (1996).

A perusal of Table 2 reveals that the null hypothesis of a unit root in stock indexes cannot be rejected using the Zivot and Andrew approach. Nonetheless, the same cannot be said about exchange rate markets.[6] In particular, the null hypotheses of a unit root are easily rejected for the markets of Hong Kong (Dollar), Indonesia (Rupiah), Korea (Won), Malaysia (Ringgit), the Philippines (Peso), and Thai (Bhat). The prevalence of lacking a unit root may be attributed to the fact that these currencies are pegged to the US dollar before the crisis.

The lack of the unit root property within the sample periods enables us to apply the statistical model readily in terms of exchange rates and rate of change in stock price. Note that from prior studies, rate of change in exchange rates was found to mirror exchange rate exposure, hence it may have better economic interpretation. It is to be pointed out that the cointegration analysis is not needed for the time periods in which the logarithmic exchange rates are of I(0). Given the outcome of the unit root test, we present the cointegration results based on the Gregory and Hansen model (1996) in Table 3.

The first line in Table 3 lists the statistics when y_{1t} (stock price) is regressed on y_{2t} (exchange rate) whereas the statistics of the reverse-order regressions are reported in the second line. In all cases of y_{2t} = stock prices, y_{1t} = exchange rates, we fail to reject the null hypothesis that no cointegration exists between exchange rates and stock prices.[7] This being the case, we employ Eq. (4) to explore the relations between exchange rates and stock prices in the Asian markets.

4. Discussion of empirical results

Despite the structural break in these Asian markets, traditional Granger causality test or Eq. (4) would suffice for studying the relations between exchange rates and stock prices. Based on the Schwarz Bayesian (SB) criterion, the optimum lag (k), except for Hong Kong (k = 3), is one for all other markets. The causality results for the three sub-periods are reported in Table 4.

During Period 1 (Jan. 3, 1986 ~ Nov. 30, 1987), there existed little interaction between currency and stock markets except for Singapore. An examination indicates changes in the exchange rates lead that of the stock prices in Singapore. Our result is in agreement with the majority of prior studies using monthly data: only a few found either exchange rates lead stock prices or there is no interaction.

Table 4
The causality test between changes in exchange rates and changes in stock prices for the nine Asian economics

H_0: exch -/→ stock	86.13 ~ 87.11.30 (1)		87.12.1 ~ 97.5.31 (2)		97.6.1 ~ 98.6.16 (3)		97.6.1 ~ 98.16[a] (4)		97.6.1. ~ 98.6.16[b] (5)		97.6.1 ~ 98.6.16[c] (6)		97.6.1 ~ 98.6.16[d] (7)	
	F Value	p Value	F Value	p Value	F Value	p Value	F Value	p Value	F Value	p Value	F Value	p Value	F Value	p Value
SHKN -/→ HKN	0.0082	0.9281	2.2231***	0.0835	3.3090**	0.0700	3.7826***	0.0528	3.8322**	0.0513	3.3467***	0.0684	3.4998***	0.0625
HKN -/→ SHKN	0.2504	0.6170	2.0227	0.1086	1.3396	0.2481	0.9360	0.3342	2.3427	0.1271	1.4824	0.2245	1.3413	0.2478
SIND -/→ IND	0.2249	0.6356	0.1163	0.7331	0.7344	0.3922	0.7351	0.3920	0.7278	0.3943	0.7388	0.3908	0.7445	0.3890
IND -/→ SIND	0.1855	0.6669	0.0001	0.9916	1.2347	0.2675	0.7009	0.4032	1.2733	0.2602	1.2338	0.2677	1.5493	0.2143
SJPN -/→ JPN	1.4411	0.2305	1.1606	0.2814	0.1392	0.7094	0.2729	0.6018	0.0887	0.7661	0.1196	0.7297	0.1430	0.7056
JPN -/→ SJPN	1.9373	0.1646	0.0010	0.9745	1.3756	0.2419	2.6785***	0.1029	1.3870	0.2399	1.3440	0.2474	1.5992	0.2071
SKOA -/→ KOA	1.0895	0.2971	0.1952	0.6587	0.0008	0.9781	0.0019	0.9650	0.0057	0.9399	0.0045	0.9466	0.0003	0.9872
KOA -/→ SKOA	0.1104	0.7399	0.0600	0.8065	17.8604*	0.0000	18.8471*	0.0000	17.9082*	0.0000	17.7958*	0.0000	19.0854*	0.0000
SMAL -/→ MAL	0.2451	0.6208	0.6395	0.4240	2.6366***	0.1056	2.5562	0.1110	2.6290***	0.1061	2.6469***	0.1049	2.7164***	0.1005
MAL -/→ SMAL	0.1948	0.6591	0.3777	0.5389	3.6616**	0.0567	2.5767	0.1096	3.6450***	0.0573	3.6592**	0.0568	3.7230***	0.0547
SPHI -/→ PHI	0.0099	0.9209	0.8305	0.3622	2.5880	0.1088	2.8004***	0.0954	2.6042	0.1078	2.2793	0.1323	2.6089	0.1074
PHI -/→ SPHI	0.0222	0.8816	1.0546	0.3046	1.2832	0.2583	1.2201	0.2703	1.6981	0.1937	1.3077	0.2538	1.3863	0.2401
SSIG -/→ SIG	1.3964	0.2379	1.4630	0.2266	5.2427**	0.0228	5.2784**	0.0224	5.0330**	0.0257	5.2446**	0.0228	6.0148**	0.0148
SIG -/→ SSIG	4.7670**	0.0295	2.7911***	0.0949	13.5657*	0.0003	12.2910*	0.0005	13.8289*	0.0002	12.0027*	0.0006	13.5391	0.0003
STHA -/→ THA	0.1370	0.7115	0.4493	0.5027	6.8189*	0.0095	6.6287**	0.0106	6.7110*	0.0101	6.6796*	0.0103	7.0278*	0.0085
THA -/→ STHA	0.8512	0.3567	0.9862	0.3208	3.7305*	0.0545	3.8431*	0.0510	3.8607*	0.0505	3.7794**	0.0529	3.6829**	0.0560
STWN -/→ TWN	1.7295	0.1891	4.2392**	0.0396	10.9414*	0.0011	11.8221*	0.0007	11.0286*	0.0010	11.1004*	0.0010	10.8029*	0.0011
TWN -/→ STWN	0.9534	0.3293	0.1995	0.6552	7.4313*	0.0068	7.1519*	0.0079	7.4039*	0.0069	7.6613*	0.0060	7.7743*	0.0057

Note: -/→ implies does not Granger cause. * = 1% significance level; ** = 5% significance level; *** = 10% significance level.
The abbreviation for each market denotes exchange rate variations (e.g. HKN = rate of changes of the Hong Kong dollar or simply Hong Kong exchange rate variations) of the market. The prefix S represents rate of changes in stock prices of the market (e.g. SHKN = rate of changes of Hong Kong stock prices or simply stock prices changes of the Hong Kong market).

[a] Changes in the U.S. stock index are included.
[b] Federal fund rates are included.
[c] Difference of federal fund rates is included.
[d] Error correction term $s_{t-1} - e_{t-1}$ is included where s_t: logrithmic stock prices, e_t: logrithmic exchange rates.

In Period 2 (or After-Crash period), there is no definitive pattern of interaction between the two markets. As is shown in Table 4, changes in exchange rates lead that in stock prices in the case of Singapore ($\alpha \geq 10\%$) again. Additionally, changes in stock prices are found to lead exchange rate changes in Taiwan ($\alpha < 5\%$), and in Hong Kong ($\alpha \geq 10\%$).

Known for its volatility, the Taiwan stock market reached its peak at 12 000 points in February 1990, and fell to as low as 4000 points in the ensuing recession. It rebounded again back to about 10 000 points, and fell to 7100 points in the midst of the Asian flu. The exchange rate of the Taiwan dollar, built on large foreign exchange surplus from export-led economic growth, had remained fairly stable (at approximately 38 to 40 dollars per US dollar) before 1988. Since then, its currency began to appreciate in value owing to the pressure from the US. The appreciation took place gradually so to minimize the impact on firms. Such a policy was indeed considered suitable to alleviate adverse effects for domestic manufacturers, but served as an open invitation to "hot money" that flowed into Taiwan's financial market. The overflow of the hot money inevitably led to the overheated stock market and overvalued prices. In addition, the overflow of the hot money noticeably increased the demand for the Taiwan dollar that in turn exerted the pressure for its currency appreciation. Fig. 2 illustrates a picture of short-term capital inflows into Southeast Asia for the past 10 years. In the first part of 1990s, before the Asian financial crisis, capital inflows were on the rise. The increased capital inflows strengthen the local currency such as the New Taiwan dollar hence its stock prices, as is anticipated under the flexible exchange rate. However, the capital flight occurred in the fourth quarter of 1994, when China lobbed missiles around the island. Consequently, the stock market experienced severe setbacks, which triggered currency depreciation. This phenomenon is consistent with the portfolio approach.

In contrast, Singapore did not experience significant amount of capital inflows. And with her current account in black frequently, its export-led economy, leads the stock market (see Fig. 3). This seems to be in agreement with the traditional approach.

No definitive pattern has been identified between currency and stock markets according to the previous literature. However, such is not the case during Period 3 (the Asian flu period). Based on our analysis, seven of the nine nations suggest significant relations between the two markets. In the case of South Korea, changes in the exchange rates lead that in stock prices.[8] The reverse direction is found in Hong Kong and the Philippines. The rest of the markets (Malaysia, Singapore, Thailand, and Taiwan) are characterized by feedback interactions in which change in exchange rates can take the lead and vice versa. Given the definitive patterns, the Granger causality test does not provide signs of these relations. To examine the short run dynamic relations of the Asian markets during Period 3, we take advantage of the impulse response (IR) functions and calculate standard deviation for each forecasting period.[9] The IR (10 periods) from shocks of each variable are shown in Table 5.

An inspection of Table 5 reveals that the results from the IR analysis are in conformity with that of the Granger causality test. This is to say, if the Granger causality test indicates change in stock prices leads that in exchange rates, the responses of stock prices from one-unit shock of exchange rates are insignificant. Such is the case for Hong Kong and the Philippines as shown in lower part of Table 5 (panel B). Overall, three distinguishing patterns can be identified from Table 5 from the IR analysis of the nine economies in Asia.

First, one-unit shocks from the exchange rates have very discernible negative responses on

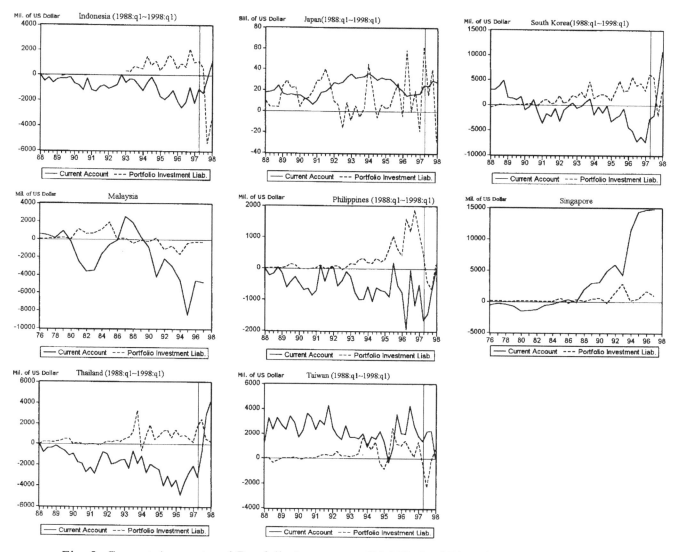

Fig. 2. Current Account and Portfolio Investment (Liability) of The Eight Asian Economies

their corresponding stock prices. This is clearly the case for South Korea. The IR analysis provides further short-run dynamic relations (sign and timing) above and beyond that obtained from the Granger causality test. For instance, it takes Korean stocks within the first 3 days to respond (negatively). The negative responses are found to be in Day 1, 2, and 3 after the one-unit shock was administered in Korea. The IR analysis for Korea confirms the validity of the traditional approach. According to the traditional approach, exchange rate leads stock price, but the correlation can go either way depending on whether the individual firm is an exporter (output) or an importer (raw material). The net effect of the aggregation (stock index) cannot be determined hence the sign is arbitrary.

In the Hong Kong and the Philippines markets, one standard deviation decrease in stock prices gave rise to the depreciation of the Hong Kong dollar and Philippines Peso in the first day and first 2 days, respectively. In the markets of Malaysia, Singapore, Thailand, and Taiwan, the feedback relations prevailed. The portfolio approach (stock price leads exchange rate) suggests a negative correlation while the traditional approach (exchange rate leads stock price) implies an arbitrary correlation. As a result, the net correlation under the feedback relation—a simultaneous combination of the two approaches—is uncertain. Barring Thai-

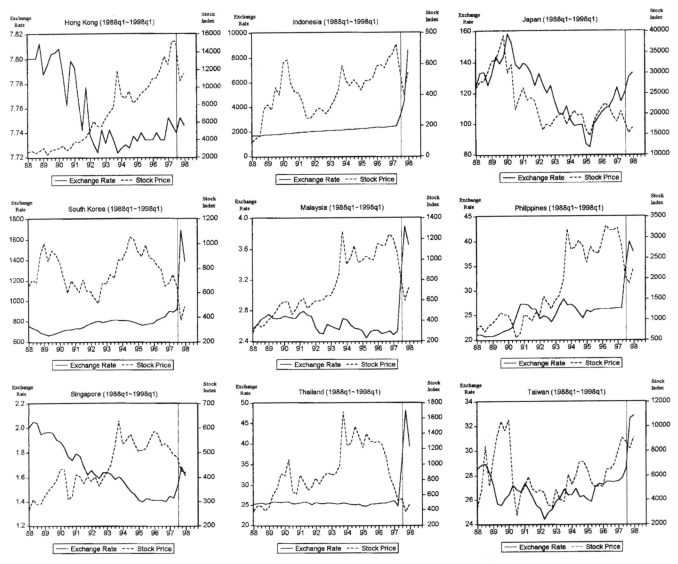

Fig. 3. Exchange Rates and Stock Indexes of the Nine Asian Economies (Quarterly Data)

land, one unit change in exchange rates has largely negative impacts on stock prices with the length of two periods (for example, Malaysia, Singapore, and Taiwan as shown in Table 5). In the case of one unit change in the stock prices, we observe the negative impacts on the corresponding exchange rates: Thai Baht (three periods), Taiwan Dollar (two periods), Malaysian Ringgit (one period), and Singapore Dollar (one period).

The result of our analysis indicates that (1) there existed no relation between the stock and the exchange rate markets for Japan and Indonesia; (2) the exchange rate led the stock price in Korea; and (3) stock prices led exchange rates to some extent for the rest of the markets. Given these results, capital flow seems to be a major cause for currency depreciations. The capital flight from Mexico in the Peso crisis (1994) went to equity markets of southeast Asia (emerging markets) high growth area. The enormous amount of capital inflow, however, does not cause exchange rates to change as most of local currencies are pegged to the US dollar. This explains why we do not observe the phenomenon consistent with the portfolio approach in Period 2. Unfortunately, trade deficit as well as inefficient banking system brought about massive capital outflow during the crisis period. Central banks, unable to

Table 5
Estimation result of impulse response function

Panel A: response of exchange rates from one-unit shock in stock price change

Period	HKN	KOA	MAL	PHI	SIG	THA	TWN
1	−0.0071***	0.0000	0.0000	−0.2139*	0.0000	−0.2073***	−0.0853*
2	−0.0060	0.0054	−0.1729***	−0.1867**	−0.0978*	−0.3353*	−0.1321*
3	0.0032	0.0014	−0.0130	−0.0659	0.0005	−0.0896*	−0.0170
4	−0.0014	0.0003	−0.0034	−0.0201	−0.0040	−0.0160	−0.0065
5	0.0006	0.0001	−0.0004	−0.0059	0.0000	−0.0020	−0.0013
6	−0.0003	0.0000	−0.0001	−0.0017	−0.0002	−0.0001	−0.0004
7	0.0001	0.0000	0.0000	−0.0005	0.0000	0.0000	−0.0001
8	0.0000	0.0000	0.0000	−0.0001	0.0000	0.0000	0.0000
9	0.0000	0.0000	0.0000	0.0000	0.0000	0.0000	0.0000
10	0.0000	0.0000	0.0000	0.0000	0.0000	0.0000	0.0000

Panel B: response of stock prices from one-unit shock in exchange rate change

Period	HKN	KOA	MAL	PHI	SIG	THA	TWN
1	0.0000	−0.7670*	−1.2877*	0.0000	−0.4740*	0.0000	0.0000
2	0.1837	−0.8979*	−0.4687*	−0.1455	−0.4343*	0.2965**	−0.2636*
3	−0.0918	−0.2258*	−0.0531	−0.0561	−0.0167	0.0844**	−0.0278
4	0.0397	−0.0488	−0.0105	−0.0175	−0.0175	0.0156	−0.0123
5	−0.0167	−0.0102	−0.0015	−0.0051	−0.0006	0.0020	−0.0023
6	0.0070	−0.0021	−0.0003	−0.0015	−0.0007	0.0001	−0.0007
7	−0.0029	−0.0004	0.0000	−0.0004	0.0000	0.0000	−0.0002
8	0.0012	−0.0001	0.0000	−0.0001	0.0000	0.0000	0.0000
9	−0.0005	0.0000	0.0000	0.0000	0.0000	0.0000	0.0000
10	0.0002	0.0000	0.0000	0.0000	0.0000	0.0000	0.0000

Note: * = 1% significance level; ** = 5% significance level; *** = 10% significance level.

maintain the pegged rates, can do nothing but let their currency values slide: a floating exchange rate system in which we witness the result predicted under portfolio approach. For detailed economic and financial descriptions readers are refereed to Kochhar et al. (1998).

5. Other related problems of the model

One of the frequently addressed problems in security price is the existence of predictable portion of changes in stock prices. Based on the Granger causality, in five of the nine nations, variations in exchange rates could be used to predict stock prices changes in Period 3 as shown in Table 6.[10] Reported in Table 6 are comparisons of adjusted R-squares ($\bar{R}^2$) between the regression in which current stock price change is the dependent variable and its lagged (k = 1) variables are independent and the regression in which current stock price change is the dependent variable and lagged (k = 1) variables of exchange rate changes and stock price changes are independent variables.[11] An examination of Table 6 indicates that past information (within a day) on exchange rate variations is useful in predicting stock price changes in the five markets. This is reflected in increased adjusted $\bar{R}^2$ from 18% to 293%. That is, investors can better predict stock price change based on exchange rate changes by adding changes of lagged (by one period) exchange rates.

23

Table 6
A comparison of predictable portion of stock price changes

	KOA	MAL	SIG	THA	TWN
(1) $\bar{R}^2$	0.0147	0.0127	0.0181	0.0474	−0.0037
(2) $\bar{R}^2$	0.0851	0.0223	0.0616	0.0569	0.0196
(2) vs. (1)	141.1%	54.9%	109.2%	18.2%	293.1%

Note: $\bar{R}^2$ = adjusted R-square. (1): $\bar{R}^2$ from the regression in which current stock price change is the dependent variable and its lagged values (k = 1) are independent variables. (2): $\bar{R}^2$ from the regression in which current stock price change is the dependent variable and lagged values (k = 1) of exchange rate variations and stock price changes are independent variables. (2) vs. (1) = percentage increases in $\bar{R}^2$ percentage = $((2) − (1))/(((2) + (1))/2) \times 100\%$.

To test for robustness of the results, we have tried several alternative specifications in which stock prices changes of the US and Japan are included. Trades with the US and Japan historically play an important role in the region. A recent study by Wei et al. (1995) indicates that US stock market exerts more impact on Asian stock markets than does the Japanese market. For this reason, we include stock price changes of the US market in Eq. (4) to reexamine the Granger causality model (Table 4, column 4).[12] As is evident from Table 4, with the exception of Japan and Malaysia, the lead-lag structures remain virtually unchanged.

During the period of Asian financial crisis, it seems worthwhile to include a variable reflecting fundamental shocks other than the US and Japanese stock prices. Because daily data of other economic variables are difficult to obtain, we use the interest rates (federal fund rate) as a proxy to reflect fundamental shocks. In addition, interest rates are very much related to exchange rates and stock prices, and as such mirror economic prospect to some extent. Often used in the literature, difference of interest rates between the US and southeast Asian economies reveals the capital flow between the two nations. For this reason, we incorporate the interest rates and the differences into Eq. (4) to perform the Granger causality test. The result shown in Table 4 [column (5) and (6)] indicates that the conclusions remain about the same except in the case of the Philippines, in which the significance level (stock price leads exchange rate) decreases somewhat. The overall lead-lag relations remain unchanged in general with added interest rate variables.

No cointegration relation between stock prices and exchange rates is found, perhaps due to the noise from using daily data, as indicated by the statistical analysis. However, by including the error correction term in Eq. (4), we are able to examine the potential short-term causality and long-term equilibrium relations. Adding the error correction term ($y_{1t-1} - y_{2t-1}$) into Eq. (4), where y_{1t} and y_{2t} denote stock price and exchange rate, we have:

$$\Delta y_{1t} = \alpha_0 + \delta_1(y_{1t-1} - y_{2t-1}) + \sum_{i=1}^{k} \alpha_{1i}\Delta y_{1t-i} + \sum_{i=1}^{k} \alpha_{2i}\Delta y_{2t-i} + \varepsilon_{1t}$$

$$\Delta y_{2t} = \beta_0 + \delta_2(y_{1t-1} - y_{2t-1}) + \sum_{i=1}^{k} \beta_{1i}\Delta y_{1t-i} + \sum_{i=1}^{k} \beta_{2i}\Delta y_{2t-i} + \varepsilon_{2t}$$

(6)

Table 7
Estimation result of impulse response function (reverse order)

Panel A: response of exchange rates from one-unit shock in stock price change

Period	HKN	KOA	MAL	PHI	SIG	THA	TWN
1	0.0000	−0.7892*	−0.7782*	0.0000	−0.2040*	0.0000	0.0000
2	−0.0091***	−0.1602	−0.1971***	−0.1540***	−0.0829***	−0.3250*	−0.1201*
3	0.0045***	−0.0329	−0.0256	−0.0593	−0.0078	−0.0926**	−0.0127
4	−0.0020	−0.0068	−0.0047	−0.0185	−0.0033	−0.0171	−0.0056
5	0.0008	−0.0014	−0.0007	−0.0054	−0.0003	−0.0022	−0.0010
6	−0.0003	−0.0003	−0.0001	−0.0016	−0.0001	−0.0002	−0.0003
7	0.0001	−0.0001	0.0000	−0.0005	0.0000	0.0000	−0.0001
8	−0.0001	0.0000	0.0000	−0.0001	0.0000	0.0000	0.0000
9	0.0000	0.0000	0.0000	0.0000	0.0000	0.0000	0.0000
10	0.0000	0.0000	0.0000	0.0000	0.0000	0.0000	0.0000

Panel B: response of stock prices from one-unit shock in exchange rate change

Period	HKN	KOA	MAL	PHI	SIG	THA	TWN
1	−0.2504	0.0000	0.0000	−0.2819**	0.0000	−0.2531***	−0.2320*
2	0.1995	−0.8285*	−0.3400***	−0.2141	−0.3773*	0.2365	−0.2613*
3	−0.0913	−0.2164*	−0.0255	−0.0745	0.0021	0.0746	−0.0358
4	0.0389	−0.0472	−0.0066	−0.0226	−0.0153***	0.0145	−0.0130
5	−0.0164	−0.0099	−0.0009	−0.0066	0.0002	0.0020	−0.0026
6	0.0069	−0.0020	−0.0002	−0.0019	−0.0006	0.0002	−0.0007
7	−0.0029	−0.0004	0.0000	−0.0005	0.0000	0.0000	−0.0002
8	0.0012	−0.0001	0.0000	−0.0002	0.0000	0.0000	0.0000
9	−0.0005	0.0000	0.0000	0.0000	0.0000	0.0000	0.0000
10	0.0002	0.0000	0.0000	0.0000	0.0000	0.0000	0.0000

Note: * = 1% significance level; ** = 5% significance level; *** = 10% significance level.

To test whether exchange rates Granger-cause stock prices, we examine the null hypothesis H_0: $\delta_1 = 0$ and $\alpha_{21} = \alpha_{22} = \cdots = \alpha_{2k} = 0$. Conversely, to test whether stock prices Granger-cause exchange rates, we examine H_0: $\delta_2 = 0$ and $\beta_{21} = \beta_{22} = \cdots = \beta_{2k} = 0$. Including the error correction terms does not alter the lead-lag relations noticeably [Eq. (7) in Table 4]. Besides, estimated δ_1 and δ_2 are insignificant (not shown), suggesting a lack of long-term equilibrium relations between stock prices and exchange rates. It is to be pointed out that these results are derived from estimating reduced form equations, and as such are open to questions of the endogeneity of the two variables. In that case, structural equations may be preferred. On the other hand, proponents of VAR models do not favor the structural form approach (Canova 1995, pp. 103–109). One problem associated with an unrestricted VAR model is the sensitivity of the results in response to different order of variable appearances. To examine the robustness of impulse responses under different contemporaneous identifying assumptions, we alter the order of variable appearances, and report the result in Table 7.[13]

An examination of Table 7 indicates the impulse response functions (mostly negative correlation) are similar to that of Table 5 save that the lengths of impact are different. In the case of Korea, as we reverse the order of the two variables, the Granger causality that exchange rate change leads stock price change becomes a feedback relation: a relation that dominated the southeast Asian economies. In general, the Granger causality remains robust in the period of the Asian flu.

6. Conclusion

Prior studies based on monthly data have found either little relation can be established between the two markets or exchange rate market leads stock market. In this paper, we apply the recently advanced statistical techniques with daily data to analyze the problem in the Asian economies. The result indicates markets were largely characterized by the phenomenon predicted under the portfolio approach. Most markets exhibit either changes in stock prices lead that in exchange rates or either market can take the lead (feedback interaction). Built on the result of the Granger causality test, the IR analysis lends further support to the importance of stock market as the leader or the existence of feedback interaction during the Asian flu period. Likewise, the inclusion of exchange rate variations (within a day) can improve predictable portion of stock price changes in the five Asian markets. In the early 1990s as barriers to capital movement were gradually removed, the role of the portfolio approach cannot be downplayed. Capital movement into and out of the Asian economies is as beneficial as it is detrimental. The Asian flu certainly has put the stock and the exchange rate markets in a spotlight that suggests financial markets in the Asian economies need an overhaul. Although the main objective of this paper is to explore short-term dynamic relations between stock prices and exchange rates, it also provides a potential avenue for future research in the spillover effects in southeast Asia.

It needs to be pointed out that although the Granger-Causality and IR may exhibit some statistical relations, it is sometimes difficult to interpret the underlying fundamental economic relation based on those results. It is likely that the results may be generated from other structure relations, i.e., via interest rate parity condition or IS-LM-related policies. For example, some recessionary shocks or unfavorable information on the country will cause a stock price decrease and an exchange rate depreciation. In this case the timing relation between the stock price and exchange rate will be generated from the relative efficiency of the stock market and foreign exchange market. In this case, both approaches may not play any role in generating the timing relation.[14]

Notes

1. For an intuitive explanation of the portfolio approach, see Krueger (1983, pp. 81–91). Other related works include Dornbusch (1975), Frenkel and Rodriguez (1975), and Boyer (1977).
2. The strength of the Zivot and Andrew approach lies in endogeneizing the breakpoint, albeit a single one. Recently, tests with multiple break points are made possible by Bai and Perron (1998), Garcia and Perron (1996), Liu et al. (1997), and Andrews et al. (1996).
3. There are three different models provided by Zivot and Andrew (1992). We adopt the

unit-root model as shown in Eq. (2) due to the existence of a trend in the time series. Similar results are obtained using the other two models proposed by them.

4. We are grateful for the generosity extended by the economics department of UCSD in providing the data.

5. After the convention in event study, the breakpoint for Period 3 would be approximately 1 month before the depreciation of the Thai Baht.

6. Note that λ values in 11 of the 18 unit root tests exceed the 95% value, indicating that the Asian financial debacles have indeed caused structural changes. Such structures break toward either end of the data could cast a doubt on the power of the Zivot and Andrew unit root test statistic. It remains an interesting research topic in the future.

7. The cointegration test is performed only on the national stock prices and exchange rates with the property of I(1), i.e., including that of Japan, Singapore, and Taiwan.

8. The significance level is greater than 10% in the case of the Philippines. The impulse response functions indicate that change in stock price leads that in exchange rate. As such, the Philippines market can be characterized by the portfolio approach.

9. As the cointegration relations are not found between the variables, we calculated the IR functions based on Eq. (4).

10. In the case of Hong Kong, the analysis indicates that exchange rate changes lead that of stock prices, and as such the predictable portion of stock price changes cannot be improved upon noticeably.

11. Harvey (1995) employs adjusted R-square to investigate the predictability of stock prices for both emerging and developed markets.

12. To conserve space, we present only the results in which changes in the US stock prices are included. The lead-lag structures remain unchanged as well if Japanese data are included in Eq. (4).

13. The variable order of the IR functions in Table 5 is based on that from the causality tests shown in Table 4. For instance, changes in stock prices lead that in exchange rates in the Hong Kong market. Hence, the variable order in which the IR functions are calculated (Table 5) is stock price changes–exchange rate changes. The order is reversed to exchange rate–stock price and the results of the IR function calculations are reported in Table 7.

14. We are indebted to one of the reviewers for pointing this possibility out.

References

Abdalla, I. S. A., & Murinde, V. (1997). Exchange rate and stock price interactions in emerging financial markets: evidence on India, Korea, Pakistan, and Philippines. *Applied Financial Economics, 7*, 25–35.

Aggarwal, R. (1981). Exchange rates and stock prices: a study of the US capital markets under floating exchange rates. *Akron Bus Econ Rev, 12*, 7–12.

Amihud, Y. (1993). Exchange Rates and the Valuation of Equity Shares. In Y. Amihud & R. Levich (Eds.) *Exchange rates and corporate performance*. Homewood: Irwin.

Andrews, D. W. K., Lee, I., & Ploberger, W. (1996). Optimal changepoint tests for normal linear regression. *J Econometrics 70*, 9–38.

Bahmani–Oskooee, M., & Sohrabian, A. (1992). Stock prices and the effective exchange rate of the dollar. *Appl Econ, 24*, 459–464.

Bai, J., & Perron, P. (1998). Estimating and testing linear models with multiple structural change. *Econometrica, 66*(1), 47–78.

Bartov, E., & Bondar, G. M. (1994). Firm valuation, earnings expectations and the exchange rate exposure effect. *J Finance, 49*, 1755–1786.

Bodnar, G. M., & Gentry, W. M. (1993). Exchange rate exposure and industry characteristics: evidence from Canada, Japan and the US. *J Int Money Finance, 12*, 29–45.

Boyer, R. S. (1977). Devaluation and portfolio balance. *Am Econ Rev, 67*, 54–63.

Canova, F. (1995). Vector autoregressive models: specification, estimation, inference, and forecasting. In M. Hashem Pesaran & M. R. Wickens (Eds.). *Handbook of Applied Econometrics: Macroeconomics*. Oxford: Blackwell Publishers.

Dickey, D. A., & Fuller, W. A. (1979). Distribution of the estimators for autoregressive time series with a unit root. *J Am Stat Assoc, 74*, 427–431.

Dornbusch, R. (1975). A portfolio balance model of the open economy. *J Monetary Econ, 1*, 3–20.

Engle, R. F., & Granger, C. W. J. (1987). Cointegration and error correction: representation, estimation, and testing. *Econometrica, 55*, 251–276.

Frenkd, P., & Rodriguez, C. A. (1975). Portfolio equilibrium and the balance of payments: a monetary approach. *Am Econ Rev, 65*, 674–688.

Garcia, R., & Perron, P. (1996). An analysis of the real interest rate under regime shifts. *Rev Econ Stat, 78*, 111–125.

Granger, C. W. J. (1969). Investigating causal relations by econometric models and cross-spectral methods. *Econometrica, 37*, 424–439.

Gregory, A. W., & Hansen, B. E. (1996). Residual-based tests for cointegration in models with regime shifts. *Econometrics, 70*, 99–126.

Hamilton, J. D. (1994). *Time Series Analysis*. Princeton: Princeton University Press.

Harvey, C. R. (1995). Predictable risk and returns in emerging markets. *Rev Finan Studies, 8*(3), 773–816.

Jorion, P. (1990). The exchange exposure of US multinational firm. *J Bus, 63*, 331–345.

Kochhar, K., Prakash, L., & Stone, M. R. (1998). The East Asian crisis: macroeconomic developments and policy lessons. *IMF Working Paper 98-128*.

Krueger, A. O. (1983). *Exchange-Rate Determination*, Cambridge: Cambridge University Press.

Krugman, P. R., & Obstfeld, M. (1997). *International Economics: Theory and Policy*, 4th ed., Reading: Addison-Wesley Inc.

Liu, J., Wu, S., & Zidek, J. V. (1997). On segmented multivariate regressions. *Statistica Sinica, 7*, 497–525.

Nelson, C. R., & Plosser, C. I. (1982). Trends and random walks in macroeconomic time series: some evidence and implications. *J Monetary Econ, 10*, 139–162.

Perron, P. (1989). The great crash, the oil price shock, and the unit root hypothesis. *Econometrica, 57*, 1361–1401.

Perron, P., & Vogelsang, T. J. (1992). Nonstationarity and the level shifts with an application to purchasing power parity. *J Bus Econ Stat, 10*(3), 301–320.

Wei, K. C. J., Liu, Y.-J., Yang, C.-C., Chaung, G.-S., (1995). Volatility and price change spillover effects across the developed and emerging markets. *Pacific-Basin Finan J, 3*, 113–136.

Zivot, E., & Andrews, D. W. K. (1992). Further evidence on the great crash, the oil-price shock, and the unit-root hypothesis. *J Bus Econ Stat, 10*(3), 251–270.

ELSEVIER

Energy Economics 24 (2002) 107–119

Energy
Economics

www.elsevier.com/locate/eneco

An analysis of factors affecting price volatility of the US oil market

C.W. Yang[a], M.J. Hwang[b,*], B.N. Huang[c]

[a]*Department of Economics, Clarion University of Pennsylvania, Clarion, PA 16214-1232, USA*
[b]*Department of Economics, College of Business and Economics, West Virginia University, P.O. Box 6025, Morgantown, WV 26505-6025, USA*
[c]*Institute of International Economics, National Chung Cheng University, Chiayi, Taiwan, ROC*

Abstract

This paper studies the price volatility of the crude oil market by examining the market structure of OPEC, the stable and unstable demand structure, and related elasticity of demand. In particular, the impacts of prosperity and recession of the world economy and the resulting demand shift on crude oil price are investigated. The error correction model is used to estimate the demand relations and related elasticity. The income effect on demand functions is evaluated to shed light on future prices. A simulation of potential oil prices under different scenarios on a cut of one million barrels per day by OPEC is evaluated. From our simulation, given the 4% cut in OPEC production, the oil price is expected to increase unless the recession is severe. The magnitude and scope of a price hike would be diminished if non-OPEC or domestic production were greatly expanded. © 2002 Elsevier Science B.V. All rights reserved.

JEL classification: 022

Keywords: Crude oil; Price volatility; Demand elasticity

1. Introduction

The volatility of crude oil prices creates uncertainty, and therefore an unstable economy for both oil-exporting and -importing countries. Higher prices result in an increase in inflation and a subsequent recession in oil-consuming nations, as oil

---

* Corresponding author. Tel.: +1-304-293-7866.
E-mail address: mjhwang@mail.wvu.edu (M.J. Hwang).

prices are negatively correlated to economic activities (Ferderer, 1996). Most post-war recessions were preceded by oil price shocks, i.e. 1974, 1980 and 1990 economic recessions in the US (Huntington, 1998). Evidence that a recession depresses prices recently occurred in 1998, when oil prices declined to approximately $10 a barrel as a result of the Asian economic setback that started in 1997. Lower oil prices would prohibit economic development and might generate political instability and social unrest in some oil-producing countries. The rollercoaster of sharp oil-price fluctuations has been remarkable in the last three decades. Thus, it is of great importance to analyze important factors affecting the volatility of crude oil prices.

The literature on price volatility of crude oil relates oil shocks either to the instability of the market structures or to the effect of the price elasticity of demand. Mork (1989) and Huntington (1998) demonstrated the asymmetric relationship, that a reduction in oil prices does not necessarily lead to noticeable output growth, while an increase can have a negative impact on output growth. The study of Ferderer (1996) points to the observation that disruptions in oil market not only give rise to higher prices, but also increase oil price volatility.

Chaudhuri (2001) recently found that non-stationary commodity prices could be attributed to the non-stationarity in oil prices. In other words, real oil prices and real commodity prices are cointegrated. Thus, he suggested that the price of oil should be another variable in the aggregate production function. Given its volatility, it comes as no surprise that the oil price is considered by the majority of researchers as being non-stationary, i.e. it may well have a unit root (e.g. Bentzen and Engsted, 1993, 1996; Jones, 1993; Hamilton, 1994; Arize, 2000), despite the conclusion by Pindyck (1999) that long-term oil prices are mean-reverting around shifting trend lines. While advances in the non-stationary time series technique are instrumental in understanding the behavior of oil prices, the mechanism of the data-generating process plays a critical role. That is, the underlying structure of oil markets, especially OPEC, can shed important light on the issue.

In his pioneering work, Griffin (1985) characterized OPEC behavior into four categories: cartel, competitive, the target revenue, and property rights models. He

---

[1] The most popular version is the one that considers Saudi Arabia or OPEC core countries (e.g. Saudi Arabia, Kuwait and Qatar) as a dominant firm (Pindyck and Hnyilicza, 1976; Tourk, 1977; Noreng, 1978; Houthakker, 1979; Aperjis, 1982; Alhajji and Huettner, 2000a). On the other hand, MacAvoy (1982) attributes price increases to supply disruptions under open market conditions. The positively sloped supply functions imply a competitive oil market. In addition, Verleger (1987) ascribes oil price disruptions to problems on the demand side. Another non-collusive model is the target revenue theory: oil-producing countries' capacity to absorb investment at an acceptable rate of return determines oil production (Ezzati, 1976; Teece, 1982; Salehi-Isfahani, 1987; Cremer and Isfahani, 1991; Alhajji and Huettner, 2000b). In addition, the production cutbacks can be viewed as a phenomenon of a backward-bending supply curve. As a result of increasing demand, the market price will increase and oil output will decrease with adequate oil revenue to cover the target investment (Alhajji and Huettner, 2000b). The property rights model (Johany, 1979; Mead, 1979) makes use of the concept of the real discount rate and real oil price in the Hotelling theory. When ownership is transferred from the international oil companies to producing nations, the result is sharp production reductions and price increases (Dasgupta and Heal, 1979).

rejected the last three models in favor of the OPEC cartel model. Recently, Alhajji and Huettner (2000a,b) employed the multi-equation estimation technique, and showed that neither OPEC nor the OPEC core could play the role of a dominant producer, because the demand for oil remains price inelastic in both cases.[1] The remarkable price volatility reflects the market structure of OPEC, unstable demand structure, and shifting demand conditions. Both the inherent unstable demand structure and non-stationary economic variables must be addressed in order to dissect the behavior of oil prices. The supply of crude oil by production decreases or increases will also be emphasized, as it has an immediate impact on prices. In particular, the impacts of changes in the world economy and its resulting demand shift on crude oil prices will be investigated.

This paper initially provides a theoretical analysis of crude oil pricing by expanding the elasticity theory developed originally by Greenhut et al. (1974). The next section presents an extended model by incorporating marginal cost (includes user cost) and highlights the unstable demand structure. Section 3 estimates and describes the demand structure and the volatility of oil prices. Section 4 presents a simulation of potential oil prices under different scenarios on a one million barrel cut per day by OPEC. The last section provides a conclusion.

2. Unstable demand structure of the oil market

Assume a well-behaved demand function for OPEC oil $P = P(q,\alpha)$ of $R^2 \to R^1$, where P and q denote price (average value of imported oil to the US) and output (import of crude oil from OPEC), respectively, and α is a shifting parameter (e.g. income level). We regard an increase in α as an increase in demand (outward shift in demand curve). A general total cost function $C(q)$, such that $C'(q) = MC(q)$, reflects the cartel's cost environment. A close-knit cartel maximizes the joint profit $\pi = Pq - TC(q)$ with the following first-order condition:

$$P(1 - 1/e) - MC(q) = 0 \tag{1}$$

---

[2] According to Griffin and Teece (1982), the marginal cost of producing crude oil includes the conventional marginal production costs and user cost. The marginal production cost consists of conventional capital, labor and material costs of producing the last unit of output. The user cost is an opportunity cost of producing an extra barrel of oil, as oil is a non-renewable product and the decision to produce a barrel of oil today precludes the possibility of producing it at some time in the future. This marginal cost is estimated between $0.10 and $0.25 per barrel in most Persian Gulf countries during the 1970s. The user cost was estimated by Pakravan (1984) to be less than $0.10 in the Middle East during the 1960s and 1970s. In addition, per barrel user cost (U_0) for OPEC may be used as a proxy for traditional marginal cost, as shown in the profit-maximizing condition $P = U_0/(1 + 1/\varepsilon)$ by Alhajji and Huettner (2000a, p. 37). By using their formulation, the user cost has three components: (i) increased security cost (military expenditure) per barrel; (ii) royalties, which account for 12% to over 50% of total revenues or 0.17-fold the real oil price; and (iii) the extraction cost $\alpha = \$0.5$ in 1970, with an increase of 3% per year. This amounts to $0.5 \cdot 1.03^{31} = \$1.25$ for year 2001. Even with the three-component cost of $5, the k value is 5.6 for a price of $28 per barrel.

or

$$P[1 - 1/e - MC(q)/P] = P(1 - 1/e - 1/k) = 0 \tag{2}$$

where $k = P/MC$. Solving for e (absolute value of the price elasticity) yields:

$$e^* = k/(k - 1) \tag{3}$$

Under a stable demand structure, the price elasticity converges toward $k/(k - 1)$. For instance, the marginal cost[2] is approximately constant at $1/10$ of the price (i.e. for $k = 10$), and the long-run price elasticity of demand should approach -1.11. However, the MC/P ratio can be quite appreciable when oil prices slide below $10 a barrel.[3]

Using the chain rule and expanding the Greenhut–Hwang–Ohta theory (Greenhut et al., 1974; the original model was well thought out by Greenhut and Ohta, 1975), we derive the following second-order condition:

$$\frac{d^2\pi}{dq^2} = \frac{dP(1 - 1/e)}{dP}\frac{dP}{dq} - \frac{dMC}{dP}\frac{dP}{dq} \tag{4}$$

$$= \frac{dP}{dq}\left[\frac{dP(1 - 1/e)}{dP} - \frac{dMC}{dP}\right]$$

$$= \frac{dP}{dq}\left[\frac{\eta* - (1 - e)}{e} - \frac{dMC}{dq}\frac{dq}{dP}\frac{q}{MC}\frac{MC}{q}\frac{P}{q}\frac{q}{P}\right]$$

$$= \frac{dP}{dq}\left[\frac{\eta* - (1 - e)}{e} - \frac{-eMC}{\beta P}\right] < 0$$

$$= \frac{dP}{dq}\left[\frac{\eta* - (1 - e)}{e} - \frac{-e}{\beta k}\right] < 0$$

where $\beta = (dq/dMC)(MC/q)$ is the output elasticity on the marginal cost curve and η^* is the elasticity of price elasticity on a given demand curve i.e. $\eta^* = (\partial e/e)/(\partial P/P)$. For a constant marginal cost curve, the second term in the parentheses drops out as $\beta \to \infty$. For an insignificant marginal cost, $1/k = MC/P$

---

[3] It is both interesting and paradoxical to note that the great majority of estimated price elasticity values (absolute value) are less than unity, except that of Alhajji and Huettner (2000a) in the case where Saudi Arabia is the dominant player. The behavior of oil price changes cannot be entirely captured by traditional microeconomic theory. Nonetheless, the long-run inelastic demand is consistent with the switch elasticity model by Greenhut et al. (1974), upon which we expand.

is trivial. For a stable case of demand structure, it is sufficient that $\eta^* > 1 - e$ holds. Note that a similar result can be derived for the leadership model with residual demand.

In the presence of a structural break in demand, that is from a competitive environment to an effective cartel scheme (e.g. OPEC), the demand function experiences a sudden change in curvature. The divergent or oscillating price elasticity of demand signals an unstable demand framework, as indicated by $d^2\pi/dq^2 > 0$ in Eq. (4) (i.e. the second-order condition is violated). A cartel can benefit from an increase in price if demand is inelastic. As the market demand changes from a relatively more elastic competitive market demand to an inelastic cartel demand (i.e. a consequence of $\eta < 1 - e$), this unstable relation will make the elasticity diverge from $e = k/(k - 1)$.

The majority of empirically estimated price elasticities of crude oil fluctuate between -0.05 and -0.4 (Griffin and Teece, 1982; Pindyck, 1978), which is far from the suggested stable long-run value of -1.11 (at $k = 10$) or -1.22 (at $k = 5.6$). Clearly, the 1973 oil embargo ushered its way to an effective oil cartel, OPEC, in which the price elasticity of demand drastically diminished in absolute value as the price of crude oil was substantially raised (i.e. η was significantly negative). Such an unstable demand structure may well contain the seed of recent great price volatility, as will be explained in the next section.

3. Volatility of crude oil price

To investigate the volatility of oil prices, we follow the generalized autoregressive conditional heteroskedasticity (GARCH) model developed by Bollerslev (1986). The estimation period is from January 1975 to September 2000.[4] The monthly crude oil price data are taken from International Financial Statistics published by the International Monetary Fund. The following relation is set forth to estimate the volatility of the price change:[5]

$$DP = \Delta \ln P_t \approx (\ln P_t - \ln P_{t-1}) \tag{5}$$

where P_t is the oil price in time period t;[6] $\Delta \ln P_t$ denotes rate of change of the oil price in logarithmic difference.

---

[4] The GARCH model used in this model extends from 1975 to 2000, instead of from 1948 to 1998 for the following reasons. First, oil prices were low and non-volatile before the 1973–1974 energy crisis. Consequently, we exclude it from the volatility analysis. Second, inclusion of 1999 and 2000 data better highlights the recent price hike. We thank one referee for this point.

[5] For small changes in price, we use $d \ln P_t = d P_t/P_t$, which reflects the rate of change in price due to logarithmic differentiation. Note that the prefix of P (or D) is only a notation, not a differentiation operator.

[6] This paper employs annual data, and as such, does not accommodate for seasonality and its potential autocorrelation problem. Models using quarterly data can directly tackle such problems and give more detailed results (Alhajji and Huettner, 2000a). We thank one of the referees for pointing out this problem.

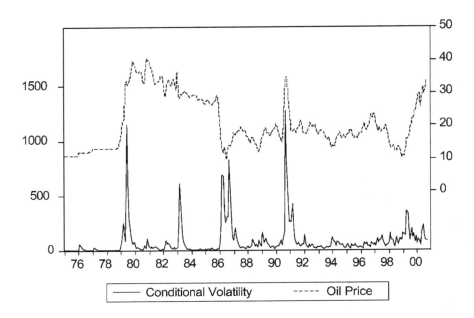

Fig. 1.   Price of crude oil and conditional variance.

According to Mandelbrot (1963), large changes in prices tend to be followed by large changes of prices in either direction, and small changes tend to be followed by small changes. As shown in Fig. 1, such volatility clustering abounds in terms of oil price change. To accommodate for this phenomenon, Engle (1982) developed the autoregressive conditional heteroscedasticity (ARCH) model. We first estimate $DP_t = a_0 + a_1 DP_{t-1} + \varepsilon_t + a_2 \varepsilon_{t-2}$ and test its residuals based on the Lagrange multiplier (LM) method with the χ^2 test statistic of 15.30 (six degrees of freedom).[7] As such, we reject the null hypothesis H_0: no ARCH phenomenon (of order six). However, estimations of the ARCH(q) model could involve long lags. To circumvent this problem, Bollerslev (1986) generalizes the model or GARCH (p,q) to include past innovations regarding an oil price change (ε_{t-q}) and its previous conditional variance (h_{t-p}). Note that the ARCH model takes only ε_{t-q} into consideration.

From estimated α and β of the variance equation [see Eq. (8) below] and the corresponding t statistics, it is evident that volatility clustered and prevailed in the sample period. In addition, $\alpha + \beta > 1$ implies that shocks to the conditional variance are persistent, in the sense that they remain important for forecasts of all horizons.

The volatility of DP_t can be estimated via the GARCH (q,p)–ARMA (m,n) model in terms of the conditional variance as shown below:

$$DP_t = a_{ij} + \sum_{i=1}^{m} a_i DP_{t-1} + \varepsilon_t + \sum_{j=1}^{n} b_j \varepsilon_{t-j} \qquad \varepsilon_t \in N(0,h) \tag{6}$$

---

[7] We based our estimates on Eq. (8) as its residuals are white noise.

$$h_t = a_{ij} + \sum_{i=1}^{p} \alpha_i \varepsilon_{ij}^2 + \sum_{j=1}^{q} \beta_j h_{t-j} \tag{7}$$

The best-fit model is the one that requires standardized residuals $u_t = \varepsilon/\sqrt{h}$ and their squares $u_t^2 = \varepsilon_t^2/h_t$ obey the white noise process. And as such, we present the estimation results in the following:[8]

$$DP = \underset{(0.5632)}{0.1205} + \underset{(3.6011)}{0.1938\,DP_{t-1}} + \varepsilon_t - \underset{(-3.1746)}{0.1798}\,\varepsilon_{t-2} \tag{8}$$

$$h_t = \underset{(6.3923)}{2.2000} + \underset{(6.3674)}{0.6909\,\varepsilon_{t-1}^2} + \underset{(9.1924)}{0.4922\,h_{t-1}} \tag{9}$$

$$\log L = -985.1621$$

$$Q_{12}(u_t) = 11.0610 \ [0.524]$$

$$Q_{12}(u_t^2) = 7.3946 \ [0.830]$$

Note that the numbers in parentheses are t statistics and those in brackets are P values. Log L is the logarithmic likelihood function value. $Q_{12}(u_t)$ are χ^2 statistics (12-month lag) to test the null hypothesis that u_t obeys the white noise process.

Evident from Eqs. (8) and (9), the rate of change in the oil price exhibits a significant ARCH phenomenon as the estimated value of $\alpha + \beta = 0.6909 + 0.4922 > 1$ and are both significant.[9] In addition, the problem of misspecification does not seem to exist, as u_t or u_t^2 is free of the serial correlation problem (up to 12-month lag). The conditional variance h_t in Eq. (9) can be used to estimate the price volatility from January 1975 to September 2000. Fig. 1 presents the trends for both the price of crude oil (dotted line) and the conditional variance (solid line).

Again, oil price trends, to a large extent, synchronize with that of conditional variances. Spikes occurred in the periods of 1979–1980, 1986 and 1990. It is noteworthy that a large conditional variance from the GARCH model is a necessary condition for leading to a recession in the US. However, it is not a sufficient condition, as some spikes did not lead to a serious economic downturn. Starting from January of 1999, spikes in terms of conditional variances appear to have surged again. Our sample period does not fully capture this volatility due to lag effects and would be more accurate if more data were included.

---

[8] Empirical results are obtained via the E-view package and are available upon request.
[9] The significance of estimated α and β in the GARCH model is also found in emerging stock markets (Huang and Yang, 2000).

Table 1
Variable definition and unit root test

Logarithmic scale	Definition of variable	ADF statistics with lag of one
ln P	P = price of oil (in cents per million BTU)	-1.44
ln PN	PN = price of natural gas (in cents per million BTU)	-1.47
ln PC	PC = price of coal (in cents per million BTU)	-1.24
ln YR	YR = real income (real GDP)	-2.05
ln Q	Q = consumption of oil (in quadrillion BTU)	-2.38

Source: The Annual Energy Review (Department of Energy, 1998) and the Economic Report of the President (United States Government Printing Office, 1998).

4. Demand and price estimation of the US oil market

The problem of forecasting future oil price is confounded due to its volatile behavior. There, a stationary time series is preferred. Following the augmented Dickey–Fuller (ADF) unit root test, we cannot reject the unit root hypothesis, even after logarithmic transformation of the variables during the sample period (1949–1998). The ADF statistics are listed in Table 1.

Invoking the Engle–Granger two-step cointegration test, we reject the null of $\varepsilon_t \sim I(1)$ of the demand function below, with the ADF statistic of -4.70:

$$\ln Q = a + b\ln P + c\ln PC + d\ln YR + e\ln PN + \varepsilon_t \tag{10}$$

In other words, a cointegration vector may reflect the long-run cointegration relation as shown below:

$$\begin{aligned} \text{ECM}_{t-1} = {} & \ln Q_{t-1} + 0.1842\ln P_{t-1} - 0.2264\ln PC_{t-1} - 0.0083\ln PN_{t-1} \\ & - 0.5006\ln YR_{t-1} \end{aligned} \tag{11}$$

It should be pointed out these variables are of $I(1)$ and, as such, t statistics cannot be used for statistical inference. We employ the error correction model (ECM) to estimate the demand relation for the US oil market with the structural break that occurred in 1975.[10] The estimated relation for 1949–1975 and 1976–1998 are reported in the following, respectively:

$$\begin{aligned} \Delta\ln Q_t = {} & -0.6716\text{ECM}_{t-1} - 0.074\Delta\ln P_t + 0.1875\Delta\ln PN_t \\ & + 0.01887\Delta\ln PC_t + 0.2141\Delta\ln YR_t + e_{1t} \end{aligned} \tag{12}$$

$\bar{R}^2 = 0.43$, $Q(1) = 0.000$ (0.999), $Q(4) = 0.552$ (0.906) and $Q(12) = 4.490$ (0.973).

$$\Delta \ln Q_t = -0.25229 \text{ECM}_{t-1} - 0.0375 \Delta \ln P_t - 0.004 \Delta \ln \text{PN}_t$$
$$+ 0.3343 \Delta \ln \text{PC}_t + 0.7105 \Delta \ln \text{YR}_t + e_{2t} \tag{13}$$

$\bar{R}^2 = 0.91$, $Q(1) = 0.887$ (0.833), $Q(4) = 3.57$ (0.312) and $Q(12) = 17.197$ (0.102), where $Q(1)$ is the Ljung–Box statistic of lag one to test the white noise process; P values are in parentheses.

However, care must be exercised in evaluating the price elasticity of demand in Eqs. (12) and (13) because the coefficients of -0.074 and -0.0375 are not the price elasticity.[11] Rather, the elasticity can be approximated by $(0.074)^{1/2} \approx 0.272$ and $(0.0375)^{1/2} \approx 0.194$. The price of oil, to a large extent, impacts production decisions, especially in industrialized economies. Given the historical evidence of business cycles, derived demand for oil is subject to great fluctuations as well. Predicting future oil prices cannot be satisfactorily carried out without considering the income effect in the demand equation. This is especially true in the presence of the recent sputtering economy. As is commonly used, a constant elasticity demand, for a given level of PC and PN, can be rewritten as:

$$Q = \alpha P^b \, \text{PC}^c \, \text{YR}^d \, \text{PN}^f$$

$$= m P^b \, \text{YR}^d \tag{14}$$

In order to simulate the impact of OPEC production reduction on the US

---

[10] The break point is chosen endogenously by seeking the minimum sum of squares of error:

$$\text{SSE}(k) = \sum_{i=1}^{k} \varepsilon_i^2 + \sum_{j=k+1}^{T} \varepsilon_j^2$$

The lagged impact of the 1973 energy embargo was fully felt in 1975.

[11] The coefficient α of first-differenced logarithmic variables can be shown as elasticity squared for infinitesimally small changes:

$$\beta = \frac{d(\ln Q_t - \ln Q_{t-1})}{d(\ln P_t - \ln P_{t-1})} = \frac{d(d\ln Q_t)}{d(d\ln P_t)}$$

$$= \frac{d\left(\dfrac{dQ_t}{Q_t}\right)}{d\left(\dfrac{dP_t}{P_t}\right)} = \frac{-\dfrac{(dQ_t)^2}{Q_t^2} + \dfrac{1}{Q_t} \times d^2 Q_t}{-\dfrac{(dP_t)^2}{P_t^2} + \dfrac{1}{P_t} \times d^2 P_t} = e^2$$

The second difference is approximately zero if the first difference is found to be stationary.

Table 2
Simulated US oil prices with a 4% cut in OPEC production

OPEC production cut %	Real GDP as % of current GDP	Simulated US oil in terms of a bench-mark price
4	1.03	1.403
4	1.02	1.345
4	1.01	1.2887
4	1.00	1.2342
4	0.99	1.1815
4	0.98	1.1305
4	0.97	1.0812
4	0.96	1.0336
4	0.95	0.9876

price,[12] we express Eq. (14) in terms of its inverse:

$$P = (Q^*/m)^{1/b} \cdot YR^{*-(d/b)} \tag{15}$$

in which $b = (0.0375)^{1/2} = 0.194$ and $d = (0.7105)^{1/2} = 0.843$ after the structural break. For a constant m and real GDP ($YR = YR^*$), a 4% production cut translates into an approximate 23% price increase. This is to say, with $Q = 0.96Q^*$, the new price, ceteris paribus, is $(0.96)^{1/-0.194} \approx 123.4\%$ of the original level (or a benchmark price). If US real GDP is estimated to decrease by 5%, the new oil price according to our estimate will be:

$$P = 0.96^{1/-0.194} \cdot 0.95^{-0.843/-0.194} \cdot P^* = 0.987P^*$$

This being the case, the future oil price, which plays a crucial role in the US economy, will actually be 1.3% lower than the current price. If the recession threat is indeed credible and serious (5% reduction in GDP), we report the simulated prices given the OPEC production cut of 4% in Table 2.

An examination of Table 2 reveals the impacts of a 4% OPEC production cut under different income scenarios. Unless the recession is severe (e.g. 5% drop in real GDP), the impact of the joint cartel production cut would most likely lead to price increases: a 40% hike is predicted if the US real GDP is up by 3% (a rather unlikely case). However, it signals the vulnerability of the US energy demand in terms of rising manufacturing cost. It is little wonder that US Energy Secretary, Spencer Abraham, warns that the nation faces a major energy supply crisis over the next two decades.

---

[12] The structural break model highlights the changing price elasticity. To fathom the effect of production cuts (actual or announced), we use the estimated demand relation after the 1973 energy crisis, as production cuts were non-existent before the formation of OPEC.

Table 3
Simulated US oil prices with a 4% cut in OPEC production and 1% output expansion from non-OPEC or domestic production

OPEC production cut (%)	Real GDP as % of current GDP	Simulated US oil in terms of % of a bench-mark price when non-OPEC or domestic production expands by:	
		1%	2%
4	1.03	1.3303	1.2618
4	1.02	1.275	1.2095
4	1.01	1.2217	1.1586
4	1.00	1.17	1.1098
4	0.99	1.12	1.062
4	0.98	1.0717	1.0165
4	0.97	1.025	0.9721
4	0.96	0.9793	0.9289
4	0.95	0.936	0.8878

If non-OPEC or domestic supply expands by 1%, the net OPEC output reduction would be reduced to 3%. The impact on oil prices can be similarly calculated (Table 3). The magnitude of higher oil prices is reduced from 40.3% to 33.03% if real GDP increases by 3%. An aggressive supply-side policy could enhance the non-OPEC oil production by 2% (i.e. the net OPEC output curtailment would be 2%). The impacts of higher production costs can be appreciably diminished, as shown in the last column of Table 3.

5. Concluding remarks

When oil prices plunged to $10 a barrel in late 1998, OPEC was on the brink of collapse. With the price reaching a 10-year high at $37 a barrel in 2000, the newly energized oil cartel has regained an assertive role on the global economic stage. It is apparent that a cartel such as OPEC tends to promote higher prices with lower production. An excessively high price would generally create conditions where there is a potential for inflation and economic recession. On the other hand, as a consequence of reduced demand and oversupply, along with relatively more elastic demand, the oil price would then drop, as indicated in our analysis. The excessive price volatility spells uncertainty for both oil-exporting countries and major consuming nations, such as the US and Japan. Consequently, the recent proposal of a price band between $22 and $28 a barrel may well be in the best interest of OPEC members. Strictly speaking, the seed of volatile oil price is contained in the switching elasticity theory by Greenhut et al. (1974). Thus, the logarithmic difference is needed to ensure a reasonable estimation of the demand structure. Unless the Western economy is recession-proof, a prediction of oil prices relying only on estimated price elasticity may be inadequate. A large conditional variance could be a harbinger for recession, a distinct possibility currently facing major economies. From our simulation given the 4% cut in OPEC production, the oil price is expected to increase unless the recession, if it exists, is severe. The magnitude and scope of a price hike would be diminished if non-OPEC or domestic production were expanded, as is planned by the Bush administration in its energy policy.

Acknowledgements

The authors would like to thank the co-editor and two anonymous referees for very helpful comments on an earlier draft.

References

Alhajji, A.F., Huettner, D., 2000a. The target revenue model and the world oil market: empirical evidence from 1971 to 1994. Energy J. 21 (2), 121–144.

Alhajji, A.F., Huettner, D., 2000b. OPEC and world crude oil markets from 1973 to 1994: cartel, oligopoly, or competitive? Energy J. 21 (3), 31–58.

Aperjis, D., 1982. The Oil Market in the 1980s, OPEC Oil Policy and Economic Development. Ballinger Publishing Company, Cambridge, MA.

Arize, A.C., 2000. US petroleum consumption behavior and oil price uncertainty: test of cointegration and parameter instability. Atl. Econ. J. 28 (4), 463–477.

Bentzen, J., Engsted, T., 1993. Short- and long-run elasticities in energy demand: a cointegration approach. Energy Econ. 15 (1), 9–16.

Bentzen, J., Engsted, T., 1996. On the estimation of short- and long-run elasticities in US petroleum consumption: comment. South. Econ. J. 62, 783–787.

Bollerslev, T., 1986. Generalized autoregressive heteroskedasticity. J. Economet. 52, 307–327.

Chaudhuri, K., 2001. Long-run prices of primary commodities and oil prices. Appl. Econ. 33, 531–538.

Cremer, J., Isfahani, D., 1991. Models of the Oil Market. Harwood, New York.

Dasgupta, P.S., Heal, G.M., 1979. Economic Theory and Exhaustible Resources. Cambridge University Press, Cambridge.

Department of Energy, 1998. Annual Energy Review. Energy Information Administration, DOE.

Engle, R.F., 1982. Autoregressive conditional heteroskedasticity with estimates of the variance of UK inflation. Econometrica 50, 987–1008.

Ezzati, A., 1976. Future OPEC price and production strategies as affected by its capacity to absorb oil revenues. Eur. Econ. Rev. 5 (3), 190–194.

Ferderer, J.P., 1996. Oil price volatility and the macroeconomy. J. Macroecon. 18, 1–26.

Greenhut, M.L., Ohta, H., 1975. Theory of Spatial Pricing and Market Area. Duke University Press, Durham.

Greenhut, M.L., Hwang, M.J., Ohta, H., 1974. Price discrimination by regulated motor carriers: comment. Am. Econ. Rev. 64 (4), 780–784.

Griffin, J.M., 1985. OPEC behavior: a test of alternative hypotheses. Am. Econ. Rev. 75 (5), 954–963.

Griffin, J.M., Teece, D.J., 1982. OPEC Behavior and World Oil Prices. Allen & Urwin, London.

Hamilton, J.D., 1994. Time Series Analysis. Princeton University Press, Princeton.

Houthakker, H., 1979. The Political Economy of World Energy. Harvard Institute of Economic Research. Discussion Paper No 617.

Huang, B.N., Yang, C.W., 2000. The impact of financial liberalization on stock price volatility in emerging markets. J. Comp. Econ. 28 (2), 321–339.

Huntington, H.G., 1998. Crude oil prices and US economic performance: where does the asymmetry reside? Econ. J. 19 (4), 107–132.

International Monetary Fund, 2001. International Financial Statistics, CD-ROM, Data Base, Washington D.C.

Johany, A.D., 1979. OPEC and the price of oil: cartelization or alteration of property rights. J. Energy Dev. 5 (1), 72–80.

Jones, C.T., 1993. A single-equation study of US petroleum consumption: the role of model specification. South. Econ. J. 59, 687–700.

MacAvoy, P.W., 1982. Crude Oil Prices: As Determined by OPEC and Market Fundamentals. Ballinger Publishing Company, Cambridge.

Mandelbrot, B., 1963. The variation of certain speculative prices. J. Bus. 36, 394–419.

Mead, W.J., 1979. The performance of government energy regulations. Am. Econ. Rev. Proc. 69, 352–356.

Mork, K.A., 1989. Oil and the macroeconomy when prices go up and down: an extension of Hamilton's results. J. Polit. Econ. 91, 740–744.

Noreng, O., 1978. Oil Politics in the 1980s. McGraw-Hill Book Company, New York.

Pakravan, K., 1984. Estimation of user's cost for depletable resource such as oil. Energy Econ. 6, 35–40.

Pindyck, R.S., 1978. Gains to producers from the cartelization of exhaustible resources. Rev. Econ. Stat. 60, 238–251.

Pindyck, R.S., 1999. The long-run evolution of energy prices. Energy J. 20 (2), 1–27.

Pindyck, R.S., Hnyilicza, E., 1976. OPEC and the monopoly price of world oil. Eur. Econ. Rev. 8, 139–154.

Salehi-Isfahani, D., 1987. Testing OPEC Behavior: Further Results. Department of Economics, Virginia Polytechnic Institute and State University. Working Paper #87-01-02.

Teece, D.J., 1982. OPEC behavior: an alternative view. In: Griffin, J.M., Teece, D.J. (Eds.), OPEC Behavior and World Oil Prices. Allen & Urwin, London.

Tourk, K., 1977. The OPEC cartel: a revival of the dominant firm theory. J. Energy Dev. 7 (2), 321–328.

United States Government Printing Office, 1998. Economic Report of the President. US GPO, Washington, DC.

Verleger, P.K., 1987. The evaluation of oil as a commodity. In: Jocoby, R., Zimmerman, M. (Eds.), M. Adelman, Energy — Markets and Regulation: Essays in Honor of M.A. Adelman. MIT Press, Cambridge, MA.

International Review of Financial Analysis
9:3 (2000) 281–297

IRFA

INTERNATIONAL REVIEW OF
Financial Analysis

Causality and cointegration of stock markets among the United States, Japan, and the South China Growth Triangle

Bwo-Nung Huang[a],*, Chin-Wei Yang[b], John Wei-Shan Hu[c]

[a]*Department of Economics, National Chung-Cheng University, Chia-Yi, 621 Taiwan, ROC*
[b]*Department of Economics, Clarion University of Pennsylvania, Clarion, PA 16214-1232, USA*
[c]*Department of Business Administration, Chung-Yuan Christian University, Chung-Li 32023, Taiwan, ROC*

Abstract

This paper explores the causality and cointegration relationships among the stock markets of the United States, Japan and the South China Growth Triangle (SCGT) region. Applying the recently advanced unit root and cointegration techniques that allow for structural breaks over the sample period (October 2, 1992 to June 30, 1997), we find that there exists no cointegration among these markets except for that between Shanghai and Shenzhen. By invoking the Granger causality test and considering the non-synchronous trading problem, we will show that stock price changes in the US have more impact on SCGT markets than do those of Japan. More specifically, price changes in the US can be used to predict those of the Hong Kong and Taiwan markets on next day. Similarly, price changes on the Hong Kong stock market lead the Taiwan market by 1 day. Furthermore, the stock returns of the US and Hong Kong markets are found to be contemporaneous. Finally, there is a significant feedback relationship between the Shanghai and the Shenzhen Stock Exchanges. © 2000 Elsevier Science Inc. All rights reserved.

JEL classification: G15; G12; C32

Keywords: Causality test; South China Growth Triangle; Structural break

1. Introduction

Based on the estimate by Masood and Gooptu (1993), the flow of portfolio investments to emerging financial markets has taken a quantum leap from $6.2 billion in 1987 to 37.2 billion

---

* Corresponding author.
E-mail address: ecdbnh@ccunix.ccu.edu.tw (B.-N. Huang)

Table 1
Basic information of SCGT stock markets

	1991	1992	1993	1994	1995	1996	1997	1998
China								
No. of listed companies	14	52	183	291	323	540	743	853
Market capitalization	2,028	18,255	40,567	43,521	42,055	113,755	206,366	231,322
Percent of total emerging market	0.22	1.82	2.40	2.27	2.18	5.01	9.38	12.12
Trading value	820	16,715	43,395	97,526	49,774	256,008	369,574	284,766
Percent of total emerging market	0.13	2.65	3.93	5.86	4.78	16.95	13.86	14.55
Turnover ratio (%)		158.9	164	233.4	115.9	329	231	130.1
Taiwan								
No. of listed companies	221	256	285	313	347	382	404	437
Market capitalization	124,864	101,124	195,198	247,325	187,206	273,608	287,813	260,015
Percent of total emerging market	13.75	10.11	11.56	12.93	9.70	12.04	13.08	13.63
Trading value	365,232	240,667	346,487	711,346	383,099	470,193	1,297,474	884,698
Percent of total emerging market	59.29	38.13	31.39	42.72	36.76	31.13	48.66	45.21
Turnover ratio (%)	330.1	209.3	235.5	323.1	174.9	204.4	440	333.2
Emerging market								
Market capitalization	907,871	1,000,319	1,688,781	1,913,273	1,929,050	2,272,184	2,200,591	1,908,258

Percent of total world market	6.95	9.36	11.13	10.84	12.65	12.05	9.16	8.00
Trading value	1,956,858	2,666,647	1,510,529	1,042,297	1,665,141	1,103,746	631,188	615,965
Percent of total world market	8.55	13.69	11.09	10.20	18.88	15.34	13.20	12.27
Hong Kong								
No. of listed companies	658	658	561	518	529	450	386	333
Market capitalization	343,394	413,323	449,381	303,705	269,508	385,247	1,721,06	1,219,86
Percent of total developed market	1.34	1.94	2.48	1.92	2.04	3.13	1.73	1.17
Trading value	205,918	489,365	166,419	106,888	1,471,58	1,315,50	78,598	38,607
Percent of total developed market	0.98	2.91	1.37	1.16	2.06	2.16	1.89	0.88
Market capitalization	25,553,855	21,317,929	18,139,951	15,859,021	13,210,778	12,327,242	9,923,024	10,434,218
Percent of total world market	93.05	90.64	88.87	89.16	87.35	87.95	90.84	92.00
Trading value	20,917,462	16,818,167	12,105,541	9,176,451	7,156,704	6,096,929	4,151,662	4,403,631
Percent of total world market	91.45	86.31	88.91	89.80	81.12	84.74	86.80	87.73
World								
Market capitalization	27,462,113	23,518,520	20,412,135	17,788,071	15,124,051	14,016,023	10,923,343	11,342,089
Trading value	22,874,320	19,484,814	13,616,070	10,218,748	8,821,845	7,194,675	4,782,850	5,019,596

Source: Emerging Stock Market Factbook 1999, International Financial Corporation (in billions of US dollars).

44

in 1992. The importance of emerging markets cannot be denied, especially for the market of the People's Republic of China (PRC). The growing PRC market consists of the Shanghai Stock Exchange and the Shenzhen Stock Exchange. The former commenced operations in December 1990 while the latter opened its market in April 1991. The rapid expansion of the PRC markets reflects China's significant economic growth. As shown in Table 1, trade volume, market capitalization, and number of listed companies have mushroomed for the 7-year period from 1991–1998. In the beginning, the market capitalization and trade volume accounted for less than 1% of that in the entire emerging markets. However, they have increased to 12.12% and 14.55%, respectively, in 1998. As the PRC gradually opens her markets to international trade, the concept of the South China Growth Triangle (SCGT) begins to take shape. Thus far, the SCGT consists of Hong Kong, Taiwan and the Southern part of the PRC.

Improved relationship between the countries, coupled with the return of Hong Kong to the PRC on July 1, 1997, has strengthened ties within the SCGT. Given that the three economies adopt export-led policies, will the interaction become stronger due to greater trade within the SCGT? Are members of the SCGT as much economically tied to the US and Japan as before? These are the questions this paper addresses.

Quite a few papers address the issue of capital markets in emerging economies (Eun & Shim, 1989; Dwyer & Hafer, 1993; Ma, 1993; Heston et al., 1995). The cointegration relationship among different markets was addressed by Wei et al. (1995) and Hu et al. (1997). Despite the analyses of the PRC market by Bailey (1994), Ma (1996) and Su and Heisher (1997), the literature on the market interactions between the PRC and other major economies are lacking. This paper intends to fill some of these gaps. It is to be noted that the three markets have begun to play more important roles in the world economy. For instance, the Taiwan market accounts for 13% and 45% in terms of market capitalization and trade volume in the emerging markets. The Hong Kong market has a share of approximately 2% of the developed markets in both market capitalization and trade volume. Altogether, the three markets of the SCGT account for about 3% to 4% in terms of world market capitalization value, and 5.3% to 11.0% in world trade volume. Owing to their growing importance, a study on interactions within the SCGT is of paramount interest.[1] Although no long-term cointegration relationships have been identified within the members of the SCGT, or between SCGT and the US, or between SCGT and Japan, we demonstrate a significant Granger causality for the markets of Taiwan–US, and Hong Kong–US. In addition, the Hong Kong market is found to lead the Taiwan market. Although the two markets in the PRC exhibit a feedback relationship, they do not appear to have any short-term relationship with other markets of the SCGT, the US or Japan.

The structure of this paper is as follows. The next section reviews the literature. The third section explains the stock markets of the PRC. The fourth section presents data and basic statistics of the U.S., Japanese and SCGT markets. The fifth section discusses unit root and cointegration results of these markets. We then explore the pairwise causality between SCGT and the US, SCGT and Japan, Shanghai and Shenzhen, and among SCGT members. A conclusion is given in the last section.

---

[1] Refer to Table 1 for basic information on the three markets.

2. Literature review

Recent studies on equity market integration and segmentation include contributions by Errunza and Losq (1987), Bekaert and Harvey (1995), and Heston et al. (1995).[2] They, to various extents, apply statistical models to study the time-varying cointegration property of different equity markets. The study on market interdependence can be traced back to as early as Granger and Morgenstern (1970). Subsequent analyses include Ripley (1973), Lessard (1974, 1976), Panto et al. (1976) and Hilliard (1979). Their conclusions generally point to the following observation: national stock indices are merely a reflection of their economies, and as such only weak correlations are found. This being the case, an international portfolio could be quite advantageous. Hilliard's paper (1979), using spectrum analysis, focused on the correlation relations (contemporaneous lagged daily stock prices) of 10 major markets. Contemporaneous correlation was found to be significant for intercontinental stock prices but weak for intracontinental prices. In other studies (Ripley, 1973; Ibbotson et al., 1982), noticeable stock price comovements were identified due to factors such as geographical proximity, institutional currency relationships, partnerships in trade and cultural or economic base similarities. Recent studies by Schollhammer and Sand (1987), Dwyer and Harter (1988), employing cross correlation techniques or the unit root model (daily or monthly returns) have indicated a lack of correlation in stock prices among the U.S. Japanese, German, and the U.K. markets. However, variations in the stock prices were found to have been correlated before the 1987 market crash. The bottom line is how one accurately estimates the correlation coefficient matrix. Eun and Resnick (1984) compare various techniques of estimating the direct and indirect correlation coefficients with the assumption that securities are correlated with one another only through common factors that respond to the world market index. Their conclusion favors the direct estimation approach. Another important question in estimating correlation coefficients is its stability in terms of long-term and intertemporal periods. For instance, intertemporal stability is refuted by Maldonado and Saunders (1981) who employ the monthly stock prices (1957–1978) in the US, Japanese, German, Canadian and U.K. markets in their study. On the other hand, intertemporal stability is supported by Watson (1980) and Philippatos et al. (1983). More recently, Longin and Solnik (1995) document the evidence of instability of both correlation and covariance coefficients via the multivariate GARCH (1,1) model. Nonetheless, in the wake of unabated and synchronized growth across various countries, there is good reason to believe that interdependence among national stock indices is on the rise.

Another key angle in investigating the interdependence of average stock returns focuses on the transmission mechanism. For instance, Eun and Shim (1989) identify some multilateral interactions via the impulse response function of a 9-country VAR model. Innovations from the U.S. market are rapidly transmitted abroad, but the innovations abroad cannot significantly impact US stock returns.[3] In addition, dynamic responses of

---

[2] Readers are referred to Adler and Dumas (1983), Cho et al. (1986), Gultekin et al. (1989), Korajczyk and Viallet (1989), Ambrose et al. (1992) and Foerster and Karolyi (1993).

[3] Eun and Shim (1989) did not explain the implications of the significance level in terms of statistical testing. Rather, they define the significance level as the one that has any positive absolute value of the impulse response function.

national markets are shown to be generally consistent with the efficient information hypothesis. In another study, Joen and Von Furstenberg (1990) analyze the interaction among major markets (the U.S. Britain, Japan and Germany) by using the impulse response functions of a VAR model. Sifting through the daily returns (January 1986 to November 1988) at market close, they examine the correlation coefficient matrix and find that the role of U.S. leadership has undergone some structural changes after the 1987 crash. That is, the leadership role by the U.S. has somewhat diminished while that of Japan has increased. In addition, the phenomenon of stock price comovement seems to be on the rise continually. Employing recently advanced unit root and cointegration techniques that accommodate structural breaks, we intend to examine lead, lag or feedback relations via the Granger causality among the U.S., Japan and SCGT members. The technique provides more accurate results (Granger et al., 2000) and the topic is potentially interesting and important: Can cultural and linguistic similarity overcome the differences in economic systems?

3. Development of the Chinese stock markets

The economic reforms started in 1978 led to the rebirth of the stock markets in China. The Shanghai Stock Exchange and the Shenzhen Stock Exchange are the two major emerging capital markets in China. The Shanghai Stock Market was officially opened in 1990 and the Shenzhen Stock Market in 1991. Two types of stocks are traded in the two markets: class A and class B. Class A shares are restricted to Chinese (PRC) citizens and denominated in Chinese currency yuan or Renminbi (RBM), while class B shares can be bought and sold only by foreigners and are settled in foreign currencies (US dollars for Shanghai, Hong Kong dollars for Shenzhen). By the end of 1995, there were 135 companies and 161 stocks listed on the Shenzhen Stock Exchange.[4] Regardless of some inevitable difficulties in its infancy stage, the rapid development of China's capital markets has generated interest among academics, investors and regulators. In this study, we focus our analysis on the relationship of share A markets with four other markets. There are at least two reasons for examining class A share markets. First, the class B shares market has been losing its appeal to foreign investors while the class A share market dominates that of class B shares in terms of the number of listed companies, trade volume and market capitalization. Second, it allows us to address an interesting issue: how the volatility of a market, which is largely closed to foreign investors, is related to the volatility of foreign markets.

4. Data and basic statistics

Sample data included in this study comprise daily stock prices at market close (close quotes October 1, 1992 through June 30, 1997). Specifically, we take (i) the Taiwan weighted

---

[4] Refer to the basic information in Table 1.

Table 2
Basic statistics of the six stock indices

	HKN	JPN	SHS	SHZ	TWN	USA
Mean	0.0817	0.0138	0.0466	0.0366	0.0747	0.0693
Median	0.0258	0.0000	0.0000	0.0000	0.0000	0.0486
Maximum	5.7072	7.5509	28.8602	27.2152	7.4081	2.6049
Minimum	−8.3484	−5.7611	−17.9047	−18.8833	−7.7818	−3.0823
Std. Dev.	1.4339	1.2151	3.4766	3.1175	1.5100	0.6771
Skewness	−0.4074	0.2715	1.4335	0.9150	0.0206	−0.3383
Kurtosis	6.5837	6.3442	15.4341	13.9098	6.1000	4.9157
JB	696.19*	591.62*	8,392.41*	6,307.30*	495.42*	212.75*
N	1,237	1,237	1,237	1,237	1,237	1,237

Daily rate of return for the six stock markets is calculated based on the conventional first difference of logarithmic prices:

$$r_{t,i} = (\log p_{t,i} - \log p_{t-1,j}) \times 100$$

where $r_{t,i}$ denotes the rate of return for the ith market on day t and $p_{t,i}$ denotes the corresponding stock price index. HKN = Hong Kong, JPN = Japan, SHS = Shanghai, SHZ = Shenzen, TWN = Taiwan. JB = Jarque–Bera Statistic. N = number of observations. Std. Dev. = Standard deviation.
* 1% significance level.

volume index (TWN), (ii) the Nikkei 225 Index of Tokyo (JPN), (iii) the Dow Jones Industrial Average of the U.S. (US), (iv) the Hang Seng Index of Hong Kong (HKN), (v) the Shanghai Index (class A shares) or SHS and (vi) the Shenzhen Index (class A shares) or SHZ. All data is obtained from the Datastream database.[5] Daily percentage changes of stock prices for the six stock markets are calculated using the conventional first difference of logarithmic prices [Eq. (1)]:

$$\triangle y_{t,i} = (\log y_{t,i} - \log y_{t-1,i}) \times 100 \tag{1}$$

where $\triangle y_{t,i}$ denotes the percentage change of stock price for the ith market on day t and $y_{t,i}$ denotes the corresponding stock price index. The descriptive statistics reported in Table 2 indicate that emerging markets tend to have greater leptokurtosis with more pronounced fluctuations. It is not surprising that the Shenzhen and the Shanghai markets have the greatest kurtosis.[6] Likewise, both the Shanghai and the Shenzhen markets have the largest standard deviations as well. These results are consistent with the finding by Bekaert and Harvey (1995) that return volatility in emerging markets is greater than that of developed markets.

The autocorrelation functions (ACF) shown in Table 3 exhibit some interesting characteristics: The coefficients of the ACFs of the U.S. and Japanese markets (barring the marginal significances at $k = 1$ and $k = 3$) are statistically insignificant for all 12 lags.

---

[5] We are grateful for the generosity extended by the economics department of UCSD.

[6] Greater kurtosis values for the Shanghai and the Shenzen markets are indicative of the clustering of stock price changes around their mean or median values with a handful of large changes. In other words, the high kurtosis values may have been related to relatively short history of the two markets in which the great majority of price changes occurred around the mean.

Table 3
Coefficients of ACF of the six stock indices

	HKN	JPN	SHS	SHZ	TWN	USA
1	−0.0270	−0.0573*	0.0059	0.0160	−0.0281	0.0404
2	0.0403	0.0020	0.0381	0.0617*	0.0312	0.0133
3	0.0483	−0.0052	0.0899*	0.0132	0.0481	−0.0634*
4	−0.0120	0.0190	0.0423	0.0955*	−0.0072	−0.0316
5	0.0337	−0.0359	0.0140	0.0108	0.0329	0.0068
6	−0.0486	−0.0125	0.0323	−0.0644*	−0.0486	−0.0101
7	−0.0238	−0.0085	−0.0067	−0.0175	−0.0278	0.0077
8	0.0469	0.0052	−0.0738*	−0.0204	0.0490	−0.0222
9	−0.0132	0.0225	0.0677*	0.0157	−0.0114	−0.0043
10	0.0648*	0.0030	−0.0209	−0.0417	0.0635*	0.0169
11	−0.0016	0.0204	0.0842*	−0.0005	−0.0008	−0.0290
12	0.0209	0.0329	0.0741*	0.0360	0.0158	0.0015

* 5% significance level.

In contrast, the Shanghai and the Shenzhen markets exhibit clear statistical significance of varying degrees in the first 12 lags. Between these two extremes, the Taiwan and the Hong Kong markets have their significant ACF coefficients in the 10th lag periods. It verifies the result by Bekaert and Harvey (1995) that the ACFs have some significant lag effects in stock returns of emerging markets.

Trend plots of the six stock indices except that of Japan (Fig. 1) reveal steady growth paths. This growth trend is most conspicuous in the US market, followed by Hong Kong. Moreover, structural changes as manifested in these indices are readily detectable, and as such, proper procedures of unit root and cointegration are necessary for valid statistical estimations.

5. Unit root and cointegration tests with regime shifts

Prior unit root analyses rest largely on the Augmented Dickey–Fuller (ADF) test developed by Dickey and Fuller (1979) in the following equation:

$$\triangle y_t = \alpha + \beta t + (\rho - 1)y_{t-1} + \sum_{i=1}^{k-1} \theta_i \triangle y_{t-i} + a_t \tag{2}$$

where $\triangle = 1 - L$, y_t = stock price at time period t, t = trend variable, and a_t obeys white noise with the null hypothesis H_0: $\rho = 1$. Failure to reject the null hypothesis implies a unit root in y_t. Popular as it is, the ADF test statistic can very likely lead to erroneous conclusions in the event of a regime shift such as a depression or an oil shock (Perron, 1989). The crux of the matter is that the traditional ADF test fails to reject the H_0 in the presence of structural break(s). To circumvent this problem Perron and Vogelsang (1992) include a dummy variable in Eq. (2). However, the use of dummy variables has its own problem as pointed out by Zivot and Andrews (1992, pp. 251). A skeptic of Perron's approach would argue that his choices of breakpoints are based on prior observations of the data, and hence problems associated with

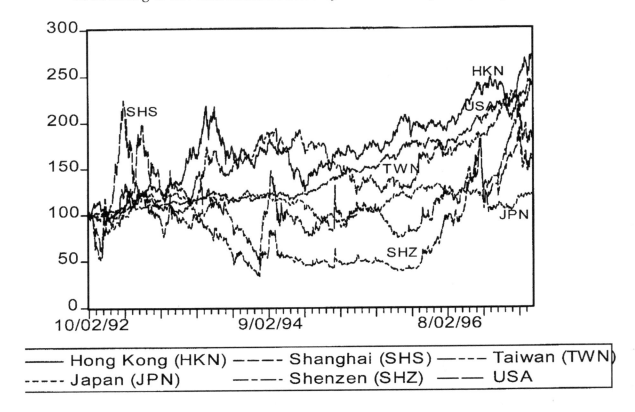

Fig. 1. Stock price trends of the US, Japan, Hong Kong, China, and Taiwan (10/05/1992 = 100).

'pre-testing' are applicable to his method. Consequently, Zivot and Andrews (1992) modify the ADF specification as shown below[7]:

$$\triangle y_t = \alpha + \beta t + \gamma DU_t(\lambda) + (\rho - 1)y_{t-1} + \sum_{i=1}^{k-1} \theta_i \triangle y_{t-i} + a_t \tag{3}$$

where $DU_t(\lambda) = 1$ for $t > T\lambda$, otherwise $DU_t(\lambda) = 1$; $\lambda = T_B/T$ represents the location where the structural break lies; T is sample size; T_B is the date when the structural break occurred. Note that estimation results hinge upon λ. As such, Zivot and Andrews (1992) simulate a set of critical value for different λ.

A perusal of Table 4 reveals that except in the case of Taiwan (with significance level slightly less than 5%), we fail to reject the null hypothesis. This is to say that the logarithmic stock prices in the study are largely of $I(1)$.[8] In their influential work, Engle and Granger (1987) pioneer the cointegration technique especially in the case of $I(1)$ variables. Despite the normalization problem, the Engle–Granger model remains popular for its simplicity and clarity. Unfortunately, it suffers from the same problem as does the ADF model. It could lead to erroneous results in the presence of a structural break. For this reason, Gregory and Hansen (1996) proposed a new test procedure that combines both the Engle and Granger (1987) and

---

[7] Owing to the presence of trends (Fig. 1), we adopt model A of Zivot and Andrews (1992) which includes a trend variable.

[8] The logarithmic stock prices of Taiwan are considered $I(1)$ and not found to be cointegrated with that of other markets.

Table 4
Unit root test

	HKN	JPN	SHS	SHZ	TWN	USA
Z&A	−3.71 [0.44]	−3.09 [0.67]	−3.66 [0.24]	−3.67 [0.76]	−3.88*[0.53]	−3.67 [0.28]

Z&A = Zivot and Andrews (1992) unit root test based on the following equation:

$$\triangle y_t = \alpha + \beta_t + (\rho - 1)y_{t-1} + \gamma DU_t(\lambda) + \sum_{t=1}^{k-1} \theta_i \triangle y_{t-i} + a_t$$

where $DU_t(\lambda) = 1$ for $t > T \lambda$; otherwise, $DU_t(\lambda) = 0$. $\lambda = T_B/T$ indicates the location of a structural break. $T =$ sample size. $T_B =$ date when a structural break occurred. Numbers in [ ] are break point λ. The sample period spans from 10/01/1992 to 06/30/1997. The smallest test statistics of the Zivot and Andrews (1992) model are used ($k = 4$) for testing hypothesis.
   * 5% significance level.

Zivot and Andrews (1992) models. Theirs is a residual-based cointegration approach that allows for regime shift. It is a two-stage model based on the following multiple regression:

$$y_{1t} = \alpha + \beta t + \gamma DU(\lambda) + \theta y_{2t} + e_t \tag{4}$$

where y_{1t} and y_{2t} are of $I(1)$; y_{2t} is a variable or a set of variables; $DU_t(\lambda)$ is the same as in Eq. (3). After estimating Eq. (4) in the first stage, we proceed to test if the residual e_t is of $I(0)$ or $I(1)$. There exists some cointegrating relation(s) between y_{1t} and y_{2t} if e_t is of $I(0)$. The result of using the Gregory and Hansen model (1996) is reported in Table 5.

The bivariate cointegration tests are performed between (1) members of the SCGT and the US, (2) members of the SCGT and Japan, (3) the Shanghai and Shenzhen Stock Exchange, and among SCGT members. The trivariate cointegration test is applied to the members of the SCGT. A moment's inspection indicates that the test statistics are greater between members of SCGT and the US than between the SCGT and Japan. However, they are not statistically significant according to the critical values provided by Gregory and Hansen (1996). This implies that there exists no long-term cointegrating relation in the stock markets between SCGT members, and either the US or Japan.

The emergence of the SCGT could profoundly tilt the balance of economic power within the region.[9] Despite the similarities in culture and language and close geographical proximity, there does not seem to exist a stable cointegration within the SCGT. In some sense it is not surprising to see three segmented capital markets, because each market has varying degrees of restrictions on capital movements (Wei et al., 1995). The Shanghai market, facing severe restrictions on foreign exchange, has the stiffest controls, followed by Taiwan. Viewed from this perspective, capital controls, more than any other factors, may have played a major role in economic integration or segmentation. As a consequence, we cannot support the notion of cointegrating relations among the capital markets in the SCGT, despite the fact that the intraregional trades have intensified. The presence of the two equity markets in the PRC along with the two-tier system (class A and B shares) offers an interesting case for studying capital market integration and segmentation effects

---

[9] The Shanghai market is taken to represent the PRC. Results remain similar if the Shenzhen market is used.

Table 5
Cointegration test results

HKN on USA	−4.67	HNK on JPN	−3.78
USA on HKN	−4.55	JPN on HKN	−3.02
SHS on USA	−4.32	SHS on JPN	−3.78
USA on SHS	−3.96	JPN on SHS	−3.09
SHZ on USA	−4.98	SHZ on JPN	−4.03
USA on SHZ	−3.96	JPN on SHZ	−3.04
TWN on USA	−3.84	TWN on JPN	−3.92
USA on TWN	−3.76	JPN on TWN	−3.58
SHS on SHZ	−5.35*	TWN on HKN	−3.91
SHZ on SHS	−5.29*	HKN on TWN	−3.68
SHS on HKN	−3.69	SHS on TWN	−3.70
HKN on SHS	−3.77	TWN on SHS	−3.93
SHZ on HKN	−4.10	SHZ on TWN	−4.02
HKN on SHZ	−3.83	TWN on SHZ	−3.89
HKZ on TWN SHS	−3.81	SHS on SHZ	−5.35*
TWN on HKN SHS	−3.95	SHZ on SHS	−5.29*

The first stage of the Gregory and Hansen cointegration model is to estimate the multiple regression that takes into consideration structural breaks:

$$\triangle y_t = \alpha + \beta_t + \gamma DU_t(\lambda) + \theta_1 y_{2t} + e_t$$

where y_{1t} and y_{2t} are of $I(1)$ and y_{2t} is a variable or a set of variables. $DU_t(\lambda) = 1$ for $t > T\lambda$; otherwise, $DU_t(\lambda) = 0$. The second stage involves a unit root test (ADF or Phillips–Perron test) on e_t. If e_t is of $I(0)$, a cointegrating relation is ascertained. The sample period spans from 10/01/1992 to 06/30/1997. The smallest test statistics of Zivot and Andrews (1992) model are used for testing hypothesis. The second stage is based on the ADF test with $k = 4$.
 * 5% significance level.

within a socialist economy. These two markets are expected to exhibit the cointegration since they reflect similar fundamental economic forces. As shown in Table 4, the test statistics indicate that we reject the null hypothesis at 5% significance level; that is, there exists a statistically significant cointegrating relation between the Shanghai and the Shenzhen markets.

6. Causality test

In the absence of long-term equilibrium (cointegration) relations between either the US or Japan and the SCGT members, a study on short-term interactions is in order. We apply the Granger (1969) model, based on the following bivariate VAR model:

$$\triangle y_{1t} = \alpha_0 + \sum_{i=1}^{k} \alpha_{1i} \triangle y_{1t-i} + \sum_{i=1}^{k} \alpha_{2i} \triangle y_{2t-i} + \varepsilon_{1t}$$

$$\triangle y_{2t} = \beta_0 + \sum_{i=1}^{k} \beta_{1i} \triangle y_{1t-i} + \sum_{i=1}^{k} \beta_{2i} \triangle y_{2t-i} + \varepsilon_{2t} \tag{5}$$

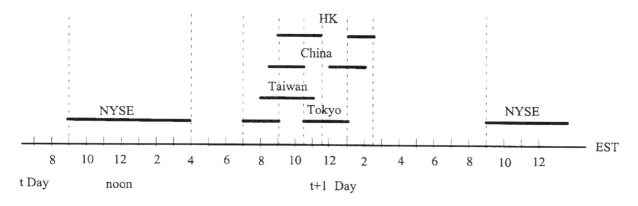

Fig. 2. Trading time pattern (EST, New York Time).

where y_{1t} and y_{2t} denote stock prices of two different countries, and ε_{1t}, ε_{2t} are assumed to be serially uncorrelated with zero mean and finite covariance matrix. When the null hypothesis $H_0:\alpha_{21} = \alpha_{22} = \ldots = \alpha_{2k} = 0$ is retained, it suggests that y_{2t} does not Granger-cause y_{1t}. Conversely, if the null hypothesis $H_0:\beta_{11} = \beta_{12} = \ldots = \beta_{1k} = 0$ is not rejected, it implies that y_{1t} does not Granger-cause y_{2t}. The main problem with the use of daily returns across countries is the nonsynchronous trading periods for different markets around the world. This institutional feature of markets has been the focus of a large literature on international returns and volatility spillovers.[10] Fig. 2 illustrates the trading hours of the five markets in terms of the Eastern Standard Time.

Evident from Fig. 2, the trading hours of all the five markets except the US occur on the same calendar day. The trading hours of the US market take place 1 calendar day before that of other markets. This is to say that we lag the US stock prices by 1 day in performing the Granger causality tests between the US and other markets. On the other hand, we use the price change data of the same calendar day for the markets of Japan and SCGT. The results of the Granger causality are reported in Table 6.

A perusal of Table 6 reveals that there exists no causal relationship between the US and the Shanghai markets. In contrast, the US market is found to lead both the Hong Kong and the Taiwan markets by 1 day in terms of price changes. However, price changes of SCGT members (except Hong Kong) do not lead the US market. Such a 1-day causality is consistent with the result from Wei et al. (1995). The result that the previous day's price movement in the US market has a positive impact on the following day's price movement in the Taiwan and Hong Kong markets is logical. The Hong Kong dollar has been pegged to the U.S. dollar since 1983. As a result, if the U.S. market movement is to reflect the expected change in the interest rate in the U.S., the movement of the U.S. market will have an immediate spillover effect to Hong Kong. On the other hand, high technology industry in Taiwan can actually serve and reflect its counterpart industry in the U.S. This is especially true as the high technology stocks account for between 50% and 60% of the total market capitalization in Taiwan. It is not unexpected, therefore, that the U.S. market has a

---

[10] See Hamao et al. (1990), Bae and Karolyi (1994), Lin et al. (1994), and Connolly and Wang (1995).

Table 6
Granger causality test results

Null hypothesis	F-statistic	Probability
USA $-/\rightarrow$ HKN	144.52	0.00
HKN $-/\rightarrow$ USA	0.55	0.46
USA $-/\rightarrow$ SHS	0.31	0.58
SHS $-/\rightarrow$ USA	2.34	0.13
USA $-/\rightarrow$ TWN	5.88	0.02
TWN $-/\rightarrow$ USA	1.31	0.25
JPN $-/\rightarrow$ HKN	0.12	0.73
HKN $-/\rightarrow$ JPN	0.00	0.96
SHS $-/\rightarrow$ JPN	1.77	0.18
JPN $-/\rightarrow$ SHS	1.37	0.24
TWN $-/\rightarrow$ JPN	0.31	0.58
JPN $-/\rightarrow$ TWN	1.63	0.20
SHS $-/\rightarrow$ HKN	0.11	0.75
HKN $-/\rightarrow$ SHS	0.01	0.91
TWN $-/\rightarrow$ HKN	0.22	0.64
HKN $-/\rightarrow$ TWN	5.75	0.02
TWN $-/\rightarrow$ SHS	0.00	0.99
SHS $-/\rightarrow$ TWN	0.12	0.73

$-/\rightarrow$ denotes does not Granger-cause. The test procedure is based on the bivariate VAR(k) model. The optimal k is based on the Schwartz Criterion (SC). $k = 1$ is found in all cases according to the SC.

major impact on Taiwan. Surprisingly, no such causal relation is found between Japan and SCGT members. The impact from the U.S. market on the SCGT markets seems to be noticeably greater than that from the Japanese market. This may be explained by the favorable trade balances the SCGT members enjoy with the US despite the most-favored-nation status dispute. From this viewpoint, an unexpected change in demand conditions of the U.S. market would certainly have a profound impact on the economic growth of SCGT members. On the other hand, stock price in Hong Kong could have an impact on the US market since it serves as the financial center in Asia. It is little wonder that the globalization of financial markets has intensified the repercussion effect, with news announcements now simultaneously transmitted across continents.[11] As for intraregional stock market interactions, it can be seen that price changes of the Hong Kong market lead that of the Taiwan market.

---

[11] For instance, during the Asian financial crisis starting October of 1997, the plummeting of the Hang Seng Index (1438 points) transmits its impact to the Dow Jones index (554.26-point fall).

Table 7
Cointegration and ECM results of the Shanghai and the Shenzen markets

ECM $_{t-1}$	C	LSHZ$_{t-1}$		
LSHS$_{t-1}$	3.6286	0.5696		
		[6.3573]		

| | | | VAR(1) | |
	C		SHS$_{t-1}$	SHZ$_{t-1}$
SHS$_t$	0.0005	−0.0197	−0.027	0.0692
	[0.4601]	[−2.7531]	[−0.7248]	[1.6737]
SHZ$_t$	0.0003	0	0.0667	−0.0319
	[0.3917]	[0.0044]	[1.9939]	[−0.8572]

LSHS and LSHZ denote logarithmic stock price indices of Shanghai and Shenzen, respectively; SHS and SHZ denotes first difference of logarithmic stock price indices of Shanghai and Shenzen, respectively; *t*-statistics are reported in brackets; ECM reflects a long-term equilibrium relationship; *C* is the intercept term.

The existence of cointegrating relations between the Shanghai and Shenzhen markets suggests the use of the error correction term in testing for causality. The causality test based on the error correction model (ECM) is shown in Eq. (6):

$$\triangle y_{1t} = \alpha_0 + (y_{1t-1} - \gamma y_{2t-1}) + \sum_{i=1}^{k} \alpha_{1i}\triangle y_{1t-i} + \sum_{i=1}^{k} \alpha_{2i}\triangle y_{2t-i} + \varepsilon_{1t} \tag{6}$$

$$\triangle y_{2t} = \beta_0 + (y_{1t-1} - \gamma y_{2t-1}) + \sum_{i=1}^{k} \beta_{1i}\triangle y_{1t-i} + \sum_{i=1}^{k} \beta_{2i}\triangle y_{2t-i} + \varepsilon_{2t}$$

where δ_1 and δ_2 reflect the speeds of adjustment to the equilibrium, and other specifications remain similar to Eq. (5). According to Engle and Granger (1987), the existence of cointegrating relations implies causality, and as such $|\delta_1| + |\delta_2| > 0$. Reported in Table 6 are long-term equilibrium relations and results of the ECM between the two stock markets.

Table 7 indicates a long-term equilibrium relation between the Shenzhen and the Shanghai stock markets: 1% price change in the former is accompanied by 0.57% price changes in the latter. Based on the transaction data (December 1994 to June 1997), volume in the Shanghai market is 2.1 times as large as that of the Shenzhen market and thus, the magnitude of price changes are within the expected range. In the price change (Shanghai) equation of the ECM, the error correction term is found to be statistically significant, and lagged price changes (by one period) of the Shenzhen market can be used to predict current price changes of the Shanghai market within 10% significance level. This implies that the Shenzhen stock market leads the Shanghai market both in the long run and short run. In the price change (Shenzhen) equation of the ECM, the error correction term is insignificant, but the lagged price changes of Shanghai stocks are good predictors for that of Shenzhen stocks with a 5% significance level. This is indicative of the leadership role of the Shanghai market in the short run. That is, the bilateral leadership role leads inevitably to a double feedback relation between the two markets.

As discussed above, the two markets of PRC exhibit a long-term cointegration relationship between them. However, they do not form any relationship with other members of SCGT, the US or Japan, be it long-run or short-run. The Hong Kong market leads the Taiwan market while the price changes (1 day before) of the US markets impact those of both markets. In short, we conclude that the PRC markets are the most segregated, followed by Taiwan. The Hong Kong market, being the most internationalized financial center in Asia, plays an important role in the analysis.

7. Concluding remarks

Despite the recent economic slumps in Association of South East Nations (ASEAN), economic growth is still on the rise for SCGT members. One may surmise that common culture (Confucianism) and language (Chinese) and other traits ought to give rise to an integrated capital market. Much to our surprise, this is not the case even though the Shanghai and the Shenzhen markets are found to be statistically cointegrated. We suggest that an ease of restrictions on capital movements among SCGT members could serve as a catalyst in forging greater market integration. In addition, Japan does not form any cointegrating relation with members of SCGT. There exists a stronger Granger causality between the US and members of SCGT. US price changes can be used to predict subsequent day price changes in the Hong Kong and the Taiwan stock markets. We have also shown that price changes in the Hong Kong market lead that in the Taiwan market. Finally, we have identified a strong feedback interaction between the Shanghai and the Shenzhen markets.

Acknowledgments

Financial assistance from the National Science Council (Taiwan) and Fulbright Scholarship Program is gratefully acknowledged. This paper is written during my tenure as a visiting scholar at UCSD. We are grateful to two anonymous referees and editor for valuable suggestions. However, we wish to absorb all the culpability.

References

Adler, M., & Dumas, B. (1983). International portfolio selection and corporation finance: a synthesis. *Journal of Finance, 38*, 925–984.

Ambrose, B. W., Ancel, E., & Griffiths, M. D. (1992). The fractal structure of real estate investment trust returns: the search for evidence of market segmentation and nonlinear dependency. *Journal of American Real Estate & Urban Economics Association, 20*, 25–54.

Bae, K.-H., & Karolyi, G. A. (1994). Good news, bad news, and international spillovers of stock return volatility between Japan and the US. *Pacific-Basin Financial Journal, 2*, 405–438.

Bailey, W. (1994). Risk and return on China's new stock markets: some preliminary evidence. *Pacific-Basin Financial Journal, 2*, 243–260.

Bekaert, G., & Harvey, C. R. (1995). Time-varying world market integration. *Journal of Finance, 50*, 403–444.

Cho, D. C., Eun, C. S., & Senbet, L. W. (1986). International arbitrage pricing theory: an empirical investigation. *Journal of Finance, 41*, 313–329.

Connolly, R., & Wang, A. (1995). *Can economic news explain the US–Japan stock market return and volatility linkages?* Working Paper, Columbia University.

Dickey, D. A., & Fuller, W. A. (1979). Distribution of the estimators for autoregressive time series with a unit root. *Journal of American Statistical Association, 74*, 427–431.

Dwyer, G. P. Jr., & Hafer, R. W. (1993). Are national stock markets linked? In S. R. Stansell (Ed.), *International Financial Market Integration,* (pp. 235–258). Cambridge, MA: Blackwell.

Dwyer, G. P. Jr., & Harter, R. W. (1988). Are national stock markets linked? *Federal Reserve Bank of St. Louis Review, 39*, 3–14.

Engle, R. F., & Granger, C. W. J. (1987). Cointegration and error correction: representation, estimation, and testing. *Econometrica, 55*, 251–276.

Errunza, V., & Losq, E. (1987). How risky are emerging markets? *Journal of Portfolio Management, 14*(1), 62–77.

Eun, C. S., & Resnick, B. G. (1984). Estimating the correlation structure of international share prices. *Journal of Finance, 39*, 1311–1324.

Eun, C. S, & Shim, S. (1989). International transmission of stock market movements. *Journal of Financial Quantitative Analysis, 24*, 241–256.

Foerster, S. R., & Karolyi, G. A. (1993). International listings of stocks: the case of Canada and the U.S. *Journal of International Business Studies, 24*, 763–784.

Granger, C. W. J. (1969). Investigating causal relations by econometric models and cross-spectral methods. *Econometrica, 37*, 424–439.

Granger, C. W. J., Huang, B.-N., & Yang, C.-W. (2000). A bivariate causality between stock prices and exchange rates: evidence from recent Asian flu. *Quarterly Review of Economics and Finance,* in press.

Granger, C. W. J., & Morgenstern, O. (1970). *Predictability of stock market prices.* Lexington, MA: Heath Lexington Books.

Gregory, A. W., & Hansen, B. E. (1996). Residual-based tests for cointegration in models with regime shifts. *Journal of Economomics, 70*, 99–126.

Gultekin, M. N., Gultekin, N. B., & Penati, A. (1989). Capital controls and international capital market segmentation: the evidence from the Japanese and American stock markets. *Journal of Finance, 44*, 849–869.

Hamao, Y., Masulis, R. W., & Ng, V. (1990). Correlations in price changes and volatility across international stock markets. *Review of Financial Studies, 3*, 281–307.

Heston, S. L., Rouwenhorst, K. G., & Wessels, R. E. (1995). The structure of international stock returns and the integration of capital markets. *Journal of Empirical Finance, 2*, 173–197.

Hilliard, J. (1979). The relationship between equity indices on world exchanges. *Journal of Finance, 4*, 103–114.

Hu, J. W.-S., Chen, M.-Y., Fok, R. C. W., & Huang, B.-N. (1997). Causality in volatility spillover effects between US, Japan and four equity markets in the South China growth triangle. *International Financial Market, Institution and Money, 7*, 351–367.

Ibbotson, R., Carr, R., & Robinson, A. (1982). International equity and stock returns. *Financial Analysts Journal, 38*(4), 61–83.

Joen, B. N., & Von Furstenberg, G. M. (1990). Growing international co-movement in stock price indexes. *Quarterly Review of Economics and Finance, 30*, 15–30.

Korajczyk, R. A., & Viallet, C. J. (1989). An empirical investigation of international asset pricing. *Review of Financial Studies, 2*, 553–585.

Lessard, D. R. (1974). World, national and industry factors in equity returns. *Journal of Finance, 29*, 379–391.

Lessard, D. R. (1976). World, country, and industry relationships in equity returns: implications for risk reduction through international diversification. *Financial Analysts Journal, 32*, 2–8.

Lin, W. L., Engle, R. F., & Ito, T. (1994). Do bulls and bears move across borders? Transmission of international stock returns and volatility. *Review of Financial Studies, 7*, 507–538.

Longin, F., & Solnik, B. (1995). Is the correlation in international equity returns constant: 1960–1990? *Journal of International Money and Finance, 14*(1), 3–26.

Ma, C. K. (1993). Financial market integration and cointegration tests. In S. R. Stansell (Ed.), *International financial market integration* (pp. 228–298). Cambridge, MA: Blackwell.

Ma, X. (1996). Capital controls, market segmentation and stock prices: evidences from the Chinese stock markets. *Pacific-Basin Financial Journal, 4*, 219–239.

Maldonado, R., & Saunders, A. (1981). International portfolio diversification and the intertemporal stability of international stock market relationships, 1957–78. *Financial Management, 10*, 54–63.

Masood, A., & Gooptu, S. (1993). Portfolio investment flows to developing coumtries. *Finance and Development, 30*(1), 9–12.

Panto, D. B, Lessig, V. P., & Joy, M. (1976). Comovement of international equity markets: a taxonomic approach. *Journal of Financial and Quantitative Analysis,* Sept., *11*(3), 415–432.

Perron, P. (1989). The great crash, the oil price shock, and the unit root hypothesis. *Econometrica, 57*, 1361–1401.

Perron, P., & Vogelsang, T. J. (1992). Nonstationarity and the level shifts with an application to purchasing power parity. *Journal of Business and Economic Statistics, 10*(3), 301–320.

Philippatos, G. C., Christofi, A., & Christofi, P. (1983). The intertemporal stability of international stock market relationships: another view. *Financial Management, 12*, 63–69.

Ripley, D. M. (1973). Systematic elements in the linkage of national stock market indices. *Review of Economics and Statistics, 55*, 356–361.

Schollhammer, H., & Sand, O. C. (1987). Lead–lag relationships among national equity markets: an empirical investigation. In S. J. Khoury & A. Ghosh (Eds.), *Recent Development in International Banking and Finance* (Vol. 1, pp. 149–168).

Su, D., & Heisher, B. M. (1997). *Risk, return and regulation in Chinese stock markets.* Working Paper, Ohio State University.

Watson, J. (1980). The stationarity of inter-country correlation coefficients: a note. *Journal of Business Finance & Accounting, 7*, 297–303.

Wei, K. C. J., Liu, Y.-J., Yang, C.-C., & Chaung, G.-S. (1995). Volatility and price change spillover effects across the developed and emerging markets. *Pacific-Basin Financial Journal, 3*, 113–136.

Zivot, E., & Andrews, D. W. K. (1992). Further evidence on the great crash, the oil-price shock, and the unit-root hypothesis. *Journal of Business and Economic Statistics, 10*(3), 251–270.

Applied Economics, 1996, **28**, 967–974

Long-run purchasing power parity revisited: a Monte Carlo simulation

BWO-NUNG HUANG and CHIN W. YANG*

Institute of International Economics, National Chung Cheng University, Chia-Yi, 621 Taiwan and *Department of Economics, Clarion University of Pennsylvania, Clarion, Pennsylvania 16214, USA*

The existence of long-run purchasing power parity (PPP) implies that a cointegration vector of nominal exchange rate, domestic price, and foreign price is expected regardless of using the Engle–Granger two-step method or Johansen maximum likelihood approach. However, this paper has found conflicting results: the Engle–Granger technique tends to reject the long-run PPP hypothesis whereas the Johansen method is generally supportive of long-run PPP. Via Monte Carlo simulations, the present paper finds that the Johansen approach has a bias toward supporting long-run PPP especially under the circumstances in which the assumption of normally or/and independently and identically distributed disturbance terms is violated.

I. INTRODUCTION

The violation of the normality assumption plays a central role in maximum likelihood estimation and hypothesis testing. An incorrectly specified likelihood function gives rise to biased, inconsistent, and inefficient maximum likelihood parameter estimates (Dorfman, 1993). As a long-standing problem in econometrics, an incorrect choice of estimation methods would, more often than not, lead to questionable conclusions. This is particularly true in the literature of international finance in which a theory may or may not be borne out by the same set of empirical data. Purchasing Power Parity (PPP) exists if currencies in different countries possess the same value (law of one price) especially in the presence of arbitrage. Some renowned theories of exchange rates are derived based on the validity of long-run PPP. For instance, the dynamic exchange rate model by Dornbusch (1976) is built upon the premise of long-run PPP, as is the asset pricing model by Lucas (1982). While the PPP hypothesis is not supported by most empirical evidence in the short run, the issue of long-run PPP remains largely unsettled. Early studies by Roll (1979), Frenkel (1981), Adler and Lehmann (1983) and Mark (1990) indicated that real exchange rate follows random walks, and as such its impact cannot be mean-reverting. It implies that the long-run PPP hypothesis may not hold. Besides, the presence of non-traded goods, transportation costs, trade restrictions and

imperfect competition can certainly weaken the result of long run PPP. In their interesting papers, Manzur (1990) and Manzur & Ariff (1995) use a methodology based on Divisia index numbers to test PPP with both short-run and long-run data for Group Seven (G7) and the Association for South East Asian Nations (ASEAN) currencies. In particular, Manzur and Ariff (1995) demonstrate a fairly close relationship between average inflation differential and average exchange rate change (logarithmic difference), as is expected from long run PPP. However, the short-run correlation between domestic price and exchange rate drops to 0.36, a value far below unity. In addition, their finding indicates that the unit root hypothesis can be rejected based on the Sims test; but it cannot be rejected by the conventional Dickey–Fuller (DF), augmented DF, and Phillips–Perron tests. In short, their result indicates that PPP does not hold at all in the short-run, but holds quite well over a longer run.

The existence of long-run PPP implies that a cointegration vector of nominal exchange rate, domestic price, and foreign price is expected regardless of using the Engle–Granger two-step method or Johansen maximum likelihood (ML) approach. However, this paper has uncovered conflicting results: The Engle–Granger technique tends to reject the long-run PPP hypothesis whereas the Johansen method generally supports it. Via Monte Carlo simulations, the present paper finds that the Johansen

0003 6846 © *1996 Routledge*

approach has a bias toward supporting long-run PPP especially in the circumstances where the assumption of normally or/and independently and identically distributed (i.i.d.) disturbance terms is violated. The paper is organized as follows. Section II provides a brief introduction of the Engle–Granger technique and Johansen maximum likelihood approach. The discussion of the violations of the Gaussian assumption for the Engle–Granger technique and the Johansen approach is given in Section III. And, Section IV contains concluding remarks.

II. ENGLE–GRANGER AND JOHANSEN MODELS AND GAUSSIAN ASSUMPTIONS

Within the framework of the Engle and Granger (1987) cointegration model, if real exchange rate does not follow a random walk, there exists a cointegration relation of $(1, -1, 1)$ for $(\alpha_1, \alpha_2, \alpha_3)$ such that

$$e_t = \alpha_1 + \alpha_2 p_t - \alpha_3 p_t^* + v_t \qquad (1)$$

where e_t, p_t and p_t^* denote nominal exchange rate, domestic, and foreign prices, respectively. However as pointed out by Davutyan and Pippenger (1985), Taylor (1988), and Taylor and McMahon (1988), the test of long-run PPP may well be clouded owing to the varying compositions in commodity baskets or/and different weights used in calculating price indices. Such measurement errors imply that long-run PPP may actually exist without having $(1, -1, 1)$ for coefficients α_1, α_2 and α_3 in Equation 1. Alternatively, one can ascertain that PPP holds as long as v_t is stationary. To test the stationarity of v_t, one normally estimates Equation 1 by applying ordinary least squares as the first step of the Engle–Granger approach. As the second step, the augmented Dickey–Fuller test is employed to test the existence of a unit root for v_t as shown below:

$$v_t = a + b_0 v_{t-1} + \sum_{i=1}^{k-1} b_i \Delta v_{t-i} + u_t \qquad (2)$$

where the value of k is chosen such that u_t follows the process of white noise. As shown via simulations (Schwert, 1989), a proper choice of k could circumvent the problem of non-normality and non-i.i.d assumptions.

Most recently, Cheung and Lai (1993a), using the Engle–Granger two-step approach, fail to reject the null hypothesis of no cointegration after taking into consideration the problem of measurement error.[1] Their result is consistent with that of Ballie and Selover (1987), Taylor (1988), Corbae and Ouliars (1988), Canarella *et al.* (1990), Mark (1990) and Flynn and Boucher (1993). Thus, it may be

viewed as supporting evidence to that the PPP hypothesis cannot be maintained since the relative prices between two countries do not form a cointegration relation. In contrast, via the multivariate ML approach developed by Johansen (1991), Cheung and Lai (1993a) are able to reject the null hypothesis of no cointegration.[2] To facilitate the comparison of these two estimation methods, we now present a brief outline of Johansen's procedure:

First, consider in general an $N \times 1$ time series vector of I(1) variables, X_t:

$$X_t = \mu + A_1 X_{t-1} + \cdots + A_k X_{t-k} + \varepsilon_t \qquad (3)$$

where ε_t is assumed to be an i.i.d. Gaussian process.

Next, defining $\Delta = 1 - L$ where L is the lag operator, we can rewrite Equation 3 as follows:

$$\Delta X_t = \mu + \Gamma_1 \Delta X_{t-1} + \cdots + \Gamma_{k-1} X_{t-k+1} + \Pi X_{t-k} + \varepsilon_t \qquad (4)$$

where

$$\Gamma_i = -(I - A_1 - \cdots - A_i), \quad i = 1, \ldots, k-1$$

$$\Pi = -(I - A_1 - \cdots - A_k)$$

Under the null hypothesis of $r(0 \leqslant r < N)$ cointegrating vector(s) with $\Pi = \alpha\beta'$, where α and β are $N \times r$ matrices of rank r such that ΠX_{t-k} is stationary, Johansen (1988) shows that ML estimation of the cointegration space spanned by β can be derived based on the least squares residuals from the following two vector regressions:

$$\Delta X_t = c_1 + \Gamma_1 \Delta X_{t-1} + \cdots + \Gamma_{k-1} X_{t-k+1} + \xi_{1t} \qquad (5)$$

$$X_{t-k} = c_2 + \Gamma_1 \Delta X_{t-1} + \cdots + \Gamma_{k-1} X_{t-k+1} + \xi_{2t} \qquad (6)$$

where c_1 and c_2 are constant terms. We can then define the product moment matrices of the residuals as

$$S_{ij} = T^{-1} \sum_{t=1}^{T} \hat{\xi}_{it} \hat{\xi}_{jt}' \qquad i, j = 1, 2 \qquad (7)$$

The likelihood ratio test statistic for the hypothesis of at most r cointegration vectors is referred to by Johansen as the trace test.

$$Trace = -2 \ln Q = -T \sum_{j=r+1}^{N} \ln(1 - \lambda_j) \qquad (8)$$

with $\lambda_{r+1}, \ldots, \lambda_N$ being the $N - r$ smallest eigenvalues of $S_{21} S_{11}^{-1} S_{12}$ with respect to S_{22}.

As an alternative likelihood ratio test statistic, one can employ the following:

$$\lambda_{max} = -2 \ln Q_{r|r+1} = -T \ln(1 - \lambda_{r+1}) \qquad (9)$$

---

[1] The problem of measurement error is discussed in detail by Cheung and Lai (1993a).

[2] Cheung and Lai also discussed the rejection frequencies with the autocorrelation coefficient less than 1 under both augmented Dickey Fuller and Johansen tests. For finite sample properties of the Johansen approach, readers are referred to Cheung and Lai (1993b).

which is also known as the maximal eigenvalue statistic. This statistic is capable of testing the null hypothesis of the existence of r cointegration vectors against the alternatives hypothesis of $r + 1$ cointegration vectors.

III. RESULTS

In order to highlight the potential consequence of the non-Gaussian distribution and to facilitate a meaningful comparison with the results by Cheung and Lai (1993a), we employ the identical data set in this paper. That is, the US is modelled as home country with Britain, France, Germany, Switzerland and Canada as foreign countries. The bilateral intercountry relations are examined during the sample period from January 1974 to December 1989. Monthly data of averages of spot exchange in US dollars (line ah), and consumers price index (line 64) are obtained from IMF's *International Financial Statistics*. All the variables are converted to logarithmic forms from the original raw data.

The unit root test by Cheung and Lai (Table 1 in their paper) indicates that the null hypothesis cannot be rejected in all cases. Since the statistical properties of variables in first-difference form play a significant role, we report the test results of normality, white noise, and ARCH in Tables 1 and 2. An examination of the tables indicates that some of the differenced variables are not normally distributed (e.g., exchange rates of the US and UK, consumer price indices of the UK and Switzerland) and some exhibit a pattern of the ARCH phenomenon (e.g. consumer price indices of the US, UK and France). This finding is of critical importance in the sense that the ARCH phenomenon of the US, being the home country, can be transmitted to other countries via the three-variable model. In addition, the UK consumer price index exhibits the ARCH effect and reveals a departure from normality as well. Thus, the basic assumptions of the UK three-variable model, to a considerable extent, remain questionable.

As for the cointegration test, we employ Johansen's ML approach to detect the existence of cointegration vector(s). In order to make the result comparable to that of Cheung and Lai (1993a), we take a lag period of eight or $k = 8$. The existence of non-Gaussian residuals suggests that their conclusion from the ML approach may be distorted owing to the invalid assumption. Much like their cointegration test result, the ML result in our paper reveals that in all cases except the UK there exists exactly one cointegration vector for the three-variable model; and there are two cointegration vectors for the UK model (Table 3). This being the case, one may claim that the long-run PPP hypothesis can be maintained from the theoretical viewpoints. To be consistent, however, the existence of such a long-run PPP should

[3] This result is also consistent with that of Cheung and Lai (1993a).

Table 1. *Tests of normality and ARCH(1) based upon first-difference variables*

		J B	ARCH(1)
CAN	e	2.4008	0.6891
	p	2.2621	0.8168
FRA	e	1.6047	0.0186
	p	0.4385	17.3584**
GMY	e	0.3265	0.0141
	p	2.0836	0.5472
SWI	e	0.4679	0.2354
	p	41.4249**	0.7450
UK	e	22.3228**	0.0398
	p	165.6013**	10.0014**
USA	p*	0.7341	31.4186**

Note: J B denotes Jarque Bera normality test. ARCH(1) indicates Engle (1982) ARCH test with $k = 1$. CAN = Canada, FRA = France, GMY = Germany, SWI = Switzerland, UK = Britain, e = nominal exchange rate in US dollars, P = consumer price index, ** denotes significant at 1% level.

Table 2. *Residual test on normality and white noise of the three-variable model*

Country	Test	Exchange rate	US price	Price
CAN	J B	0.3849	34.2205**	4.8099
	Q (12)	14.1233	10.5838	10.5409
FRA	J B	4.8890	1.9204	8.0839*
	Q (12)	2.7412	7.7028	12.5834
GMY	J B	0.3124	8.0251*	15.9304**
	Q (12)	1.8553	13.8965	12.6292
SWI	J B	0.5255	9.1305*	9.0923*
	Q (12)	3.3438	6.7522	19.4517
UK	J B	5.3578	5.5610	159.8049**
	Q (12)	10.7806	11.9247	21.8713**

Note: J B denotes Jarque Bera normality test. Q(12) is the Box Pierce Q. test with $k = 12$, * and ** are 5% and 1% significance levels respectively.

be borne out via the Engle–Granger two-step method as well. Surprisingly, as shown in Table 4, this is not the case. There does not seem to exist a cointegration relation of the three-variable model for any country.[3] Is there some significance in the finding? Since it is not likely a result of sampling fluke, the inconsistency of the two cointegration

Table 3. *The cointegration results of the Johansen ML approach*

Country	Trace			λ_{max}		
	$r \leqslant 2$	$r \leqslant 1$	$r = 0$	$r = 2$	$r = 1$	$r = 0$
CAN	10.40	5.05	31.09*	0.40	4.65	26.05*
FRA	4.80	17.15	43.12**	4.80	12.35	25.97*
GMY	4.17	15.05	50.37**	4.17	10.88	35.31**
SWI	1.49	6.46	33.32**	1.49	4.98	26.86*
UK	5.27	23.74*	53.93**	5.27	18.46*	30.20**

Note: 95% critical values (three variables) for Trace and λ_{max} are obtained from Table 1 of Osterwald Lenum (1992). ** and * denote significance at 99% and 95% level, respectively.

Table 4. *The cointegration results of the Engle Granger approach*

Country	ADF
CAN	− 0.8882
FRA	− 1.6321
GMY	− 1.3691
SWI	− 1.8396
UK	− 1.6241

Note: Results are derived from using the ADF model with $k = 4$. The 5% critical value (three variables) for the ADF test with $k = 4$ is − 3.78 (Engle and Yoo, 1987).

tests suggests that the robustness of these two approaches needs to be examined carefully, i.e. the sensitivity of the assumptions to the test results, especially in the presence of ARCH and non-normal residuals. In the following section, we examine the impact of these factors in terms of the statistical power under both methods via Monte Carlo simulations.

Violations of Gaussian assumptions

Well known in the literature, the critical values of the cointegration tests are derived under the assumption of normally, and independently and identically distributed residuals. Unfortunately, such assumptions on the whole are not satisfied in many empirical studies especially in price data. Thus, we focus the analysis on the properties of robustness and its corresponding test power. Consider a bivariate error correction model (ECM):

$$\Delta y_t = \theta(y_{t-1} - x_{t-1}) + 0.5\Delta x_t + 0.6\Delta y_{t-1} + u_{1t} \quad (10)$$

$$\Delta X_t = u_{2t} \qquad u_{2t} \sim N(0, 1) \quad (11)$$

where x_t follows a Gaussian random walk; y_t may be characterized by a Gaussian, or a non-normal random walk, or a random walk with a GARCH process; $\theta \in \{0, -0.2\}$ in which $\theta = 0$ implies that there exists no cointegration relation and $\theta = -0.2$ implies that there exists exactly one cointegration relation with the cointegration vector of $(1, -1)$. Given this preliminary information of the model, we can employ the following data-generating procedures. First, assuming that u_{1t} follows a GARCH-normal process, we have

$$u_{1t}|\Omega_{t-1} \sim N(0, h_t) \quad (12)$$

$$h_t = 0.002 + 0.2u_{1t-1}^2 + 0.3h_{t-1} \quad (13)$$

Alternatively, u_{1t} may be distributed as GARCH-$\chi^2_{(1)}$ or

$$u_{1t}|\Omega_{t-1} \sim \chi^2_{(1)}(0, h_t) \quad (14)$$

$$h_t = 0.002 + 0.2u_{1t-1}^2 + 0.3h_{t-1} \quad (15)$$

In addition, we assume that u_{1t} obeys $\chi^2_{(1)}$ distribution. Finally, u_{1t} and u_{2t} of Equations 10 and 11 are assumed to follow a GARCH-normal process characterized by Equations 12 and 13 in conjunction with the following:

$$u_{2t}|\Omega_{t-1} \sim N(0, ha_t) \quad (16)$$

$$ha_t = 0.004 + 0.4u_{2t-1}^2 + 0.3ha_{t-1} \quad (17)$$

In order to explain the inconsistent test results arising from the Johansen ML and the Engle–Granger two-step methods, a replication of 10 000 simulations are performed for each of the four different data generating processes. To minimize the sensitivity of the initial values on the test results, we delete the first 20 observations. Thus, the sample size in general is about 200.[4]

Reported in the left half of Table 5 is the simulation result under $\theta = 0$ (no cointegration relation) while that under $\theta = -0.2$ (cointegration relation) is shown in the right half.

---

[4] The sample size used in the previous section is about 160 (January 1974 to December 1986) and 200 if the sample period is extended to December of 1989.

Table 5. *Rejection frequencies using Johansen's ML approach (in 10 000)*

k		$\theta = 0$ (no cointegration)					$\theta = -0.2$ (with cointegration)				
		(1)	(2)	(3)	(4)	(5)	(1)	(2)	(3)	(4)	(5)
2	1%	0.0109	0.9892	0.6019	0.9635	0.2026	1.0000	0.8011	0.9606	0.9067	1.0000
	5%	0.0506	0.9987	0.8547	0.9954	0.4603	1.0000	0.9056	0.9916	0.9815	1.0000
	10%	0.1048	1.0000	0.9443	0.9989	0.6477	1.0000	0.9525	0.9980	0.9953	1.0000
3	1%	0.0110	0.8101	0.7496	0.8761	0.3480	1.0000	0.3066	0.5040	0.7150	1.0000
	5%	0.0515	0.9571	0.9341	0.9802	0.5956	1.0000	0.4708	0.7105	0.8870	1.0000
	10%	0.1065	0.9883	0.9806	0.9965	0.7538	1.0000	0.5847	0.8353	0.9544	1.0000
4	1%	0.0112	0.6007	0.6201	0.7562	0.2537	1.0000	0.1914	0.3771	0.5479	1.0000
	5%	0.0544	0.8491	0.8718	0.9388	0.4878	1.0000	0.3484	0.5863	0.7488	1.0000
	10%	0.1103	0.9413	0.9513	0.9830	0.6497	1.0000	0.4633	0.7233	0.8626	1.0000
5	1%	0.0110	0.4388	0.4963	0.6460	0.1837	1.0000	0.1151	0.2807	0.3830	1.0000
	5%	0.0536	0.7193	0.7791	0.8812	0.3826	1.0000	0.2477	0.4760	0.5925	1.0000
	10%	0.1093	0.8571	0.8996	0.9576	0.5454	1.0000	0.3564	0.6094	0.7262	1.0000
6	1%	0.0116	0.3245	0.3965	0.4976	0.1289	1.0000	0.0872	0.2130	0.3055	0.9998
	5%	0.0529	0.5974	0.6857	0.7860	0.3105	1.0000	0.1955	0.3999	0.5076	1.0000
	10%	0.1134	0.7618	0.8386	0.9023	0.4650	1.0000	0.2990	0.5375	0.6405	1.0000
7	1%	0.0118	0.2339	0.3283	0.4111	0.1148	1.0000	0.0536	0.1553	0.2211	0.9930
	5%	0.0550	0.4785	0.6277	0.7102	0.2831	1.0000	0.1456	0.3271	0.4068	0.9998
	10%	0.1149	0.6559	0.7918	0.8520	0.4320	1.0000	0.2325	0.4588	0.5440	1.0000
8	1%	0.0129	0.1925	0.2844	0.3410	0.0910	1.0000	0.0409	0.1208	0.1621	0.9629
	5%	0.0556	0.4137	0.5706	0.6362	0.2439	1.0000	0.1129	0.2703	0.3343	0.9979
	10%	0.1174	0.5936	0.7444	0.8022	0.3902	1.0000	0.1947	0.4042	0.4664	0.9998
9	1%	0.0133	0.1455	0.2359	0.2929	0.0799	1.0000	0.0311	0.0836	0.1181	0.9193
	5%	0.0561	0.3495	0.5006	0.5762	0.2214	1.0000	0.0957	0.2153	0.2707	0.9889
	10%	0.1179	0.5172	0.6834	0.7567	0.3578	1.0000	0.1675	0.3357	0.4052	0.9982
10	1%	0.0152	0.1274	0.2007	0.2433	0.0636	1.0000	0.0276	0.0763	0.0967	0.8490
	5%	0.0592	0.3164	0.4549	0.5141	0.1991	1.0000	0.0827	0.2005	0.2308	0.9728
	10%	0.1219	0.4857	0.6406	0.6920	0.3326	1.0000	0.1521	0.3122	0.3558	0.9947

Note: (1) = rejection frequencies under the Gaussian assumption; (2) = rejection frequencies if one of the two variables is from GARCH (1, 1) normal; (3) = rejection frequencies if one of the two variables is generated from $\chi^2_{(1)}$ GARCH (1, 1); (4) = rejection frequencies if one of the two variables is consistent with $\chi^2_{(1)}$; (5) = rejection frequencies if both variables are generated from GARCH (1, 1).

Test results on normality and i.i.d. under different assumptions are reported in column 1. It is worth noting that an interesting phenomenon surfaces readily in Table 5: if the residual of one of the two variables deviates from normality or i.i.d. assumption, the power of the ML method is seriously distorted. That is, when the data are generated from the assumption of no cointegration ($\theta = 0$), the ML method has unusually high rejection frequencies under all four different scenarios. The result makes a strong prima facie case for an inappropriate use of the ML model especially in the presence of non-Gaussian and ARCH residuals. Needless to say, the ML method tends to reject the null hypothesis of no cointegration. As a consequence, it produces the spurious results to support the existence of a cointegration relation as shown in Table 5 (columns 2–5). At the other end of the spectrum, when the data are generated from $\theta = -0.2$

(with cointegration) with lag period of $k = 2$, the ML approach has reasonably good power. However, its rejection frequency decreases with increase of k. On close inspection, there are several interesting results from Table 5. First, the distortions of the ML method are relatively severe if one of the variables has violated the normality or i.i.d. assumption (see column 2 and 4, respectively). Surprisingly, such distortions are less serious if both assumptions are violated (column 3). Second, the rejection frequencies are comparable to that under the Gaussian assumption if both variables do not have Gaussian residuals (column 5) in the scenario where data are generated from $\theta = -0.2$. In sharp contrast, such distortions become relatively less severe when the data are generated from $\theta = 0$. As indicated from the simulation results, the ML method tends to support the existence of cointegration especially in the case where the assumption of normality or i.i.d. is violated. Consequently, the issue of the existence of cointegration in the population is very much clouded in the presence of departures from the Gaussian and/or i.i.d. assumptions.

Engle–Granger or Johansen approach

Depending on the choice of the models and nature of the data, the result can be vastly different. Thus, it is important to explore the relationship between the data generation process and the estimation procedure. Table 5 and Table 6 report the statistical power of the Engle–Granger two-step and ML approaches under identical data generating processes. When the data are generated from a no cointegration relation (via the ECM), the Engle–Granger two-step method possesses greater power in testing the existence of

a cointegration relation than does the Johansen ML approach in the bivariate model if one of the variables is found to violate the normality assumption (Table 6). That is to say, the Engle–Granger approach tends not to reject the null hypothesis of no cointegration when data are generated from $\theta = 0$ (no cointegration). However, this moderate advantage of the Engle–Granger method over the ML approach decreases as lag period k increases. When the data are generated from $\theta = -0.2$ (with cointegration), the Engle–Granger method possesses satisfactory power for small k values, but will less likely reject the null hypothesis of no cointegration for larger k values. In either case, the distortion from applying the Engle–Granger approach is less than that of the ML method. One important difference between the two approaches is that the Engle–Granger method has slightly more rejection frequency than does the

Table 6. *Rejection frequencies using the Engle Granger two-step cointegration approach*

k		$\theta = 0$ (no cointegration)					$\theta = -0.2$ (with cointegration)				
		(1)	(2)	(3)	(4)	(5)	(1)	(2)	(3)	(4)	(5)
2	1%	0.0123	0.0098	0.0101	0.0102	0.0078	1.0000	0.9620	0.9960	0.9986	0.4791
	5%	0.0509	0.0399	0.0412	0.0440	0.0337	1.0000	0.9981	0.9995	0.9998	0.8563
	10%	0.1000	0.0798	0.0819	0.0879	0.0712	1.0000	0.9998	1.0000	1.0000	0.9542
3	1%	0.0112	0.0098	0.0101	0.0107	0.0069	1.0000	0.8654	0.9699	0.9874	0.1535
	5%	0.0536	0.0434	0.0418	0.0409	0.0372	1.0000	0.9787	0.9968	0.9989	0.5412
	10%	0.1039	0.0838	0.0817	0.0831	0.0725	1.0000	0.9951	0.9994	0.9996	0.7754
4	1%	0.0131	0.0102	0.0094	0.0081	0.0088	0.9994	0.6656	0.8728	0.9280	0.0255
	5%	0.0553	0.0432	0.0401	0.0390	0.0350	1.0000	0.9018	0.9811	0.9926	0.2160
	10%	0.1075	0.0808	0.0796	0.0780	0.0718	1.0000	0.9656	0.9959	0.9987	0.4664
5	1%	0.0141	0.0106	0.0099	0.0094	0.0078	0.9986	0.4826	0.7345	0.8116	0.0073
	5%	0.0523	0.0406	0.0431	0.0422	0.0320	1.0000	0.7597	0.9384	0.9630	0.0906
	10%	0.1047	0.0797	0.0852	0.0870	0.0684	1.0000	0.8897	0.9828	0.9891	0.2468
6	1%	0.0138	0.0089	0.0079	0.0077	0.0062	0.9882	0.3346	0.5703	0.6655	0.0015
	5%	0.0538	0.0370	0.0362	0.0383	0.0311	0.9996	0.5861	0.8508	0.9027	0.0275
	10%	0.1071	0.0787	0.0761	0.0800	0.0650	0.9998	0.7461	0.9422	0.9688	0.0978
7	1%	0.0125	0.0088	0.0082	0.0086	0.0073	0.9744	0.2473	0.4391	0.5110	0.0003
	5%	0.0561	0.0393	0.0381	0.0373	0.0332	0.9971	0.4570	0.7481	0.8098	0.0096
	10%	0.1090	0.0780	0.0762	0.0800	0.0677	0.9995	0.6180	0.8850	0.9205	0.0461
8	1%	0.0126	0.0083	0.0083	0.0092	0.0079	0.9303	0.1718	0.3276	0.3884	0.0000
	5%	0.0502	0.0394	0.0387	0.0373	0.0303	0.9937	0.3492	0.6299	0.6954	0.0038
	10%	0.1072	0.0813	0.0780	0.0774	0.0653	0.9988	0.4972	0.7920	0.8474	0.0241
9	1%	0.0129	0.0073	0.0091	0.0112	0.0075	0.8747	0.1308	0.2607	0.3097	0.0000
	5%	0.0528	0.0370	0.0386	0.0389	0.0347	0.9826	0.2735	0.5347	0.6004	0.0017
	10%	0.1024	0.0801	0.0803	0.0786	0.0675	0.9968	0.4011	0.7222	0.7678	0.0122
10	1%	0.0098	0.0091	0.0071	0.0055	0.0053	0.8059	0.1054	0.1981	0.2298	0.0000
	5%	0.0472	0.0350	0.0365	0.0327	0.0309	0.9638	0.2294	0.4597	0.4972	0.0010
	10%	0.0957	0.0775	0.0761	0.0747	0.0622	0.9917	0.3488	0.6302	0.6757	0.0064

Note: (1) = rejection frequencies under the Gaussian assumption; (2) = rejection frequencies if one of the two variables is from GARCH (1, 1) normal; (3) = rejection frequencies if one of the two variables is generated from $\chi^2_{(1)}$ GARCH (1, 1); (4) = rejection frequencies if one of the two variables is consistent with $\chi^2_{(1)}$; (5) = rejection frequencies if both variables are generated from GARCH (1, 1).

ML method (see column 5 of Table 6) if (1) data are generated from $\theta = 0$ (no cointegration), and (2) both variables exhibit a GARCH process. Conversely if data are generated from $\theta = -0.2$ (with cointegration), the ML method seems to be superior to the Engle–Granger approach.

In a nutshell, the Engle–Granger method in testing the existence of cointegration relation is superior to the ML method if (1) the variable does not have Gaussian and i.i.d. residuals and (2) the data are generated from $\theta = -0.2$. Viewed in this perspective, the ML method is more likely to lead us to wrong conclusions because some assumptions in the three-variable model are not met. Given that the US consumer price index clearly follows the ARCH(1) process (Table 1),[5] the ML method has a greater probability of rejecting the null hypothesis of no cointegration while the data do not contain cointegration relations. Seen in this light, it is not surprising that the same result, that there exists a cointegration relation for each of the five countries, is found in both our paper (Table 3) and Cheung and Lai (1993a). However, such findings, according to the simulations of Table 5, can be attributed to the distortion of the ML method, which is in turn caused by the invalid assumptions on residuals. Furthermore, the existence of two cointegration relations for the UK may well be accounted for by the departure from both normality and i.i.d. assumptions.

To further illustrate the impact of the statistical assumptions on the test result of long-run PPP, we report the critical values of the Johansen ML approach based on relations (12) and (13) using Monte Carlo simulations with a replication of 10 000 ($T = 200$). Clearly the cointegration results obtained by using the ML critical values of Table 7 suggest that the null hypothesis of no cointegration cannot be rejected. Viewed in this perspective, the long-run equilibrium PPP relationship in the case of five countries does not exist. On the other hand, the simulation results of Table 6 suggest that in the absence of the Gaussian and i.i.d. residuals, the Engle–Granger two-step approach is more robust and hence dependable. That is, with some caveats, the long-run PPP hypothesis should be rejected as was evidenced by the majority of previous empirical studies.

IV. CONCLUSIONS

The empirical studies on long-run PPP have been prolific, yet the issue has remained unsettled. No clear consensus has emerged regarding the hypothesis. In their celebrated work, Engle and Granger pioneer the cointegration theory which offers an excellent playground for examining the long-run PPP hypothesis. Early studies, based upon their cointegration theory, tend more to reject the existence of long-run PPP. In contrast, the multivariate ML method developed

Table 7. *Selected critical values of the Johansen ML model with disturbance terms following GARCH(1,1)*

$n-r$	99%	95%	90%
1	6.52 (6.52)	3.96 (3.96)	2.88 (2.88)
2	81.47 (80.10)	66.55 (65.09)	59.30 (57.81)
3	141.26 (131.34)	123.17 (114.20)	114.22 (105.23)
4	194.60 (172.56)	174.56 (153.85)	164.10 (143.04)
5	244.03 (206.02)	223.28 (185.13)	212.16 (174.71)

Note: The results are based upon the Monte Carlo method with a replication of 10 000 ($T = 200$). The figure in parenthesis is λ_{max}; and the figure not in parenthesis is the trace statistic.

by Johansen leans more towards the support of long-run PPP. From the theoretical viewpoint, one would expect an identical conclusion regardless of the choice of the two approaches. That is, there exists a conintegration relation among nominal exchange rate, domestic and foreign prices if long-run PPP is indeed a tenable economic theory. Conflicting results are therefore indicative of a potential error in reaching the conclusion from using inappropriate models. Since many empirical data exhibit departures from the Gaussian or i.i.d. assumptions, the importance in choosing an appropriate approach cannot be downplayed. From the Monte Carlo simulation results, this paper has found that the ML approach tends to reject the null hypothesis of no cointegration where there does not exist one. The most striking result is perhaps that such a bias becomes more profound in the situation where residuals deviate from the normality or i.i.d. assumption. In the three-variable and five-country model, at least one variable is found to depart from the Gaussian assumption or to follow the ARCH(1) process. Thus, the empirical finding that there exists a cointegration relation in some of these models can be attributed to the spurious result emanated from using the ML approach. Quite differently, the Engle–Granger two-step approach has better statistical power under the same data generating procedures. It should be pointed out that based upon the critical values from the Monte Carlo simulations, the null hypothesis of no conintegration cannot be rejected in the presence of a GARCH process. Our paper suggests that the existence of long run PPP needs to be evaluated with great caution especially if the normality or i.i.d. assumption is violated.

[5] In addition, the normality assumption on some of variables is not satisfied (Table 2) in the three-variable VAR model with $k = 8$. The UK consumer price index is neither normally distributed nor consistent with the white noise assumption.

ACKNOWLEDGEMENTS

The authors are grateful to an anonymous referee and the editor of this journal for useful comments. Part of this study is supported by the National Science Council of Taiwan under grant No. NSC82-0301-H-194-043.

REFERENCES

Adler, M. and Lehmann, B. (1983) Deviations from purchasing power parity in the long run, *Journal of Finance*, **38**, 1471 87.

Ballie, R. and Selover, D. (1987) Cointegration and models of exchange rate determination, *International Journal of Forecasting*, **3**, 43 51.

Canarella, G., Pollard, S. K. and Lai, K. (1990) Cointegration between exchange rates and relative prices: another view, *European Economic Review*, **34**, 1303 22.

Cheung Yin-Wong and Lai, K. S. (1993a) Long-run purchasing power parity during the recent float, *Journal of International Economics*, **34**, 181 91.

Cheung Yin-Wong and Lai, K. S. (1993b) Finite-sample sizes of Johansen's likelihood ratio tests for cointegration, *Oxford Bulletin of Economics and Statistics*, **55**, 313 28.

Corbae, D. and Ouliars, S. (1988) Cointegration and tests of purchasing power parity, *Review of Economics and Statistics*, **70**, 508 11.

Davutyan, N. and Pippenger, J. (1985) Purchasing power parity did not collapse during the 1970's, *American Economic Review*, **75**, 1151 58.

Dorfman, J. H. (1993) Should normality be a normal assumption, *Economics Letters*, **42**, 143–7.

Dornbush, R. (1976) Expectations and exchange rate dynamics, *Journal of Political Economy*, **84**, 1161 76.

Engle, R. and Granger, C. (1987) Cointegration and error correction: representation, estimation, and testing, *Econometrica*, **36**, 143 59.

Engle, R. F. (1982) Autoregressive conditional heteroscedasticity with estimates of the variance of United Kingdom inflation, *Econometrica*, **50**, 987 1007.

Engle, R. F. and Yoo, B. S. (1987) Forecasting and testing in co-integrated systems, *Journal of Econometrics*, **35**, 143 59.

Flynn, N. A. and Boucher (1993) Test of long-run purchasing power parity using alternative methodologies, *Journal of Macroeconomics*, **15**, 109 22.

Frenkel, J. A. (1981) The collapse of purchasing power parities during the 1970's, *European Economic Review*, **16**, 145 65.

Johansen, S. (1988) Statistical analysis of cointegration vectors, *Journal of Economic Dynamics and Control*, **12**, 231 54.

Johansen, S. (1991) Estimation and hypothesis testing of cointegration vectors in Gaussian vector autoregressive models, *Oxford Bulletin of Economics and Statistics*, **52**, 169 210.

Lucas, R. (1982) Interest rates and currency prices in a two-country world, *Journal of Monetary Economy*, **83**, 335 60.

Mark, N. (1990) Real and nominal exchange rates in the long run: an empirical investigation, *Journal of International Economics*, **28**, 115 36.

Manzur, Meher (1990) An international comparison of prices and exchange rates: a new test of purchasing power parity, *Journal of International Money and Finance*, **9**, 75 91.

Manzur, Meher and Mohamed Ariff (1995) Purchasing power parity: new methods and extensions, *Applied Financial Economics*, **5**, 19 26.

Osterwald-Lenum, M. (1992) A note with fractiles of the asymptotic distribution of the maximum likelihood cointegration rank test statistics: four cases, *Oxford Bulletin of Economics and Statistics*, **54**, 461 72.

Roll, R. (1979) Violations of purchasing power parity and their implications for efficient international commodity markets, in M. Sarnat and G. Szego (eds), *International Finance and Trade* (Ballinger, Cambridge, MA).

Schwert, G. W. (1989) Tests for unit roots: a Monte Carlo investigation, *Journal of Business and Economic Statistics*, **7**, 147 59.

Taylor, M. P. (1988) An empirical examination of long run purchasing power parity using co-integration techniques, *Applied Economics*, **20**, 1369 82.

Taylor, M. P. and McMahon, P. C. (1988) Long run purchasing power parity in the 1920s, *European Economic Review*, **32**, 179 97.

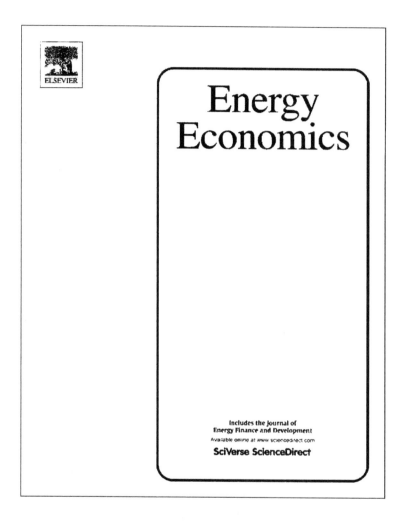

Energy
Economics

Includes the Journal of
Energy Finance and Development
Available online at www.sciencedirect.com
SciVerse ScienceDirect

(This is a sample cover image for this issue. The actual cover is not yet available at this time.)

Energy Economics 34 (2012) 1284–1300

Contents lists available at SciVerse ScienceDirect

Energy Economics

journal homepage: www.elsevier.com/locate/eneco

Oil price movements and stock markets revisited: A case of sector stock price indexes in the G-7 countries ☆

Bi-Juan Lee [a], Chin Wei Yang [b,c], Bwo-Nung Huang [d,*]

[a] Ministry of National Defense, Taipei, Taiwan
[b] Department of Economics, Clarion University of Pennsylvania, Clarion, PA 16214-1232, United States
[c] Department of Economics, National Chung Cheng University, Chia-Yi 621, Taiwan
[d] Department of Economics & Center for IADF, National Chung Cheng University, Chia-Yi 621, Taiwan

ARTICLE INFO

Article history:
Received 31 May 2010
Received in revised form 3 June 2012
Accepted 4 June 2012
Available online 12 June 2012

JEL classification:
Q42
Q43
G12

Keywords:
Oil prices
Stock returns
Industry sectors
Sector stock return

ABSTRACT

Applying sector stock prices and oil prices in 1991:01–2009:05 from the G7 countries we find oil price shocks do not significantly impact the composite index in each country. However, stock price changes in Germany, the UK and the US were found to lead oil price changes.

As for the interaction between oil price changes and sector stock price changes, we find short-run negative causal relationships: 4 of 7 sector index returns were impacted by oil price changes in Germany, 2 in the US and 1 in France. In particular, stock returns of information technology and consumer staples sectors were found to be impacted most by oil price shocks, followed by financial, utilities and transportation sectors. In terms of causality from sector stock price changes to oil price changes, we find stock price changes lead oil price changes in 8 of 9 sectors in Germany, most in the G7 countries followed by the UK, Italy, France, Canada and the US. No such a causal relationship, however, is found for Japan. With respect to specific sectors, stock price changes in consumer staples and materials sectors were impacted most significantly by oil price changes followed by transportation, financial, energy, health care, industrials, utilities, information technology and telecommunication sectors with the exception of consumer discretionary sector. In addition, short term stock price changes are found to lead positively oil price changes.

© 2012 Elsevier B.V. All rights reserved.

1. Introduction

Until the late 2008, two features regarding stock prices and oil prices stood out. First, stock market had grown impressively around the world. Second, oil prices had surged and hit a new record of US $147 per barrel in 2008. It appeared that the equity market was impacted by the high oil prices. Logically speaking, there is one good reason why this might be the case. That is the higher the oil prices, the greater the production cost, which translates into lower profits. As a consequence, it is quite logical that increased oil price could impact stock markets. Nonetheless, each sector absorbs different magnitudes of shocks in terms of production cost and as such exhibits different responses.

However, the existing research on the relationship between oil price and stock market primarily focuses on the behavior of broad-based equity prices (i.e. the composite stock index), giving rise to the problem that the "averages" tend to hide what's really going on.

For instance, sectors comprising oil companies are expected to benefit from oil price increase. If, however, this sector contains industries that are adversely affected by the oil price increase, the result may show no or negative relationship. As a matter of fact, sector equity indexes may well be more suitable for examining the impact of oil price on equity markets because oil price may affect different sectors in different ways. This being the case, a sector-based classification is appropriate for investigating the difference in responses among sectors in the presence of an oil price change. Furthermore, examining the oil price-stock market linkages via sector indexes helps us determine whether or not oil price changes are more closely linked to sector equity indexes than to national composite equity indexes.

Although oil price changes constitute a source of systematic risk in terms of aggregate market returns, the exposure to this risk varies across sectors. To explore the possibility, this paper attempts to find out whether oil prices provide additional information beyond what the market portfolio suggests in terms of sector stock returns. As many different financial assets are traded based on sector stock returns, it is important for traders to understand the oil price transmission mechanism across the sectors in order to make optimal portfolio decisions. The purpose of this paper is to investigate the transmission from oil price changes to various sector stock returns and to national market (composite) stock returns. Unlike most of

☆ Many thanks to two reviewers for their useful comments. However all remaining errors are authors' only.
  * Corresponding author.
  E-mail addresses: bijuanlee@gmail.com (B.-J. Lee), Yang@mail.clarion.edu (C.W. Yang), ecdbnh@ccu.edu.tw (B.-N. Huang).

0140-9883/$ – see front matter © 2012 Elsevier B.V. All rights reserved.
doi:10.1016/j.eneco.2012.06.004

Table 1
Summary of market and sector indexes and sample period of the G-7 nations.

Country	Canada	France	Germany	Italy	Japan	UK	US
sector	S&P/TSX	Euronext Paris CAC	CDAX	Milan SE	TOPIX	UK FTSE	S&P500
Composite	91:01–09:05	91:01–09:05	91:01–09:05	91:01–09:05	91:01–09:05	91:01–09:05	91:01–09:05
Consumer discretionary						91:01–09:05	91:01–09:05
Consumer staples			91:01–09:05	91:01–09:05	91:01–09:05	91:01–09:05	91:01–09:05
Energy	91:01–09:05				91:01–09:05	91:01–09:05	91:01–09:05
Financial	91:01–09:05	91:01–09:05	91:01–09:05	91:01–09:05	91:01–09:05	91:01–09:05	91:01–09:05
Health care			91:01–09:05		91:01–09:05	91:01–09:05	91:01–09:05
Industrials		91:01–09:05	91:01–09:05	91:01–09:05	91:01–09:05	91:01–09:05	91:01–09:05
Information Technology		91:01–09:05	91:01–09:05			91:01–09:05	91:01–09:05
Materials	91:01–09:05	91:01–09:05	91:01–09:05			91:01–09:05	91:01–09:05
Utilities	91:01–09:05		91:01–09:05	91:01–09:05	91:01–09:05	91:01–09:05	91:01–09:05
Transportation			91:01–09:05			91:01–09:05	91:01–09:05
Telecommunications			91:01–09:05		91:01–09:05	91:01–09:05	91:01–09:05

the previous literature, which mainly focused on the national market indexes, we emphasize the effects of changes in oil prices on different sector stock indexes in the G-7 nations: Canada, France, Germany, Italy, Japan, the UK, and the US. Our results would be useful to the individual investors and arbitrageurs, who buy stocks of different sectors, to comprehend how the shares in different sectors react to oil price changes.

The organization of this paper is as follows. Section 2 reviews the related literature. Section 3 briefly presents the methodology used in this paper. Section 4 describes the data and presents the empirical estimations before discussing the results. Section 5 concludes and provides some policy implications.

2. Literature review

There have been quite a few papers investigating the links between oil price movements and stock market. Prior studies about oil prices and stock returns can be classified into two categories, one at the market and another at the industry (sector) levels. At the market level, the research on the impact of oil price changes on stock markets was pioneered by Jones and Kaul (1996), who used 1947–1991 quarterly data for Canada, Japan, the U.K. and the U.S., to find that the oil price had a negative impact on the aggregate stock returns. Huang et al. (1996) used a VAR (Vector Autoregressive) model to examine the relationship between daily oil futures returns and daily U.S. stock returns. They discovered that oil futures returns did not have much impact on the broad-based market indexes such as the S&P 500. In addition, Sadorsky (1999) presented an unrestricted VAR model using 1947–1996 monthly data of oil prices, stock returns, short-term interest rate, and industrial production to examine the links among these variables. In contrast to the results of Huang et al. (1996), Sadorsky (1999) identified that oil price shocks played an important role in explaining US broad-based stock returns. Similarly, Papapetrou (2001), using 1989–1999 monthly data of Greek stock market, also found that oil price was an important component in explaining stock price movements, and positive oil price shocks depressed real stock returns. Applying a multivariate threshold regression model, Huang et al. (2005) found oil price changes could negatively impact stock returns only when the price increase in previous period exceeded a threshold value. The ensuing impact lasts for one period for Canada, Japan and the US. Recently Bjørnland (2008) used a structural VAR model to find that higher oil prices had a stimulating effect on the Norwegian economy as is expected of an oil exporting country. In contrast, Park and Ratti (2008) pointed out that the impacts of oil price shocks on oil-importing countries' stock market were negative in direction while those on oil-exporting countries' stock market were positive. Miller and Ratti (2009) employ a structural change-robust VEC (Vector Error Correction) model to investigate the long-run relationship between world oil price and national stock indexes of the 6 OECD countries. Their

results indicate stock market indexes respond negatively to increases in the oil price in the long run.

At the industry level, Faff and Brailsford (1999), making use of several industry returns in Australian stock market, found that oil, gas and diverse resources industries had positive responses from oil price increase, while papermaking, packing, and transportation industries reacted negatively. Also, Sadorsky (2001) found evidence that a rise of the stock market index and oil price had a positive effect on oil industries' returns. However, Hammoudeh and Li (2005) discovered a negative relationship between returns of US transportation industry and oil prices. Employing a multifactor (mainly two-factor) model, El-Sharif et al. (2005) explored the relationship between the sector returns of oil and gas industry and oil price. As expected, they found a positive relationship.

In a paper, Nandha and Faff (2008) investigated the oil price-sector index relationship via a market factor market model using Data Stream global equity indices (April 1983 to September 2005) and 35 global industry indices and oil prices. With the exception of mining and oil-gas industries, there exist significantly negative relationships between stock returns and oil price change. In a similar vein, Elyasiani et al. (2011) studied the causal relationship between oil price shocks and 13 industry stock returns of the US from the perspective of the volatility of a GARCH model.

Despite the proliferation of the above literature, the focus is limited largely to a specific country or some region (e.g., US or global) or to specific industries (e.g., oil and gas industry). Rarely was the study made on the relationship between a specific industry and oil price change across different countries. Furthermore, many prior studies analyzed the relationship based on multifactor models, which preassume the direction of causality: market return or oil price change causes industry stock return. They obviously ignored the causality from stock price returns to oil price changes. Beyond it, multifactor model being static in nature, cannot address the dynamic relationship between the variables. To improve upon the multifactor model, this paper applies the VAR or VEC model to 12 industry stock indexes of the G7 countries to re-examine the short-run interactions between oil prices and sector stock index returns. [1]

3. Model and source of data

Different from the multifactor models, the empirical framework for investigating the interactions between oil price changes and stock prices in this paper is built upon an unrestricted VAR model.[2] A VAR model has been frequently used to analyze the impact of oil

---

[1] To facilitate comparisons on sectors between countries, we follow the classification system by Global Industry Classification Standard (GICS) developed by MSCI and Standard & Poor's.

[2] In the presence of cointegration relations among model variables, one would have to opt for a VEC model for analysis.

Table 2
ADF Unit Root Test of the G-7 Nations.

Country sector	Canada S&P/TSX level	diff.	France Euronext Paris CAC level	diff.	Germany CDAX level	diff.	Italy Milan SE level	diff.	Japan TOPIX level	diff.	UK UK FTSE level	diff.	US S&P500 level	diff.
Composite	-2.95	-10.51***	-0.91	-13.54***	-2.18	-11.26***	-1.09	-11.71***	-2.32	-12.08***	-1.71	-11.15***	-0.65	-11.82***
Consumer discretionary	n.a.	n.a.	n.a.	n.a.	n.a.	n.a.	n.a.	n.a.	n.a.	n.a.	-1.69	-12.78***	-1.24	-12.37***
Consumer staples	n.a.	n.a.	n.a.	n.a.	-2.33	-10.34***	-0.61	-12.31***	-2.97	-11.57***	-1.94	-10.87***	-0.94	-11.28***
Energy	-2.52	-11.12***	-2.48	-10.87***	n.a.	n.a.	n.a.	n.a.	-1.79	-12.29***	-2.15	-13.04***	-2.14	-12.49***
Financial	-2.57	-10.78***	n.a.	n.a.	-2.27	-9.62***	-2.29	-11.24***	-2.09	-11.20***	-1.07	-10.16***	-1.93[a]	-11.15***
Health care	n.a.	n.a.	-2.41	-11.04***	-2.22	-12.37***	n.a.	n.a.	-2.23	-14.39***	-1.15	-14.18***	-0.18	-12.87***
Industrials	n.a.	n.a.	-1.26	-10.88***	-1.89	-10.29***	-2.87	-12.12***	-1.52	-11.19***	-2.64	-9.70***	-0.39	-11.92***
Information Technology	n.a.	n.a.	-2.48	-10.68***	-2.84	-10.40***	n.a.	n.a.	n.a.	n.a.	-1.64	-9.66***	-1.08	-11.42***
Materials	-1.62	-11.51***	n.a.	n.a.	-1.57	-10.29***	n.a.	n.a.	-2.22	-12.33***	-2.89	-9.69***	-3.04	-11.23***
Utilities	-2.45	-10.94***	n.a.	n.a.	-2.06	-11.70***	-0.29	-12.57***	n.a.	n.a.	-1.58	-13.06***	-2.01	-11.98***
Transportation	n.a.	n.a.	n.a.	n.a.	-1.73	-11.43***	n.a.	n.a.	-2.61	-11.31***	-2.33	-9.40***	-2.77	-12.42***
Telecommunications	n.a.	n.a.	n.a.	n.a.	-2.32	-11.40***	n.a.	n.a.	n.a.	n.a.	-1.69	-11.41***	-1.27	-10.90***
Lroilp	-3.20*	-11.16***	-3.15*	-11.18***	-3.11	-11.19***	-3.09	-11.15***	-3.11	-10.97***	-3.13	-11.30***	-3.14*	-11.10***
Ly	-2.07[a]	-5.22***	-1.44[a]	-5.30***	-1.53[a]	-6.28[a]	-1.43[a]	-6.11***	-3.52***[a]	-.687***	-.077[a]	16.51[a]	-2.38[a]	-3.46***
R	-3.02**[a]	-10.98***	-1.71[a]	-11.31***	-1.58[a]	-5.99***	-1.91[a]	-8.46***	-5.63***	-3.66***	-3.06**[a]	-21.95***	-1.85[a]	-5.40***

Note: *,**,*** represent 10%, 5% and 1% significance levels respectively; diff denotes 1st-differenced variable; superscript a indicates the unit root test with a drift term, and other unit root tests are performed with drift term and trend variable; n.a. = sector data not available; logarithmic transformation is made on all stock price; lroilp = logarithmic real oil price ; ly = logarithmic industrial production index; r = interest rate.

price shocks on economic activity since the work by Darby (1982) and Hamilton (1983). The main advantage of this model is the ability to capture the dynamic relationships among the economic variables of interest. That is, we don't need to provide a priori assumptions regarding which variables are response variables and which variables are explanatory variables. Thus, our VAR model consists of a system of equations that expresses each variable in the system as a function of its own lagged variable and lagged variables of all other variables in the system.

Following Sadorsky (1999) and Huang et al. (2005), we use monthly data of the four variables in our empirical model. The real oil price (roilp) is obtained from the average world crude oil prices deflated by the consumer price index. In the same manner, real stock price (rstkp) is taken from the equity indexes deflated by the consumer price index. Moreover, the industrial production is chosen because the equity market is fundamentally related to changes in output level. Hence the industrial production is a proxy for output (y). Additionally, it is important to include an interest rate because it could affect stock returns through the discount rate on companies' expected earnings. Hence, the money market rate is used as a proxy for the interest rate (r). Furthermore, natural logarithms are taken of each data series except interest rates. As such, the study on the reaction of real stock returns to the oil price changes includes logarithm or log of real oil prices ($lroilp_t$), log of real stock prices ($lrstkp_t$), log of industrial production (ly_t), and interest rate (r_t). Monthly data from 1991:01 to 2009:05 of the crude oil price, interest rate, and industrial production are collected from International Financial Statistics. The data for all stock indexes are acquired from Global Financial Data (GFD). The GFD follows the Global Industry Classification Standard (GICS) system and therefore makes it possible to compare sector indexes across different countries. The GFD consists of 11 sectors. Table 1 lists the names of the markets, sectors available in the G-7 nations.

An inspection of Table 1 indicates that of the G-7 nations only the US and the UK have all the 11 sector indexes, Germany has 9, Japan has 7, and Canada, France and Italy have 4 sector indexes.[3] Let $\mathbf{y}' = [r, lroilp, ly, lrstkp]'$ is a vector of 4×1. Before deciding on either VAR or VEC model, we need to test if $\mathbf{y}$ follows I(1) or I(0) process. If vector $\mathbf{y}$ follows I(0), we can build a VAR model using vector $\mathbf{y}$. If $\mathbf{y}$ or some components of $\mathbf{y}$ follow I(1), a cointegration test is to be performed on the variables that are of I(1). In this case, the VEC model is estimated as shown below:

$$\Delta \mathbf{y}_t = \mathbf{A}_0 + \mathbf{B}\mathbf{y}_{t-1} + \sum_{i=1}^{p-1} \mathbf{A}_i \Delta \mathbf{y}_{t-i} + \boldsymbol{\varepsilon}_{1t}, \tag{1}$$

where Δ is 1st difference operator, $\mathbf{y}_{t-i}$ is a vector of error correction terms and $\mathbf{B}$ is the matrix denoting speed of adjustment needed toward the equilibrium. $\mathbf{A}_i$ is 4×4 matrix of unknown coefficients, $\mathbf{A}_0$ is a 4×1 column vector of deterministic constant terms, $\boldsymbol{\varepsilon}_{1t}$ is a 4×1 column vector of error with the properties of $E(\boldsymbol{\varepsilon}_{1t}) = \mathbf{0}$ for all t, $E(\boldsymbol{\varepsilon}_{1s}\ \boldsymbol{\varepsilon}'_{1t}) = \Omega$ if $s = t$ and $E(\boldsymbol{\varepsilon}_{1s}\ \boldsymbol{\varepsilon}'_{1t}) = 0$ if $s \neq t$, where Ω is the variance-covariance matrix. If the I(1) variables in $\mathbf{y}$ do not exhibit cointegration relations, we opt for the following VAR model for analysis:

$$\Delta \mathbf{y}_t = \mathbf{A}_0 + \sum_{i=1}^{p-1} \mathbf{A}_i \Delta \mathbf{y}_{t-i} + \boldsymbol{\varepsilon}_{1t}. \tag{2}$$

It is to be pointed out that structural change(s) may have occurred during the sample period (1991:01–2009:05). Such a structural change (Invasion of Kuwait by Iraq, Asian Financial Crisis, World Financial Tsunami) may very well produce a different causal relationship. In other words, failing to take it into consideration can render

---

[3] In order to compare data across different countries and sectors, the starting time is from 1991:01. Stock price data may not be available in 1991:01 for some sectors.

Table 3
Johansen cointegration test of the G-7 nations.

	Canada	France	Germany	Italy	Japan	UK	US
Composite	54.94*** [0.01]	34.04 [0.50]	39.10 [0.26]	45.35* [0.08]	7.18 [0.56]	53.72*** [0.01]	78.53 c, *** [0.00]
Consumer discretionary	n.a.	n.a.	n.a.	n.a.	n.a.	51.74** [0.02]	66.88 c, *** [0.00]
Consumer staples	n.a.	n.a.	58.79*** [0.00]	60.56 c, * [0.09]	11.38 [0.19]	55.74*** [0.01]	80.99 c, *** [0.00]
Energy	70.78*** [0.00]	n.a.	n.a.	n.a.	5.22 [0.79]	45.51* [0.08]	81.19 c, *** [0.00]
Financial	59.94*** [0.00]	45.55* [0.08]	36.62 [0.37]	81.86 a,c, *** [0.00]	6.11 [0.68]	54.44*** [0.01]	86.19 c, *** [0.00]
Health care	n.a.	n.a.	48.04** [0.05]	n.a.	8.61 [0.40]	68.38 a, *** [0.00]	77.79 c, *** [0.00]
Industrials	n.a.	36.28 [0.38]	43.78 [0.11]	54.10*** [0.01]	5.77 [0.72]	45.32* [0.08]	79.67 c, *** [0.00]
Information technology	n.a.	39.05 [0.26]	54.65*** [0.01]	n.a.	n.a.	44.47* [0.10]	81.14 a,c, *** [0.00]
Materials	46.53* [0.07]	47.86 [0.14]	48.35** [0.05]	n.a.	n.a.	58.59 a, *** [0.00]	81.74 b,c, *** [0.00]
Utilities	68.46 a, *** [0.00]	n.a.	56.54** [0.00]	59.89 c, * [0.10]	5.54 [0.75]	57.87*** [0.00]	75.77 c, *** [0.00]
Transportation	n.a.	n.a.	38.61 [0.28]	n.a.	n.a.	51.00** [0.02]	79.75 a,c, *** [0.00]
Telecommunications	n.a.	n.a.	45.31* [0.09]	n.a.	17.94** [0.02]	44.90* [0.09]	73.75 a,c, *** [0.00]

Note: Trace statistics by Johansen (1988) are reported in the table. Null hypothesis H_0 is there exists no cointegration relation. Rejection of H_0 implies there exist at least one set of cointegration relation. Superscripts a and b indicate there exist 2 and 3 sets of cointegration relations respectively. Numbers in [ ] are p values. For Japan, we test the cointegration relation between oil price and stock price only. Superscript c indicates intercept and trend terms are included in the cointegration test, and the rest without superscript c indicate the cointegration test including intercept only. n.a. = sector prices of stocks are not available. *,**,*** represent 10%, 5% and 1% significance levels respectively.

Table 4
Choice of models and optimal lag periods for the G7 nations.

	Canada	France	Germany	Italy	Japan	UK	US
Composite	VEC(12)	VAR(10)	VAR(3)	VAR(3)	VAR(4)	VEC(10)	VEC(6)
Consumer discretionary	n.a.	n.a.	n.a.	n.a.	n.a.	VEC(10)	VEC(9)
Consumer staples	n.a.	n.a.	VEC(4)	VAR(11)	VAR(4)	VEC(10)	VEC(10)
Energy	VEC(8)	n.a.	n.a.	n.a.	VAR(4)	VAR(10)	VEC(3)
Financial	VEC(6)	VAR(7)	VAR(5)	VEC(10)	VAR(4)	VEC(10)	VEC(6)
Health care	n.a.	n.a.	VEC(2)	n.a.	VAR(3)	VEC(10)	VEC(3)
Industrials	n.a.	VAR(3)	VAR(3)	VEC(7)	VAR(4)	VAR(10)	VEC(12)
Information technology	n.a.	VAR(10)	VEC(5)	n.a.	n.a.	VAR(10)	VEC(3)
Materials	VAR(1)	VAR(3)	VEC(3)	n.a.	n.a.	VEC(10)	VEC(10)
Utilities	VEC(8)	n.a.	VEC(12)	VAR(11)	VAR(5)	VEC(10)	VEC(9)
Transportation	n.a.	n.a.	VAR(4)	n.a.	n.a.	VEC(10)	VEC(12)
Telecommunications	n.a.	n.a.	VAR(2)	n.a.	VEC(3)	VAR(12)	VEC(10)

Note: Numbers in ( ) are optimal lag period using likelihood ratio test. VAR = vector antoregressive; VEC = vector error correction.

Table 5
Sup F test results on the multivariate model with structural change.

	Canada	France	Germany	Italy	Japan	UK	US
Composite	5.6867 [1998:09]	6.0067 [1999:01]	9.8095** [1992:09]	12.3420** [2007:06]	12.6252** [2007:06]	14.9312*** [2003:08]	9.8390** [2006:08]
Consumer discretionary	n.a.	n.a.	n.a.	n.a.	n.a.	19.0785*** [2003:05]	8.5805** [2008:01]
Consumer staples	n.a.	n.a.	5.4882 [1999:03]	8.1052* [1996:01]	12.5370** [2007:06]	11.2053** [2007:06]	11.7485** [1996:09]
Energy	16.6493*** [1999:03]	n.a.	n.a.	n.a.	12.9226** [2007:06]	15.8167*** [2007:08]	20.3602*** [2008:07]
Financial	14.5784*** [1993:07]	10.2547** [2007:05]	4.5850 [2002:11]	4.6594 [1996:01]	13.3631*** [2007:07]	8.5119* [2007:02]	17.6798*** [1994:12]
Health care	n.a.	n.a.	11.8202** [1993:03]	n.a.	9.7280** [2007:06]	16.1352 [1999:03]	15.8888*** [2008:08]
Industrials	n.a.	15.4894*** [2007:07]	10.5132** [1992:09]	10.5535** [2003:05]	13.1435** [2007:06]	11.8944 [2003:07]	9.3840** [2002:04]
Information Technology	n.a.	5.7066 [1997:01]	8.9875* [1993:03]	n.a.	n.a.	14.1066 [2007:08]	29.1587*** [2008:08]
Materials	58.2800*** [1992:01]	15.5044*** [2007:07]	18.9759*** [2010:09]	n.a.	n.a.	13.5422*** [1999:01]	15.7351*** [2000:02]
Utilities	10.9954** [1995:04]	n.a.	7.7627* [2003:04]	8.1711* [1996:01]	14.5727*** [2007:03]	19.2389*** [2001:04]	9.6240** [2008:01]
Transportation	n.a.	n.a.	7.5592 [1993:03]	n.a.	n.a.	15.3707*** [2003:04]	10.5672** [2002:04]
Telecommunications	n.a.	n.a.	12.9144*** [1993:03]	n.a.	14.5834 [1995:03]	6.4368 [1998:08]	14.9269*** [2006:08]

Note: Numbers in the table are sup F statistics and number in [ ] are dates of structural break. *, **, *** denote 10%, 5% and 1% significance level. Multivariate models used are four-variable VAR or VEC model (see Table 4).

Table 6
Results of 4-variable composite index causality model.

Canada	France	Germany	Italy	Japan	UK	USA
		$\Delta lroilp \to \Delta r$	$\Delta lroilp \to \Delta r$		$\Delta lroilp \to \Delta r$	
				$\Delta lroilp \to \Delta ly$		
$\Delta ly \to \Delta r$	none	$\Delta ly \to \Delta r$			$\Delta ly \to \Delta r$	$\Delta ly \to \Delta r$
		$\Delta lrstkp \to \Delta r$			$\Delta lrstkp \to \Delta r$	
		$\Delta lrstkp \to \Delta lroilp$			$\Delta lrstkp \to \Delta lroilp$	$\Delta lrstkp \to \Delta lroilp$
		$\Delta lrstkp \to \Delta ly$	$\Delta lrstkp \to \Delta ly$		$\Delta lrstkp \to \Delta ly$	$\Delta lrstkp \to \Delta ly$
$\Delta r \to \Delta ly$		$\Delta r \to \Delta ly$		$\Delta r \to \Delta ly$	$\Delta r \to \Delta ly$	
					$\Delta r \to \Delta lroilp$	

Note: The causality results are reported with significance level 5% or less; $a \to b$ implies a Granger causes b; $\Delta =$ first difference; $lroilp$, $lrstkp$ and ly represent logarithmic real oil price, real stock price and real industrial production index, r is interest level.

a biased estimation. Chow proposed an F statistic in order to test the stability of estimated coefficients in regressions as early as 1960 if the point of structural break is known a priori. To overcome this limitation, Hansen (1991) developed a procedure to examine the stability of estimated coefficients in the univariate model. However, Hansen's method cannot locate the position where a structural break takes place. To remedy the shortcoming, Bai and Perron (1998) proposed a univariate model in which multiple structural breaks can be located and tested. For multivariate models, Bai et al. (1998) developed a Sup F procedure to locate and test multiple structural break points, which we use in the following VEC model:

$$\Delta y_t = A_{01} + B_1 y_{t-1} + \sum_{i=1}^{p-1} A_{i,1} \Delta y_{t-i}$$
$$+ d_t(k) \left(A_{02} + B_2 y_{t-1} + \sum_{i=1}^{p-1} A_{i,2} \Delta y_{t-i} \right) + \varepsilon_t, \tag{3}$$

where $\Delta y_t' = (\Delta r, \Delta lroilp, \Delta ly, \Delta lrstkp)'$ is a vector of 4×1; $d_t(k) = 0$ for $t \leq k$ but $d_t(k) = 1$ for $t > k$. Stacking Eq. (3) leads to

$$\Delta y_t = (V_t' \otimes I)\theta_1 + d_t(k)(V_t' \otimes I)\theta_2 + \varepsilon_t, \tag{4}$$

where $V_t' = (1, y_{t-1}, \Delta y_{t-1}, \Delta y_{t-2}, ..., \Delta y_{t-p-1})'$, $\theta_1' = vec(A_{01}, B_1, A_{1,1}, A_{2,1}, ..., A_{p-1,1})'$, $\theta_2' = vec(A_{02}, B_2, A_{1,2}, A_{2,2}, ..., A_{p-1,2})'$, I is an identity matrix of 4 x 4 indicating the presence of a full structure

change (intercept and slope). In the case of a partial structural change (a mean shift or changes on part of estimated coefficients), its test power is known to be superior to that of the full change. Thus, we opt for the model with partial structural changes as shown below:

$$\Delta y_t = (V_t' \otimes I)\theta_1 + d_t(k)(V_t' \otimes I)S'S\theta_2 + \varepsilon_t. \tag{5}$$

where S is the selection matrix consisting of 0 and 1 with full row rank. The rank of S depends on the number of coefficients that are allowed to change. If $S = I$, it indicates Eq. (4) represents a full structural change. If $S = (s \otimes I)$ with $s = (1, 0, ...0)$, Eq. (5) can be expressed as

$$\Delta y_t = (V_t' \otimes I)\theta_1 + \theta_2 d_t(k) + \varepsilon_t. \tag{6}$$

Eq. (6) represents the model with the structural change in intercept only, which can be written more compactly as

$$\Delta y_t = Z_t'(k)\beta + \varepsilon_t, \tag{7}$$

where $Z_t'(k) = [(V' \otimes I), d_t(V_t' \otimes I)S']$, $\beta = (\theta_1'. (S\theta_2)')'$, and ε_t obeys multivariate Gaussian distribution. The F statistic can be used to

Table 7
Causality results from oil price changes to stock price changes.

	Canada	France	Germany	Italy	Japan	UK	US	% of significant causal relations
Composite	8.6573 [0.73]	7.2370 [0.70]	3.1099 [0.38]	3.4285 [0.33]	6.9991 [0.14]	11.5475 [0.32]	9.5109 [0.15]	0/7 (0%)
Consumer discretionary	n.a.	n.a.	n.a.	n.a.	n.a.	15.1461 [0.13]	2.5229 [0.98]	0/2 (0%)
Consumer staples	n.a.	n.a.	9.5411** [0.05]	13.9622 [0.24]	6.6293 [0.16]	15.3616 [0.12]	19.6277** [0.03]	2/5 (40%)
Energy	3.0582 [0.93]	n.a.	n.a.	n.a.	9.1489* [0.06]	13.5312 [0.20]	2.2582 [0.52]	0/4 (0%)
Financial	8.5485 [0.20]	7.9913 [0.33]	11.8735** [0.04]	15.9576* [0.10]	4.8644 [0.30]	10.1324 [0.43]	4.3744 [0.63]	1/7 (14.3%)
Health care	n.a.	n.a.	0.1183 [0.94]	n.a.	3.3918 [0.34]	13.2120 [0.21]	4.3127 [0.23]	0/4 (0%)
Industrials	n.a.	2.4199 [0.49]	1.5319 [0.67]	12.7422* [0.08]	5.3041 [0.26]	11.0357 [0.35]	20.5429* [0.06]	0/4 (0%)
Information technology	n.a.	18.3393** [0.05]	18.8701*** [0.00]	n.a.	n.a.	16.6319* [0.08]	4.5830 [0.21]	2/4 (50%)
Materials	0.1500 [0.70]	0.6267 [0.89]	3.8511 [0.28]	n.a.	n.a.	13.7975 [0.18]	10.3769 [0.41]	0/5 (0%)
Utilities	11.1976 [0.19]	n.a.	31.2573*** [0.00]	15.2038 [0.17]	2.8827 [0.72]	13.0693 [0.22]	5.1375 [0.82]	1/6 (16.7%)
Transportation	n.a.	n.a.	8.6049* [0.07]	n.a.	n.a.	16.9522* [0.08]	23.7025** [0.02]	1/3 (33.3%)
Telecommunications	n.a.	n.a.	3.7668 [0.15]	n.a.	3.4102 [0.33]	18.4532* [0.10]	12.7893 [0.24]	0/4 (0%)
% of significant causal relations	0/5 (0%)	1/5 (20%)	4/10 (40%)	0/5 (0%)	0/8 (0%)	0/12 (0%)	7/57 (16.7%)	7/57 (12.3%)

Note: The null hypothesis is oil price changes do not Granger-cause ($\to$) stock price changes; Number reported in the Table are χ^2, statistics and number in [ ] are p values; *, **, *** represent 10%, 5% and 1% significant levels; % of significant causal relations = % of significant Granger causal relation (5% significance) for a given country or a given sector.

locate and test structural breaks in the multivariate model as shown below:

$$\hat{F}(k) = T \cdot \left\{ \boldsymbol{R}\hat{\boldsymbol{\beta}}(k) \right\}' \left\{ \boldsymbol{R} \left(T' \sum_{t=1}^{T} \boldsymbol{Z}_t \hat{\Sigma} \boldsymbol{Z}_t' \right) \boldsymbol{R}' \right\}^{-1} \left\{ \boldsymbol{R}\hat{\boldsymbol{\beta}}(k) \right\}, \qquad (8)$$

in which $\hat{\boldsymbol{\beta}}(k) = \left\{ \sum_{t=1}^{T} \left(\boldsymbol{Z}_t \hat{\Sigma}^{-1} \boldsymbol{Z}_t' \right) \right\}^{-1} \sum_{t=1}^{T} \boldsymbol{Z}_t \hat{\Sigma}^{-1} \Delta \boldsymbol{y}_t.$

Note that $\boldsymbol{R} = (\boldsymbol{0}, \boldsymbol{I})$ such that $\boldsymbol{R}\boldsymbol{\beta} = \boldsymbol{S}\boldsymbol{\theta}_2$. Failure to reject $H_0 : \boldsymbol{S}\boldsymbol{\theta}_2 = \boldsymbol{0}$ implies an absence of a structural break, the variance covariance relation ($\hat{\Sigma}$) can be estimated via an OLS estimator for a given k and alternative hypothesis. Within a given region, the greatest F statistic (Sup F) is used to test and locate the structural break. If such structural break(s) took place in the data, we would segment the model into different regimes.

4. Empirical results and discussion

The analysis of the relationship between oil price changes and stock returns requires that all the time series are to be test for stationarity (unit root) and cointegration tests. A unit root test on individual time series is conducted first before examining the existence of a cointegration relation. Such a test investigates the presence of a stochastic trend in all the individual series. If the time series have a stochastic trend, then a shock that impacts them will have permanent effects. The unit root test employed in this investigation is the Augmented Dickey-Fuller (ADF) test. The unit root test results on composite indexes, sector indexes, oil price, industry output and interest of the G-7 countries are listed in Table 2.

An inspection of Table 2 indicates that we cannot reject the null of a unit root for both composite and sector indexes in level. However, we can reject the null of I(1) for these indexes in first difference,

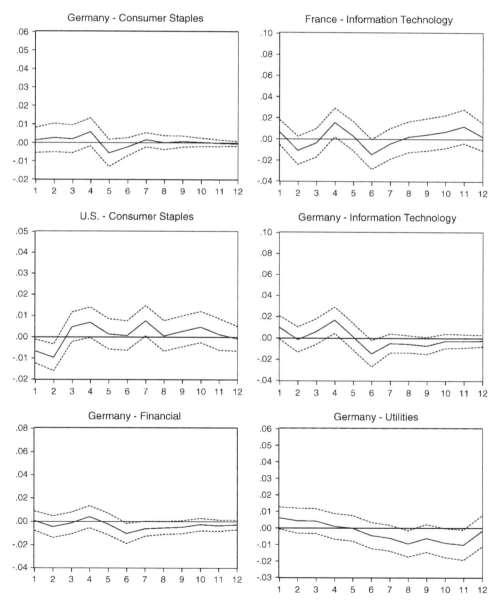

Fig. 1. Impulse response functions for sectors with significant Granger causality ($\Delta lroilp \rightarrow \Delta lrstkp$). Notes: Solid line represents the response function of stock prices change to a one standard deviation change in oil prices. Dotted lines are 95% confidence intervals.

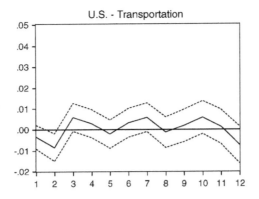

U.S. - Transportation

Notes: Solid line represents the response function of stock prices change to a one standard deviation change in oil prices. Dotted lines are 95% confidence intervals.

Fig. 1. (*continued*).

Table 8
Causality results from stock price changes to oil price changes.

	Canada	France	Germany	Italy	Japan	UK	US	% of significant causal relations
Composite	20.3073* [0.06]	9.4624 [0.49]	14.6777*** [0.00]	3.0464 [0.38]	5.9355 [0.20]	24.6016*** [0.01]	14.4563** [0.03]	3/7 (42.9%)
Consumer discretionary	n.a.	n.a.	n.a.	n.a.	n.a.	7.5277 [0.67]	12.5424 [0.18]	0/2 (0%)
Consumer staples	n.a.	n.a.	14.4081*** [0.01]	21.4602** [0.03]	7.4623 [0.11]	19.4837** [0.04]	20.6388** [0.02]	4/5 (80%)
Energy	26.5921*** [0.00]	n.a.	n.a.	n.a.	3.2464 [0.52]	21.7000** [0.02]	3.3570 [0.34]	2/4 (50%)
Financial	4.6168 [0.59]	10.2950 [0.17]	12.8133** [0.03]	18.6414** [0.05]	3.2510 [0.52]	25.7384*** [0.00]	16.9621*** [0.01]	4/7 (57.1%)
Health care	n.a.	n.a.	5.6930* [0.06]	n.a.	2.0883 [0.55]	23.8715*** [0.01]	9.7910** [0.02]	2/4 (50%)
Industrials	n.a.	14.9977*** [0.00]	14.3024*** [0.00]	8.5243 [0.29]	2.3175 [0.68]	20.7221** [0.02]	19.8059* [0.07]	3/6 (50%)
Information technology	n.a.	9.1688 [0.52]	17.2255*** [0.00]	n.a.	n.a.	12.7785 [0.24]	4.5242 [0.21]	1/4 (25%)
Materials	8.2532*** [0.00]	23.9839*** [0.00]	22.4225*** [0.00]	n.a.	n.a.	33.2957*** [0.00]	15.9390* [0.10]	4/5 (80%)
Utilities	13.2935* [0.10]	n.a.	23.8866** [0.02]	21.2634** [0.03]	9.3438* [0.10]	15.9142* [0.10]	30.8874*** [0.00]	3/6 (50%)
Transportation	n.a.	n.a.	17.5728*** [0.00]	n.a.	n.a.	19.7570** [0.03]	10.4106 [0.58]	2/3 (66.7%)
Telecommunications	n.a.	n.a.	3.7561 [0.15]	n.a.	0.2352 [0.97]	31.4171*** [0.00]	14.9845 [0.13]	1/4 (25%)
% of significant causal relations	2/5 (40%)	2/5 (40%)	8/10 (80%)	3/5 (60%)	0/8 (0%)	9/12 (75%)	5/12 (41.7%)	29/57 (50.9%)

Note: Null hypothesis is stock price changes do not Granger cause (↛) oil price changes. See Table 8 for other explanations. *,**,*** represent 10%, 5% and 1% significance levels respectively.

indicating they follow an I(1) process.[4] Test results for other 3 macroeconomic variables suggest that we cannot reject the null of I(1) in level but can reject the null of I(1) in first difference with the exception of *ly* and *r* of Japan. As such we employ Johansen's trace statistics (1988) on two variables (*lroilp* and *lrstkp*) for Japan and all the four variables for 6 other countries. The test results are listed in Table 3.

An examination of Table 3 shows that there exists a cointegration relation in the composite index, energy, financial and utilities sectors (except materials sector) for Canada. In contrast, there is a lack of 4-varable cointegration relations in the composite index or 4 sector indexes in France. For Germany the 4-variable cointegration exists in the 5 sectors of consumer staples, heath care, information technology, materials and utilities, but a cointegration relation is not found for composite index and other 4 sectors. In Italy, the cointegration

relation exists in financial and industrial sectors, but is missing in the composite index, consumer staples and utilities sectors. For Japan, the cointegration relation is found only in telecommunication sector. For the UK, the cointegration relation exists in all the sectors except energy, industrials, IT and telecommunications sectors. Finally, the cointegration relation is found to exist for all sector indexes and the composite index for the US. Based on the results of Table 3, Table 4 illustrates the appropriate models for each sector in the G7countries.

As shown in Table 4, a VEC model is used for sectors with a cointegration relation but a VAR model is chosen instead for sectors without a cointegration relation. The optimal delay period based on the likelihood ratio (LR) test is shown in parentheses.

After deciding on appropriate models and optimal lags for G7 countries, we need to test the possibility of the structural break based on Eq. (8). As shocks are often emanated exogenously, we opt for mean shift as the structural change. Table 6 reports test results on the structural break in terms of Sup F.

---

[4] We can reject the null of I(1) at $\alpha = 5\%$ for interest rate in level for Canada. As is well known in the literature, the statistical power is not strong in unit root tests. Thus, we choose $\alpha = 1\%$ as significance level. A 5% significance level is used in other tests.

Listed in Table 5 are Sup F statistic values along with the potential dates of structural break shown in brackets. When a Sup F test statistic exceeds its critical value, it implies that a structural break occurred on the date as shown in the bracket. We then reject the null of no structural break. For instance, except for the composite index, all remaining 4 sector indexes exhibited the structural break in Canada. In a similar vein, values of Sup F indicate financial, industrials and materials sector indexes in France experienced structural breaks in 2007. Of the 10 indexes in Germany, we reject the null of no structural break for 5 sector indexes: composite, health care, industrials, materials and telecommunications. Of the 5 indexes in Italy, only composite index and industrial sector index showed the structural break. In the case of Japan, all indexes except for the telecommunications sector exhibit the structural break: we reject the null for 7 of the 8 indexes. In the UK, we fail to reject the null for 5 sector indexes: financial, health care, industrials, IT and communications sectors. However, we reject the null for the remaining sectors. In the US, we reject the

null for all 12 indexes based on Sup F test statistics. Based on the above test results, majority of stock price indexes exhibited significant structural breaks during the period. Failing to take it into consideration will render estimates biased. Given that our structural change analysis is built upon mean shift, we add dummy variable (intercept) D into the model. When $t \geq \tau$, D = 1, otherwise D = 0. Note that τ represents location of the structural break (dates of structural break in brackets in Table 6). After taking structural shift into consideration, the VEC and VAR models take the following forms:

$$\Delta y_t = A_{01} + A_{02}D + By_{t-1} + \sum_{i=1}^{p-1} A_i \Delta y_{t-i} + \varepsilon_{1t}, \tag{1'}$$

$$\Delta y_t = A_{01} + A_{02}D + \sum_{i=1}^{p-1} A_i \Delta y_{t-i} + \varepsilon_{1t}, \tag{2'}$$

where $\Delta y_t' = (\Delta r, \Delta lroilp, \Delta ly, \Delta lrstkp)'$ is a vector of 4×1; y_{t-1}' is a vector denoting $(r_{t-1}, lroilp_{t-1}, ly_{t-1}, lrstkp_{t-1})'$ in level; and D is the intercept dummy vector reflecting the mean-shift effect. With the

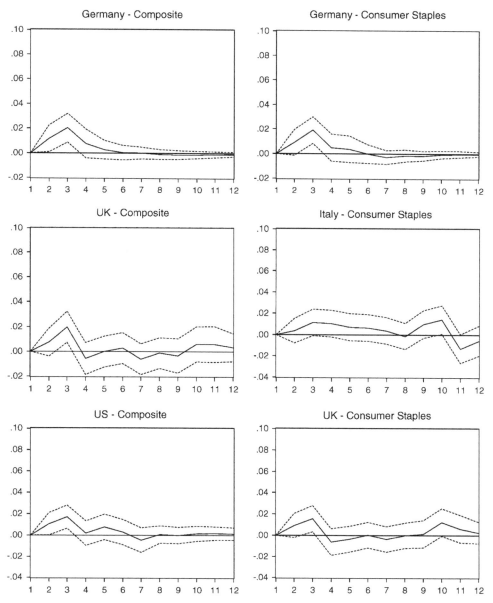

Fig. 2. Impulse response functions for sectors with significant Granger causality ($\Delta lrstkp \rightarrow \Delta lroilp$). Notes: Solid line represents the response function of oil prices to a one standard deviation change in stock prices. Dotted lines are 95% confidence intervals.

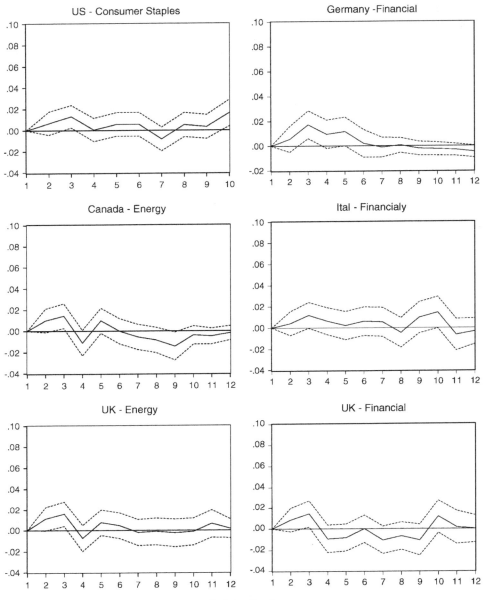

Fig. 2. (continued).

structural break included in the model, we are ready to examine the Granger causality between oil price changes and stock price changes. The causality results are reported in Table 6.[5]

An examination of Table 6 suggests readily that oil price changes do not Granger-cause composite stock price changes in all of the G7 countries. Nontheless, composite stock price changes are found to lead oil price changes in Germany, the UK and the US.[6]

We discover more interactions among the 4 variables of the composite index model in Germany and the UK. On the contrary, no

significant interaction among the 4 variables is found in France. Within the G7 nations, significant feedback effects exist between industrial output and interest rate for Canada. Oil price changes lead interest rate changes while stock price changes lead industrial output changes in Italy. In Japan, both oil price changes and interest rate changes lead changes in industrial output. Stock price changes in the US lead both oil price changes and changes in industrial output and changes in industrial output lead interest rate changes. The interaction pattern becomes more complicated in Germany: oil price changes lead interest rate changes while significant feedback effects exist between interest rate changes and changes in industrial output. It is found that stock price changes lead not only oil price changes, they also lead changes in interest rate and industrial output as well. The same six interactions in Germany are found in the UK as well. In addition, interest rate changes are found to lead oil price changes. This translates into a strong feedback relationship between changes in oil price and

---

[5] As our emphasis is on the impact of oil price changes on stock price changes especially from the viewpoint of sector indexes, we report the results only on if oil price Granger-causes stock price in Table 7. Other Granger-causality results, however, are available upon request.

[6] Empirical estimates are reported in Table 8 regarding whether stock price changes Granger-cause oil price changes.

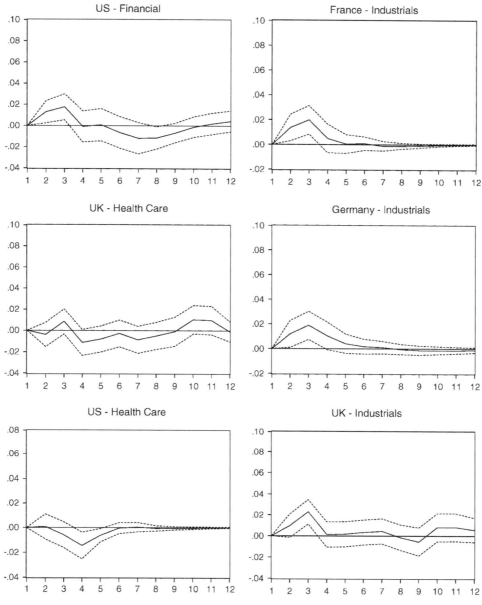

Fig. 2. (*continued*).

interest rate in the UK. In the case of the US, stock price changes are found to lead oil price changes. The other two Granger causality relations are consistent with Sadorsky's finding (1999): stock price changes lead changes in industrial output which in turn lead changes in interest rate. The difference may be attributed to different sample periods used. That is, early findings rarely pointed to the results that stock price changes Granges-cause oil price changes. Our finding may point to a new trend.

Generally speaking, high oil price impacts economic growth negatively, but it influences stock price via other factors. First, if oil is used as input (consumers), an expectation of oil price increase may well exert a negative impact on its stock price depending on the degree of oil dependence of the industry. In contrast if oil is produced as output (suppliers), an expectation of oil price increase tends to boost the company's stock price. Next, stock prices are more or less impacted by

an oil price increase even if oil is not directly employed as input in the production process. This is because an oil price increase, more often than not, gives rise to inflation, which may slowdown economic growth. If the sector is pro-cyclical (counter-cyclical), its stock prices are expected to decline (increase). Depending on the extent of oil dependence and its relationship with respect to business cycle, the impact of an oil price change on stock prices also hinges on fiscal and monetary policies implemented in fine-tuning the economy. Table 7 reports various impacts of oil price changes on stock prices for each country.

A perusal of Table 7 reveals that stock price in 6 of 11 sector indexes generally are not Granger-caused by oil price changes. That is, sector stock prices of consumer discretionary, energy, health care, industrials, materials and telecommunications were oil price-resistant. Except for energy industry, the other 5 sectors are oil

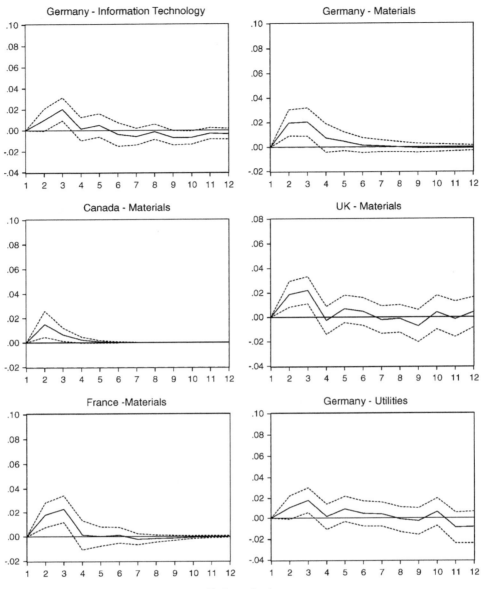

Fig. 2. (continued).

consumers. The result may also be explained by the data frequency and period of data; that is, monthly data cannot reflect the short term dynamics if an oil price increase exerts only a temporary influence on stock-prices.[7] For the remaining 5 sector (consumers staples, financial, information technology, utility and transportation), we find that oil price changes Granger-caused sector stock price changes for at least one country. Impulse response functions (IRF) are reported in Fig. 1.[8]

Of the 5 countries that have consumers staples sector, only 2 exhibit significant IRF: Germany and the US. Despite the Granger causality result for Germany, the IRF does not show a clear-cut pattern (Fig. 1). On the other hand, the IRF for the US indicates that (i) sector stock price of consumer staples (i) reacted negatively in periods 1–2 as one unit of oil price change is administered positively to the economy and (ii) bounced back positively in period 7, but tapered off after

the 8th period. Note that consumer staples sector consists of foods and personal (household) products, which are directly affected by oil price change. As such, oil price change is expected to impact the sector stock price to various extents depending on the magnitude of pass-through effect from oil price to sector stock prices. It is small wonder that the stock price in some countries (e.g. Italy) experienced relatively smaller changes.

Of the 7 countries with financial sector stock prices, only in Germany, do we find oil price changes Granger-cause changes in stock price. As shown in the IRF of Fig. 1, the stock price reacted negatively in periods 6, 7 and 8 and tapered off afterwards. As oil is not a direct input in financial sector, its stock prices were influenced indirectly via inflation and resulting economic slowdown. The pass-through effect might have been more pronounced in Germany with a longer lag effect (after 6, 7 and 8 months).

As illustrated by Sadorsky (2003), oil price change can have two diametrically different impacts on IT stock prices. The first view recognizes that oil price changes have very little impact on IT

---

[7] We thank a reviewer for pointing out this possibility.
[8] Fig. 1 provides only IRFs with significant Granger causality.

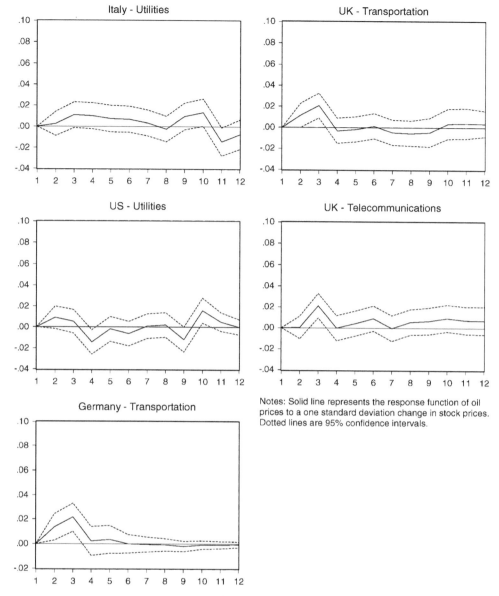

Notes: Solid line represents the response function of oil prices to a one standard deviation change in stock prices. Dotted lines are 95% confidence intervals.

Fig. 2. (*continued*).

stock prices because IT companies engage in business activities that are not very energy intensive. Another view, however, clinched to the idea that oil price changes do have an impact on IT stock prices because oil price increase fuels inflation and inflation leads to change in business cycle conditions and/or economic down turns. IT companies are known to be very sensitive to the overall business cycle and thus generally do not fare well with oil price increases. This said we expect oil price change may or may not impact stock prices of IT sectors. That is, some IT companies that produce hardware are likely to be more oil-intensive in production process than companies that supply service (software venders). The procyclical nature of oil price changes tends to dominate in Germany and France, in which oil price change Granger-causes IT stock prices. As can be noted from the IRF of Germany, IT stock price changes reacted positively in period 4 but negatively in period 6

as one unit of shock is administered. The IRF of France echoed that of Germany: IT stock price changes responded positively in period 4 but negatively in period 6.

Given that both utility and transportation sectors are notably related to oil products, oil price changes are expected to impact their sector stock prices. The extent to which stock prices are impacted depends on the pass-through effect and speed of adjustment. Our empirical results indicate that only in Germany do we find the Granger causality between oil price changes and utility stock price: utility stock prices reacted negatively to oil price change in periods 8, 10 and 11. The other 4 countries did not exhibit any significant causality. Among 3 countries (Germany, the UK, the US) with transportation sector index data, only in the US do we find Granger causality: transportation stock price responded negatively to oil price change in period 2 as one unit of shock were given to model.

France - Industrials

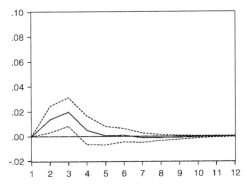

Germany - Industrials

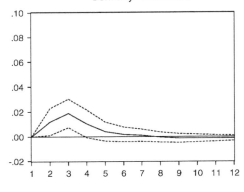

UK - Industrials

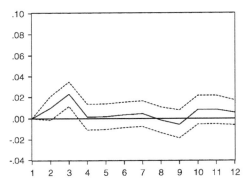

Fig. 2. (continued).

Based on the above Granger causality results, it is evident that oil price changes cannot be used to predict changes in composite indexes. Our result is at odds with that by Sadorsky (1999) and Papapetrou (2001) in which a negative relationship was found. On the other hand, a positive relationship was proposed by Bjørnland (2008) and Park and Ratti (2008). The difference in conclusion can be explained away by the fact that we use a more recent sample period on the G7 nations. As mentioned before, recent research (Blanchard and Gali, 2007; Hooker, 2002) indicates that the transmission effect from oil price increase to inflation has been weakened and not surprisingly, it impacts composite index only moderately in our model. However oil price increases may well have greater effect (negatively) on some specific sectors and this is borne out in our analysis. [9]

Our result echoing that by Hammoudeh and Li (2005) indicates that oil price shocks Granger-cause US transportation sector index negatively. However the positive relationship between oil price and oil and/or gas industry (El-Sharif et al., 2005; Sadorsky 2001) is not supported with the exception of Japan ($\alpha = 6\%$) by our model.[10] This may well be the result of different sample period and definition of the energy sector in our paper, which includes oil gas and energy equipment and services.

The Granger causality model is used here to investigate the short-run dynamic relationships between oil price and stock price. On the other hand, the multifactor model analyzes the problem from the supply side perspective: the rise in production cost due to an oil price increase leads to reduced profit, which in turn depresses the stock price. Thus an oil price increase is expected to impact stock

[9] The main focus of our paper is not on why some sector index responds more significantly than others to oil price changes. However it remains an interesting topic in the future.

[10] As shown in the IRF, a unit increase in oil price shock impacted the energy sector of Japan negatively in periods 5 and 6.

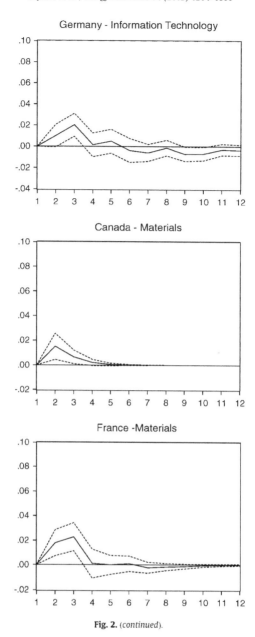

Fig. 2. (continued).

price negatively. However, one can approach the same problem from the demand-side perspective: As high stock prices reflect a booming economy, the demand for oil increases. As a result, price of crude oil rises as well. That is to say, one expects stock price changes lead (Granger-cause) oil price changes positively as show in Table 8.

A close examination of Table 8 indicates that the frequency with which stock price changes lead oil price changes far exceed that of oil price changes lead stock price changes: 29 to 7, about half of the 57 relationships. Such a causality was found most frequently in Germany: 8 of 10 sectors (8/10 = 80%), followed by the UK's 75% (9/12), Italy's 60% (3/5), US's 42% (5/12), Canada's and Frances 40% (2/5). Note that stock price changes did not lead oil price changes at all in Japan. In terms of frequencies in which sector stock price leads oil price, we find sectors of consumer staples and materials exhibit

such causality in 4 of 5 countries (80%), followed by the transportation sector in 2 of 3 countries or 67%, financial sector (4 of 7 countries or 57%), health care sector (2 of 4 or 50%), industrial and utility sectors (3 of 6 or 50%), composite index (3 of 7 or 43%), and IT and telecommunication sector (1 of 4 or 25%). None of the two consumer discretionary sector exhibits the stock-price-leads-oil-price causality. Fig. 2 displays the IRF of 29 such causalities.

Consistent with our expectation, Fig. 2 illustrates largely positive responses to stock price changes. Of the 29 causal relationships, 24 showed positive responses. The minor exceptions are found in (i) financial sector of Italy showed no significant response, (ii) utility sector of Italy showed positive response in the 10th period but turned negative in the 12th period, (iii) the health sector of the UK displayed no significant responses, and (iv) both health and utility sectors in the

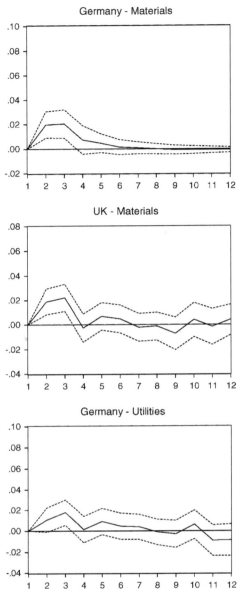

Fig. 2. (continued).

US responded negatively in the 4th period. Average length of time of the response time to stock price changes is about one quarter (3 months) except for Italy where oil price took 10 months to responds to stock price changes.

5. Conclusions

Since Jones and Kaul (1996) ushered in the oil price–stock price relationship, many more followed along the line. Most of the literature deals with the impact of oil price on nation's composite index. For example Faff and Brailsford (1999) were among the first to extend the use of composite index to industry (sector) for a given country (Australia). Recently, focus has been shifted to oil price–sector stock price relationship based on a given country or specific sectors (e. g., gas and oil sector). These studies fail to consider how sector stock prices respond to oil price change across countries. Furthermore,

their analyses are built on multifactor model, which pre-assume the causality direction: oil price changes give rise to sector stock returns. Besides, the multifactor model is essentially a static model in the sense that it lacks dynamic interactions. To overcome these limitations, using sector stock prices from 1991:01 to 2009:05 and the GICS definition to facilitate comparison, we establish a four-variable (interest, oil price, industrial production index and stock prices), we perform the unit root and cointegration tests in order to choose between VAR and VEC models. Further, to take potential structural breaks into consideration in the 20-year period, we employ the Sup F testing procedure by Bai et al. (1998). The result indicates the existence of many a structure break. Failure to take it into consideration will likely lead to biased estimates.

Our analysis finds that oil price shocks did not impact composite indexes of the G7 economies in terms of the Granger causality. However, from the perspective of individual sectors, oil price shocks did

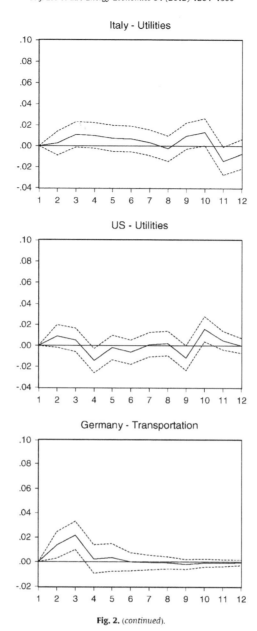

Fig. 2. *(continued).*

exert significant influences on some sector indexes for some countries. As far as individual countries are concerned, 4 of the 9 sector indexes of Germany are significantly affected by oil price shocks, followed by France and the US. Sector indexes of the other four economics were not significant impacted. In terms of sectors that were impacted more frequently, the IT sector takes the lead: 2 (France and Germany) of the 4 (France, Germany, the UK and the US) countries in which oil price shocks impacted IT sectors, followed by the consumer staples sector: 2 (Germany and the US) of the 5 countries in which consumer staples sector index was impacted by oil price shocks. As for the transportation sector, only in the US do we find that it was impacted by oil price shocks. Similarly oil price shocks were found to impact utility and financial sector indexes only in Germany. Further, oil price shocks did not significantly impact price indexes of consumer discretionary, energy, health care, industrials, materials and telecommunication sectors. Beyond that our results indicate that the

frequency in which stock price changes lead oil price changes exceeds that of oil price changes lead stock price changes. The direction of change is found to be positive. That is, higher stock prices reflecting the booming economy increase the demand for oil, and hence higher oil prices.

More often than not, financial assets are traded based on sector stock returns; therefore, it is important for trader to understand the oil price transmission mechanism across the sectors and across the countries in order to make optimal portfolio decisions. By using sector stock prices of the G7 nations, this paper provides useful information for traders to execute optimal portfolio across both different countries and different sectors. In particular, investors might have to restrain themselves from buying sectors stocks in Germany, France and the U.S. in the presence of an oil price increase. At the same time, it may be to an investor's best interest to shun stocks from IT, consumer staples, transportation, and utilities sectors.

B.-J. Lee et al. / Energy Economics 34 (2012) 1284–1300

UK - Transportation

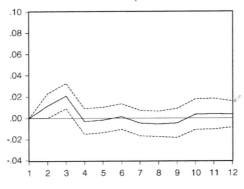

UK - Telecommunications

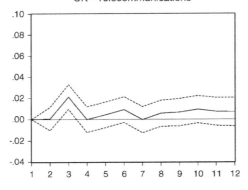

Notes: Solid line represents the response function of oil prices to a one standard deviation change in stock prices. Dotted lines are 95% confidence intervals.

Fig. 2. *(continued).*

References

Bai, J., Perron, P., 1998. Estimating and @@testing linear models with multiple structural changes. Econometrica 66 (1), 47–78.

Bai, J., Lumsdaine, R.L., Stock, J.H., 1998. Testing for and Dating Common Breaks in Multivariate Time Series. Rev. Econ. Stud. 65, 395–432.

Bjørnland, H.C., 2008. Oil Price Shocks and Stock Market Booms in an Oil Exporting Country. Scot. J. Polit. Econ. 56, 232–254.

Blanchard, O.J., Gali, Jordi, 2007. The Macroeconomic Effects of Oil Price Shocks: Why Are the 2000s so Different from 1970s? NBER Working Paper No. 13368.

Darby, M.R., 1982. The Price of Oil and World Inflation and Recession. Am. Econ. Rev. 72, 738–751.

El-Sharif, I., Brown, D., Burton, B., Nixon, B., Russell, A., 2005. Evidence on the Nature and Extent of the Relationship between Oil Prices and Equity Values in the UK. Energy Econ. 27, 819–830.

Elyasiani, E., Mausur, I., Odusami, B., 2011. Oil Price Shocks and Industry Stock Returns. Energy Econ. 33 (5), 966–974.

Faff, R.W., Brailsford, T.J., 1999. Oil Price Risk and the Australian Stock Market. J. Energy Finance Dev. 4, 69–87.

Hamilton, J.D., 1983. Oil and the Macroeconomy since World War II. J. Polit. Econ. 91, 228–248.

Hammoudeh, S., Li, H., 2005. Oil Sensitivity and Systematic Risk in Oil-sensitive Stock Indices. J. Econ. Bus. 57, 1–21.

Hansen, B., 1991. Parameter Instability in Linear Models. J. Policy Model. 14 (4), 517–533.

Hooker, Mark A., 2002. Are Oil Shocks Inflationary? Asymmetric and Nonlinear Specifications versus Changes in Regime. J. Money, Credit, Bank. 34 (2), 540–561.

Huang, R., Masulis, R., Stoll, H., 1996. Energy Shocks and Financial Markets. J. Futur. Mark. 16, 1–27.

Huang, Bwo-Nung, Hwang, M.J., Peng, Hsiao-Ping, 2005. The Asymmetry of the Impact of Oil Price Shocks on Economic Activities: An Application of the Multivariate Threshold Model. Energy Econ. 27, 455–476.

Johansen, S., 1988. Statistical Analysis of Cointegration Vectors. J. Econ. Dyn. Control. 12, 231–254.

Jones, C.M., Kaul, G., 1996. Oil and the Stock Market. J. Finance 51, 463–491.

Miller, J.I., Ratti, R.A., 2009. Crude Oil and Stock Markets: Stability, Instability, and Bubbles. Energy Econ. 31, 559–568.

Nandha, M., Faff, R., 2008. Does Oil Move Equity Prices? A Global View. Energy Econ. 30, 986–997.

Papapetrou, E., 2001. Oil Price Shocks, Stock Market, Economic Activity and Employment in Greece. Energy Econ. 23, 511–532.

Park, J., Ratti, R.A., 2008. Oil Price Shocks and Stock Markets in the US and 13 European Countries. Energy Econ. 30, 2587–2608.

Sadorsky, P., 1999. Oil Price Shocks and Stock Market Activity. Energy Econ. 21, 449–469.

Sadorsky, P., 2001. Risk Factors in Stock Returns of Canadian Oil and Gas Companies. Energy Econ. 23, 17–28.

Sadorsky, P., 2003. The Macroeconomic Determinants of Technology Stock Price Volatility. Rev. Financ. Econ. 12, 191–205.

INTERNATIONAL JOURNAL OF FINANCE AND ECONOMICS
Int. J. Fin. Econ. **7**: 37–50 (2002)
Published online in Wiley InterScience (www.interscience.wiley.com). DOI: 10.1002/ijfe.177

VOLATILITY OF CHANGES IN G-5 EXCHANGE RATES AND ITS MARKET TRANSMISSION MECHANISM

BWO-NUNG HUANG[1][*][†] and CHIN WEI YANG[2]

[1] *Department of Economics, National Chung-Cheng University, Taiwan*
[2] *Department of Economics, Clarion University of Pennsylvania, Clarion, USA*

ABSTRACT

This paper studies the transmission mechanism of G-5 exchange rate changes within each market and across the three major markets: London, New York and Tokyo. It is found that the volatility in both the London and New York markets leads that of Tokyo. In addition, the New York market slightly leads the London market in its volatility. After the Euro monetary system crisis, the frequencies of both the volatility spillover effect from London to New York and mutual feedback phenomena have increased. Furthermore, the volatility spillover effects from both London and New York to Tokyo have been on the rise after the Asian financial debacle. Within the framework of the causality model, we find better forecasting performance in predicting G-5 exchange rates across the three markets. It outperforms the traditional ARMA model in terms of both in- and out-sample forecasting. Copyright © 2002 John Wiley & Sons, Ltd.

JEL CODE: C32; F32; G15

KEY WORDS: Volatility spillover; causality in variance; G-5 exchange rates

1. INTRODUCTION

Predicting volatility is important for determining the cost of capital and for evaluating direct investment and asset allocation decisions. Generally, higher volatility implies higher capital costs and, as such, may also increase the option value of waiting. Hence it delays major investment projects. This paper finds increasing linkage effects in terms of volatility in G-5 exchange rates after the Euro monetary system (EMS) crisis, the Peso crisis and the Asian financial debacle. Such an enhanced linkage phenomenon does not mesh well with the risk diversification of international portfolio theory. Engle, Ito and Lin (1990) pioneer the concept of volatility spillover by the analogy of a heat wave. Volatility has location-specific autocorrelation so that a volatile day in, for example, New York is likely to be followed by another volatile day in New York (intra-market volatility), but not typically a volatile day in Tokyo. On the other hand, their meteor shower hypothesis claims that intra-day volatility of foreign exchanges can spillover from one trading centre to another so that a volatile day in New York is likely to be followed by a volatile day in Tokyo (inter-market volatility spillover effect). The spillover phenomenon is explained generally by two approaches. First, the model by Kyle (1985) and Admanti and Pfleiderer (1988) emphasizes the trading strategy of uniformed liquidity traders and optimizing traders with private information. They show that private information is only gradually incorporated into prices, with the price at the end of the relevant trading interval finally reflecting all private information. Therefore, such market dynamics may well extend the volatility to the next period. Second, spillover can result from stochastic policy coordination among industrial countries. Suppose the volatility in one country is caused by a policy change or announcement. Then the government of another country would respond, giving rise to volatility spillovers. Such a volatility spillover represents a challenge to an efficient market hypothesis: the exchange rate (yen/$) in the London

---

*Correspondence to: Bwo-Nung Huang, Department of Economics, National Chung-Cheng University, Chia-Yi, Taiwan 621.
[†]E-mail: ecdbnh@ccunix.ccu.edu.tw

market could be used to speculate in the Tokyo market if the former is found to lead the latter statistically. Prior studies on volatility spillover emphasize (i) the exchange rate markets (Lin, 'The source of intraday volatility in the foreign exchange: a multivariate factor GARCH approach; unpublished manuscript, 1989; Engle *et al.*, 1990; Baillie and Bollerslev, 1991), (ii) the stock market (Hamao *et al.*, 1990; King and Wadhwani, 1990; Hu *et al.*, 1997); and (iii) interest rate volatility (Edwards, 1998; Edwards and Susmel, 2000; Fleming and Lopez, 1999).

In general, most of the literature seems to support volatility spillover in all the three markets. However, the process of volatility transmissions are rather arbitrarily modelled; that is, the direction of spillover is determined *a priori* without resorting to an appropriate statistical testing procedure.[1] Some studies simply assume the direction of spillover from an industrialized economy to emerging markets, even though such a causal direction is intuitively acceptable. Cheung and Ng (1996) propose a two-stage causality-in-variance model in which dynamic specification (including the lagged relations) of the volatility spillover can be statistically tested first. Such an approach of causality in second moment is of importance to both academicians and practitioners because a change in variance reflects the arrival of information and, to some extent, provides a valuable insight concerning the characteristics and temporal dynamics of financial data. Causality-in-variance is of great importance in forecasting volatility, and, to the best of our knowledge, has not been applied to the exchange rate market. This paper intends to analyse such a phenomenon in three major markets: New York, Tokyo and London. The G-5 exchange rates are expressed in Canadian dollars/US \$, Mark/US \$, British Pounds/US \$ and Yen/US \$. Section 3 presents data and basic statistics. Section 4 discusses the estimated results from a causality-in-variance model and examines its forecasting accuracy. Conclusions are given in Section 5.

2. A CAUSALITY TEST IN VARIANCE

2.1. The two-stage procedure for testing the causality

In this section we briefly introduce the model proposed by Cheung and Ng (1996) since it is quite involved. Readers are referred to their paper for details. First, consider two stationary and ergodic time series X_t and Y_t, and let I_t and J_t be two information sets such that $I_t = \{X_{t-j}; j \geqslant 0\}$ and $J_t = \{Y_{t-j}; j \geqslant 0\}$. The following causality relations are defined before starting the model:

(1) Y_t is said to cause X_{t+1} in variance if

$$E\{(X_{t+1} - \mu_{x,t+1})^2 | I_t\} \, /= E(X_{t+1} - \mu_{x,t+1})^2 | I_t, J_t\} \tag{1}$$

where $\mu_{x,j+1}$ is the mean of X_{t+1} conditional on I_t.

(2)   Feedback in variance occurs if X causes Y and Y in turn causes X. That is,

$$E\{(X_{t+1} - \mu_{x,t+1})^2 | I_t\} \, /= E\{(X_{t+1} - \mu_{x,t+1})^2 | I_t, J_t\} \tag{2}$$

and

$$E\{(X_{t+1} - \mu_{y,t+1})^2 | J_t\} \, /= E\{(X_{t+1} - \mu_{x,t+1})^2\} | I_t, J_t\} \tag{3}$$

(3)   The contemporaneous causality-in-variance exists if

$$E\{(X_{t+1} - \mu_{x,t+1})^2 | I_t, J_t\} \, /= E\{(X_{t+1} - \mu_{x,t+1})^2 | I_t, J_t, Y_{t+1}\} \tag{4}$$

For estimation purpose, we rewrite X_t and Y_t as:

$$X_t = \mu_{x,t} + h_{x,t}^{1/2} \varepsilon_t \tag{5}$$

$$Y_t = \mu_{y,t} + h_{y,t}^{1/2} \eta_t$$

where $\{\varepsilon_t\}$ and $\{\eta_t\}$ are two independent white-noise processes with zero mean and unit variance with their conditional means and variances given by

$$\mu_{z,t} = \sum_{i=1}^{\infty} \varphi_{z,i}(\theta_{z,\mu})Z_{t-i} \tag{6}$$

$$h_{z,t} = \varphi_{z,o} + \sum_{i=1}^{\infty} \varphi_{z,i}(\theta_{z,h})\left[(Z_{t-i} - \mu_{z,t-i})^2 - \varphi_{z,o}\right] \tag{7}$$

where $\theta_{z,w}$ is a column parameter vector of dimension $P_{z,w}$; W_t assumes the values of Z and h; $\varphi_{z,i}(\theta_{z,h})$ are uniquely defined functions of $\theta_{z,h}$ and $Z = X, Y$. Note that the specifications of $\mu_{z,t}$ and $h_{z,t}$ encompass the time-series models such as the commonly used ARMA models and (generalized) autoregressive conditional heteroscedastic ((G)ARCH) processes.

Let $\gamma_{uv}(k)$ be the sample cross-correlation of lag k:

$$\gamma_{uv}(k) = \frac{c_{uv}(k)}{\sqrt{c_{uu}(0)}\sqrt{c_{vv}(0)}} \tag{8}$$

and U_t and V_t are the squares of standardized innovations or

$$U_t = \left[(X_t - \mu_{x,t})^2/h_{x,t}\right] \tag{9}$$

$$V_t = \left[(Y_t - \mu_{y,t})^2/h_{y,t}\right]$$

and $c_{uv}(k)$ is the sample cross-covariance with the kth lag given by

$$c_{uv}(k) = T^{-1}\sum(U_t - \bar{U})(V_t - \bar{V}_1) \quad \text{for} \quad k = 0, \pm 1, \pm 2, \dots \tag{10}$$

and $c_{uu}(0)$ and $c_{vv}(0)$ are the sample variance of U and V respectively.

The asymptotical behaviour of $\sqrt{T}(\hat{\gamma}_{uv}(k_1), \dots \hat{\gamma}_{uv}(k_m))$ enables us to construct a normal or Chi-square statistic to test the null hypothesis of no causality (theorem 1 from Cheung and Ng, 1996) as will be described in the next section.

2.2. Test procedures for causality-in-variance

Following Cheung and Ng (1996), we compute $\sqrt{T}\hat{\gamma}_{uv}(k)$ to test the causal relationship at a specific lag k via a standard normal distribution. To test the null hypothesis of no causality from lag j to k, Cheung and Ng (1996) suggest a Chi-square test statistic S with the degree of freedom of $k - j + 1$:

$$S = T\sum_{i=j}^{k} \hat{\gamma}_{uv}(i)^2 \tag{11}$$

The choice of j and k depends on the specification of alternative hypotheses. In the case where no *a priori* information regarding the direction of causality is available, one may set $-j = k = m$ for a large enough m value to allow for a sufficient number of non-zero lags (Cheung and Ng, 1996, p. 37). To detect a cross-correlation pattern, Koch and Yang (1986) suggest the following improved statistic:

$$S^* = T\sum_{k=-m}^{m-j}\left[\sum_{i=0}^{j}\hat{\gamma}_{uv}(k+i)\right]^2, i = 0, 1, \dots, m-1 \tag{12}$$

The causality-in-variance tests via the cross-correlation of standardized residuals indeed possess some desirable properties over the GARCH model in that it takes into consideration both changing variances and means, and as such is found to have satisfactory statistical power.

3. SOURCE OF DATA AND SOME BASIC STATISTICS

Five exchange rates of G-5 countries in terms of per US dollar are employed for the three major markets: New York, London and Tokyo. The sample spans from 30 June 1995 to 6 September 1999, a total of 1092 data points for the Tokyo market; it starts from 2 January 1987 to 6 September 1999, a total of 3307 data points for both New York and London. All the data are obtained from Datastream International. In the US market, the quote price is taken at market close in Eastern Standard Time at 18:00 (Global Treasury Information Service of New York); the price in the UK is quoted by WMR/Reuters at 16:00 London time; and the price in Japan is quoted at market close by the Bank of Tokyo.[2] Note that the samples are lengthy enough to encompass three major events that led to substantial exchange rate volatility: the Eurodollar Monetary System (EMS) crisis (September 1992) the Mexican Peso crisis (December 1994) and the Asian financial crisis (starting July 1997). These events are normally conducive to enhanced and more integrated volatility in unique ways. To determine impacts on exchange rate volatility, we divide the samples into four segments. Period I runs from 2 January 1987 to 31 August 1992, the pre-EMS crisis period; Period II from 1 September 1992 to 30 November 1994, the EMS period; Period III from 1 December 1994 to 30 June 1997, the Peso crisis period; Period IV from 1 July 1997 to 6 September 1999, the Asian financial debacle period. Figure 1 shows time-series plots for each of the three markets.[3]

We label the times when the three financial crises occurred. Before the EMS crisis, the British pound, German mark and French franc experienced a fair amount of volatility. As shown in Figure 1, their values depreciated rapidly during the EMS crisis. Moreover, the Japanese yen, German mark and the French franc were noticeably impacted during the Peso crisis. On the other hand, the Canadian dollar and the Japanese yen witnessed appreciable changes from the Asian crisis. While Figure 1 reveals explicitly trends of currency appreciation and depreciation, the magnitudes of currency volatility can be better detected through the absolute value of rate of return for each currency.

Following convention, we first define the rate of return or percentage change in foreign exchange by taking the first difference of five logarithmic prices (per US dollar) in three different markets:

$$r_{i,j,t} = (\log P_{i,j,t} - \log P_{i,j,t}) \times 100\% \tag{13}$$

The subscript i denotes the currencies of Canada, France, Germany, Japan and Britain respectively; the subscript j denotes the markets of Tokyo, London and New York, respectively; and the subscript t denotes the price of day t. Absolute values of $r_{i,j,t}$ are shown in Figure 2.

An examination of Figure 2 indicates the British pound underwent upward volatility during the EMS crisis as did the franc and the mark. However, both the franc and the mark had experienced relatively large volatility before the EMS crisis. In addition, G-5 currency prices had their shares of volatility during the Peso crisis. Finally, the Canadian dollar and the Japanese yen saw greater volatility in their values during the Asian financial turmoil. Table 1 reports the magnitudes of volatility G-5 currencies during the three major events.

The volatility of the G-5 exchange rates except the Japanese yen and the British pound in the London market is slight greater than that in New York during the EMS crisis (Table 1). Moreover, the average volatility of the G-5 exchange rates in New York is the greatest of the three markets in the Peso crisis. It is to be noted that the volatility in the Tokyo market does not differ noticeably from the other two markets during the Asian financial debacle. Clearly, the spillover phenomena of exchange rate volatility are different in the three markets during each crisis. The primary contribution of this study lies in that we do not assume either the direction of volatility transmission, i.e. from a major currency (e.g. the Japanese yen) to a secondary currency (e.g. the Canadian dollar), or an arbitrary route (e.g. from New York to Tokyo). By employing the causality-in-variance test developed by Chung and Ng (1996), we examine the spillover relations of exchange rate volatility.

Int. J. Fin. Econ. 7: 37–50 (2002)

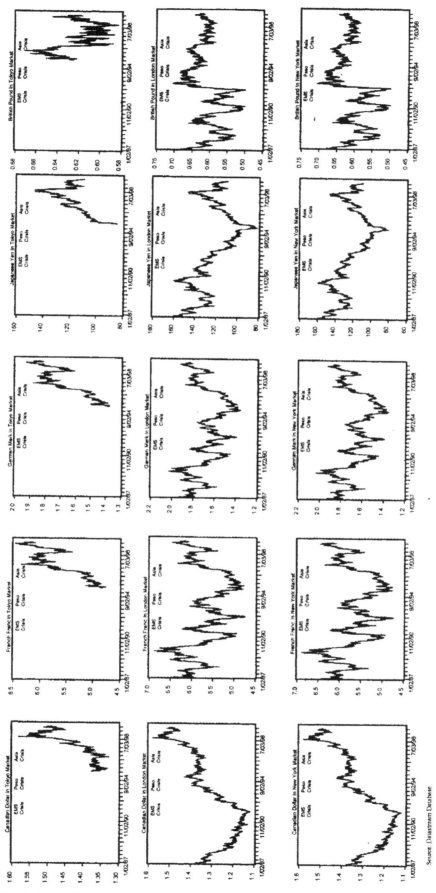

Figure 1. Time series of G-5 currency versus US dollar in three different markets.

Int. J. Fin. Econ. **7**: 37–50 (2002)

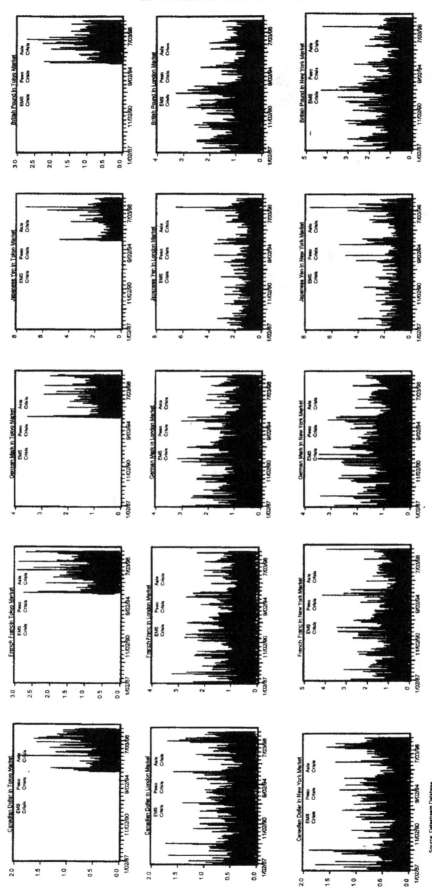

Figure 2. Time series of absolute return of G-5 currency in three different markets.

Int. J. Fin. Econ. **7**: 37–50 (2002)

Table 1. Absolute values of G-5 exchange rate changes and basic statistics

		EMS crisis					Peso crisis					Asia crisis				
		CAN	FRA	GMY	USA	UK	CAN	FRA	GMY	USA	UK	CAN	FRA	GMY	USA	UK
Tokyo	Mean						0.1983	0.3630	0.3987	0.4452	0.2995	0.2718	0.4371	0.4392	0.6397	0.3519
	S.D.						0.2166	0.3401	0.3793	0.4766	0.3254	0.2758	0.4393	0.4400	0.6572	0.3314
	N						521	521	521	521	521	570	570	570	570	570
London	Mean	0.2605	0.4929	0.5198	0.4641	0.4956	0.1929	0.3739	0.4125	0.4701	0.3099	0.2630	0.4384	0.4405	0.6660	0.3522
	S.D.	0.2213	0.4372	0.4693	0.4502	0.4965	0.1877	0.3788	0.4154	0.4945	0.3073	0.2515	0.3668	0.3681	0.6540	0.3096
	N	587	587	587	587	587	673	673	673	673	673	570	570	570	570	570
New York	Mean	0.2279	0.4882	0.5138	0.4721	0.5309	0.2010	0.3917	0.4230	0.4893	0.3302	0.2663	0.4352	0.4311	0.6665	0.3703
	S.D.	0.2078	0.4599	0.4827	0.4878	0.5526	0.1880	0.4448	0.4477	0.5492	0.3527	0.2537	0.4169	0.3855	0.6805	0.3866
	N	587	587	587	587	587	673	673	673	673	673	570	570	570	570	570

Note: S.D. = Standard deviation. N = number of observations. CAN = Canada, FRA = France, GMY = Germany, UK = United Kingdom.

Int. J. Fin. Econ. 7: 37–50 (2002)

4. CAUSALITY-IN-VARIANCE TEST

In order to test the causality-in-variance, an appropriate GARCH model is needed before applying the bivariate causality model via standardized residuals (Cheung and Ng, 1996).[4] The leptokurtic distribution of exchange rate returns combined with the clustering phenomenon make the ARCH (Engle, 1982) or GARCH (Bollerslev, 1986) model a reasonable candidate for describing the arrival process. Some recent research specifically focuses on GARCH(1,1)–MA(1) as it allows for the autocorrelation induced by discontinuous trading in exchange rates that make up an index. Note that the MA(1) phenomenon may be considered as the non-trading effect for the markets.[5] Well-known in the literature, an appropriate GARCH model requires that standardized residuals $\varepsilon_t/h^{1/2}$ or their squares ε^2/h_t be free of serial correlation problems. For this reason, we employ the generalized GARCH(r, m)-ARMA(p, q) model:

$$r_{i,j,t} = b_0 + \sum_{k=1}^{p} b_k r_{i,j,t-k} + \sum_{l=1}^{q} c_l \varepsilon_{i,j,t-1} + \varepsilon_{i,j,t} \quad \varepsilon_{i,j,t} | I_{i,j,t-e} \sim N(0, h_{i,j,t})$$

$$h_{i,j,t} = \alpha_0 + \sum_{k=1}^{m} \alpha_k \varepsilon_{i,j,t}^2 + \sum_{l=1}^{\gamma} \beta_l h_{i,j,t-1} \tag{14}$$

where $r_{i,j,t}$ is the percentage change of the ith exchange rate in the jth market and is conditional on the past information $I_{i,j,t}$. The optimal GARCH(r, m)-ARMA(p, q) required that $\varepsilon_{i,j,t}/h_{i,j,t}^2$ and $\varepsilon_{i,j,t}^2/h_{i,j,t}$ be free of serial correlation before using them in the causality-in-variance test.[6] As in the Granger causality test, an optimal lag is required as part of the testing procedure. In order to capture the dynamics of daily exchange rate change, we take 5 days (a week) as the optimal lag.[7] The resulting lead–lag relations are reported in Table 2.

Within each market, it appears that lead–lag relations are less frequent than that across the markets. For instance, within the London market, both the franc and the mark led the Canadian dollar in Period I; the mark and the British pound led the Canadian dollar in Period II; the yen and the Canadian dollar led the British pound in Period III; and the British pound led the Japanese yen in Period IV. In the New York markets, the frequencies of such lead–lag relations are 3, 3, 6, 3 in the four periods respectively while they are 0, 0, 2, 4 in the Tokyo markets (Table 2). In the case of the spillover effect in New York and London, we found significantly more lead–lag relations due to potential effect of currency combinations. We summarize them in Table 3.

An examination of Table 3 indicates that the New York market leads the London market (New York → London) with the greatest frequencies in each of the four periods, followed by feedback relations (New York ↔ London). After the EMS crisis (Period II), the frequency of London → New York seems to have increased. From the viewpoint of the London and Tokyo markets, 80% of the spillover phenomena are from London to Tokyo (Period III), and it becomes 96% in Period IV. In the case of the New York and Tokyo combination, about 67% of the cross-market effects are from New York to Tokyo (Period III). It increases to 75% in Period IV. In terms of cross-market spillover effect, it seems that (i) both the New York and the London markets lead the Tokyo market and (ii) the New York market also leads the London market to a lesser extent. Based on the time when a market opens, the London exchange rate market opens first until noon, when the New York market starts the day. Soon after the market closes in New York, the Tokyo market opens (see Figure 3). It is not surprising that both markets in New York and London lead the Tokyo market. While it is a little harder to comprehend the leadership role of the New York market versus that of London, one must take the existence of overlapping trading hours into consideration. This may explain the abundance of the feedback relations (approximately 1/3 of the frequencies) between the two markets, especially for a given currency. For instance, there exist a feedback relation of the yen rate between the New York and the London markets. At fast glance, Figure 3 appears to indicate that the London market leads the US market. However, this is not supported by the empirical result. Just like the stock market, the US exchange rate market dominates other major markets; an asymmetrical relation. The same phenomenon is also found between the New York and the Tokyo markets, but with less frequency.

Table 2. Results of the causality-in-variance model

Market	Period I	Period II	Period III	Period IV
Tokyo			franc → pound mark → pound	dollar → mark dollar → pound franc → yen mark → yen
London	franc → dollar mark → dollar	mark → dollar pound → dollar	yen → pound dollar → pound	pound → yen
New York	mark → franc franc → yen mark → yen	franc → dollar mark → dollar pound → dollar	franc → dollar yen → dollar pound → dollar mark → franc yen → franc yen → mark	dollar → yen dollar → pound franc → mark
London versus New York	dollar/usa → dollar/uk franc/usa ↔ franc/uk mark/usa → franc/uk yen/usa → franc/uk pound/usa → franc/uk dollar/usa ← mark/uk franc/usa ↔ mark/uk mark/usa → mark/uk yen/usa → mark/uk pound/usa → mark/uk franc/usa → yen/uk mark/usa → yen/uk yen/usa ↔ yen/uk pound/usa → yen/uk franc/usa → pound/uk mark/usa → pound/uk yen/usa → pound/uk pound/usa ↔ pound/uk	dollar/usa → dollar/uk franc/usa → dollar/uk mark/usa → dollar/uk yen/usa → dollar/uk pound/usa → dollar/uk dollar/usa ← franc/uk franc/usa ↔ franc/uk mark/usa ↔ franc/uk yen/usa → franc/uk pound/usa ↔ franc/uk dollar/usa ← mark/uk franc/usa ↔ mark/uk mark/usa ↔ mark/uk yen/usa → mark/uk pound/usa ↔ mark/uk franc/usa → yen/uk yen/usa ↔ yen/uk dollar/usa ← pound/uk franc/usa → pound/uk mark/usa → pound/uk yen/usa → pound/uk pound/usa ↔ pound/uk	dollar/usa ↔ dollar/uk franc/usa → dollar/uk yen/usa → dollar/uk dollar/usa ← franc/uk franc/usa ↔ franc/uk mark/usa → franc/uk yen/usa ↔ franc/uk pound/usa → franc/uk franc/usa ↔ mark/uk mark/usa ↔ mark/uk yen/usa → mark/uk pound/usa → mark/uk dollar/usa ← yen/uk franc/usa → yen/uk mark/usa → yen/uk yen/usa ↔ yen/uk franc/usa → pound/uk mark/usa → pound/uk yen/usa → pound/uk pound/usa ↔ pound/uk	dollar/usa ↔ dollar/uk franc/usa ← dollar/uk pound/usa ← dollar/uk franc/usa ↔ franc/uk mark/usa ↔ franc/uk yen/usa → franc/uk pound/usa → franc/uk franc/usa ↔ mark/uk mark/usa ↔ mark/uk yen/usa → mark/uk pound/usa → mark/uk franc/usa → yen/uk mark/usa → yen/uk yen/usa → yen/uk dollar/usa → pound/uk franc/usa → pound/uk mark/usa → pound/uk pound/usa ↔ pound/uk

Table 2 (continued)

Tokyo versus New York

Period III

- dollar/usa → dollar/jpn
- pound/usa → dollar/jpn
- dollar/usa ↔ franc/jpn
- franc/usa ↔ franc/jpn
- mark/usa → franc/jpn
- yen/usa → franc/jpn
- pound/usa ↔ franc/jpn
- dollar/usa → yen/jpn
- franc/usa → yen/jpn
- mark/usa → yen/jpn
- yen/usa → yen/jpn
- pound/usa ↔ yen/jpn
- dollar/usa → mark/jpn
- franc/usa ↔ mark/jpn
- mark/usa → mark/jpn
- yen/usa → mark/jpn
- pound/usa ↔ mark/jpn
- franc/usa → pound/jpn
- mark/usa → pound/jpn
- yen/usa → pound/jpn
- pound/usa ↔ pound/jpn

Period IV

- dollar/usa → dollar/jpn
- pound/usa ← dollar/jpn
- franc/usa → franc/jpn
- mark/usa → franc/jpn
- yen/usa ↔ franc/jpn
- pound/usa → franc/jpn
- franc/usa → pound/jpn
- mark/usa → pound/jpn
- yen/usa → pound/jpn
- pound/usa → pound/jpn
- dollar/usa ← yen/jpn
- franc/usa → yen/jpn
- mark/usa → yen/jpn
- yen/usa → yen/jpn
- pound/usa → yen/jpn
- franc/usa → mark/jpn
- mark/usa → mark/jpn
- yen/usa ↔ mark/jpn
- pound/usa → mark/jpn
- dollar/usa ← pound/jpn

Tokyo versus London

Period III

- dollar/uk → dollar/jpn
- mark/uk ← dollar/jpn
- franc/uk → franc/jpn
- mark/uk → franc/jpn
- yen/uk → franc/jpn
- pound/uk ↔ franc/jpn
- dollar/uk ← mark/jpn
- franc/uk → mark/jpn
- mark/uk → mark/jpn
- yen/uk → mark/jpn
- pound/uk ↔ mark/jpn
- franc/uk → yen/jpn
- mark/uk → yen/jpn
- yen/uk → yen/jpn
- dollar/uk ← pound/jpn
- franc/uk → pound/jpn
- mark/uk → pound/jpn
- yen/uk → pound/jpn
- pound/uk ↔ pound/jpn
- dollar/uk → yen/jpn

Period IV

- dollar/uk → dollar/jpn
- franc/uk → dollar/jpn
- mark/uk → dollar/jpn
- pound/uk ← dollar/jpn
- dollar/uk → franc/jpn
- franc/uk → franc/jpn
- mark/uk → franc/jpn
- yen/uk → franc/jpn
- pound/uk → franc/jpn
- franc/uk → yen/jpn
- mark/uk → yen/jpn
- yen/uk → yen/jpn
- pound/uk → yen/jpn
- dollar/uk → mark/jpn
- franc/uk → mark/jpn
- mark/uk → mark/jpn
- yen/uk → mark/jpn
- pound/uk → mark/jpn
- dollar/uk → pound/jpn
- franc/uk → pound/jpn
- mark/uk → pound/jpn
- yen/uk → pound/jpn
- pound/uk → pound/jpn

Notes: → = lead; and ↔ = feedback.
Period I runs from 2 January 1987 to 31 August 1992.
(pre-EMS period)
Period II runs from 1 September 1992 to 30 November 1994.
(EMS period)
Period III runs from 1 December 1994 to 30 June 1997.
(peso crisis period)
Period IV runs from 1 July 1997 to 6 September 1999. (Asian Hu period)

Table 3. Summary of frequencies of lead–lag relations based on causality-in-variance model

Relations	Period I	Period II	Period III	Period IV
LD → NY	1/18 (5.56%)	4/22 (18.18%)	2/20 (10%)	2/18 (11.11%)
NY → LD	13/18 (72.22%)	10/22 (45.45%)	11/20 (55%)	10/18 (55.56%)
NY ↔ LD	4/18 (22.22%)	8/22 (36.36%)	7/20 (35%)	6/18 (33.33%)
LD → TK			16/20 (80%)	22/23 (95.65%)
TK → LD			2/20 (10%)	1/23 (4.35%)
LD ↔ TK			2/20 (10%)	0/23 (0%)
NY → TK			14/21 (66.67%)	15/20 (75%)
TK → NY			0 (0%)	3/20 (15%)
NY ↔ TK			7/21 (33.33%)	2/20 (10%)

Note: These frequencies are obtained from Table 2. NY = New York, LD = London, TK = Tokyo; → = lead, ↔ = feedback.

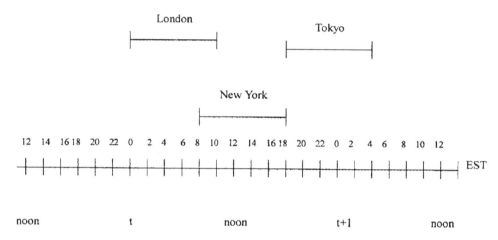

Figure 3. Trading hours of the three exchange rate markets.

Again, the transmission of volatility is not symmetrical. In other words, return volatility is generally from the US to other stock markets. In the exchange rate market, however, London as well as New York is an important centre. In terms of currency volatility, it can be seen that the feedback phenomena between New York and London increased substantially during the EMS crisis as compared to that before the crisis. The pattern remained roughly the same during both the peso and Asian crises as that after the EMS crisis. In addition, we find greater frequencies of the spillover effect from either New York or London to Tokyo. For a given exchange rate, the lead–lag relation appears to depend on the different times when a market opens. Table 4 summarizes these relationships.

An inspection of Table 4 shows that the feedback relations prevail between the two overlapping markets: New York and London. Except for the Canadian dollar in Periods I and II, in which the New York market led the London market, the feedback relations dominated for the other four currencies. In Period III, the exchange rate of the mark in New York led that of London while feedback relations prevail between the two markets for other currencies. Similarly, in Period IV, the yen rate in New York is found to lead that of London, and feedback relations abound for other currencies. Generally, the feedback phenomenon can be attributed to the existence of the overlapping trading hour. After close study, the New York market could be said to have led the London market.[8] The results of Table 4 also shows that the New York or London markets lead the Tokyo market. In addition, lead–lag relations do not seem to have undergone significant changes before and after these crises.

The lead–lag relations of different exchange rates within a market or that of a same currency across different markets via the causality-in-variance test can shed important light on forecasting. Generally, prior

Int. J. Fin. Econ. **7**: 37–50 (2002)

Table 4. Lead–lag relations of an exchange rate in the three markets based on causality-in-variance model

Market	Period I	Period II	Period III	Period IV
LD versus NY	dollar/usa → dollar/uk franc/usa ↔ franc/uk mark/usa → mark/uk yen/usa ↔ yen/uk pound/usa ↔ pound/uk	dollar/usa → dollar/uk franc/usa ↔ franc/uk mark/usa ↔ mark/uk yen/usa ↔ yen/uk pound/usa ↔ pound/uk	dollar/usa ↔ dollar/uk franc/usa ↔ franc/uk mark/usa → mark/uk yen/usa ↔ yen/uk pound/usa ↔ pound/uk	dollar/usa ↔ dollar/uk franc/usa ↔ franc/uk mark/usa ↔ mark/uk yen/usa → yen/uk pound/usa ↔ pound/uk
TK versus NY			dollar/usa → dollar/jpn franc/usa → franc/jpn mark/usa → mark/jpn yen/usa → yen/jpn pound/usa → pound/jpn	dollar/usa → dollar/jpn franc/usa → franc/jpn mark/usa → mark/jpn yen/usa → yen/jpn pound/usa → pound/jpn
TK versus LD			dollar/uk → dollar/jpn franc/uk → franck/jpn mark/uk → mark/jpn yen/uk → yen/jpn pound/uk → pound/jpn	dollar/uk → dollar/jpn franc/uk → franc/jpn mark/uk → mark/jpn yen/uk → yen/jpn pound/uk → pound/jpn

Note: → = lead; and ↔ = feedback.

Table 5. A comparison of forecasting accuracy

	Within the Tokyo market model					
	In-sample				Out-sample	
Model	RMSE	Theil U	Correlation	RMSE	Theil U	Correlation
UARMA	0.0591	0.1374	0.1613	0.1562	0.3073	0.4773
OUR	0.0586	0.1361	0.2105	0.1617	0.3192	0.4056
	Across the markets model					
	In-sample				Out-sample	
Model	RMSE	Theil U	Correlation	RMSE	Theil U	Correlation
UARMA	0.0252	0.1147	0.9127	0.0254	0.1132	0.0550
OUR	0.0201	0.0905	0.9460	0.0149	0.0650	0.7409

Note: In-sample performance results are shown in the left half; out-sample in the right half.
SUARMA = univariate ARMA model
OUR = causality-in-variance model.
RMSE = root mean square error.
Correlation = correlation coefficient of actual and forecast values.

studies choose explanatory variables arbitrarily. According to the causality-in-variance model, once a causality is established, the past values of a variable can be used to predict the current or future values of another variable. Ashley *et al.* (1980) show that a genuine causality of a model can be ascertained if the in- and out-sample performances are better than those of other models. To this end, we resort to the results of Table 2 and 4 for in- and out-sample testings. Given that the franc, the yen and the British pound were found to lead the Canadian dollar in Period III in New York, we may use these currencies to forecast the Canadian dollar. Similarly, the franc and the mark can be explanatory variables for predicting the British pound in Tokyo. In the case of inter-market forecasting, the exchange rates of the Canadian dollar in New York and London are explanatory variables to forecast the exchange rate in Tokyo. These relations are expressed in the following equations:

$$v_{\text{pound/jpn},t} = a_0 + a_1 v_{\text{franc/jpn},t-1} + a_2 v_{\text{mark/jpn},t-1} + \varepsilon_{1t} \tag{15}$$

$$v_{\text{dollar/jpn},t} = b_0 + b_1 v_{\text{dollar/usa},t-1} + b_2 v_{\text{dollar/uk},t-1} + \varepsilon_{2t} \tag{16}$$

Equation (15) is set to forecast the volatility of the British pound via that of the mark and franc rates in Tokyo. Equation (16) employs the volatility of the Canadian dollars in New York and London to predict that in Tokyo. Note the volatility variable v's are obtained from the optimal $GARCH(r, m)$-$ARMA(p, q)$ model: its conditional variance $h_{i,j,t}$ with the in-sample period from 1 December 1994 to 30 May 1997 and the out-sample period from 1 June 1997 to 30 June 1999. The results in terms of $v_{pound/jpn,t}$ and $v_{dollar/jpn,t}$ are compared with that of a univariate ARMA (UARMA) model(benchmark model).[9] The criteria based on the correlation coefficient, root mean square error and the Theil U statistic are reported in Table 5.

Study of Table 5 reveals that the variance-in-causality model (Tokyo market) outperforms the UARMA model for in-sample period in terms of the three criteria. It is, however, slightly less accurate than the UARMA model for the out-sample period. As for an inter-market forecast, our model fares much better than the UARMA model especially in the correlation coefficient between the actual and predict values within the out-sample period (0.74). According to Ashley *et al.*'s criterion (1980), there exists indeed some causality-in-variance relations across the three exchange rate markets.

5. CONCLUSIONS

Transmission of exchange rate volatility is of great concern to academicians and practitioners alike. Transmissions of different exchange rates in a given market and that of a given exchange rate across different markets are analysed on G-5 currencies from 2 January 1987 to 6 September 1999. Applying the causality-in-variance test of Cheung and Ng (1996), we find a significant amount of spillover effect among different currencies in a given market and that across different markets. In particular, the lead–lag relations in three different markets depend to a large extent on the time when a market opens, that is, the exchange rates in New York and London lead that in Tokyo while feedback relations prevail between New York and London. According to Ashley *et al.* (1980), a genuine causality exists if a model dominates the benchmark mode (in our case, the UARMA model) in terms of in- and out-sample performance. To this end, we first compare the forecasting results (pound/dollar) using franc and mark values in Tokyo. Second, by making use of the volatility of the Canadian dollar rates (per US dollar) in both New York and London, we are able to better predict the volatility of the Canadian dollar in the Tokyo market. In the case of the intra-market volatility spillover model, the causality-in-variance model outperforms the UARMA benchmark model in terms of in-sample accuracy, but is slightly less accurate in terms of the out-sample. However, in the case of the inter-market volatility spillover model, our model dominates the UARMA model in all three criteria for either in- or out-sample comparison. Summarizing, the causality-in-variance model seems to have predicted more accurately in the exchange rate markets where rate volatility abounds.

NOTES

1. For instance, Edwards (1998) assumes that interest rate volatility in Mexico transmits to Argentina, not vice versa.
2. By priced quoted at market close, we calculate the average of bid rates and offer rates.
3. All the G-5 exchange rates are calculated based on the number of local currencies per US dollars in each market.
4. By an appropriate model, we mean standardized residuals or their squares are free from serial correlation problem.
5. Scholes and Williams (1977) first pointed out the moving average phenomenon.
6. Note that the MA(1) process could be attributed to discontinuity in stock market trading.
7. See Hamilton (1994) and Hu *et al.* (1997).
8. In each of the four periods, there exist some, though not many, leadership roles of the New York market.
9. We employ the optimal conditional variance $\hat{h}_{i,j,t}$ from the $GARCH(r,m)$-$ARMA(p,q)$ to replace $v_{i,j,t}$, and $v_{i,j,t-1}$ to perform in-sample and out-sample forecast.

REFERENCES

Ashley R, Granger CWJ, Schmalensee R. 1980. Advertising and aggregate consumption: an analysis of causality. *Econometrica* **48**(5): 1149–1167.

Admati AR, Pfleiderer P. 1988. Selling and trading on information in financial markets. *American Economic Review* **78**(2): 96–103.

Baillie RT, Bollerslev T. 1991. Intra-day and inter-market volatility in foreign exchange rates. *Review of Economic Studies* **58**(3): 565–585.

Bollerslev T. 1986. Generalized autoregressive conditional heteroscedasticity. *Journal of Econometrics* **31**: 307–327.

Byers JD, Peel DA. 1995. Evidence on volatility spillovers in the interwar floating exchange rate period based on high/low prices. *Applied Economics Letters* **2**: 394–396.

Cheung YW, Ng LK. 1996. A causality-in-variance test and its application to financial market prices. *Journal of Econometrics* **72**: 33–48.

Edwards S. 1998. Interest rate volatility, contagion and convergence: an empirical investigation of the cases of Argentina, Chile and Mexico. *Journal of Applied Economics* **1**(1): 58–86.

Edwards S, Susmel R. 2000. Interest rate volatility and contagion in emerging markets: evidence from the 1990s. NBER Working Paper No. 7813.

Engle RF. 1982. Autoregressive conditional heteroscedasicity with estimates of the variance of U.K. inflation. *Econometrica* **50**: 987–1008.

Engle RF, Ito T, Lin W. 1990. Meteor showers or heat waves? Heteroskedastic intra-daily volatility in the foreign exchange market. *Econometrica* **58**: 525–542.

Fleming MJ, Lopez Jose A. 1999. Heat waves, meteor showers and trading volume: An analysis of volatility spillovers in the U.S. Treasury Market. Federal Reserve Bank of New York, Staff Reports No. 82.

Hamao Y, Masulis RW, Ng V. 1990. Correlations in price changes and volatility across international stock markets. *Review of Financial Studies* **3**: 281–307.

Hamilton J. 1994. *Time Series Analysis*. Princeton University Press: Princeton, NJ.

Hu JW-S, Chen M-Y, Fok RCW, Huang B-N. 1997. Causality in volatility and volatility spillover effects between US, Japan and four equity markets in the South China growth Triangle. *International Financial Markets, Institution and Money* **7**: 351–367.

King M, Wadhwani S. 1990. Transmission of volatility between stock markets. *Review of Financial Studies* **3**: 5–33.

Koch PD, Yang SS. 1986. A method for testing the independence of two time series that accounts for a potential pattern in the cross-correlation function. *Journal of the American Statistics Association* **81**: 533–544.

Kyle AS. 1985. Continuous auctions and insider trading. *Econometrica* **53**(6): 1315–1335.

Scholes M, Williams J. 1977. Estimating betas from nonsynchronous data. *Journal of Financial Economics* **5**: 309–327.

Applied Economics Letters, 2001, **8**, 725–729

Stock market integration – an application of the stochastic permanent breaks model[1]

BWO-NUNG HUANG* and ROBERT C. W. FOK‡

*Department of Economics, National Chung Cheng University, Chai-Yi, Taiwan 621
and ‡Department of Finance, Shippensburg University, Shippensburg, PA 17257, USA
E-mail: ecdbnh@ccunix.ccu.edu.tw and CWFok@wharf.ship.edu*

Using the Stochastic Permanent Breaks (STOPBREAK) model, this study examines the relationships of the US stock market with the Japanese and eight European stock markets. The evidence indicates that the US stock market is temporally cointegrated with the markets in Japan, Germany, Netherlands and Switzerland. However, cointegration relationship exists only between the US and Netherlands market when the Johansen cointegration test is used. In other words, some sort of cointegrating relationships may exist between two markets even if the standard cointegration test indicates that the two markets are not cointegrated. According to the STOPBREAK model, when two markets are temporally cointegrated, the movement of the two markets does not follow a random walk and market inefficiency is implied.

I. INTRODUCTION

Long-term relationships among different stock markets (industries) may exist when stock prices in various markets are affected by common factors. Granger (1986) claims that the hypothesis of market efficiency does not hold when stock prices of different markets are cointegrated. According to the Granger representation Theorem (Engle and Granger, 1987) when two data series are cointegrated, at least one Granger causality relationship exists. In this case, we can predict the value of one variable by using the past/or current information of other variables. The validity of efficiency market is challenged when stock prices in one market can be explained by additional information such as price movement in other stock markets. As Bekaert and Harvey (1997) state, when the impact of foreign factors on domestic market accelerates, the domestic and foreign markets become more and more integrated.

Most recent studies on market efficiency using cointegration tests focus on foreign exchange markets (Hakkio and Rush, 1989; MacDonald and Taylor, 1989; Copeland, 1991). Relatively few studies investigate whether a cointe-

gration relationship exists among stock prices. MacDonald and Taylor (1991) find no cointegration exists among a sample of 40 stocks in the UK classified by industry. On the other hand, Chelly-Steeley and Pentecost (1994) find that stock prices of small firms are more likely to be cointegrated than those of large firms. Although stocks in the same industry are affected by identical fundamental factors, the price of different stocks may deviate from each other due to external shocks. It is possible that stock prices move together most of the time but jump apart from each other occasionally. In this case, cointegration relations are not likely to be detected by conventional testing techniques. Considering the possibility of structural breaks, Engle and Smith (1998) developed the Stochastic Permanent Break (STOPBREAK) model to investigate if temporary cointegration exists among stock prices. The model provides a new approach to the modelling process where the effect on stock fluctuates between permanent and transient. Engle and Smith find that the prices of stocks in the same industry are temporally cointegrated.

Due to the globalization of most stock markets, and advances in technology, both academics and practitioners

* Corresponding author.

Applied Economics Letters ISSN 1350–4851 print/ISSN 1466–4291 online © 2001 Taylor & Francis Ltd
http://www.tandf.co.uk/journals
DOI: 10.1080/13504850011003633 7

725

are interested in whether international stock markets are cointegrated. However, such studies ignore the possibility of temporary cointegration. As suggested by Engle and Smith (1998), a cointegrating relationship may exist over short periods of time when imperfect information exists.

The purpose of this study is to fill the gap by examining the dynamics of the relationship between the US stock market and other major stock markets using Engle and Smith's STOPBREAK model. In addition, the results are compared with those obtained by Johansen's (1988) cointegration test. Our results indicate that the market index in the US is temporally cointegrated with that in Japan, Germany, Netherlands and Switzerland. This implies that the STOPBREAK model is more appropriate than the conventional test in analysing relationships among different stock markets when structural breaks exist in the data series.

The paper is organized as follows: the following section describes the methodology – the STOPBREAK model; Section III describes the data and reports and discusses the empirical results. Section IV concludes the paper.

II. STOCHASTIC PERMANENT BREAKS MODEL (STOPBREAK)

The Stochastic Permanent Breaks Model (STOPBREAK) considers structural breaks in the data series. Different from conventional tests for structural changes (e.g. Chow's test, 1960), it does not require dividing the series into different regimes. Instead, the model allows the process to predict whether part or all of a shock will be permanent or transitory. For some time series y_t, the STOPBREAK process can be written as

$$y_t = m_t + \varepsilon_t \qquad (1)$$

$$m_t = m_{t-1} + q_{t-1}\varepsilon_{t-1}$$

where m_t and ε_t denote a conditional forecast and a martingale difference sequence respectively. The series becomes a random walk process when $q_{t-1} = 1$ and a constant mean process when $q_{t-1} = 0$. Allowing for intermediate values of q_t, the process allows a permanent shock range between zero and one, and large shocks indicate permanent break. The STOPBREAK process is generated from the following general breaking process:

$$A(L)B(L)(y_t - x_t\delta) = z_{t-1}A(L)\varepsilon_t + (1 - Z_{t-1})B(L)\varepsilon_t$$

$$t = 1, 2, \ldots, T$$

$$A(L) = 1 - a_1L - a_2L^2 - \cdots - a_pL^p \qquad (2)$$

$$B(L) = 1 - \beta_1L - \beta L^2 - \cdots - \beta_sL^2$$

where x_t denotes a vector of explanatory variables, ε_t an innovation term, Z_{t-1} an information function up to $t - 1$,

and L the lag operator. The basis STOPBREAK process is obtained by setting $\delta = 0$, $B(L) = 1 - L$, $A(L) = 1$, and $Z_{t-1} = q_{t-1}(\gamma_0)$, i.e.

$$\Delta Y_t = \varepsilon_t - (1 - q_{t-1}(\gamma_0))\varepsilon_{t-1} \qquad (3)$$

Engle and Smith (1998) suggest specifying q_t as the following continuous function:

$$q_t(\gamma_0) = \frac{\varepsilon_t^2}{\gamma_0 + \varepsilon_t^2} \qquad \gamma_0 > 0 \qquad (4)$$

Equation 4 allows for partial permanent breaks in the process. It allows large shocks be more likely to have a permanent effect than small ones.

The formulation (Equation 2) can generate a number of possible specifications of the simple STOPBREAK process. For example, when setting $\delta = 0$, $B(L) = 1 - L$, $A(L) = 1 - \alpha_0L$, and $Z_{t-1} = q_{t-1}(\gamma_0)$, a moving average representation is implied:

$$\Delta y_t = \alpha_0\Delta y_{t-1} + \varepsilon_t - \theta_{t-1}\varepsilon_{t-1} \qquad (5)$$

where $\theta_{t-1} = 1 - (1 - \alpha_0)q_{t-1}(\gamma_0)$ and $0 \leq \alpha_0 < 1$. In this case, y_t follows either an AR(1) or random walk process. In this process, some temporal correlation may exist during the 'non-breaking' periods.

Including explanatory variables (i.e. x_t) considers a type of temporary cointegration since it implies that a linear combination is approximately stationary for periods of time. The cointegrating coefficients do not change. This is analogous to the intercept correction widely used in forecasting (Hendry and Clements, 1996). Shifts in the intercept correct mean shifts, while maintaining the fundamental relationships among variables unchanged.

To examine if the data series follows a random walk process, the null $q_t = 1$ is tested using Equation 3 and the null $\gamma_0 = 0$ based on Equation 4. Rejections of the nulls indicate that the data series have a STOPBREAK process. As pointed out by Engle and Smith (1998, p. 14), the test statistics depend on the values of the parameters. As a result, no uniformly most-powerful test of a random walk against a STOPBREAK process exists. Nevertheless, Engle and Smith suggest three alternative tests for the existence of a STOPBREAK process:

(1) Using a t-test for Ho: $\phi = 0$ against a negative alternative in the following regression:

$$\Delta y_t = \varphi \frac{\Delta v_{t-1}}{\bar{\gamma} + \Delta y_{t-1}^2} + \mu_t \qquad (6)$$

(for the choice of $\bar{\gamma}$, see Table 2 in Engle and Smith (1998)).

The standard distribution theory applies to the t-statistic. Since the process is never exactly stationary, there are no unit moving average roots for Δy_t under the null and the alternative hypothesis.

(2) Some data series may have temporal correaltion in all periods. When Equation 5 is used, the testing procedure

becomes more complicated since α_0 is undefined under the null of a random walk process. As a result, there is a need to determine the value of α_0 even if it does not exist under the null. Let $\bar{\alpha}$ be the chosen value.

Engle and Smith compute the Neyman Pearson test statistic as the t-statistic on φ for the following regression:

$$\Delta y_t = \varphi \sum_{i=1}^{l} \bar{\alpha}^{i-1} \frac{\Delta y_{t-1}}{\bar{\gamma} + \Delta y_{t-1}^2} + u_t \qquad (6')$$

The test statistic is asymptotic normally distributed. According to Engle and Smith, the value of $\bar{\alpha}$ can be chosen from the range 0 to 0.9.

(3) The test can be approximated by regressing Δy_t on $\Delta y_{t-1}/(\bar{\gamma} + \Delta y_{t-i}^2)$, $i = 1, 2, \ldots, p$ where p is some predetermined number. Let R^2 and T be the r-square and number of observations of the regression. Under the null hypothesis, $T^* R^2$ of the regression is distributed as $\chi_{(p)}^2$. This procedure is relatively simple to implement, however, the testing power is low since it is only an approximation. In addition, the procedure cannot test against a one-sided alternative.

III. DATA AND EMPIRICAL RESULTS

In this study the daily closing market index in the USA relative to the stock market indices of Japan and eight European countries from January 1990 to June 1998 is examined. The market indices are obtained from the DataStream database. The three statistical tests detailed in Section II are used to determine if the STOPBREAK process exists for the sample. If the null hypothesis of a random walk is rejected for the market indexs of USA relative to that of another stock market, temporary cointegration exists between the two markets and the cointegration vector is $(1, -1)$.* First, logarithms of a pair of market indices (e.g. Japan and USA) are taken and denoted them as y_t and x_t. Referring to equation (2), it is specified $\delta = 1$, $B(L) = 1 - L$, $A(L) = 1 - \alpha_0 L$, and $Z_{t-1} = q_{t-1}(\gamma_0)$ and the following STOPBREAK model is obtained, which considers temporal correlation and temporary cointegration:

$$(1 - L)(y_t - x_t) = (1 - \alpha_0 L)\varepsilon_t + (1 - q_{t-1})(\varepsilon_t - \varepsilon_{t-1}) \quad (7)$$

In addition, it can be specified the conditional log likelihood function for Equation 7 and estimate it using Quasi-maximum likelihood estimation (QMLE). In this study, conditional heteroscedasticity and specify the conditional variance h_t are incorporated as follows:

$$h_t = \beta_0 + \beta_1 \varepsilon_{t-1}^2 + \beta_2 h_{t-1} + \kappa d_{t-1}(\varepsilon_{t-1} < 0)\varepsilon_{t-1}^2$$

Table 1. *Testing the null of a simple random walk against STOP-BREAK*

	$\mathrm{Sup}_\alpha t_\gamma$ Equation 6	$\alpha = 0.8$ Equation 6'	χ_5^2	χ_{10}^2
Belgium/USA	−1.8595**	−0.3473	6.3561	7.8823
Germany/USA	−5.5098*	−1.6295**	30.4605*	34.8208*
Japan/USA	−2.4902*	−2.6681*	3.7232	5.4904
Netherlands/USA	−4.3708*	−1.5178**	21.6627*	24.5478*
Swiss/USA	−2.0422**	−1.8842*	9.2880**	14.8157
Denmark/USA	−0.4059	0.1151	1.3509	5.0310
Ireland/USA	−0.5015	−1.4347	3.3420	10.7386
Norway/USA	−0.2359	0.7357	5.7818	8.6925
Spain/USA	−0.9172	−0.3443	1.5973	9.5630
10% critical value	−1.72	−1.28	9.23	15.99
5% critical value	−2.07	−1.65	11.07	18.31

Note: * indicates significance at 5% and ** significance at 10%. Critical values other than χ^2 are taken from the Table 4 in Engle and Smith (1998).

Variable d_{t-1} is defined to be unity if ε_{t-1} is negative, and zero otherwise.

As shown in Table 1, among the nine market pairs, the market index in the USA is temporally cointegrated with the market index in Belgium, Germany, Japan, Netherlands and Switzerland. Compared with the other four markets, the test statistic for the relationship between USA and Belgium is less significant. In other words, temporary cointegration exists between Germany and USA; Japan and USA; Netherlands and USA; Swiss and USA. No evidence of temporary cointegration is found for other stock markets. The results of QMLE are reported in Table 2.

Using the STOPBREAK model can take account of the existence of temporal correlation and temporary cointegration among stock prices in different markets. In addition, the possibility of a GARCH(1,1) process is allowed. In Table 2, the null hypothesis of $\gamma = 0$ is rejected for Japan/USA, Germany/USA, Netherlands/USA and Switzerland/USA. However, the null is not rejected for Belgium/USA. The maximum value of γ is 0.3664 for Japan/USA, and the minimum value of is 0.0278 for Switzerland/USA. The coefficient of temporal correlation (α_0) is statistically different from zero only for Japan/USA. This implies that temporal correlation does not exist for most of the pair-wise relationship between the USA and other markets. For the GARCH (1,1) estimation, both the coefficient of β_1 and β_2 are consistent with the null hypothesis, i.e. both β_1 and β_2 are greater than zero. In sum, the above results suggest that temporary cointegration exists between the US market and markets in Japan, Germany, Netherlands and Switzerland. The existence of temporary

---

* Referring to Equation 7, when taking logarithms for the two market indices, $y_t - x_t = \log(Y_t/X_t)$, where Y_t and X_t denotes market index for a particular market and the US market. Therefore, when the null cannot be rejected, temporary cointegration exists between the two markets and the cointegration vector is $(1,1)$.

Table 2. *Quasi-maximum likelihood estimation*

	Belgium/USA	Germany/USA	Japan/USA	Netherlands/USA	Swiss/USA
β_0	0.0111*	0.1975*	0.0627*	0.0183*	0.0826*
β_1	0.9528*	0.6794*	0.8921*	0.9224*	0.8467*
β_2	0.0491	0.2082*	0.0363*	0.0971*	0.0955*
κ	−0.0327*	−0.1319*	0.0880*	−0.0830*	−0.0011
γ	0.0082	0.1079*	0.3664*	0.0355**	0.0740*
α_1	−0.4590	0.0000	0.4901*	−0.1336	0.000
log L	−749.28	−1092.70	−1950.40	−758.88	−1356.89

Note: The STOPBREAK MODEL is defined as:

$$\Delta y_t = \alpha_1 \Delta y_{t-1} + \varepsilon_t - \theta_{t-1}\varepsilon_{t-1} \qquad \varepsilon_t \sim (0, h_t)$$

where $\theta_{t-1} = 1 - (1 - \alpha_1)q_{t-1}(\gamma)$ and

$$0 \leqslant \alpha_1 < 1, q_{t-1}(\gamma)\frac{\varepsilon_{t-1}^2}{\gamma + \varepsilon_{t-1}^2}, \gamma > 0.$$

$$h_t = \beta_0 + \beta_1\varepsilon_{t-1}^2 + \beta_2 h_{t-1} + \kappa d_{t-1}(\varepsilon_{t-1} < 0)\varepsilon_{t-1}^2.$$

cointegration may imply arbitrage opportunity as Engle and Smith (1998) claimed

The STOPBREAK model predicts the direction that the price ratio will move, i.e., it forecasts whether prices will move towards or away from each other. If they are predicted to move apart, the investor will buy the higher valued stock and sell the lower stock short. In a STOPBREAK framework, such an investor is expected to make small gains regularly and then to make either large gains or large losses when the unexpected permanent shocks occur. On average, these large profits and losses will cancel each other out, leaving accumulated wealth with no money down.

Does the existence of temporary cointegration imply that regular cointegration does not exist? To address this interesting question, the Johansen (1988) cointegration test is used to examine if the stock price in the US market is cointegrated with the stock prices in the nine markets in the conventional sense. The results of Johansen's test for the market pairs are reported in Table 3.†

From Table 3, no support was found for the hypothesis of a cointegration relationship between the US market and the markets indicating temporary cointegration except for Netherlands. This result implies that finding a temporary cointegration relationship between the US market and Netherlands market does not improve the prediction of future stock prices for these two markets. For the other three markets (Germany, Japan and Switzerland), the existence of temporary cointegration, but not regular cointegration, implies that investors can enhance their forecasts of stock price in one market using information about stock prices in another market. However, no

Table 3. *Results of Johansen's cointegration test*

	$r \leqslant 1$	$r = 0$
Germany/USA	0.77	10.73
Japan/USA	0.86	12.36
Netherlands/USA	0.00	18.72**
Switzerland/USA	0.15	9.31
Belgium/USA	0.92	18.61**
Denmark/USA	0.12	8.46
Ireland/USA	0.04	13.89
Norway/USA	0.18	6.17
Spain/USA	0.19	10.62
5% Critical value	3.76	15.41
1% Critical value	6.65	20.04

Note: Critical values for trace test are obtained from Johansen (1988), r denotes the number of cointegrating vectors, ** denotes the 5% level of significance.

arbitrage profits are possible based on the results of a Johansen test.

IV. CONCLUSION

Previous studies tend to find no cointegration relationships among stock prices. However, stock prices may be cointegrated in a way that cannot be detected by the conventional cointegration test. It is possible that stock prices may move apart sometimes but are cointegrated most of the time. Applying the Stochastic Permanent Breaks (STOPBREAK) model developed by Engle and Smith (1998) to the Japanese and eight European stock markets, it was found that the relative market index of USA with respect

† Similar results are obtained when the Engle and Granger (1987) two-stage approach is used.

to Germany, Netherlands, Japan and Switzerland follows a STOPBREAK process. This means that the US stock market is temporally cointegrated with the markets in Germany, Netherlands, Japan and Switzerland. However, the US market is cointegrated only with the market in Netherlands according to the Johansen (1998) cointegration test. When two stock price series move together in some time period but move apart when external shocks occur, investors can still predict stock prices in one market using information of stock prices in the other market. The findings indicate that the conventional cointegration test may not be sufficient when trying to investigate the dynamic relations between two markets.

ACKNOWLEDGEMENTS

The authors are grateful to the Editor and an anonymous referee for valuable suggestions.

REFERENCES

Bekaert, G. and Harvey, C. R. (1997) Emerging equity market volatility, *Journal of Financial Economics*, **43**, 29–77.

Chelley-Steeley, P. L. and Pentecost, E. M. (1994) Stock market efficiency, the small firm effect and cointegration, *Applied Financial Economics*, **4**, 405–11.

Chow, G. C. (1960) Tests of equality between sets of coefficients in two linear regressions, *Econometrica*, **28**, 591–605.

Copeland, L. S. (1991) Cointegration tests with daily exchange rate data, *Oxford Bulletin of Economics and Statistics*, **53**, 185–98.

Engle, R. and Granger, C. N. J. (1987) Cointegration and error correction: representation, estimation, and testing, *Econometrica*, **55**, 251–76.

Engle, R. and Smith, A. D. (1998) Stochastic permanent breaks, *UCSD Discussion Paper 98-3*.

Granger, C. W. J. (1986) Developments in the study of cointegrated economic variables, *Oxford Bulletin of Economics and Statistics*, **48**, 213–28.

Hakkio, C. S. and Rush, M. (1989) Market efficiency and cointegration: an application to the sterling and Deutschemark exchange markets, *Journal of International Money and Finance*, **8**, 75–88.

Hendry, D. F. and Clements, M. P. (1996) Intercept corrections and structural change, *Journal of Applied Econometrics*, **11**: 475–94.

Johansen, S. (1988) Statistical analysis of cointegration vectors, *Journal of Economics Dynamics and Control*, **12**, 231–54.

MacDonald, R. and Taylor, M. (1991) Stock prices, efficiency and cointegration: some evidence from recent float. Working Paper, Dundee University, Department of Economics.

MacDonald, R. and Taylor, M. (1989) Foreign exchange market efficiency and cointegration: some evidence from the recent float, *Economics Letters*, **29**: 63–8.

JOURNAL OF ECONOMIC DEVELOPMENT
Volume 24, Number 2, December 1999

State Dependent Correlation and Lead-Lag Relation when Volatility of Markets is Large: Evidence from the US and Asian Emerging Markets

Bwo-Nung Huang, Soong-Nark Sohng and Chin Wei Yang[*]

By using the filtered probability calculated from the SWARCH model (Hamilton and Susmel (1994)), this paper examines the state of volatility of the equity markets. More specifically, we explore the nature of correlation coefficients and lead-lag relations between the US and the emerging economies. Such relations and correlation are found to have intensified during the Asian financial crisis. In the case when the volatility was great, US stock prices clearly led the emerging markets. Furthermore, stock prices of Japan, Hong Kong and Singapore also led the Asian emerging markets.

I. Introduction

The flow of portfolio investments to emerging financial markets experienced a significant increase from $6.2 billion in 1987 to $37.2 billion in 1992 (Gooptu (1994)). Although debt instruments - bonds, certificates of deposit and commercial paper - are still the main components of such flows, portfolio investors have shown increasing interest in equities of developing countries. For example, Claessens and Gooptu (1994) estimate that the flow of foreign capital to the equity markets of emerging economies almost doubled from $7.6 billion in 1991 to $13.1 billion in 1992. Needless to say, the revival of emerging financial markets after the debt crisis of the early 1980s represents a new challenge to researchers. However, the recent turmoil in Asian markets has cast a dark cloud over the stability of international portfolio investment, especially in the presence of large volatility. Since July of 1997, the 'Asian Flu' spread out rapidly to Hong Kong, South Korea and Japan. On October 27, the Hang Seng index plummeted 1438 points as a result of sky rocketing short term interest rate in an effort to prevent the Hong Kong dollar from depreciating. Triggered by the free fall of the Hang Seng index, the Dow Jones Industrial Average (DJIA) suffered a 554.26 point loss. As shown in Table 1, all nine major Asian stock markets experienced serious set backs, ranging from 13.4%

* Professor of Economics, National Chung-Cheng University, Chia-Yi, Taiwan 621, Professor of Economics, Clarion University of Pennsylvania, Clarion, PA 16214-1232 and Professor of Economics, Clarion University of Pennsylvania, Clarion, PA 16214-1232, respectively.
Generous financial assistances from the National Science Council (Taiwan) and Fulbright Scholarship Program are gratefully acknowledged. We thank anonymous referees for their valuable comments; however, all errors are ours.

in Thailand to 40.3% in Indonesia. It is to be pointed out that average rate of return in the Asian markets is higher than that of developed nations, so is their volatility.[1] Admittedly, low correlation between the emerging and the developed equity markets is of particular interest to portfolio managers who use international diversification to reduce risk. Nonetheless, major events like the so-called 'Asian Flu' would undoubtedly add profound volatility to the markets. As such a question often raised is whether the international correlation (or interdependence) increases in period of high turbulence. The presence of volatility in markets is precisely the culprit when the benefit of risk diversification is needed most. It is therefore to the disadvantage of international portfolio managers if large volatility accompanies high correlation.

Table 1 A Comparison of Daily Stock Prices and Exchange Rates Between July 1 and November 14, 1997

stock index	HKN	IND	JPN	KOA	MAL	PHI	SIG	THA	TWN
11/14/97	9957.33	436.84	15082.52	520.01	677.47	1844.95	425.89	456.87	7482.92
7/1/97	15196.79	731.61	20175.52	758.03	1078.90	2815.54	494.00	527.28	9030.28
%Change	−34.48%	−40.29%	−25.24%	−31.40%	−37.21%	−34.47%	−13.79%	−13.35%	−17.14%

Note: HKN = Hong Kong, IND = Indonesia, JPN = Japan, KOA = Korea, MAL = Malaysia, PHI = Philippines, SIG = Singapore, THA = Thailand, TWN = Taiwan. All prices are based on daily market close; all exchange rates are expressed as number of local currencies per US dollar.

Prior studies largely fall into two categories. First, like many others, Ratner (1992) claims that the international correlations remain relatively constant over the period of 1973-89. Hence a set of constant correlation coefficients is indicative of stable interaction among equity markets. Second, the correlation relation is found to be evolving through time. For example, employing daily data for the eight markets over the three years (1972, 1980 and 1987), Koch and Koch (1991) conclude from a simple Chow test that international markets have recently grown more interdependent. Moreover, Longin and Solnik (1995) discover via a multivariate GARCH model that the correlation is increasing in periods of high market volatility. The phenomenon is reinforced by Solnik et al. (1996). Similar results may be found in King and Wadhwani (1990), King, Sentana and Wadhwani (1994) and Karolya and Stulz (1996), Erb, Harvey, and Viskanta (1994). In particular, their findings demonstrate that correlations are higher in bear markets and during recession. One stylized fact found in Bekaert and Harvey (1997) reveals again that the correlation between markets rises in the periods when the volatility of markets is large. It seems that the international correlation increases when global factors dominate domestic ones and affect all financial markets. The dominance of global factors tends to be associated with major events (e.g., the oil crises, the Gulf war etc.). Viewed in this perspective, we expect a stronger correlation (interaction) among various national stock markets during the period of high volatility than in a period of low volatility.

Generally speaking, previous analyses consider primarily unconditional correlation computed over different subperiods. In this paper, we propose an explicit model that

---

1. See Bekaert and Harvey (1997).

provides conditional correlation coefficients in different states of volatility. Beyond that, we apply the Granger-causality model to analyze the interaction between different states (low and high volatility) among various national stock markets. Low coefficients are found to exist between most of the Asian emerging markets and the US in terms of unconditional correlation. However, after taking different states into consideration via the filtered probability of the Markov-switching model, asymmetrical state-dependent correlation coefficients begin to surface. This is to say, while in state one (low volatility), the correlation between the US and the Asian emerging markets is low, becoming stronger and significant in state two (high volatility). Applying the Granger-causality test, we find stronger interactions in a high volatility period, but only weak interdependence when the volatility is low. The result is in general agreement with Bekaert and Harvey (1995). The next section discusses the Markov Switching model (MS); Section III provides a description of data and results from the MS model; Section IV presents (i) the state-dependent correlation coefficients conditional on the probabilities estimated from the MS model and (ii) the Granger-causality results. Section V illustrates that cumulative abnormal profit in state 1 is generally positive but becomes negative in state 2. A conclusion is given in the last section.

II. Switching ARCH (SWARCH) Model[2]

Consider an AR(p)-GARCH(p,q) process for variable y_t:

$$y_t = \beta_0 + \beta_1 y_{t-1} + \dots + \beta_p y_{t-p} + e_t; \quad e_t \sim N(0, h_t), \tag{1}$$

$$e_t = \sqrt{h_t} w_t,$$

$$h_t = \alpha_0 + \sum_{i=1}^{p} \alpha_i e_{t-1}^2 + \sum_{j=1}^{q} \gamma_j h_{t-j}, \tag{2}$$

where w_t is assumed to be i.i.d. and N(0,1). This model has found a wide variety of applications in the finance literature and its appeal lies in the ability to capture the time varying nature of volatility. Notwithstanding its strength, such a model, however, fails to capture structural shifts in the data caused by low probability events (e.g., the crash of 1987, recession or recent Asian financial crisis). Diebold (1986), as well as Lamoreux and Lastrapes (1990) argues that the persistence frequently found in the ARCH models is due to the presence of structural breaks. Cai (1994), Brunner (1991) and Hamilton and Susmel (1994) modify the ARCH specification to account for such structural changes in data and propose a switching ARCH (SWARCH) model where the variance Equation (2) is revised to be:

$$e_t = \sqrt{g_{s_t}} \cdot u_t,$$

2. The model is primarily based on Hamilton and Lin (1996).

$$u_t = \sqrt{h_t} \cdot w_t, \tag{3}$$

$$h_t = a_0 + \sum_{i=1}^{p} a_i e_{t-1}^2.$$

Note w_t is assumed to be i.i.d. and N(0,1),[3] ; s_t is an unobserved latent variable that represents the volatility phases of a stock market. In absence of such phases, the parameter g_{s_t} would simply equal unity for all t. In that case Equation (3) describes stock returns with an autoregression whose residual e_t follows a pth-order ARCH process.

More generally, for g_{s_t} not identically equal to unity, the latent ARCH process u_t is multiplied by a scale factor $\sqrt{g_{s_t}}$. It represents the current phase s_t which in turn characterizes overall stock volatility. Hamilton and Susmel (1994) normalize $g_1 = 1$, in that case g_2 has the interpretation as the ratio of the average variance of stock returns when $s_t = 2$ to that when $s_t = 1$. To coefficients of Equation (3) may be estimated via a maximum likelihood approach.

As a byproduct of the maximum likelihood approach, Hamilton (1989) shows that we can make inferences about a particular state of the stock return at any date. The filter probability, $p(s_t, s_{t-1} \mid y_t, y_{t-1}, ...)$, denotes the conditional probability with the state at date t being represented by s_t and that at date $t-1$ by s_{t-1}. The smooth probabilities, $p(s_t \mid y_r, y_{t-1}, ...)$, on the other hand are inferences about the state at date t based on data available through some future date T (end of sample). Smooth probabilities reflect ex post evaluation as they encompass entire sample period. On the other hand, the filter probability evaluates the likelihood at time t whether the rate of change in stock returns belong to state 1 (small volatility) or state 2 (large volatility).

III. Data and Results of the SWARCH Model

Eight Asian along with the US stock markets are included in the present study. Stock prices taken from Datastream Data Bank span from Jan. 3 1986 to Jan. 5 1998 (3132 observations in total) for the following markets: Hong Kong (HKN), Japan (JPN), South Korea (KOA), Malaysia (MAL), Philippines (PHI), Singapore (SIG), Thailand (THA), and Taiwan (TWN).[4] To start the analysis, we first define the rate of change of stock prices as

$$y_t = (\log Y_t - \log Y_{t-1}) \times 100 \tag{4}$$

in which Y_t is stock prices in each markets at time t. The estimation results based on Equation (4) are represented in Table 2.

---

3. To better estimate the model parameters in the presence of profound leptokurtic return distributions in the Asian markets, we replace the normality assumption with a student t distribution.

4. We are very grateful for the generosity extended by the economics department of the University of California, San Diego, in providing the data.

Table 2 Estimates of the SWARCH Model for the Nine Asian and the US Stock Markets

	HKN	JPN	KOA	MAL	PHI	SIG	THA	TWN	USA
β_0	0.0984	0.0630	0.0123	0.0459	0.0284	0.0235	0.0751	0.0937	0.0726
	0.0185[a]	0.0160	0.0392	0.0146	0.0219	0.0123	0.0177	0.0244	0.0127
β_1	0.0846	0.0033	0.0675	0.1771	0.2066	0.1619	0.1750	0.0523	0.0144
	0.0190[a]	0.0182	0.0214	0.0187	0.0187	0.0193	0.0192	0.0181	0.0139
α_0	0.9486	0.4433	1.1634	0.5409	0.8269	0.3783	0.6542	1.2479	0.4809
	0.0562[a]	0.0362	0.0712	0.0516	0.0854	0.0303	0.0536	0.0930	0.0345
α_1	0.1218	0.1033	0.0854	0.2407	0.2033	0.2217	0.1822	0.0777	0.0527
	0.0300[a]	0.0296	0.0344	0.0423	0.0424	0.0465	0.0338	0.0266	0.0186
α_2	0.0844	0.1515	0.0772	0.1890	0.1295	0.1025	0.2084	0.2221	0.0321
	0.0268[a]	0.0339	0.0438	0.0397	0.0366	0.0402	0.0380	0.0415	0.0192
g_2	5.6558	4.9928	5.8094	3.7135	5.1863	4.4844	5.0400	5.4489	2.6529
	0.6639[a]	0.4063	0.8746	0.4292	0.5630	0.7660	0.5149	0.5239	0.2041
v	2.9708	3.5913	5.4927	1.8710	2.1905	2.3114	2.8650	2.7664	2.1662
	0.4576[a]	0.5924	1.2004	0.3085	0.4006	0.4136	0.4846	0.4962	0.3367
p_{11}	0.9965	0.9915	0.9913	0.9963	0.9867	0.9925	0.9936	0.9948	0.9984
p_{22}	0.9842	0.9914	0.9401	0.9955	0.9830	0.9692	0.9885	0.9920	0.9982
$(1-p_{11})^{-1}$	285.71	117.65	114.94	270.27	75.19	133.33	156.25	192.31	625.00
$(1-p_{22})^{-1}$	63.29	116.28	16.69	222.22	58.82	32.47	86.96	125.00	555.56
π_1	0.8170	0.5011	0.8729	0.5487	0.5607	0.8033	0.6416	0.6060	0.5220
π_2	0.1830	0.4989	0.1271	0.4513	0.4393	0.1967	0.3584	0.3940	0.4780
log L	−5029.04	−4756.27	−5291.35	−4659.49	−5603.96	−3671.37	−5145.43	−6188.37	−3801.41

Note: HKN = Hong Kong, JPN = Japan, KOA = South Korea, MAL = Malaysia, PHI = Philippines, SIG = Singapore, THA = Thailand, TWN = Taiwan. The estimates are based on the SWARCH model:

$$y_t = \beta_0 + \beta_1 y_{t-1} + e_t \qquad e_t \sim t_v(0, h_t) \qquad e_t = \sqrt{g_{s_t}} \cdot u_t \qquad u_t = \sqrt{h_t} \cdot w_t \qquad h_t = \alpha_0 + \sum_{i=1}^{2} \alpha_i e_{t-1}^2$$

where v = the degrees of freedom of the t distribution; g_2 = average variance of state 2 (high volatility) with $g_1 = 1$ by design; p_{11} and p_{22} are transition probability from state 1 (2) to state 1 (2); π is average ergodic probability; logL = likelihood function value; rows with superscript a denote standard errors of the estimates. a = estimated standard error.

As indicated in Table 2, significant coefficients of AR(1) except in the markets of the US and Japan are manifestly present especially in the markets of Malaysia, the Philippines, Singapore and Thailand where the coefficients exceed 18%. As Harvey (1995) points out, the AR(1) coefficients are generally greater than 10% for emerging stock markets. As such it represents a noticeable portion of predictability in their future prices. The estimation based on ARCH(2) is found to be statistically significant for all the markets indicating the existence of clustering phenomenon of stock returns. Note that the index of persistence $(\alpha_1 + \alpha_2)$ from the SWARCH after considering Markov-switching process is significantly below 1. This is very much in agreement with what Lamoureux and Lastrapes (1990) have warned: "The extent to which persistence in variance may be overstated is because of the existence of, and failure to take account of, deterministic structural shift in the model."

Are the two states justified in analyzing the stock returns?[5] An examination of Table 2 provides an affirmative answer. As the size of variance in state 1 is set at unity, the average value of variance in state 2 or g_2 throws some light on the structural change on the stock markets. In the US market, for instance, the variance in state 2 is 2.67 times as large as that of state 1; 5.55 times as large as in the Hong Kong market indicating a significant difference. Bekaert and Harvey (1995) find greater volatility in emerging markets than in developed markets. However, we discover that the developed Asian markets such as Hong Kong, Japan and Singapore have no less volatility than that of the emerging markets. As our sample period encompasses the Asian Flu period, it indicates the volatile nature even for developed markets in the presence of major events.

As a result of the leptokurtic return distributions, v values (the degrees of freedom from the t distributions) are found to be significant, ranging from the smallest v value of 3.90 (Malaysia) to the largest one of 8.73 (South Korea). Also reported in Table 2 are both transition probability and ergodic probability. Transition probability measures the magnitude of persistency observed in which data stay in one state; that is, higher values suggest length of stay is more likely to be longer. The length of stay can be calculated as $(1 - p_{ii})^{-1}$ $i = 1$ or 2. Ergodic probability reflects the proportion of time (probability) the sample data stay in a particular state. For instance, the ergodic probabilities are less than 20% for the Hong Kong, South Korea and Singapore markets to stay in state 2 (high volatility). In contrast, the Japanese stock market experienced the high volatility state in half of the sample period. The average length of stay in state 1 (low volatility) is longest (625 days) for the US, but lasted only 75 days in the Philippines market. Similarly, the average length of stay in state 2 is 556 days for the US market with the shortest one of 17 days in South Korea. These estimates from the Markov switching technique provide valuable pieces of information and as such are instrumental in deciphering the lead-lag relations discussed in the next section.

IV. State Dependency and Granger-Causality Between the US and the Asian Stock Markets

Previous studies have successfully identified large correlation coefficients and stronger intermarket dynamics (integration) during the periods of greater volatility (e.g., Bekaert and Harvey (1995, 1997)). However, the segmentation of sample periods was entirely event-based. For example, 'Black Monday' is used as a demarcation date. While convenient, it ignores the information of small volatility during the period of great volatility and vice versa. Hence a statistical approach to determining the state of volatility in terms of the filtered probability may be considered appropriate. We first calculate the correlation coefficients between states and test the Granger-causality among different national stock indices.

---

5. Hamilton and Lin (1996) suggest that the parsimony in estimating parameters of the SWARCH model is important. Besides, it does not seem necessary to have more than two states in the present paper, and hence, we use two states.

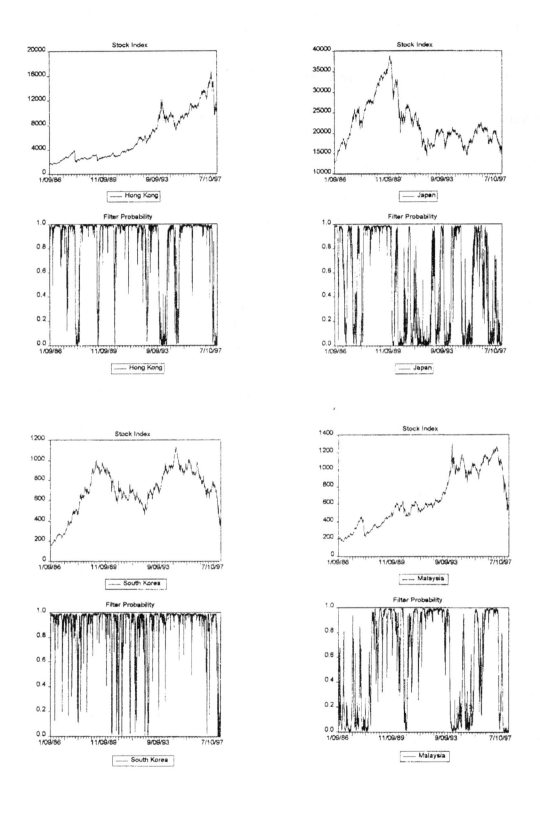

Figure 1-1 Stock Price Trend and the Filter Probability of the Nine National Stock Markets

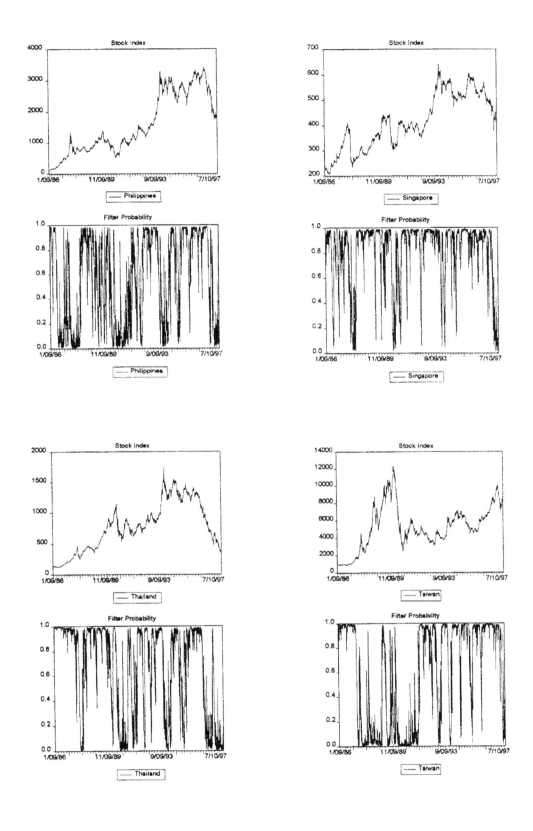

Figure 1-2 Stock Price Trend and the Filter Probability of the Nine National Stock Markets

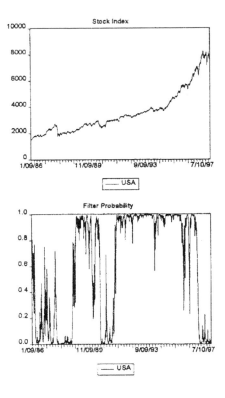

Figure 1-3 Stock Price Trend and the Filter Probability of the Nine National Stock Markets

Stock price trends and filter probabilities from the SWARCH model (Figure 1) reveal several interesting phenomena. First, the Asian stock markets exhibited a great deal of volatility in response to the US market crash of 1987. Its persistence was not long except for the markets of Philippines, Taiwan and Malaysia. During the recent Asian financial crisis, the magnitude of volatility increases noticeably as mirrored by the filter probability (less than 0.5) shown on the right hand side of graphs in Figure 1.

Table 3 Correlation Coefficients of Nine Stock Indices Under Two Different Volatility States

Panel A	HKN	JPN	KOA	MAL	PHI	SIG	THA	TWN	USA
HKN	1.0000								
JPN	0.2401	1.0000							
KOA	0.0710	0.0558	1.0000						
MAL	0.3943	0.2683	0.1090	1.0000					
PHI	0.1811	0.0991	0.0403	0.1987	1.0000				
SIG	0.3995	0.2701	0.1201	0.6300	0.1695	1.0000			
THA	0.2494	0.1529	0.1209	0.3225	0.1780	0.3187	1.0000		
TWN	0.0824	0.1245	0.0845	0.1133	0.0831	0.1348	0.1489	1.0000	
USA	0.1808	0.1018	0.0287	0.1182	−0.0217	0.1714	0.0225	−0.0291	1.0000

Table 3   (Continued)

Panel B	HKN	JPN	KOA	MAL	PHI	SIG	THA	TWN	USA
HKN	1.0000								
JPN	0.2062	1.0000							
KOA	0.0514	0.0136	1.0000						
MAL	0.3343	0.1946	0.0423	1.0000					
PHI	0.1224	0.0048	0.0411	0.0917	1.0000				
SIG	0.2974	0.1669	0.0357	0.6424	0.1969	1.0000			
THA	0.2204	0.1292	0.0531	0.2888	0.1022	0.2374	1.0000		
TWN	0.0647	0.0141	0.0528	0.0360	0.0148	0.0361	0.0564	1.0000	
USA	0.0655	0.0870	0.0252	0.0118	0.0799	0.0444	0.0031	0.0187	1.0000

Panel C	HKN	JPN	KOA	MAL	PHI	SIG	THA	TWN	USA
HKN	1.0000								
JPN	0.3656	1.0000							
KOA	0.2008	0.1185	1.0000						
MAL	0.5111	0.3580	0.3684	1.0000					
PHI	0.3489	0.1953	0.1460	0.2251	1.0000				
SIG	0.5431	0.4475	0.3541	0.6912	0.2161	1.0000			
THA	0.3628	0.2129	0.3599	0.4281	0.2815	0.4494	1.0000		
TWN	0.1035	0.2427	0.2707	0.1842	0.1061	0.2770	0.2987	1.0000	
USA	0.3048	0.1147	0.0886	0.1504	−0.0592	0.3152	0.0291	−0.0327	1.0000

Note: Panel A, B and C represent the correlation coefficients for the entire sample period (01/03/1986 ~ 01/05/1998), state 1 (low volatility) and state 2 (high volatility), respectively.

Reported in the first part of Table 3 are correlation coefficients for the entire sample period for the nine national stock markets. These low coefficients with respect to the US market are consistent with the results from prior studies. The strongest correlation is found in the Hong Kong market (0.18), followed by Singapore, Malaysia and Japan. In contrast, Taiwan and the Philippines markets witnessed negative correlation with the US market. Insignificant but positive coefficients are found for the markets of South Korea and Thailand. Of the emerging markets, the Malaysia market has the strongest correlation with the US market.[6] It appears that Hong Kong and Japan exert positive correlation with the rest of markets while Taiwan and the Philippines exhibit negative correlation with the other markets. As is well-known in econometric estimations, lumping together observation of inherently different states can leave out important information, and lead to inaccurate conclusion. For this reason, we partition the sample period into low and high volatility periods to obtain better estimates of the nine national stock indices. During the low volatility period, insignificant correlation coefficients between the Asian and the US markets suggest that investment in the Asian markets could well reduce the risk in international portfolio. The greatest one is found between the US

6. By using monthly data from 1976 through 1992, Harvey (1995) found a strong correlation coefficient (0.53) between the markets of Malaysia and the US.

and Japan (0.087) and the smallest one is between the US and Thailand (0.0031) during the low volatility period. The correlation coefficients among the Asian stock markets are also found to be less than those for the entire sample period. On the contrary, the correlation coefficients between the US and the Asian markets are greater than those for the entire sample period with the largest correlation coefficients (0.3152) between the US and Singapore and the smallest coefficient (0.0291) between the US and Thailand during the high volatility period. In addition, markets of Taiwan and the Philippines had negative correlation with the US market. Similar results are found as well within the Asian markets: greater correlation coefficients during the high volatility period. For instance, the correlation coefficients between the Hong Kong and other Asian markets during the high volatility period are two or three times as great as those during the low volatility period. The fact that correlation coefficients are relatively high during the high volatility period implies that risk-reduction via international diversification may hold true only in low volatility period.

On one hand, recent studies have provided some evidence highlighting the positive correlation between correlation and integration of capital markets. On the other hand, as Harvey (1995b, p.809) puts it: "... However, there is no necessary link between correlation and integration. A country can have zero correlation with the world market and be perfectly integrated into world capital markets. The low correlation could be caused by the weighted average of the firm betas ..." But, market integration is too important a topic for analyzing risk to be ignored. As both the single-factor and the multiple-factor CAPMs represent some measures of risk, the specification of CAPMs is of utmost importance in the statistical estimations. Needless to say, integration relations become stronger if some foreign explanatory variables can better explain the regression structure especially in the era of high volatility.

Strictly speaking, the strength of the Granger-causality does not lie in the causality per se. Rather, it can be used in improving predictive power via using historical data of the explanatory variables. A significant Granger-causality in international equity markets implies intensified integration of the markets. The Granger-causality model can be formulated as shown below:

$$\triangle y_{1t} = \alpha_0 + \sum_{i=1}^{k} \alpha_{1i} \triangle y_{1t-i} + \sum_{i=1}^{k} \alpha_{2i} \triangle y_{2t-i} + \varepsilon_{1t},$$

$$\triangle y_{2t} = \beta_0 + \sum_{i=1}^{k} \beta_{1i} \triangle y_{1t-i} + \sum_{i=1}^{k} \beta_{2i} \triangle y_{2t-i} + \varepsilon_{2t},$$

(5)

in which y_{1t} and y_{2t} represent stock prices of country 1 and 2 at time t. Failure to reject the $H_0 : \alpha_{21} = \alpha_{22} = ... = \alpha_{2k} = 0$ implies that the stock price of nation 2 does not Granger cause that of nation 1. Likewise, failure to reject $H_0 : \beta_{11} = \beta_{12} = ... = \beta_{1k} = 0$ suggests that the stock price of nation 1 does not Granger cause that of nation 2. Before applying the causality test based on Equation (5), one must examine the time series properties of these variables. That is, should the bivariate-VAR model (first difference) or that with the error correction term (VAR-VECM) be employed. This common procedure is first to apply a unit root test before conducting cointegration analysis. After that, one may perform the Granger-causality test. In our analysis, we adopt the Phillips and Perron approach (1988) that can handle the serial correlation problem in testing unit roots of national stock indices. There exist noticeable time trends in these indices as shown in Figure 1. As a result, we use the Phillips and Perron's τ_τ test statistics for the hypothesis test (Table 4).[7]

Table 4 Unit Root Test of the Nine National Stock Indices

	y	$\triangle y$
HKN	-2.5755	-57.2188^*
JPN	-2.8500	-56.2788^*
KOA	-1.5862	-52.6301^*
MAL	-0.2025	-47.3484^*
PHI	-2.4897	-47.4762^*
SIG	-1.7954	-48.3632^*
THA	-2.3422	-46.9511^*
TWN	-2.0974	-51.7245^*
USA	-2.2262	-55.1771^*

Note the Phillips-Perron or PP test is adopted with the null of a unit root. $\triangle y = \log y_t - \log y_{t-1}$ and y are logarithmic stock price indices. We employ the PP test with time trend or τ_τ test. $^* = 1\%$ significant level.

An examination of Table 4 indicates that we cannot reject the mull hypothesis (unit root) for logarithmic stock price indices of the nine nations. However, we are able to reject the mull hypothesis easily using the first difference. According to the interpretation of the Engle-Granger cointegration technique, the linear combination of national stock indices - which are I(1) - could be I(0). To test the cointegration, we apply the two-stage Engle-Granger model.[8] First, we perform the following regression analysis:

$$y_{1t} = a + bt + y_{2t} + e_t. \tag{6}$$

To test whether e_t is of I(1), we then make use of the Phillips and Perron test (or as an alternative, one could apply augmented Dickey Fuller test).

7. Readers are referred to Hamilton (1994, pp.506-515) for details about the Phillips and Perron unit root test.

8. One could also use the Johansen's maximum likelihood model, but the residual-based approach is convenient to apply.

Table 5 Pairwise Cointegration Test Results

x	y	x on y	y on x	x	y	x on y	y on x
HKN	JPN	−2.8312	−2.8949	KOA	THA	−3.1380	−2.0688
HKN	KOA	−2.4711	−1.3422	KOA	TWN	−1.4041	−1.6836
HKN	MAL	−2.7468	−0.2605	KOA	USA	−3.0134	−3.2128
HKN	PHI	−3.0498	−3.1754	MAL	PHI	−2.4135	−3.0934
HKN	SIG	−2.7020	−1.9139	MAL	SIG	−2.1291	−3.1232
HKN	THA	−2.5789	0.3409	MAL	THA	−2.6907	−2.9903
HKN	TWN	−2.9314	−2.4285	MAL	TWN	−0.1447	−1.8156
HKN	USA	−2.3044	−1.9126	MAL	USA	−1.0633	−2.8153
JPN	KOA	−2.5997	−1.7086	PHI	SIG	−3.9827**	−3.3030
JPN	MAL	−2.9833	−0.5195	PHI	THA	−3.1065	−1.1208
JPN	PHI	−2.4251	−2.2185	PHI	TWN	−2.1323	−1.6334
JPN	SIG	−2.9436	−1.8621	PHI	USA	−2.8531	−2.5259
JPN	THA	−2.6577	0.0913	SIG	THA	−2.9333	−1.8160
JPN	TWN	−3.7485**	−3.3348	SIG	TWN	−1.7409	−1.9863
JPN	USA	−2.7454	−2.1328	SIG	USA	−1.9916	−2.3997
KOA	MAL	−3.4331**	−1.9648	THA	TWN	0.7760	−1.4201
KOA	PHI	−2.3930	−3.0041	THA	USA	−2.6258	−3.8107**
KOA	SIG	−2.6236	−2.5303	TWN	USA	−2.1678	−2.2804

The two-stage residual-based test by Engle and Granger is applied. With the Phillips-Perron (1988) model used in the second stage. Failure to reject the null hypothesis implies a lack of cointegration between the variables. ** = 5% significant level.

An inspection of Table 5 suggests that we cannot reject the null hypothesis for 36 pairwise cointegration relations between national stock indices. That is, there exists no cointegration relations and as such Equation (5) is sufficient for the Granger-causality test. The Granger-causality results during different volatility states (panel A and B) are reported in Table 6.

Table 6 The Granger Causality Test Results under Two Different States

Panel A	HKN	JPN	KOA	MAL	PHI	SIG	THA	TWN	USA
HKN		0.34	0.02	1.39	3.91**	5.30**	5.35**	0.82	0.05
JPN	0.67		2.63***	0.01	0.91	0.17	0.03	0.53	1.85
KOA	0.28	0.79		0.38	0.77	3.02***	0.66	1.61	0.00
MAL	5.28**	1.21	5.69**		3.58**	0.09	0.38	0.34	0.08
PHI	0.00	0.04	0.88	1.54		3.69**	0.21	1.10	0.03
SIG	0.01	1.12	3.93**	4.00**	5.57*		5.57**	1.94	3.17***
THA	0.48	0.17	0.62	1.06	0.88	1.27		6.53**	4.02**
TWN	0.73	1.32	0.52	0.42	0.91	0.44	1.60		0.00
USA	197.39*	38.63*	0.69	141.94*	23.10*	208.00*	69.36*	6.65*	

Table 6 (Continued)

Panel B	HKN	JPN	KOA	MAL	PHI	SIG	THA	TWN	USA
HKN		3.18**	0.00	0.20	4.76**	0.53	8.78*	8.15*	6.83**
JPN	1.51		0.98	1.00	13.15*	5.39*	9.52*	6.05**	14.90*
KOA	0.00	0.70		4.22**	3.40***	0.91	2.77***	3.08***	0.01
MAL	12.46*	2.17	1.65		28.01*	1.44	35.58*	8.98*	3.68***
PHI	0.03	9.19*	0.00	3.01***		1.62	5.37**	0.38	2.20
SIG	13.13*	22.21*	0.01	22.15*	33.11*		34.70*	18.46*	0.21
THA	1.15	0.00	0.79	0.19	6.09**	0.02		5.68**	0.38
TWN	7.57	1.63	0.00	1.91	2.75***	0.00	3.31**		0.40
USA	29.34*	180.81*	10.17*	137.50*	63.91*	14.75*	70.91*	27.31*	

Note: The numbers in Panel A and B represents F statistics of the Granger-causality tests under state 1 (low volatility) and state 2 (high volatility) respectively. *, **, *** are significance level at $\alpha = 1\%$, 5% and 10% respectively. The Granger-causality model is based on the following:

$$\triangle y_{1t} = \alpha_0 + \sum_{i=1}^{k} \alpha_{1i} \triangle y_{1t-i} + \sum_{i=1}^{k} \alpha_{2i} \triangle y_{2t-i} + \varepsilon_{1t},$$

$$\triangle y_{2t} = \beta_0 + \sum_{i=1}^{k} \beta_{1i} \triangle y_{1t-i} + \sum_{i=1}^{k} \beta_{2i} \triangle y_{2t-i} + \varepsilon_{2t},$$

in which y_{1t} and y_{2t} represent stock prices of nation 1 and 2 respectively. Failure to reject the $H_0: \alpha_{21} = \alpha_{22} = \cdots = \alpha_{2k} = 0$ implies that change in stock price of nation 2 (column) does not Granger-cause that of nation 1 (row). Likewise, failure to reject $H_0: \beta_{11} = \beta_{12} = \cdots = \beta_{1k} = 0$ suggests that change in stock price of nation 1 (row) does not Granger-cause that of nation 2 (column).

Note that the first column lists the national stock indices that lead the price movement (rate of change in stock prices) while the first row provides the national stock indices that lag behind in price movement. For example, the numbers in the third row of Panel A in Table 4 represent F statistics for the null hypothesis of Japanese stock price movement does not Granger cause that of the other eight markets. On the contrary, the numbers in the third column reports the F statistics for the null hypothesis that price movements of the other eight stock markets do not Granger-cause that in the market of Japan. During the state 1 of low volatility (Panel A of Table 4), there exist 23 significant F statistics for at least $\alpha = 10\%$. This is to say we reject the non-existence of the Granger-causality among the nine national stock indices. Interestingly enough, such significant F's increase to 40 in state 2 of high volatility. The result is consistent with prior studies in that degree of market interaction and integration intensifies during the period of high volatility. It is well known in the literature (e.g., Wei et al. (1995) and Hu et al. (1998)) that the US exerts a far greater impact on the Asian emerging markets than does Japan. From Table 4, it is found that the US market leads all other Asian markets in both states. Furthermore, the market of Japan is found to lead the South Korea market in the state of low volatility with $\alpha = 10\%$. In the state of high volatility Japanese stock prices (past) can be used to improve the predictive power of the stock prices in current period of the market of the Philippines, Taiwan, Thailand and Singapore.

Beyond that, price movements of Hong Kong and Singapore are beneficial in forecasting that of other markets (see Panel B of Table 4). It is to be pointed out that the US market basically assumes the leader's role especially over the markets of Taiwan and Thailand during the low volatility period. However, all the markets except Philippines are found to lead the Taiwan and the Thailand markets during the high volatility period. As shown in Panel B of Table 4, there exist some feedback relations among the US, Japan, Hong Kong and Malaysia markets with a strong Japan-Hong Kong feedback relation ($\alpha < 5\%$). This finding is significant: while there is no feedback relation from the Asian markets to the US in the low volatility period, there indeed exists a feedback relationship from the Asian markets to the US in the high volatility period. Little wonder that such a feedback relation was borne out in the recent Asian financial crisis.

V. Event Study

Section IV indicates that lead-lag and correlation relations are relatively weak in state I (low volatility), and as a practical matter, one may like to examine the abnormal returns in each state. To do so, we employ the event-study approach for the 12-year sample period. The major events are (i) the market crash on October 19, 1987 in which DJIA plummeted nearly 500 points (event 1); (ii) the great depreciation of the Thai baht on July 2, 1998 triggered the financial debacle in Asia (event 2); and (iii) the Hang Seng index suffered a major landslide drop (1438 points) on October 27, 1997 through raising the short term interest rates substantially in order to peg its currency value to the US dollar (event 3). Event 3 in turn triggered a 544.26-point decrease in DJIA[9] in the US. Now we define the abnormal return as follows:

$$\epsilon_{it}^* = R_{it} - E[R_{it} \mid X_t],\qquad(7)$$

where ϵ_{it}^*, R_{it} and $E[R_{it}]$ denote abnormal, real, and normal returns respectively. Two models are typically used to estimate normal returns: constant-mean-return model and market model. In the case of constant mean model, X_t is a constant; it becomes market returns in the market model. Since we are more interested in market volatility, the market model is used to estimate normal returns.

As in these event analyses, an estimation window is needed before the event took place. One can then calculate normal and abnormal profits based on Equation (7). Figure 2 illustrates the process.

9. Other events could also be important; however, we choose these three events for their major impacts on national stock markets.

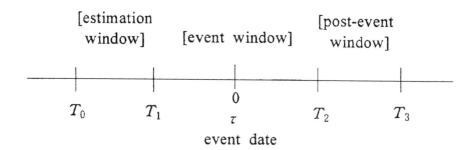

Source: Campbell, Lo and Mackinlay (1997, p.157)

Figure 2   Time Line for an Event Study

In the case of three different events, the relationship among cumulative abnormal returns of the national stock markets is depicted in Figure 3.

A perusal of Figure 3 points out that only Thailand and South Korea have positive cumulative abnormal market returns during event-1 period. Within 19 days of event 1, the abnormal returns are lowest for the US; followed by the Philippines, Taiwan, Japan, Singapore, Malaysia and Hong Kong. Beyond that period, the abnormal returns are lowest in the Philippines market. In terms of the filter probability of the SWARCH model (Table 7), all the markets except South Korea, had switched from state 1 to state 2 (e.g., Hong Kong, Japan, Singapore). From Figure 3, it can be seen that only Thailand's cumulative abnormal return was positive in state 2. In state 1, however, all the cumulative abnormal returns were negative except in the Korean market.

In the period of event 2 (Table 7), the markets of the US, Thailand, Japan were in state 2; markets of Hong Kong, South Korea, Singapore and Taiwan stayed in state 1; and those of Malaysia and the Philippines were in transition state (from state 1 to 2). In terms of cumulative abnormal returns, markets of Taiwan, Hong Kong, South Korea, the Philippines and Japan experienced positive returns while those of Singapore and Thailand were negative. The cumulative abnormal returns hovered around zero for the US and Malaysian market, with greater volatility for the US market. In general, the cumulative abnormal returns were less significant during event-2 period than those of event-1. In a similar vein, we could see positive cumulative abnormal returns during state 1 (lower volatility) but negative ones during state 2 (greater volatility). In the period of event 3, when the 'Asian flu' exerted its influence on the world financial markets, we can readily identify negative cumulative abnormal returns (Figure 2), with the lowest cumulative abnormal returns in the markets of Hong Kong, followed by Malaysia, Taiwan, the Philippines, South Korea, Singapore, Japan, Thailand and the US. The result is consistent with the filter probability approach in which all nine markets are in high volatility state.

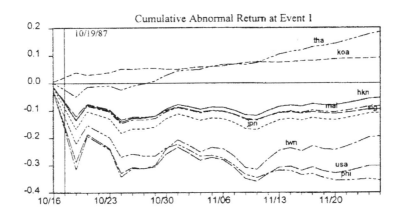

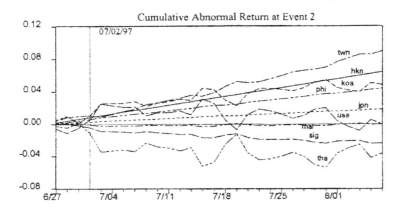

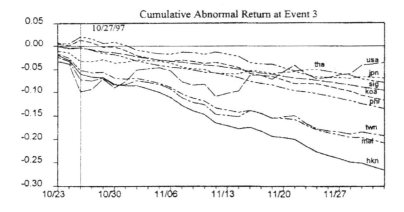

Note: hkn = Hong Kong, jpn = Japan, koa = South Korea, mal = Malaysia,
phi = Philippines, sig = Singapore, the = Thailand, twn = Taiwan

**Figure 3 Path of Cumulative Abnormal Returns of the Nine Stock Markets
(3 Events)**

Table 7   Changes of the Filter Probability of the Nine National Stock Indices

Date	HKN	JPN	KOA	MAL	PHI	SIG	THA	TWN	USA
10/16/87	0.9865	0.9637	0.9886	0.0498	0.0910	0.7493	0.6708	0.0124	0.0008
10/19/87	**0.5851**	**0.7332**	**0.9910**	**0.0126**	**0.0066**	**0.3693**	**0.3004**	**0.0017**	**0.0006**
10/20/87	0.6174	0.1465	0.9460	0.0126	0.0046	0.3976	0.0565	0.0016	0.0016
10/21/87	0.6632	0.1387	0.9685	0.0189	0.0214	0.4535	0.0497	0.0034	0.0013
10/22/87	0.8208	0.1522	0.9715	0.0146	0.0069	0.0703	0.0567	0.0029	0.0017
10/23/87	0.9151	0.1335	0.9828	0.0144	0.0091	0.0696	0.0878	0.0024	0.0050
10/26/87	0.1273	0.0837	0.9890	0.0159	0.0130	0.0987	0.0207	0.0220	0.0013
10/27/87	0.1353	0.0814	0.9920	0.0238	0.0389	0.1562	0.0175	0.0030	0.0019
10/28/87	0.1691	0.0960	0.9556	0.0153	0.0920	0.0357	0.0430	0.0224	0.0052
10/29/87	0.0975	0.0530	0.9729	0.0156	0.0070	0.0964	0.0181	0.0259	0.0014
7/1/97	0.9659	0.2794	0.9727	0.6772	0.9600	0.9122	0.1576	0.9891	0.0450
7/2/97	**0.9839**	**0.4026**	**0.9194**	**0.7390**	**0.8706**	**0.8949**	**0.0063**	**0.9920**	**0.0362**
7/3/97	0.9826	0.4998	0.9551	0.7804	0.9102	0.9336	0.0058	0.9941	0.0214
7/4/97	0.9672	0.4801	0.9696	0.7966	0.9270	0.9474	0.0090	0.9827	0.0350
7/7/97	0.9820	0.2830	0.9717	0.7175	0.9542	0.9308	0.0089	0.9800	0.0409
7/8/97	0.9882	0.3370	0.9809	0.8014	0.8395	0.9449	0.0153	0.9872	0.0218
7/9/97	0.9909	0.3571	0.9865	0.5349	0.5268	0.8254	0.0108	0.9906	0.0100
7/10/97	0.9909	0.5140	0.9892	0.4396	0.4889	0.7415	0.0219	0.9922	0.0139
7/11/97	0.9533	0.6059	0.9920	0.4382	0.2247	0.7494	0.0110	0.9934	0.0208
7/14/97	0.9693	0.2628	0.9922	0.5023	0.2566	0.7599	0.0137	0.9753	0.0357
10/24/97	0.0026	0.0106	0.0188	0.0097	0.0268	0.0833	0.0217	0.2255	0.0034
10/27/97	**0.0033**	**0.0074**	**0.0043**	**0.0198**	**0.0741**	**0.1859**	**0.0536**	**0.3217**	**0.0011**
10/28/97	0.0015	0.0012	0.0043	0.0037	0.0068	0.0126	0.0127	0.0393	0.0013
10/29/97	0.0016	0.0017	0.0374	0.0054	0.0072	0.0183	0.0388	0.0534	0.0044
10/30/97	0.0107	0.0025	0.0119	0.0150	0.0376	0.0757	0.0318	0.0496	0.0030
10/31/97	0.0299	0.0190	0.0280	0.0322	0.1019	0.0697	0.0528	0.1001	0.0044
11/3/97	0.0081	0.0555	0.0017	0.0034	0.2074	0.0149	0.1055	0.0338	0.0014
11/4/97	0.0054	0.1227	0.0057	0.0120	0.0375	0.0754	0.0054	0.0436	0.0048
11/5/97	0.0378	0.2088	0.0632	0.0245	0.0849	0.1449	0.0286	0.0728	0.0102
11/6/97	0.0180	0.3052	0.2180	0.0340	0.1435	0.1827	0.0301	0.0319	0.0183

Event on 10/19/87 = Black Monday; Event on 7/2/97 = The Great debacle of Thai Baht; Event on 10/27/97 = Major free fall on the Hang Seng index which trigger 554.26-point fall in DJIA the next day; state 1 (lower volatility) pertains to the state with the filter probability greater than 0.5; and state 2 (high volatility) corresponds to that with the filter probability less than 0.5.

In a nutshell, there seems to exist a tendency that greater volatility corresponds to negative cumulative abnormal returns while lower volatility matches with the positive returns. In the transition period (from state 1 to 2), the abnormal returns can go either way or even become zero.

VI. Conclusion

Stock price movements mirror the impact of information arrivals; larger movements reflect stronger momentum of the information content. Arbitrary segmentation of sample period (e.g., the 1987 Black Monday stock crash) can mask important interaction effects and hence lead to inaccurate results. To circumvent this problem, we make use of the SWARCH model developed by Hamilton and Susmel (1994) to determine the state of volatility via the filter probability. It is found that even mature markets such as Hong Kong exhibit a great deal of volatility during the recent Asian financial turmoil. From the calculated state-dependent correlation coefficient via the filter probability, we have found noticeable asymmetry between the two states: The correlation is much higher during the period of high volatility. As such international portfolio diversification may not be satisfactorily achieved from investing in the Asian market.

Also discovered is the increased integration among the national markets in the high volatility period. The existence of a strong integration is important in model specification (e.g., CAPM). It also provides the possibility of using prices of foreign markets to better predict the domestic stock price movements. In terms of the Granger-causality test, such causality effect becomes much stronger during the high volatility period. In addition to the existing knowledge that the US market frequently leads the Asian emerging markets, we have found that other mature markets in Asia such as Hong Kong, Japan and Singapore can play a leader's role as well during the high volatility period.

In addition, via the event analysis, there seemed to be a connection between (i) greater volatility and negative cumulative abnormal returns and (ii) lower volatility and positive cumulative abnormal returns. To a multinational portfolio manager, risk diversion and high cumulative abnormal returns are nearly mutually exclusive in the period of great market volatility. As such, forecasting market volatility plays a pivotal role in order to minimize investment losses.

Reference

Bekaret, G., and C.R. Harvey (1995), "Time-varying World Market Integration," *Journal of Finance*, 50, 403-444.

_____ (1997), "Emerging Equity Market Volatility," *Journal of Financial Economics*, 43, 29-77.

Brunner, A.D. (1991), "Testing for Structural Breaks in US Post-war Inflation Data," Unpublished Manuscript, Board of Governors of the Federal Reserve System, Washington, D.C..

Cai, J. (1994), "A Markov Model of Unconditional Variance in ARCH," *Journal of Business and Economic Statistics*, 12, 309-316.

Campbell, J.Y., A.W. Lo, and A.C. Mackinlay (1997), *The Econometrics of Financial Markets*, Princeton University Press, New Jersey.

Claessens, S., and S. Gooptu (1994), "Introduction," in *Investing in Emerging Markets*, ed. by M.J. Howell, *Euromoney Books*, London.

Diebold, F.X. (1986), "Modeling the Persistence of Conditional Variances: A Comment," *Econometric Reviews*, 5, 51-56.

Erb Claude, H.C., and Viskanta Tadas (1994), "Forecasting International Correlation," *Financial Analysts Journal*, 50, Nov./Dec., 32-45.

Gooptu, S. (1994), "Portfolio Investment Flows to Emerging Markets," in *Investing in Emerging Markets*, ed. by M.J. Howell, *Euromoney Books*, London.

Hamilton, J.D. (1989), "A New Approach to the Economic Analysis of Nonstationary Time Series and the Business Cycle," *Econometrica*, 57, 357-384.

Hamilton, J.D., and Lin, Gang (1996), "Stock Market Volatility and the Business Cycle," *Journal of Applied Econometrics*, 11, 573-593.

Hamilton, J.D., and R. Susmel (1994), "Autoregressive Conditional Heteroscedasticity and Changes in Regime," *Journal of Econometrics*, 64, 307-333.

Harvey, C.R. (1995), "The Risk Exposure of Emerging Equity Markets," *World Bank Economic Review*, 9, 19-50.

_____ (1995b), "Predictable Risk and Returns in Emerging Markets," *Review of Financial Studies*, 8, 773-816.

Hu, J.W.S., M.Y. Chen, R.C.W. Fok, and B.N. Huang (1998), "Causality in Volatility and Volatility Spillover Effects between US, Japan and Four Equity Markets in the South China Growth Triangular," *Journal of International Financial Markets, Institution and Money*, 7, 351-367.

Karolyi, G.A. and R.M. Stulz (1996), "Why Do Markets Move Together?: An Investigation of US-Japan Stock Return Comovements," *Journal of Finance*, 51, 951-986.

King, M., E. Sentana, and S. Wadhwani (1994), "Volatility and Links between National Stock Markets," *Econometrica*, 62, 901-33.

King, M., and S. Wadhwani (1990), "Transmission of Volatility between Stock Markets," *Review of Financial Studies*, 3, 5-33.

Koch, P.D., and T.W. Koch (1991), "Evolution in Dynamic Linkages across National Stock Indexes," *Journal of International Money and Finance*, 10, 231-51.

Lamoureux, C.G., and W.D. Lastrapes (1990), "Persistence in Variance, Structural Change and the GARCH Model," *Journal of Business and Economics Statistics*, 5, 121-29.

Longin, F., and B. Solnik (1995), "Is the Correlation in International Equity Returns Constant: 1960-1990?" *Journal of International Money and Finance*, 14, 3-26.

Ratner, M. (1992), "Portfolio Diversification and the Inter-temporal Stability of International Indices," *Global Finance Journal*, 3, 67-78.

Solnik, B., C. Boucrelle, and Y.L. Fur (1996), "International Market Correlation and Volatility," *Financial Analysts Journal*, 52, Sept./Oct., 17-34.

Wei, K.C.J., Y.J. Liu, C.C. Yang, G.S. Chaung (1995), "Volatility and Price Change Spillover Effects across the Developed and Emerging Markets," *Pacific-Basin Finance Journal*, 3, 113-36.

Oil Price Volatility

M. J. HWANG
West Virginia University
Morgantown, West Virginia, United States

C. W. YANG
Clarion University of Pennsylvania
Clarion, Pennsylvania, United States

B. N. HUANG
Providence University
Shalu, Taiwan

H. OHTA
Aoyama Gakuin University
Tokyo, Japan

Glossary

crude oil price Crude oil is sold through many contract arrangements, in spot transactions, and on futures markets. In 1971, the power to control crude oil prices shifted from the United States to OPEC when the Texas Railroad Commission set prorating at 100% for the first time.

demand elasticity The relative responsiveness of quantity demanded to changes in price. The price elasticity of demand is important in affecting the pricing behavior of OPEC.

economic impact The study of oil price shocks and their effects on economic activities.

oil price volatility The volatility of crude oil prices is influenced more by both demand structure and shifting demand and supply conditions and less by the cost of producing crude oil. Oil price volatility creates uncertainty and therefore an unstable economy.

Organization of Petroleum Exporting Countries (OPEC) Formed in 1960 with five founding members: Iran, Iraq, Kuwait, Saudi Arabia, and Venezuela. By the end of 1971, six other countries had joined the group: Qatar, Indonesia, Libya, United Arab Emirates, Algeria, and Nigeria. OPEC effectively controlled oil prices independent from the United States during the Arab oil embargo.

The world oil price has been extremely volatile in the past three decades. The cartel pricing is largely affected by the aggregate demand the cartel faces and related elasticity. The stable and unstable cases of price elasticity of demand are investigated in this article to shed light on the seemingly mysterious Organization of Petroleum Exporting Countries pricing behavior. We estimate the elasticity of oil demands in the U.S. market (the world's largest energy consumer) and use this information to probe and predict movements in the market price of crude oil. The volatility of crude oil prices creates uncertainty and therefore an unstable economy. Employing recent data, our empirical results indicate that a higher oil price seems to have a greater impact on the stock market than on the output market.

1. INTRODUCTION

The world oil price has been extremely volatile in the past three decades. It was as low as $2.17 per barrel in 1971 but spiked to $34 in 1981. It soared to approximately $40 per barrel toward the end of February 2003. The Organization of Petroleum Exporting Countries (OPEC) price increases of the

1970s drove Western industrialized economies into recession. During the 1950s and 1960s, many competitive independent producers characterized the crude oil industry. The demand facing the individual producer is relatively elastic, even though the demand curve for the entire industry is rather inelastic. The price of crude oil was close to the marginal cost during that period of time. OPEC emerged as an effective cartel in approximately 1971 when it successfully raised the pattern of world prices with the Tehran and Tripoli agreements. With the Arab–Israeli war of 1973, its consequent oil embargo, and the nationalization of oil production in member countries, the structure of the cartel provided the means to raise prices substantially from $4.10 per barrel in 1973 to $11.11 in 1974 to reap a monopoly profit.

With the beginning of political problems in Iran, another large increase in oil prices occurred in 1978 when Iranian production declined from a peak of 6 million barrels per day to 0.5 million barrels. Even though half of the reduction in Iranian oil was offset by expanded production by other OPEC producers, the effect was immediate and substantial, causing the price to increase to $13.49 in 1978, because the elasticity of demand was rather inelastic in the short term. Because of the war between Iraq and Iran, Iraq's crude oil production declined by 2.7 million barrels per day and Iran's production declined by 600,000 barrels per day. As a result, the price of crude oil more than doubled from $13.49 per barrel in 1978 to $34 in 1981.

Higher oil prices create inflationary pressure and slow worldwide economic activities. The recession of the early 1980s reduced demand for oil. From 1982 to 1985, OPEC tried to stabilize the world oil price with low production quotas. These attempts were unsuccessful because various members of OPEC produced beyond their quotas, causing crude oil prices to decrease below $10 per barrel by mid-1986. In particular, Saudi Arabia's increases in oil production frequently depressed the oil price. The price of crude oil remained weak until the start of the Gulf War in 1990, when it eclipsed $40 per barrel. Because of the uncertainty associated with the invasion of Kuwait by Iraq and the ensuing Gulf War, the oil price spiked again to $34 in late 1990. After the war and the recession of the early 1990s, the crude oil price began a steady decline. The economy, however, started to turn around in 1994. With a strong economy in the United States and a booming economy in Asia, increased demand led to a steady price recovery well into 1997. The financial crisis and subsequent economic setbacks in Asia started in 1997 and oil prices plummeted to approximately $10 in late 1998 and early 1999. With the recovery in Asia and the decrease in oil quotas by OPEC, the price has recovered to approximately $30 a barrel in recent years. In 2002, for the first time, Russia's oil production surpassed that of Saudi Arabia, signaling a more complicated pricing scheme for oil.

The literature on the volatility of crude oil prices relates oil price changes either to the effect of the price elasticity of demand or to the instability of the market structures. It is apparent that the stable price is established through the equilibrium of total world demand and supply, including OPEC and non-OPEC production. In the short term, the price change is largely impacted by the immediate substantial increase or decrease in oil production from OPEC members and political events. The cartel's pricing policy is largely affected by the aggregate demand it faces and related elasticity.

The volatility of crude oil prices in turn creates uncertainty and therefore an unstable economy. In his pioneering work, Hamilton indicated that oil price increases have partially accounted for every U.S. depression since World War II. Many researchers, using different estimation procedures and data, have tested the relationships between the oil price increases and many different macroeconomic variables. Using a multiequation statistical approach incorporating the U.S. interest rate, oil price, industrial production, and real stock returns with daily data from 1947 to 1996, Sadorsky found that oil price volatility does have an impact on real stock returns. Recently, emphasis has shifted to the asymmetry of the impact of oil price shocks on economic activities and on stock markets. Mork was the first to provide the asymmetry of oil price shocks or its volatility on economic activities. Using data from industrial nations, Mork and Olson verified that there is a negative relationship between an oil price increase and national output, whereas no statistical significance can be attributed to them when the oil price declines. Lee et al. estimated oil price shocks from a generalized econometric model and investigated the impacts of positive and negative oil price shocks on economic activities. They came to the same conclusion that positive shocks have a statistically significant impact on economic activities, whereas negative shocks have no significant impact.

This article examines the volatility of crude oil prices by first determining the potential maximal price that OPEC can extract based on the microeconomic foundation of the elasticity theory proposed by

Greenhut *et al.* The market structure of OPEC, the stable and unstable demand structure, and related elasticity of demand are discussed. In particular, the theory of unstable price elasticity of demand helps explain some of the pricing behavior of OPEC. The price elasticity of demand is then estimated to shed light on the volatility of oil prices. This article further investigates the significance of changing oil prices on the economy by examining the relationship between oil price volatility and industrial production and/or the stock market.

2. MICROECONOMIC FOUNDATION OF THE PRICE ELASTICITY OF DEMAND AND ITS ESTIMATION

Consider a cartel (e.g., OPEC) whose objective is to maximize joint profit:

$$\pi = PQ - \sum_{i=1}^{n} TC_i(q_i) = PQ - TC(Q). \quad (1)$$

where P is a uniform cartel price with $P' < 0$, $Q = \Sigma q_i$ is the sum of outputs of n cartel members (q_i), and TC_i is the total cost function of cartel member i. Note that OPEC behaves much like a monopoly or an effective cartel despite occasional squabbles over output quotas. Under recent OPEC arrangements, if the price becomes too low, OPEC would reduce output by half a million barrels per day ($\Delta Q = 500,000$) at one time. A 500,000-barrel decrease (increase) in Q is approximately 2% of the total cartel output, and each member must accept a 2% decrease (increase) in its quota in the case of a tightly controlled cartel. Within this framework, it is the total cartel output Q, instead of q_i, that plays a crucial role in the pricing decision.

The first-order condition requires

$$\pi' = QP' + P - MC = 0. \quad (2)$$

where $\pi' = d\pi/dQ$ is the marginal profit.

Equation (2) states that the marginal revenue (MR) equals the common cartel marginal cost (MC, or horizontal summation of marginal cost curves for all cartel members) in equilibrium. Note that if some MC_i's exceed the going market price, these members have no role to play in the pricing decision. However, this is not likely the case for OPEC because production costs fall far short of the market price. Substituting $MR = P(1-1/e)$ into Eq. (2) yields

$$P(1 - 1/e) = MC$$

or

$$P(1 - 1/e) = P/k, \quad (3)$$

where $k = P/MC$. Solving for the profit maximizing price elasticity leads to

$$e^* = k/(k - 1). \quad (4)$$

The second-order condition plays a critical role in describing the switching demand conditions. Here, we classify two major demand cases via expanding the elasticity theory formulated by Greenhut *et al.* in terms of the second-order condition.

By using the chain rule, we have

$$\frac{d^2\pi}{dQ^2} = \frac{d\pi'}{dQ}$$
$$= \frac{dP}{dQ}\left(\frac{\eta - (1 - e)}{e} - \frac{-e}{\beta k}\right) < 0, \quad (5)$$

where $\beta = (dQ/dMC)(MC/Q)$ is the output elasticity on the cartel marginal cost curve, and the price elasticity of elasticity $\eta = (de/dP)(P/e)$ measures the percentage change in price elasticity from the percentage price change. Alternatively, by taking the derivative of the marginal profit π' with respect to price P, the second-order condition in Eq. (5) appears as

$$\frac{d\pi'}{dP} = \frac{\eta - (1 - e)}{e} + \frac{e}{\beta k} > 0. \quad (6)$$

For a given output elasticity β, if the marginal cost is an insignificant portion of price (i.e., a large k) or the marginal cost is constant ($\beta \to \infty$, as is true in a large-scale crude oil production with a sizable fixed cost), the second term on the right side of Eq. (6) plays a trivial role and can be ignored. It follows that $d\pi'/dP > 0$ for $d\pi'/dQ < 0$, and it is sufficient that relation A holds:

$$\frac{d\pi'}{dP} > 0 \qquad \text{if } \eta > 1 - e, \eta = 1, \text{ or } \eta > 1. \quad (A)$$

The residual term $e/\beta k$ would reinforce the relation, but its magnitude may be insignificant, with k and/or β being large enough.

On the other hand, the unstable relation B would follow if the second-order condition is not satisfied:

$$\frac{d\pi'}{dP} < 0 \qquad \text{if } \eta < 1 - e. \quad (B)$$

Relation A indicates that no matter what the value of e is, the marginal profit π' will increase with price less than, proportionately equal to, or more rapidly than price if and only if $1-e < \eta < 1$, $\eta = 1$, or $\eta > 1$, respectively. If prior market conditions establish a price at a level at which the elasticity of demand is less than unity and MR is below MC (i.e., the marginal profit being negative), an OPEC price hike could be favorable. Under relation A, an increasingly elastic demand at higher prices (e.g., $\eta > 0$) would generally create conditions in which alternative fuels become more competitive following the price hike. The desired result under profit-maximizing principles is to have the marginal profit increase to 0 and elasticity close to $e^* = k/(k-1)$ depending on the ratio k ($k = P/\mathrm{MC}$).

Since the marginal cost for OPEC countries is very low relative to price, and k is accordingly large, e^* should be very close to 1. Under the condition $\eta = 1$ or $\eta > 1$, the increase in elasticity at a higher price is proportionately equal to or greater than the increase in price, and we reach the point at which negative marginal profit increases to zero more rapidly than under $1 - e < \eta < 1$. On the other hand, if the price elasticity of demand is greater than unity and MR is greater than MC, it becomes advantageous for OPEC to lower its price. In general, the elasticity of demand would converge toward $e^* = k/(k-1)$ and the marginal profit to zero. That is, the market system would adjust itself under stable relation A.

Relation B indicates that if $\eta - (1-e) < 0$, the marginal profit will decline (increase) as the price of crude oil is raised (lowered). This relation is an unusual and basically unstable case. Relation B also implies that the second-order condition in Eqs. (5) or (6) is not satisfied. The elasticity will therefore diverge from $e^* = k/(k-1)$. A cartel can benefit from an increase in price if demand is inelastic. Under the unstable relation B, the elasticity of demand decreases markedly as the price is raised. Inasmuch as marginal profit decreases and is negative, output will be curtailed and the price will be raised further. This can occur only in unusual circumstances (e.g., the energy crisis of 1973–1974). It is not a mathematical curiosity to have relation B—that is, an unusually strong convex demand curve. This could be a direct result of a structural break in the world oil market from a preembargo competitive oil market to an after-embargo cartel market. For example, the market demand changed from an elastic competitive market demand (before the 1973 embargo) to an inelastic cartel demand (after the embargo, from point F to C in Fig. 1).

The regime shift occurred when OPEC emerged as an effective cartel in 1971 after the Tehran Agreement. The ensuing Arab–Israeli War of 1973, subsequent oil embargo, and the nationalization of oil production in member countries enabled the cartel to raise the price of crude oil from $2 per barrel in 1971 to $34 in 1981. The price skyrocketed again to $37 in 2000 and $39.99 in late February

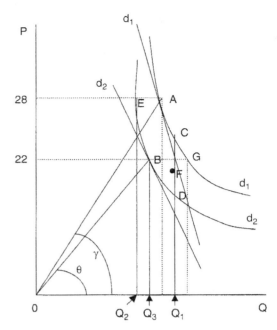

FIGURE 1 Switch in demand structure for oil. F, preembargo point; C, after-embargo point; A and E, upper limit of the price band; B and G, lower limit of the price band; A to B, market positions due to recession; C to D, market positions due to recession without a production cut.

2003. The market changed from a competitive one to one in which OPEC was able to exercise a considerable degree of monopoly power. Demand, in turn, changed from a relatively elastic individual market demand curve to a relatively inelastic industry cartel demand curve. The OPEC countries could therefore raise the oil price even further under this structural break in demand. This was the direct result of the unstable relation B.

In order to estimate the demand relations, we use data from "The Annual Energy Review" and "The Economic Report of the President." The price of coal (PC) is measured in cents per million Btu (cost, insurance, and freight price to electric utility power plants), as are prices of petroleum (P) and natural gas (PN). The quantity of petroleum consumption (Q) is in quadrillion Btu (10^{15}) and real gross domestic product is used to reflect income (Y). The sample period extends from 1949 to 1998. As with other economic variables, the unit root (unstable variables or not stationary) property needs to be examined before the estimations. Note that we fail to reject the null hypothesis of a unit root for all the variables even after the logarithmic transformation. That is, these variables are not stationary for the ordinary least squares technique: A technique is needed that relates a set of independent variables and dependent variables in terms of a linear equation. Since all the variables in our model are not stationary, we apply the two-step technique by Engle-Granger to explore the cointegration relation (comovements of unstable variables). The cointegration model, used in the presence of unstable variables, suggests the use of the error correction model (ECM) in estimating the demand relation. That is, demand for petroleum can be appropriately formulated as

$$dQ_t = \text{ECM}_{t-1} + dP_t + dPC_t$$
$$+ dPN_t + dY_t + v_t. \tag{7}$$

where the prefix d denotes first difference of the variables (in logarithm); Q_t denotes consumption of crude oil; ECM_{t-1} is representing past disequilibrium; P_t denotes the price of crude oil; Y_t is income; PC_t is the price of coal; and PN_t represents the price of natural gas. Despite its complexity, the model is more general because it allows past disequilibrium or perturbations to be included in explaining oil consumption. The estimated results for the entire sample period are shown as follows:

$$dQ_t = \underset{(-1.56)}{-0.1645}\,\text{ECM}_{t-1} \underset{(-3.22)}{-0.0502}\,dP_t$$

$$+ \underset{(1.02)}{0.0275}\,dPN_t \underset{(-0.58)}{-0.0210}\,dPC_t$$

$$+ \underset{(6.89)}{0.7664}\,dY_t + \underset{(5.33)}{0.7462}\,\text{AR}(1)$$

$$\underset{(-0.45)}{-0.0054} + e_{1t} \tag{8}$$

$$\bar{R}^2 = 0.71.$$

where $\bar{R}^2$ is the adjusted R square, which reflects general fitness of the model; t statistics in parentheses indicate the significance of estimated coefficients. Within the framework of the partial adjustment model (or model with lagged dependent variable) with ECM, the price elasticity tends to converge approximately to $-0.198 = [-0.0502/(1-0.7426)]$ when estimated in the level of first difference. Similarly, we report the estimation results for the two-subsample periods (1949–1975 and 1976–1998):

$$dQ_t = \underset{(-2.72)}{-0.1182}\,\text{ECM}_{t-1} \underset{(-2.52)}{-0.1778}\,dP_t$$

$$\underset{(-1.22)}{-0.060}\,dPN_t + \underset{(1.92)}{0.1150}\,dPC_t$$

$$+ \underset{(4.26)}{0.6097}\,dY_t + \underset{(2.89)}{0.0230} + e_{2t} \tag{9}$$

$$\bar{R}^2 = 0.70$$

and

$$dQ_t = \underset{(-2.83)}{-0.1643}\,\text{ECM}_{t-1} \underset{(-3.06)}{-0.0625}\,dP_t$$

$$+ \underset{(2.23)}{0.0673}\,dPN_t + \underset{(0.26)}{0.0403}\,dPC_t$$

$$+ \underset{(6.61)}{1.3127}\,dY_t \underset{(-5.11)}{-0.0366} + e_{3t} \tag{10}$$

$$\bar{R}^2 = 0.76,$$

The price elasticity of demand for petroleum within the entire sample period tends to approach -0.198 $[-0.0502/(1-0.7462)]$ in the long term, as indicated by Eq. (8). Thus, there appears to be enough room for further price hikes because the estimated price elasticity is not close to $k/(k-1)$ or approximately -1.05 in the empirical estimation as predicted by our theory.

The short-term price elasticities before and after the structural break can thus be estimated as -0.1778 and -0.0625, respectively. The empirical results indicate that (i) the elasticity after the energy crisis decreased in absolute value from 0.1778 to 0.0625 (a 65% decrease), clearly an unstable case of $\eta < 0$ (i.e., the price elasticity decreases as price increases), and (ii) there seems to be considerable room for a price hike because the long-term elasticity from Eq. (8) of -0.198 falls far short of $k/(k-1)$.

Neither the long-term price elasticity for the entire sample period (-0.198) nor the short-term price elasticity after the structural break (-0.0625) are close to the theoretical limit: $k/(k-1) \approx 1.05$ as is expected by a profit-maximizing cartel. The discrepancy may be explained by the significant income elasticity of 1.3127 in Eq. (10). Since the business cycle is both inevitable and unpredictable, a recession could certainly shift the demand curve to the left (Fig. 1). Continuous and gradual price hikes without disruptions in demand would render price elasticity toward $k/(k-1)$ in the long term. A recession would generally depress both price and quantity, as shown in Fig. 1 from point A to B. Given that the price elasticity can be geometrically measured as $e_A = \tan\gamma/\text{slope}\,(d_1)$ at point A, the size of the price elasticity does not appear to have changed appreciably. That is, $e_A = \tan\gamma/\text{slope}\,(d_1)$ and $e_B = \tan\theta/\text{slope}\,(d_2)$ are similar because both $\tan\theta$ and the slope of the line d_2 have decreased. Unless the U.S. economy is recession-proof in the long term, it does not seem possible that the long-term price elasticity would approach $k/(k-1)$ as implied by the theory. The wild swings in oil price after 1973 speak to the fact that demand for and supply of crude oil are not independent of political events. The significant change in the oil price was and can be detrimental to the suppliers with limited capacity and relatively higher marginal cost. In contrast, producers with lower marginal cost and greater capacity (e.g., Saudi Arabia) would benefit from greater output quota. To prevent such violent price changes, it is advantageous to OPEC to have a price band in which the price elasticity is not too low. However, can $k/(k-1) \approx 1.05$. a theoretical limit developed previously, ever be reached? There are at least three reasons in favor of this argument. First, switching to alternative fuels becomes more plausible at high oil prices and thereby tends to push long-term price elasticity close to $k/(k-1)$. Second, noncartel output (e.g., 5–7 million barrels per day from Russia) in the long term can present a credible threat when the price is high enough. This renders OPEC's residual demand curve flatter for every high price. Third, significant *ad valorem* tariffs in G8 countries would effectively pivot the demand curve downward.

Substantial price fluctuations have been witnessed for the past three decades, especially since the 1973 energy crisis. Based on our estimate, the average price elasticity was very low, −0.1778 before 1975, but the change in market structure has given rise to the unstable demand case. It has been borne out with the short-term elasticity of −0.0625 after 1975. Notice that the negative and significant ECM coefficient (−0.1643) in Eq. (10) suggests a convergent trend toward equilibrium values after the structural break. In the absence of major war, it does not seem plausible that the price of oil will significantly exceed $40 due to the income effect in the demand structure. The price elasticity, hovering at −0.0625 after the structural break, suggests that the oil price would more likely approach the upper limit of the price band ($28) than the lower limit ($22) if OPEC members strictly adhered to the production cut (e.g., from Q_1 to between Q_2 and Q_3; Fig. 1) in the presence of a recession. The result is borne out because oil prices lingered around $28 at the beginning of May 2001, 1 month after the production cut. As the recession deepened after the September 11, 2001, terrorist attacks, the price of crude oil dropped to $20 in November 2001 when the OPEC cartel was reluctant to further reduce production. Failing to do so would result in a drastic price reduction (from point C to point D in Fig. 1) as occurred before. On the other hand, the upward pressure on price could be too irresistible because −0.0625 is far below the theoretical limit of −1.05, as inelastic demand promotes price increase.

3. VOLATILITY OF AN OIL PRICE CHANGE AND MACROECONOMIC ACTIVITY

Given that oil is of great importance in the production process, its impact on industrial output cannot be ignored. Furthermore, the growing magnitude of oil price changes reflects the major role of uncertainty in statistical modeling. For instance, Lee *et al.* found that uncertainty in oil price (conditional variance of an oil price change) could significantly impact economic growth. In particular, one unit of positive normal shock (a large price increase) gives rise to a decreased output, whereas one unit of negative shock does not necessarily lead to an increased output. Similarly, the impact of an oil price change on the financial market has received increased attention. Sadorsky was among the first to apply a formal statistical model to investigate such an impact. To capture the uncertainty inherent in the macroeconomic foundation, we need to first define the real oil price after adjusting for the consumer price index or $roilp_t$. To ensure stationarity property, let $\Delta roilp_t$ represent the first difference of the logarithmic transformation on $roilp_t$. One of the common ways to measure the volatility of oil price is to apply an econometrical model called autoregressive and conditional heteroskedastic (ARCH) by Engle or its generalized version, the GARCH model, by Bollerslev. The GARCH model shown below is fairly general in describing the behavior of a volatile time series such as oil price in macroeconomics:

$$\Delta roilp_t = a_0 + \sum_{i=1}^{p} \phi_i \Delta roilp_{t-1} + \varepsilon_t - \sum_{j=1}^{k} \theta_j \varepsilon_{t-j}$$

$$\varepsilon_t = z_t \sqrt{h_t} \quad z_t \sim N(0,1)$$

$$h_t = \alpha_0 + \beta_1 h_{t-1} + \alpha_1 \varepsilon_{t-1}^2, \tag{11}$$

where h_t is the conditional variance often used to represent volatility, and $z_t = \varepsilon_t / \sqrt{h_t}$ denotes a standardized disturbance term. Note that optimal lags of the GARCH $(1,1) - ARMA(p, q)$ are selected based on the criterion that a series correlation (correlation via residuals) is not found in z_t and z_t^2. Finally, z_t is normally distributed with zero mean and unit variance or $N(0,1)$. Even though h_t is frequently chosen to measure the volatility, the major problem is that variance masks the direction of changes to which shocks are administered. The asymmetric impact—only positive shocks (or price increases) slow economic activity and/or the stock market—begs the use of z_t instead of h_t as a proxy for volatility. As such, we use z_t as the proxy in our research model.

Monthly data of industrial production (ip_t), real stock return ($\Delta rstkp_t$), consumer price index (cpi_t), exchange rate ($exch_t$), interest rate (r_t), and stock indices from July 1984 to March 2002 were obtained from the International Financial Statistics (IFS) data bank. Data for oil prices are from "Energy Prices and Taxes" published by the International Energy Agency. Note that we include exchange rates because oil is exported to many industrialized economies. In addition, the interest rate is incorporated to account for monetary policy. Major industrialized countries included in the study are the United States, Canada, Italy, Japan, Germany, and the United Kingdom.

As in many other time series analyses, we need to examine the existence of a unit root (unstable) on the variables of lip_t, $lrstkp_t$, lr_t, and $lexch_t$. The prefix l indicates the logarithmic transformation of the original variables, used to damp unstable variables. If these variables are $I(0)$ or integrated of order zero indicating stationarity or stability of the variable, we could add z_t to form a five-variable VAR model. However, if these variables are $I(1)$ or integrated of order 1 (not stationary), we need to first examine the potential existence of any cointegration relation (or comovements of unstable variables). A five-variable vector error correction model can be formulated in the presence of such a cointegration relation. Otherwise, the four-variable (first difference) model along with z_t would suffice to analyze the direction of causality among the variables.

The techniques for examining variable stationarity by Phillips and Perron and by Kwiatkwoski *et al.* are applied to the four variables for the six countries. The result is unanimous in that all the variables are of $I(1)$ or not stationary. Consequently, we examine the potential existence of cointegration relations via the trace method developed by Johansen. An examination of the trace statistics suggests that no cointegration relation exists for the five countries except Japan, and two sets of cointegration relations exist for Japan. That is, we can include the past disturbance term ECM_{t-1} in the model to explain industrial production for Japan as follows:

$$\Delta lip_t = a_4 + r_4 ECM_{t-1} + \sum_{i=1}^{k} b_{4i} \Delta lr_{t-i}$$
$$+ \sum_{i=1}^{k} c_{4i} \Delta lexch_{t-i} + \sum_{i=1}^{k} d_{4i} z_{t-i}$$
$$+ \sum_{i=1}^{k} f_{4i} \Delta lip_{t-i} + \sum_{i=1}^{k} g_{4i} \Delta lrstkp_{t-1} + e_{4t}.$$

(12)

where ECM_{t-1} is the error correction term representing the long-term cointegration. On the other hand, the five-variable vector autoregression model is applied to the other five countries as follows (only the stock return equation is shown):

$$\Delta lrstkp_t = a_5 + \sum_{i=1}^{k} b_{5i} \Delta lr_{t-i} + \sum_{i=1}^{k} c_{5i} \Delta lexch_{t-i}$$
$$+ \sum_{i=1}^{k} d_{5i} z_{t-i}$$
$$+ \sum_{i=1}^{k} f_{5i} \Delta lip_{t-i} + \sum_{i=1}^{k} g_{5i} \Delta lrstkp_{t-i} + e_{5t}.$$

(13)

The optimal lag lengths are determined based on the Akaike's Information Criterion, and directions of the Granger causality can be analyzed by the impulse response function.

To purge serial correlation in z_t of Eq. (11), for more accurate estimation, a lag of 5 on autoregression and a lag of 1 on moving average are needed.

TABLE I

Results of the Granger Causality[1]

H_0	Canada	Germany	Italy	Japan	United Kingdom	United States
z_t not $\to \Delta lr_t$	Cannot reject	Cannot reject	(+) Reject[*]	Cannot reject	Cannot reject	Cannot reject
z_t not $\to \Delta lexch_t$	Cannot reject	Cannot reject	Cannot reject	Cannot reject	Cannot reject	Cannot reject
z_t not $\to \Delta lip_t$	Cannot reject	Cannot reject	(−/+) Reject[**]	Cannot reject	(+) Reject[***]	Cannot reject
z_t not $\to lrstkp_t$	Cannot reject	(−) Reject[**]	(−) Reject[*]	Cannot reject	Cannot reject	(−) Reject[***]

[1]"not $\to$" means "does not Granger cause." The Granger causality carries the information that variable x causes variable y, whereas the ordinary regression model provides x associates y. A rejection of H_0 implies the existence of a Granger causality. [*], [**], and [***] denote 1, 5, and 10% significance level, respectively. The signs in parentheses denote the direction of the causality from the impulse response analysis. −/ + indicates the impact is first negative and then turns positive.

The result indicates that the volatile behaviors of oil prices are best described by an ARCH model for Canada due to its insignificant β_1. However, the GARCH model is preferred for the remaining countries, of which Japan has the strongest persistence in oil price volatility (i.e., $\alpha_1 + \beta_1 > 1$). The United States has the smallest persistence, with $\alpha_1 + \beta_1 = 0.7866$. Except for Japan, in which $\alpha_1 > \beta_1$ (unexpected shock impacts current conditional variance more than past conditional variance), the reverse is true for other countries. Note that the GARCH model is selected so that z_t and z_t^2 are free of serial correlation, an undesirable property in a regression analysis.

It is not surprising that the greatest z_t (greatest change in oil price) value occurred during the Gulf War in 1990. Table I reports the causality results from the statistical models. The lengths of impacts can be determined from the analysis. In the case of Germany, the volatility of a real oil price change exerts a negative impact on stock returns for three periods before it tapers off to zero. For Italy, the volatility of a real oil price change leads the monthly interest rate to change positively for 7 months, the monthly stock return to change negatively for approximately 4 months, and the monthly industrial production to change negatively in the first month but positively during the next 3 months.

Surprisingly, the same volatility exerts positive impacts on industrial production (3 months) for the United Kingdom before leveling off to zero. The favorable impact of higher oil prices on industrial output can be attributed to the fact that Britain is one of the non-OPEC producers: A higher oil price commands the greater revenue ($\eta < 0$) from export, which can in turn be reinvested. For the United States, the volatility of a real oil price change leads to negative stock returns for 3 months before diminishing gradually to zero. The same volatility has no appreciable impacts on either industrial production or stock returns for Canada and Japan. It seems that the volatility of oil price changes leads to negative stock returns in three of the six countries. It affects industrial production in only two countries, including a positive impact in the United Kingdom.

4. CONCLUSION

Recent studies have highlighted the role of oil price uncertainty. The results seem to favor the long-held conjecture that higher oil prices have a greater impact on the stock market than on the output market. This is supported in our model for the United States. Higher oil prices seem to have a lower impact on major economies such as those of the United States and Japan due to the presence of strategic oil reserve. A strategic oil reserve is essential to absorb the shocks from an excessive price hike. At the heart of the volatility analysis is the concept of oscillating price elasticity at the microeconomic foundation developed by Greenhut et al. The existence of unstable price elasticity indeed sows the seed of inherent volatile oil price changes, which can readily inflict shocks to the economy, especially a relatively small economy whose demand for oil depends solely on import. Our volatility model, employing the most recent data, supports this conclusion. The impacts on the interest rate and other monetary variables remain a challenging research avenue for the future.

SEE ALSO THE FOLLOWING ARTICLES

Business Cycles and Energy Prices • *Energy Futures and Options* • *Inflation and Energy Prices* • *Markets for Petroleum* • *Oil and Natural Gas: Economics of Exploration* • *Oil and Natural Gas Leasing* • *Oil Crises, Historical Perspective* • *Oil Industry, History of* • *Oil-Led Development: Social, Political, and Economic Consequences*

Further Reading

Alhajji, A. F., and Huettner, D. (2000). OPEC and world crude oil market from 1973 to 1994: Cartel, oligopoly, or competitive? *Energy J.* 21(3), 31–58.

Bollerslev, T. (1986). Generalized autoregressive heteroskedasticity model. *J. Econometrics* 31, 307–327.

Burbridge, J., and Harrison, A. (1984). Testing for the effect of oil price rises using vector autogressions. *Int. Econ. Rev.* 25(1), 459–484.

Dargay, J., and Gately, D. (1995). The response of world energy and oil demand to income growth and changes in oil prices. *Annu. Rev. Energy Environ.* 20, 145–178.

Department of Energy, Energy Information Administration (1998). *Annu. Energy Rev.*

Engle, R. F. (1982). Autoregressive conditional heteroskadasticity with estimates of variance of United Kingdom inflation. *Econometrica* 50, 987–1007.

Engle, R. F., and Granger, C. W. J. (1987). Cointegrantion and error correction: Representation, estimation, and testing. *Econometrica* 55, 251–276.

Ferderer, J. (1996). Oil price volatility and the macroeconomy. *J. Macroecon.* 18(1), 1–26.

Gately, D. (1993). The imperfect price—Reversibility of world oil demand. *Energy J.* 14(4), 163–182.

Greene, D. L. (1991). A note on OPEC market power and oil prices. *Energy Econ.* **13**, 123–129.

Greene, D. L., Jones, D. W., and Leiby, P. N. (1998). The outlook for U.S. oil dependence. *Energy Policy* **26**(1), 55–69.

Greenhut, M. L., Hwang, M. J., and Ohta, H. (1974). Price discrimination by regulated motor carriers: Comment. *Am. Econ. Rev.* **64**(4), 780–784.

Griffin, J. M. (1985). OPEC behavior: A test of alternative hypotheses. *Am. Econ. Rev.* **75**(5), 954–963.

Griffin, J. M., and Teece, D. J. (1982). "OPEC Behavior and World Oil Prices." Allen & Unwin, London.

Hamilton, J. D. (1983). Oil and the macroeconomy since World War II. *J. Political Econ.* **99**(2), 228–248.

Huang, R. D., Masulis, R. W., and Stoll, H. R. (1996). Energy shocks and financial markets. *J. Future Market* **16**(1), 1–27.

Hwang, M. J. (1982). Crude oil pricing in the world market. *Atlantic Econ. J.* **10**(2), 1–5.

Johansen, S. (1988). Statistical and hypothesis testing of cointegration vectors. *J. Econ. Dynamic Control* **12**, 231–254.

Kwiatkwoski, D., Phillips, C. B., Schmidt, P., and Shin, Y. (1992). Testing the null hypothesis of stationary against the alternative of a unit root. *J. Econometrics* **54**, 159–178.

Lee, K., Ni, S., and Ratti, R. A. (1995). Oil shocks and the macroeconomy: The role of price variability. *Energy J.* **16**(4), 39–56.

Mork, K. A., and Olson, M. H. T. (1994). Macroeconomic responses to oil price increases and decreases in seven OECD countries. *Energy J.* **15**(4), 19–35.

Papapetrou, E. (2001). Oil price shocks stock market: Economic activity and employment in Greece. *Energy Econ.* **23**, 511–532.

Phillips, P. C. B., and Perron, P. (1989). Testing for unit root in time series regression. *Biometrika* **75**, 335–346.

Sadorsky, P. (1999). Oil price shocks and stock market activity. *Energy Econ.* **21**, 449–469.

U.S. Government Printing Office (1998). "Economic Report of the President." U.S. Government Printing Office, Washington, DC.

Yang, C. W., Hwang, M. J., and Huang, B. N. (2002). An analysis of factors affecting price volatility of the U.S. oil market. *Energy Econ.* **24**, 107–119.

II. Granger Causality Models Using Panel Data

Panel data or longitudinal data regression model has advantages over the cross-section models in that it has much larger sample size (NxT) and may in some cases detects causal relationship. The observed persistent difference may arise from unobserved individual-specific effect (fixed effect model that allows for unobserved individual heterogeneity to be correlated with regressors) or past habit (dynamic panel model with lagged dependent variable). On the other hand, however, panel data such as pooled technique may produce overly optimistic result in terms of low standard error and larger t statistics than they should be. In the era of big data where both cross section size N and time period T have increased, panel data regression models are playing a far more important role in applied econometric modeling.

A working paper at UCSD by Levin and Lin (1992) and Levin et al. (2002) ushered in the testing procedure on panel unit root. Pesaran and Smith (1995) investigated the cointegration properties. Phillips and Moon (1999) and Pedroni (2004) provided general theory for inference with non-stationary panel data. The panel Granger causality model without doubt has gained the momentum in applied social science modeling.

The paper "Causal Relationship between Energy Consumption and GDP Growth Revisited: A Dynamic Panel Data Approach" applied the dynamic panel data model by Blundell and Bond (1998) to 82 countries for the period of 1972-2002. The GMM-SYS approach was employed to estimate the panel VAR in each of the four groups: low income, lower middle income, upper lower income and high income groups. The results indicated that: (1) there existed no causal relationship between energy consumption and economic growth in the low income panel (2) In the lower and upper middle income panel, economic growth was found to lead or Granger-cause energy consumption. (3) In the higher income panel, energy consumption Granger-causes economic growth. (4) In the upper middle income panel after the energy crisis, it was found the energy efficiency declined and releases of carbon dioxide increased. As a result, a proper energy policy should be targeted at fuel conservation.

Would military spending of developing countries crowd out investment in private sectors and hence harm badly needed economic growth? The paper "Military Expenditure and Economic Growth across Different Groups: A Dynamic Granger Causality Approach" answers this question convincingly. This paper applies GMM by Arellano and Bond (1991) to a large panel data of 90 countries over 1992-2006 to investigate the dynamic causality between military spending and economic growth.

It was found in general for low income countries, military spending Granger-caused negatively economic growth. For the developing countries in Africa, Europe, Middle East, South Asia and Pacific Rim, such a causal relationship held true with a p value of 3%. The dynamic panel model within the framework of the VAR or vector autoregressive model does not require a priori theory on causality; it is data-driven adding yet another flexibility dimension to the Granger causality methodology.

Changing social landscape has made smoking habit not only a financial but also an ethical issue. That is, people in low income bracket remain a majority of cigarette smoking group. This paper "New Evidence on Demand for Cigarette: A Panel Data Approach" employed a panel of 42 states and Washington D.C. for the period of 1961-2002 to estimate the elasticity. First, by using the Im, Pasaran and Shin (1997) panel unit root method , which takes cross-section heterogeneity into consideration, authors showed the panel variables are stationary. As a result, the pooling technique (fixed effect model) was employed to estimate the elasticity. Price elasticity, income elasticity and price elasticity of neighboring states were found to be -0.41, 0.06 and 0.09 respectively. It is noteworthy that transfer income recipients had very low income elasticity indicating they had tough time kicking the habit, hence the social inequality and vicious cycle may very well persist in the near future.

ECOLOGICAL ECONOMICS 67 (2008) 11-54

available at www.sciencedirect.com

ScienceDirect

www.elsevier.com/locate/ecolecon

ECOLOGICAL
ECONOMICS
THE TRANSDISCIPLINARY JOURNAL OF
THE INTERNATIONAL SOCIETY FOR
ECOLOGICAL ECONOMICS

ANALYSIS

Causal relationship between energy consumption and GDP growth revisited: A dynamic panel data approach[☆]

Bwo-Nung Huang[a,b,1], M.J. Hwang[c,*], C.W. Yang[d]

[a]Department of Economics and Center for IADF, National Chung Cheng University, Chia-Yi 621, Taiwan
[b]Department of Economics, West Virginia University, Morgantown, WV 26505-6025, United States
[c]Department of Economics, College of Business and Economics, West Virginia University, P. O. Box 6025, Morgantown, WV 26505-6025, United States
[d]Department of Economics, Clarion University of Pennsylvania, Clarion, PA 16214-1232, United States

ARTICLE INFO

Article history:
Received 30 September 2007
Received in revised form
8 November 2007
Accepted 9 November 2007
Available online 20 February 2008

Keywords:
Panel VAR
GMM
Environmental Kuznets Curve
Energy consumption
Economic growth

JEL classification:
Q43

ABSTRACT

This paper uses the panel data of energy consumption and GDP for 82 countries from 1972 to 2002. Based on the income levels defined by the World Bank, the data are divided into four categories: low income group, lower middle income group, upper middle income group, and high income group. We employ the GMM-SYS approach for the estimation of the panel VAR model in each of the four groups. Afterwards, the causal relationship between energy consumption and economic growth is tested and ascertained. We discover: (a) in the low income group, there exists no causal relationship between energy consumption and economic growth; (b) in the middle income groups (lower and upper middle income groups), economic growth leads energy consumption positively; (c) in the high income group countries, economic growth leads energy consumption negatively. After further in-depth analysis of energy related data, the results indicate that, in the high income group, there is a great environmental improvement as a result of more efficient energy use and reduction in the release of CO_2. However, in the upper middle income group countries, after the energy crisis, the energy efficiency declines and the release of CO_2 rises. Since there is no evidence indicating that energy consumption leads economic growth in any of the four income groups, a stronger energy conservation policy should be pursued in all countries.

1. Introduction

The crude oil price hit a historical high of $77.05 per barrel on August 2006. Practitioners and academics alike are again concerned about the economic impact of high oil prices. Similarly, could the economic growth resulting from the increase in oil consumption, at the same time, offset the negative externality inflicting on environments? This has been the focus of debate in the last two decades. If the benefit in economic growth outweighs the cost of environmental damage, it is worth increasing energy use to accelerate economic growth. On the other hand, if energy consumption

☆ The valuable comments from the editor and two referees are greatly appreciated. Any errors are our own.
* Corresponding author.
  E-mail address: mjhwang@mail.wvu.edu (M.J. Hwang).
[1] Financial support from the National Science Council (NSC95-2415-H-194-002) is gratefully acknowledged.

0921-8009/$ – see front matter © 2007 Elsevier B.V. All rights reserved.
doi:10.1016/j.ecolecon.2007.11.006

does not increase or even adversely impacts economic growth, a conservation energy consumption policy is needed to avoid the adverse impacts on the economy.

The literature on the relationship between energy consumption and income dates back to the late 1970s. Kraft and Kraft (1978), in their pioneering work, used U.S. data from 1947–1974 to discover that GNP leads energy consumption. Using U.S. monthly data from 1973 to 1979, Akarca and Long (1979) showed instead that energy consumption leads employment (in the literature, some economists use employment or production to substitute for economic growth). More inconsistencies ensued. Akarca and Long (1980), Erol and Yu (1987a), Yu and Choi (1985), and Yu and Hwang (1984) found no relationship between the two. Erol and Yu (1987b, 1989), Yu and Jin (1992), and Yu et al. (1988) went one step further to test the neutrality hypothesis and found a neutrality relation (i.e., no causal relationship between the two).

The main reason for the discrepancy in results in the previous research comes from the use of different econometric methods. In most cases, the OLS model of log-linear was used to estimate parameters and to conduct statistical tests without taking into consideration the special features of time series data. As is well known, a spurious regression in the analysis could exist (Granger and Newbold, 1974) and as a result, the previous statistical results might well be misleading.

The statistical method in time series has made important advances in the past decade. As in many other economic fields, the relationship between energy consumption and economic growth was revisited and statistically tested again using newer time series analysis. Yu and Jin (1992) applied the Engle–Granger technique to 1974.01–1990.04 data and found no long-term cointegration relation and no causal relationship between the two. The neutrality relation is, therefore, established. Masih and Masih (1996, 1997) used the Johansen cointegration algorithm to test the existence of cointegration between real GDP and total energy consumption using data in the period of 1955–1990 for India, Pakistan, Indonesia, Malaysia, Singapore, the Philippines, Taiwan, and South Korea. Next, either the vector error-correction model (VEC) or the vector autoregressive model (VAR) was used to test the relative causal relationship. The test results indicate no cointegration relation exists in Malaysia, Singapore, and the Philippines and, therefore, the neutrality hypothesis was supported. The rest of the other five countries do have a cointegration relation between energy consumption and economic growth. In particular, the test results show that: in India, energy consumption leads economic growth as a causal relationship; in Indonesia, GNP leads energy consumption; and Pakistan, Taiwan, and South Korea show a bi-directional relationship.

Cheng and Lai (1997) employed Engle-Granger's cointegrating test for Taiwan during 1955–1993 to investigate the relationship between energy and GDP, and between energy consumption and employment. They used the FPE (Final Prediction Error) version of Hsiao (1981), rather than AIC or SBC, to determine the optimal lag in Granger's causality test. They discovered that GDP leads energy as a uni-directional causal relationship in Taiwan. Their test result is in contrast to that of Masih and Masih (1997) and Hwang and Gum (1992) (bi-directional relationship). Interestingly, Yang (2000) updated the data of Taiwan to 1997 (1954–1997) and used the same

Engle and Granger (1987) cointegration method along with the FPE of Hsiao (1981) to discover a bi-directional relationship between energy consumption and economic growth.

Glasure and Lee (1997) applied the cointegrating technique and error-correction model to test the relationship between energy consumption and economic growth for South Korea and the Philippines. Based on the Granger cointegrating causality test, they discovered a bi-directional relationship in these two countries. Without considering the cointegration among variables, South Korea shows no Granger causal relationship, and the Philippines indicates a uni-directional causal relationship running from energy consumption to GDP.

In the bivariate model, Asafu-Adjaye (2000) added the price factor (using the consumer price index, i.e., CPI, to represent energy price) and applied Johansen's cointegration technique and the Granger causality test to investigate energy dependency and the relationship between energy consumption and economic growth in four countries in Asia: India, Indonesia, the Philippines, and Thailand. Both Thailand and the Philippines show a bi-directional relationship, while India and Indonesia show a uni-directional causality with energy consumption leading economic growth. Hondroyiannis et al. (2002) employed a trivariate model (energy consumption, real GDP, and price) and applied Johansen's cointegration technique and error-correction model to test the causality relationship in Greece during 1960–1996. They found no relationship among the three variables in the short run and some relationship in the long run. They concluded that the adoption of suitable structural policies aiming at improving economic efficiency can induce energy conservation without impeding economic growth.

Using cointegration and vector error-correction techniques, Soytas and Sari (2003) examined the causal relationship between GDP and energy consumption from 1950 to 1992 in the top 10 emerging countries (China excluded) and the G-7 countries. They discovered bi-directional causality in Argentina, uni-directional causality with energy consumption leading GDP in Turkey, France, West Germany and Japan, and the causality with GDP leading energy consumption in Italy and Korea.

Altinay and Karagol (2004) employed Hsiao's criterion to investigate the causal relationship between the GDP and energy consumption in Turkey during the period of 1950–2000. They concluded that there was no evidence of causality between the two and the data were trend stationary with a structural break.

Oh and Lee (2004a) used four variables (energy consumption, GDP, capital, and labor) from the supply side and three variables (energy consumption, GDP, and price) from the demand side in their multivariate Granger causality analysis to investigate the relationship between energy consumption and GDP in South Korea during the period of 1981:1–2004:4. They also employed the VEC model to distinguish between a long run and short run relationship among the variables and to identify the source of causation. In the short run, no causality was detected; however, GDP led energy consumption in the long run. Therefore, the government in South Korea can pursue conservation energy policy in the long run without compromising economic growth. Using the same techniques, with different periods during 1970–1999, Oh and Lee (2004b)

137

indicated a short run uni-directional causality running from energy consumption to GDP and a long run bi-directional causal relationship.

Paul and Bhattacharya (2004) applied the Johansen multi-variate cointegration technique on four variables (energy consumption, GDP, capital, and labor) and found bi-directional causality between energy consumption and economic growth. Lee (2005) employed panel cointegration and panel error-correction models to investigate the causal relationship between GDP and energy consumption in 18 developing countries during the period of 1975 to 2001. There is evidence of a short run and long run uni-directional causal relationship running from energy to GDP. Consequently, energy conservation may harm economic growth in those developing countries. Lee (2006) used the Granger causality cointegration test suggested by Toda and Yamamoto (1995) to investigate the relationship between energy consumption and GDP for 11 industrialized countries from 1960 to 2001. He discovered that: (i) there is no causal relationship between the two for the UK, Germany, and Sweden; (ii) U.S. data indicate a bi-directional causal relationship; (iii) Canada, Belgium, Netherlands, and Switzerland show a uni-directional causal relationship running from energy consumption to GDP; and finally (iv), France, Italy, and Japan show the relationship with GDP leading energy consumption. However, Lee and Chang (2007) applied the panel data to 22 developed countries and 18 developing countries to investigate the causal relationship between energy consumption and GDP using the bivariate model under the panel VAR framework of Holtz-Eakin et al. (1998). They discovered a uni-directional causal relationship running from GDP growth to energy consumption in the developing countries. In the developed countries, however, a bi-directional (or feedback) causality exists between the two.

Based on the literature review in Table 1, the causal relationship using the same country data could be different due, in part, to differences in research periods or in research methodologies. The most probable reason for the discrepancy is the insufficient number of observations in the data. It is manifest from the literature that most data are in the 30 to 40 years span. For the unit root or Johansen cointegration test, the 30–40 data points are few and as such, low statistical testing power is expected. Thus, the inconsistency in results is not unexpected.

In order to compensate for the deficiency in an inadequate sample size, the panel data approach is needed to reevaluate the relationship between energy consumption and income. The Granger causality test is mostly used in the time series data to investigate the relationship between energy consumption and economic growth. However, the dynamic panel estimation (DPE) approach needs to be used to identify the causal relationship for the panel data. Holtz-Eakin et al. (1998) and Arellano and Bond (1991) first suggested using all of the available lags as instruments to estimate the equation in first difference from dynamic panel data (DPD). With the availability of macroeconomic panel data, more DPE are used to investigate the causal relationship among macroeconomic variables. Though the DPE approach has not been widely used to investigate the relationship between energy consumption and economic growth in the literature, it is beginning to be used in recent years in other research areas such as defense spending and military growth

Table 1 – Summary of Literature review on the causal relationship between energy consumption and income

Authors	Countries	Results
Kraft and Kraft (1978)	US	y→ec
Akarca and Long (1979)	US	ec→employment
Akarca and Long (1980)	US	Neutral
Erol and Yu (1987a)	Japan	ec→y
Yu and Choi (1985)	S. Korea	y→ec
Yu and Hwang (1984)	Philippines	ec→y
Yu and Jin (1992)	US	Neutral
Masih and Masih (1996)	India	ec→y
	Pakistan	y→ec
	Indonesia	y→ec
	Malaysia, Singapore, Philippines	Neutral
Masih and Masih (1997)	Taiwan	y↔ec
	S. Korea	y↔ec
Cheng and Lai (1997)	Taiwan	y→ec
Yang (2000)	Taiwan	y↔ec
Glasure and Lee (1997)	S. Korea	y↔ec
	Singapore	y↔ec(ec→y)
Asafu-Adjaye (2000)	India, Indonesia	ec→y
	Philippines, Thailand	y↔ec
Stern (2000)	US	ec→y
Hondroyiannis et al. (2002)	Greece	Neutral
Soytas and Sari (2003)	Argentina	y→ec
	Italy, S. Korea	y→ec
	Turkey, France, Germany, Japan	ec→y
	Brazil, India, Indonesia, Mexico, Poland, South Africa, U.S., U.K., Canada	Neutral
Altinay and Karagol (2004)	Turkey	Neutral
Oh and Lee (2004b)	S. Korea	y↔ec(ec→y)
Paul and Bhattacharya (2004)	India	y↔ec
Jumbe (2004)	Malawi	y↔ec
Lee (2006)	U.K., Germany	Neutral
	Sweden, U.S.	y↔ec
	Canada, Belgium, Netherlands, Switzerland.	ec→y
	France, Italy, Japan.	y→ec
Lee and Chang (2007)	Developing Countries (18)	y→ec
	Developed Countries (22)	y↔ec

Notes: → denotes leads, ↔ denotes bi-directional causality or feedback, ec = energy consumption, and y = per capita real GDP.

(Yildirim et al., 2005), public finance (Fiorito and Kollintzas, 2004; Feeny et al., 2005), finance (Alessie et al., 2004), and labor supply (Baltagi et al., 2005). Due to the econometric deficiency of inadequate sample size in time series data, and the greater availability of macropanel data, the literature is on the rise in the use of panel data in the macrorelated research. As was indicated by Bond (2002), "Dynamic models are of interest in a wide range of economic applications, including Euler equations for household consumption, adjustment cost models for firm's factor demands, and empirical models of economic growth. Even when coefficients on lagged dependent variable are not of direct interest, allowing for dynamics in the underlying process may be crucial for recovering consistent estimator of other parameters (p.142)".

It is, therefore, necessary that the DPE approach be used to investigate the dynamic relation between energy consumption and economic growth. As such, this is one of our major contributions.

However, the use of panel data also creates another problem, in which different countries as a whole are treated as an entity, not as a separate unit. As a result, we cannot identify the difference in the dynamic relationship between energy consumption and income among countries. As the degree of economic development in each country is different, the relationship between energy consumption and economic growth will be different as well. For example, a developed country may use more resources to increase the efficiency of energy use and to better regulate environmental protection, while a developing country may put more resources in industrial production rather than energy efficiency and environmental protection. As a result, the relationship between energy consumption and economic growth should be different in two countries with different degrees of economic development (e.g., Lee, 2006). Another contribution of this paper is to partially resolve the "lump-together" problem in using panel data; we classify the panel data into four sub-panels based on the difference in income levels before further estimation.[2] Our results indicate that the dynamic relationship between income and energy consumption is indeed different in each income group.

If we use the panel data as a whole for 82 countries from 1972 to 2002, there is a bi-directional (feedback) relationship between energy consumption and economic growth. However, by grouping the data into four income groups based on the income levels defined by the World Bank (low income group, lower middle income group, upper middle income group, and high income group), we discover: (a) in the low income group, there exists no causal relationship between energy consumption and economic growth; (b) in the middle income groups (lower and upper middle income groups), economic growth leads energy consumption positively; (c) in the high income group countries, economic growth leads energy consumption negatively. This paper is organized as follows. Section 1 discusses the research motives and a review

of related literature. Section 2 introduces data and econometric methods. Section 3 analyzes and discusses empirical results. Section 4 is the policy implications derived from this study. The final Section 5 gives concluding remarks.

2. Model specification, econometric method, and data

In reference to the often-used explanatory variables in the literature (Oh and Lee, 2004a,b), we specify $lec_{i,t}$ (log of energy consumption), $ly_{i,t}$ (log of per capita real GDP), and other controlling variables as $liy_{i,t}$ (log of the share of capital formation to GDP to represent capital stock),[3] $lf_{i,t}$ (log of population to represent labor force), and $lp_{i,t}$ (log of GDP deflator). This is a 5-variable VAR model, where the subscripts are ith country and tth period. By taking into consideration the individual effect, the 5-variable panel VAR model can be shown as:

$$y_{i,t} = \sum_{j=1}^{p} \alpha_j y_{i,t-j} + \beta'(L)x_{i,t} + \eta_i + v_{i,t}. \tag{1}$$

η_i represents unobserved country-specific and time-invariant effect with $E(\eta_i)=\eta$ and $Var(\eta_i)=\sigma_\eta^2$. The $v_{i,t}$ are assumed to be independently distributed across countries with zero mean, but arbitrary forms of heteroskedasticity across units and times are possible. $y_{i,t}$ is $lec_{i,t}$ or $ly_{i,t}$; $x_{i,t}$ are predetermined variables as $liy_{i,t-j}$, $lp_{i,t-j}$, $lf_{i,t-j}$, $ly_{i,t-j}$ or $lec_{i,t-j}$, where $j=1,...,p$. Since η_i is assumed to follow a stochastic process of an individual effect, $E(y_{i,t-1} \eta_i) \neq 0$ and $E(x_{i,t} \eta_i) \neq 0$. $\beta(L)$ is a polynomial lag operator. To avoid the bias from the OLS estimate as a consequence of the country specific effect, we take the first difference of Eq. (1) suggested in the literature as

$$\Delta y_{i,t} = \sum_{j=1}^{p-1} \alpha_j^* \Delta y_{i,t-j} + \beta^{*'}(L)\Delta x_{i,t} + \Delta v_{i,t}, \tag{2}$$

where Δ is the first-difference operator. Eq. (2) may take care of the OLS estimation problem due to a correlation between individual effect and explanatory variables, but it also gives rise to another problem: the correlation between the lagged dependent and error term, that is, $E(\Delta y_{i,t-1} \Delta v_{i,t}) \neq 0$. Thus, the estimation of Eq. (2) by OLS will render a biased and inconsistent result. Arellano and Bond (1991) employed lagged dependent variables ($y_{i,t-s}$ for $s \geq 2$) in level as instrument in the GMM (Generalized Method of Moment) to overcome the problem of $E(\Delta y_{i,t-1} \Delta v_{i,t}) \neq 0$. Then, the corresponding optimal instrument matrix Z_i with predetermined regressors x_{it} correlated with the individual effect is given by

$$Z_i = \begin{pmatrix} y_{i1} & x_{i1} & x_{i2} & 0 & 0 & 0 & 0 & 0 & \cdots & 0 & \cdots & 0 & 0 & \cdots & 0 \\ 0 & 0 & 0 & y_{i1} & y_{i2} & x_{i1} & x_{i2} & x_{i3} & \cdots & 0 & \cdots & 0 & 0 & \cdots & 0 \\ \vdots & \vdots & \vdots & \vdots & \vdots & \vdots & \vdots & \vdots & & \vdots & & \vdots & \vdots & & \vdots \\ 0 & 0 & 0 & 0 & 0 & 0 & 0 & 0 & \cdots & y_{i1} & \cdots & y_{i(T-2)} & x_{i1} & \cdots & x_{i(T-1)} \end{pmatrix} \tag{3}$$

---

[2] If we pool every country's data together as a whole, the statistical testing power of estimation is greatly enhanced, but the heterogeneity among countries is neglected. On the other hand, if each country's data is separately estimated, there could be small sample bias in estimation due to inadequate data points. Owing to the difference in the degree of economic development, the relationship between energy consumption and economic growth may well be different. We classify the data into four categories according to different income levels. As a result, we solve the problem of inadequate data points in each country and partially solve the problem of not tackling the homogeneity when combining all 82 countries.

---

[3] Since the share of capital formation to GDP is a flow variable and capital stock is a stock variable, the use of the share of capital formation to GDP to represent capital stock may seem inappropriate. In reality, capital stock is difficult to estimate, and a proxy variable is needed. Most related literature uses the share of capital formation to GDP to represent capital stock (see Ram, 1986). We thank greatly one of the referees for pointing out this problem.

where rows correspond to the first-difference equation (Eq. (2)) for periods $t=3, 4,..., T$ for individual i, which exploit the moment conditions

$$E[Z_i' \, \Delta v_i] = 0 \quad \text{for } i = 1, 2..., N, \tag{4}$$

where $\Delta v_i = (\Delta v_{i3}, \Delta v_{i4},..., \Delta v_{iT})'$. In general, the asymptotically efficient GMM estimation based on this set of moment conditions minimizes the criterion.

$$J_N = \left(\frac{1}{N}\sum_{i=1}^{N} \Delta v_i' Z_i\right) W_N \left(\frac{1}{N}\sum_{i=1}^{N} Z_i' \Delta v_i\right). \tag{5}$$

Using the weight matrix

$$W_N = \left[\frac{1}{N}\sum_{i=1}^{N} \left(Z_i' \, \widehat{\Delta v_i} \, \widehat{\Delta v_i}' Z_i\right)\right]^{-1},$$

where the $\widehat{\Delta v_i}$ are consistent estimates of the first-differenced residuals obtained from a preliminary consistent estimator. Hence, this is known as a two-step GMM estimator. Under the assumption of homoskedasticity v_{it}, the particular structure of the first-differenced model implies that an asymptotically equivalent GMM estimator can be obtained in one-step, using instead the weight matrix

$$W_{1N} = \left[\frac{1}{N}\sum_{i=1}^{N} (Z_i' H Z_i)\right]^{-1},$$

where H is a $(T-2)$ square matrix with 2's on the main diagonal, -1's on the first off-diagonals and zeros elsewhere. Notice that W_{1N} does not depend on any estimated parameters.[4]

As to the use of the one-step or two-step estimator, Bond (2002) mentioned that "In fact, a lot of applied work using these GMM estimators has focused on results for the one-step estimator rather than the two-step estimator. This is partly because simulation studies have suggested very modest efficiency gains from using the two-step version, even in the presence of considerable heteroskedasticity (see Arellano and Bond, 1991; Blundell and Bond, 1998; Blundell et al., 2000), but more importantly because the dependence of the two-step matrix on estimated parameters makes the usual asymptotic distribution approximations less reliable for the two-step estimator (p.147)". For this reason, in our estimation, the robust one-step estimator is employed.[5]

Ever since Nelson and Plosser (1982) pointed out the unit root problem in aggregate time series data, the procedure of a unit root test has become one of the necessary procedures in econometric estimation. Bound, Jaeger and Baker (1995) stated that "When the individual series have near unit root properties, the instruments available for the equations in first-difference are likely to be weak. Instrument variable estimator can be subject to serious finite sample biases where the instruments used are weak".

To solve the problem of estimating the first-difference equation, the use of an instrument variable in level form is non-stationary and, therefore, is a weak instrument. Blundell and Bond (1998) suggested the use of the system GMM (GMM-SYS) model by Arellano and Bover (1995). In other words, lagged difference instead of the level form is used as possible instruments to solve the statistical problem of unit root or near unit root. Their simulation results indicate that when the coefficient on the lagged dependent variable is close to 1, the efficiency of using the GMM-SYS estimator is greatly improved.[6] The estimation of the GMM-SYS is to stack another instrument variable of the first difference to the original level instrument variable matrix (Eq. (3)) as follows:

$$Z_i^+ = \begin{pmatrix} Z_i & 0 & 0 & \cdots & 0 \\ 0 & \Delta y_{i2} & 0 & \cdots & 0 \\ 0 & 0 & \Delta y_{i3} & \cdots & 0 \\ \cdot & \cdot & \cdot & & \cdot \\ 0 & 0 & 0 & \cdots & \Delta y_{i(T-1)} \end{pmatrix}, \tag{6}$$

where Z_i is defined as in Eq. (3). The computation of the one-step or two-step GMM-SYS is as earlier shown. The only difference is the substitution of Z_i^+ for Z_i in the instrument variable matrix.

Since the coefficient of lagged dependent variable from yearly macrodata is close to 1, the robust one-step GMM-SYS of Blundell and Bond (1998) is used to estimate the relation in Eq. (3) and test the Granger causality between energy consumption and economic growth.

The kilograms (kg) of oil equivalent per capita are used to represent energy consumption. The data are obtained from the Energy Balance CD published by the International Energy Agency (IEA). The real GDP in terms of U.S. dollars based on the 2000 price index is used to represent income data. In addition, other control variables such as liy(gross capital formation as % of GDP), lf(population), and lp(GDP deflator) are all collected from the World Development Indicators (WDI) of the World Bank database. The data span 32 years from 1971 to 2002, including 82 countries from the poorest country (Congo) based on GNI of 2000 to the richest country (Luxembourg). Among these 82 countries, 19 countries are classified by the World Bank as low income countries, 22 countries are lower middle income group, 15 countries are upper middle income group, and 26 countries are high income countries.[7]

3. Analysis and discussion of results

Before Eq. (3) can be estimated, an optimal lag period p needs to be determined. There is a certain standard procedure such as AIC or SBC to determine the optimal lag period under the VAR model in time series data. However, the panel VAR model does not have a similar procedure to identify the

---

[4] This portion of the discussion on the GMM methodology is mainly based on Bond (2002).

[5] For the applied work using the one-step GMM estimator, see the related literature by Arai et al. (2004), Yao (2006) and Falk (2006) etc.

[6] In addition to deal with the weak instrument problem, the GMM-SYS can also handle the problems related to measurement error and time-invariant country specific effect (see Felbermayr, 2005).

[7] Appendix Table 1 displays the names of 82 countries, income groups classified from the GNI in 2000 by the World Bank.

optimal lag. So far, two approaches in the literature are available to select the optimal lag. First, the likelihood ratio test is used to select the optimal lag (Holtz-Eakin et al., 1998). Second, the *mj* statistics suggested by Arellano and Bond (1991) (where *j* is the order of autocorrelation) is employed to identify the most appropriate optimal lag. That is, under different lag periods, the selection is based on the existence of no serial correlation in the panel VAR residuals.[8] The *mj* statistic is a standardized residual autocovariance, which are asymptotically $N(0,1)$ under the null of no autocorrelation. If the disturbance v_i is not serially correlated, there should be evidence of significant and negative first order $(j=1)$ serial correlation in the difference form (i.e. $\hat{v}_{i,t} - \hat{v}_{i,t-1}$), and no evidence of second order $(j=2)$ serial correlation in the differenced residuals (Doornik et al., 2006). The advantage of using *mj* statistic for an optimal lag is that the panel VAR model will also be free of misspecification from serial correlation with the optimal lag. Table 2 displays the estimated results in four different income groups from the panel VAR model using one-step GMM-SYS.[9]

The m1 and m2 of Table 2 display the first order and second order serial uncorrelated test results from the panel VAR residuals. The selection of the three lag periods is needed for the 82 countries as a whole (fifth column) and the high income group (fourth column) in order to rid the serial correlation of panel VAR residuals. For the lower middle income group (second column), the use of VAR(1) is sufficient to satisfy the assumption. Yet, the low income and the upper middle income group countries (first and third columns) require a lag of 2 periods for the economic growth equation and a delay of 1 period for the energy consumption equation in order to satisfy the assumption. Further, in all models, the Sargan statistics indicate that we cannot reject the null hypothesis, H_o: over-identifying restrictions are valid. It is apparent that the instrument variables used in the GMM-SYS estimation in our model are appropriate.

Looking at the estimated results of the panel data from 82 countries as a whole (fifth column), the test results of Granger causality indicate that we reject the null hypothesis of $\Delta lec_{i,t-j} \nrightarrow$ (does not Granger cause) $\Delta ly_{i,t}$ at the 5% significance level and also reject $\Delta y_{i,t-j} \nrightarrow \Delta lec_{i,t}$ at the 1% significance level. That is, the estimated dynamic panel data (DPD) from the GMM-SYS show that there is a feedback relationship between energy consumption and economic growth. Further analysis reveals a positive feedback relationship. In other words, an increase in energy consumption may bring about economic growth and an increase in economic growth may also bring about further increase in energy consumption. As is expected, most other explanatory variables under this 82-country category do not have significant explanatory power. The only exception is that

there is a negative causal relationship between capital stock variable and economic growth.[10]

The advantage of using the panel data approach is the increase in data points and hence the power of statistical estimation. The disadvantage is that all 82 countries, as a whole, are treated as a unit, and we neglect the difference among countries. In past research using time series data for individual countries, only a few researchers discovered a bi-directional causal relationship. Most of these bi-directional relationships occurred in developing countries (Pakistan, as indicated by Masih and Masih, 1996; Taiwan and South Korea, by Masih and Masih, 1997; the Philippines and Thailand, by Asafu-Adjaye, 2000; Argentina, by Soytas and Sari, 2003; and India, by Paul and Bhattacharya, 2004). The U.S. is the only industrialized nation exhibiting the bi-directional relationship (Lee, 2006). As indicated in the introduction, the deficiency of the time series approach is the small sample size for statistical analysis, and the estimated results are not as reliable.

To investigate the difference among country blocks with sufficient sample size, groups of countries are classified based on their income characteristics (as a proxy for economic development). The panel data approach is then used to test the causal relationship between energy consumption and economic growth under different characteristics of countries in groups. The national income (representing the living standard in a country) is often used in the literature as a way to classify panel data into different groups. For example, De Gregorio and Guidotti (1995) classify data into three groups based on different income levels to investigate the correlation between banking development and economic growth. In addition, based on Fig. 1, the relationship between the energy consumption growth (average) and the economic growth (average) under different income groups from 1972 to 2002 is clearly different. There seems to be a strong positive relationship in the low income and middle income groups, but no such relationship appears in the high income group. Given the correlation coefficients between energy consumption growth and economic growth from the low to high income groups are calculated to be 0.7524, 0.6791, 0.5401, and 0.1050 respectively, the relationship between the two tends to decrease as income increases. If the data are not classified into four income groups, the correlation coefficient will be 0.5072, and the weak relationship between the two in the high income group cannot have been detected. Finally, an Environmental Kuznet Curve (EKC) indicates that there is an inverted U relation between the level of pollution and the level of income. Since the source of pollution is from energy consumption, it is reasonable to investigate energy consumption based on income levels.

According to the World Bank definition of GNI (Gross National Income) in 2000, these 82 countries are classified as low income (19 countries), lower middle (22 countries), upper

---

[8] In general, an optimum lag period is determined by rendering the panel VAR residual free of serial correlation. Therefore, the optimal lag is selected until no serial correlation in residual is obtained (Arellano, 2003, p.123).

[9] All of our estimations in this paper employ the DPD package under Ox. (see Doornik et al., 2006 for the use of the package). We thank the free package of DPD under Ox provided by the web-site, www.doornik.com/download.htm.

---

[10] Since we use the "first-difference" approach to solve the existence of an individual effect (η_i) problem in the model, the capital stock variable represents a change in capital stock. Also, the VAR model does not take into consideration the change in capital stock in the period (t). As a result, there may not be a positive relationship between economic growth and the change in the capital stock of a lagged period.

Table 2 – The estimated results from the dynamic panel GMM-SYS (four different groups per 2000 GNI)

Independent	Low income (1)		Lower middle (2)		Upper middle (3)		High income (4)		World (5)	
	$\Delta ly_{i,t}$	$\Delta lec_{i,t}$	$\Delta ly_{i,t}$	$\Delta lec_{i,t}$	$\Delta ly_{i,t}$	$\Delta lec_{i,t}$	$\Delta ly_{i,t}$	$\Delta lec_{i,t}$	$\Delta ly_{i,t}$	$\Delta lec_{i,t}$
$\Delta lec_{i,t-1}$	0.0091 (0.17)	0.9589* (73.0)	−0.0033 (−0.10)	0.8932* (60.6)	0.0280 (0.69)	0.8929* (41.8)	0.0103 (0.48)	0.7593* (11.4)	0.0372*** (1.90)	0.8863* (12.4)
$\Delta lec_{i,t-2}$	−0.0078 (−0.16)				−0.0360 (−0.98)		−0.0291 (−1.11)	−0.1376 (0.11)	−0.0567 (−1.50)	−0.0187 (−0.18)
$\Delta lec_{i,t-3}$							0.0076 (0.27)	0.0198 (0.38)	−0.0150 (−0.37)	0.0076 (0.86)
$\Delta ly_{i,t-1}$	1.1220* (14.1)	0.0020 (0.23)	0.9531* (24.5)	0.0358* (2.53)	1.3884* (21.8)	0.0852* (6.34)	1.4185* (23.5)	0.4378* (4.07)	1.3484* (27.0)	0.1790* (3.63)
$\Delta ly_{i,t-2}$	−0.1560** (−1.97)				−0.4195* (−7.19)		−0.5550* (−6.14)	−0.4832** (−2.23)	−0.3255* (−4.14)	−0.1138*** (−1.83)
$\Delta ly_{i,t-3}$							0.1372 (1.36)	0.0739 (0.48)	−0.0150 (−0.37)	0.0071 (0.17)
$\Delta liy_{i,t-1}$	0.1879* (3.26)	0.1139* (3.26)	0.0862* (2.53)	0.1703* (4.46)	−0.2625* (−4.66)	0.1208 (1.10)	−0.4294* (−5.57)	−0.2558 (−1.45)	−0.1804*** (−1.85)	−0.0626 (−0.79)
$\Delta liy_{i,t-2}$	−0.0600 (−0.94)				0.1809 (3.47)		0.3212* (2.51)	0.2729 (1.40)	0.1513*** (1.78)	0.1057 (1.15)
$\Delta liy_{i,t-3}$							−0.1057 (−1.07)	−0.1203 (−0.77)	−0.0350 (−0.72)	−0.0168 (−0.27)
$\Delta lp_{i,t-1}$	−0.0134** (−2.21)	0.0002 (0.48)	0.0004 (0.51)	0.0005 (0.90)	0.0108 (1.50)	0.0025** (2.04)	−0.0559 (−1.30)	−0.0965* (−3.04)	0.0006 (0.10)	−0.0075 (−0.87)
$\Delta lp_{i,t-2}$	0.0133** (2.19)				−0.0099 (−1.44)		0.0925 (1.25)	0.2154* (3.85)	−0.0007 (−0.06)	0.0130 (0.87)
$\Delta lp_{i,t-3}$							−0.0374 (−1.17)	−0.1095* (−4.57)	0.0009 (0.14)	−0.0031 (−0.43)
$\Delta lf_{i,t-1}$	−0.3278 (−0.66)	0.0054* (1.76)	0.0030 (0.33)	0.0045 (0.90)	−0.0174 (−0.09)	−0.0108 (1.10)	0.3572 (1.24)	−0.7027*** (−1.89)	0.0500 (0.19)	−0.0116 (0.03)
$\Delta lf_{i,t-2}$	0.3356 (0.67)				0.0151 (0.08)		−0.4874 (−1.03)	1.3807* (2.68)	−0.2485 (−0.56)	−0.6618 (−0.69)
$\Delta lf_{i,t-3}$							0.1219 (0.53)	−0.6779* (−2.60)	0.1979 (0.74)	0.6512 (1.16)
N	19	19	22	22	15	15	26	26	82	82
NT	551	551	660	660	435	435	754	754	2296	2296
Sargan test p-value	0.79	0.16	1.00	1.00	1.00	1.00	1.00	1.00	1.00	1.00
m1	−3.48*	−2.85*	−2.59*	−4.05*	−3.35*	−1.82***	−4.04*	−3.32*	−6.03*	−5.19*
m2	−1.39	−1.25	−0.52	−0.97	−0.25	0.86	−0.37	−0.75	−1.19	1.50
$\Delta lec_{i,t-j} \nrightarrow \Delta ly_{i,t}$	0.03 [0.99]		0.01 [0.92]		3.06 [0.22]		4.63 [0.20]		8.26** [0.04]	
$\Delta ly_{t,t-j} \nrightarrow \Delta lec_{t,-i}$		0.05 [0.82]		6.42* [0.01]		40.25* [0.00]		22.10* [0.00]		69.71* [0.00]

Note: N = no of countries; NT = no of observations; Sargan statistics are used to test H₀: over-identifying restriction are valid; number inside ( ) are t statistics; number inside [ ] are p-values; Δ = first difference; ly, lec, liy, lp and lf represent log of per capital income, log of energy consumption, log of capital formation to output ratio, log of price level and log of labor force (population), respectively. m1 and m2 denote the statistics of serial uncorrelated residuals of the first and second order in the testing of the panel model; ⇸ represents "does not Granger cause"; *, **, and *** represent respectively 1%, 5% and 10% significance levels.

middle (15 countries), and high income (26 countries) groups.[11] The data span from 1971 to 2002 and year 2000 is used as a base year for both classification for income groups and per capita real GDP computation. The detailed grouping of countries under different GNI levels is shown in Appendix

Table 1.[12] The estimated results from the GMM-SYS of panel data in four income groups are shown in Table 2.

For the low income group countries (column 1 in Table 2), the Granger causality test indicates that an increase in energy consumption does not lead economic growth and an increase in economic growth also does not bring about increase in energy consumption. The energy policy in this income group

[11] Following the World Bank definition for classification based on GNI in 2000, countries are classified as low income if GNI is lower than $826, as lower middle income countries if $826≤GNI≤$3255, as upper middle income countries if $3256≤GNI≤$10,065, and as high income countries if GNI is greater than $10,065.

[12] The standard for grouping based on GNI may have changed over the years, but very few countries have moved from one group to the other.

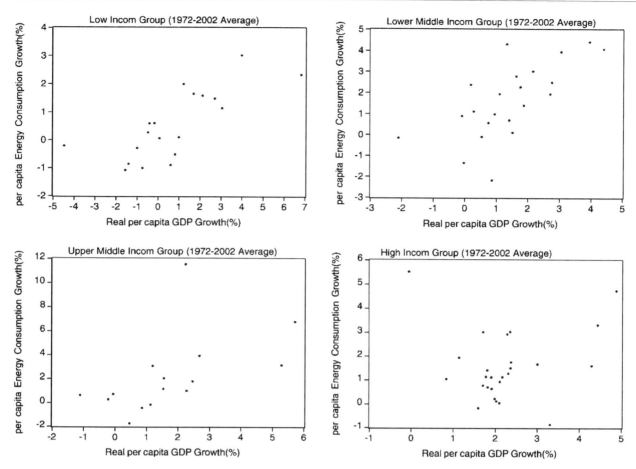

Fig. 1 – Per capita energy consumption growth vs. per capita GDP growth.

is difficult to determine because energy consumption may not bring about economic growth, according to our empirical evidence. For the lower middle income group countries (column 2 of Table 2), the energy consumption does not Granger cause economic growth, but an increase in economic growth may bring about increase in energy consumption at the 1% significance level. For the upper middle income group, we have a causal relationship similar to the lower middle income group. That is, an increase in economic growth will bring about an increase in energy consumption, but not vice versa. The only difference between those two groups is that it is statistically more significant for the upper middle income group than is the lower middle income group. For the middle income group countries (includes the lower middle income and upper middle income groups), it is appropriate to take a more aggressive energy conservation policy. Finally, for the high income group countries, the Granger causality test indicates that income change leads energy consumption change. The estimated results of the energy consumption equation further reveals that the overall effect of economic growth on energy consumption is negative (the coefficient for one-period lag is 0.4378 and for two-period is −0.4832). In other words, the economic growth may bring about a decrease in energy consumption. It seems to imply that the high income group countries have undertaken an energy conservation policy. After we group data into four categories on the basis of

income levels in these 82 countries, we do not find the same causal relationship in each group as we found when 82 countries are pooled as a whole. In the 82-country data as a whole, energy consumption leads economic growth positively. It is apparent that the classification of countries into different income groups is conducive to a better and finer understanding of causal relationship between energy consumption and economic growth under different income levels.

As indicated previously, energy use may bring about both economic growth and the externality of environmental pollution. The question is whether energy consumption can result in greater benefits in economic growth relative to the cost of environmental pollution. This simple benefit–cost relation is implied in the causal relationship between energy consumption and income. When energy consumption can bring about economic growth, it suggests that the benefit of energy use to the economy is greater than the cost of environmental damage. Conversely, if economic growth leads energy consumption positively, it may suggest the cost of using energy is greater than the benefit it brings. After we classify these 82-country data into four different income groups, we do not find any group where energy consumption leads economic growth as a uni-directional causal relationship. Conversely, the economic growth leads energy consumption in the middle income groups (lower middle income group and upper middle income group). For those two middle

143

Table 3 – Mean and standard deviation of some energy related data characteristics in four different income groups

Group	ec/y	Δy	Δec	CO_2	ind/y	$\Delta ec / \Delta y$
a. 1971–2002						
Low	0.3199	0.1478	0.2125	0.2991	25.4005	1.44
	(0.2036)	(5.1368)	(4.2003)	(0.2108)	(8.9467)	
Lower Middle	0.1969	1.6745	1.5582	0.4816	32.7056	0.93
	(0.0986)	(4.8444)	(5.6429)	(0.3514)	(8.4108)	
Upper Middle	0.2200	1.9182	2.8025	0.6002	39.2199	1.46
	(0.1206)	(5.2602)	(8.7131)	(0.4499)	(10.9005)	
High	0.2067	2.3463	1.4914	0.4834	33.1352	0.64
	(0.0733)	(2.6544)	(4.9375)	(0.2027)	(3.7049)	
b. 1971–1980						
Low	0.3294	0.5915	0.6701	0.2919	24.2093	1.13
	(0.2165)	(6.1130)	(3.7839)	(0.2335)	(9.1846)	
Lower Middle	0.2086	2.9430	2.7936	0.4934	32.2813	0.95
	(0.1662)	(5.9702)	(6.6326)	(0.4753)	(9.8380)	
Upper Middle	0.1843	3.6108	4.5266	0.5850	42.7159	1.25
	(0.0787)	(5.4993)	(9.6935)	(0.3800)	(14.6010)	
High	0.2328	2.8495	1.9795	0.6068	37.0352	0.69
	(0.0943)	(3.1386)	(6.1689)	(0.3204)	(3.8466)	
c. 1981–2002						
Low	0.3172	−0.0337	0.0252	0.3014	26.0839	−0.75
	(0.2060)	(4.6753)	(4.3497)	(0.2078)	(9.5750)	
Lower Middle	0.1938	1.1556	1.0529	0.4780	32.8985	0.91
	(0.0852)	(4.1978)	(5.1061)	(0.3242)	(7.9429)	
Upper Middle	0.2297	1.2257	2.0972	0.6053	37.9504	1.71
	(0.1381)	(5.0050)	(8.1903)	(0.4809)	(9.3390)	
High	0.1996	2.1405	1.2917	0.4483	31.5493	0.60
	(0.0698)	(2.4022)	(4.3247)	(0.1747)	(3.7918)	

Notes: ec/y = energy use per PPP GDP (kg of oil equivalent per constant 2000 PPP $); Δy = per capita real GDP growth (%); Δec = energy use (kg of oil equivalent per capita) growth (%); CO_2 = CO_2 emissions (kg per 2000 PPP $ of GDP); ind/y = % of value added in industry to GDP; numbers inside ( ) are standard deviations and numbers above ( ) are means; $\Delta ec / \Delta y$ represents the % increase in energy consumption resulting from a 1% increase in GDP.

groups, an increase in economic growth may enhance energy consumption and may bring about the externality of environmental damage without the benefit of economic growth from energy consumption. For the high income group countries, energy consumption does not bring about economic growth. However, as income increases, they begin to pay attention to the possible cost of environmental pollution and try to reduce energy consumption. For the low income countries, there is no evidence that energy consumption may bring about economic growth, or that an increase in income may bring about energy consumption. When energy consumption cannot bring about economic growth, the implication is that those countries should adopt a conservation policy to avoid damage to the environment and a waste of resources.

To understand our contributions, it is necessary to compare our econometric models and empirical results with the most recent publication using DPD by Lee and Chang (2007). There are many differences between our analyses notwithstanding the fact that they also applied the DPD model to investigate the causal relationship between energy consumption and economic growth. First, we include more countries (82 countries vs. 40 countries). Our data are classified into four income group countries rather than two, as in their study. We also take into consideration the measurement error, weak instrument and

time-invariant country specific effect from the GMM-SYS instead of the GMM model. Our VAR model includes other control variables, while Lee and Chang employ the bivariate model. We use EKC and related energy data to further investigate the causality results. Lee and Chang (2007) do not include such discussion. Our discovery of economic growth leading positively the energy consumption in the middle income group countries is the same as those of developing countries by Lee and Chang (2007). Lee and Chang indicated that, for those 18 developed countries, economic growth leads positively the energy consumption and energy consumption leads negatively the economic growth (a bi-directional feedback relation). Our results of economic growth leading energy consumption negatively for the developed countries (high income group countries) seems to be more consistent with the recent energy policy adopted in those developed countries.

Our estimated results can further be explained by the concept of EKC. EKC is an "inverted U" relationship between the level of economic development and the pollution level. In the low income countries, there are not many industrial activities to pollute the environment. As the economy improves, pollution gradually increases. Furthermore, as the industrial potential expands, it offers location advantages for high-pollution industries. Sooner or later, the pollution problem becomes a major concern which calls for remedial actions. Generally speaking, as the income increases beyond some threshold, there is a tendency towards producing low-pollution products. More resources are devoted to environmental protection. Therefore, we expect the causal relationship that economic growth leads energy consumption negatively. As a result, pollution falls as income grows. The pioneering work of Grossman and Krueger (1995) sets the path to investigate the existence of EKC in which they discovered the highest EKC point at $8000 per capita income (1985 price level). This corresponds to the income level of Mexico and Malaysia in 1994. Although the main purpose is not to investigate the EKC relation, our results, using the panel data analysis of four income groups, are quite consistent with those of the EKC prediction.[13] For the middle income group countries (lower middle income group and upper middle income group), we discovered that economic growth leads positively energy consumption. The implication is that, as income begins to increases, a negative externality (e.g., pollution) of energy consumption starts to increase. Once a country achieves high income group status, an increase in income may reduce the negative externality of energy consumption as is shown by the EKC relation. For the low income group countries in which basic industry and transportation systems are insufficient, and energy use is low, these countries are unable to generate much output (income). It is no wonder there exists no causal relationship between energy consumption and economic growth.

In order to analyze further the Granger causality in each income group, the pollution-related calculations need to be included in the statistical analysis. In each of these four income group countries, we collect additional information, such as average CO_2 emissions (pollution level) per $1 real GDP, the share of value added in industry to GDP (ind/y, the weight of industrial production), the share of energy used per $1 real GDP (ec/y, the efficiency of energy use), average per capita real GDP growth (Δy), and average growth of energy use (Δec).[14] These calculations are shown in Table 3.

As seen in the CO_2 column of Table 3a when all the data (1972–2002) are considered, the most serious pollution appears in the upper middle income group countries. The least pollution happens to be in the low income group countries, followed by the lower middle and high income group. These results indicate that there is indeed an EKC relation. By separating the data into during-energy crisis (1972–1980) and post energy crisis (1981–2002) periods, we found that the order of pollution levels in the four groups is slightly different. During the energy crisis, the most serious polluters of CO_2 are

in the high income group, followed by the upper middle income group, the lower middle income group, and the low income group. After the energy crisis, the most serious polluters of CO_2 are in the upper middle income group, followed by the lower middle income group, and high and low income groups. Viewed from the prospect of pollution, the EKC relation in fact appears in the post energy crisis period. In other words, after the energy crisis, the high income group countries made great strides to reduce the pollution. Yet the upper middle income group and low income group countries increased rather than decreased their emission of CO_2. The lower middle income group countries tended to decrease slightly the pollution of CO_2 after the energy crisis.

The effort made by the high income group countries to reduce pollution and to increase the efficiency of energy use can be seen from the column of ec/y in Table 3. The variable for energy efficiency (ec/y) represents the required unit of energy use per $1 increase in GDP. It is difficult to see the trend of relative energy efficiency in four different income groups when the whole period (1972–2002 average) is used. The most efficient use of energy is in the lower middle income group countries followed by the high income group, the upper middle income group, and finally the low income group countries. If the data are delineated into energy crisis period and post energy crisis period, the ec/y ratio in the high income group decreases from 0.2328 of energy crisis to 0.1996 of post energy crisis (the most among those four groups). That is, the increase in the efficiency of energy use is the most for the high income group. For the lower middle income group and the low income group countries, there is a small increase in the efficiency of energy use. The upper middle income group is the only group with efficiency getting worse not better (from the energy crisis of 0.1843 to the post crisis of 0.2297). As was pointed out by Cleveland et al. (2000), in some industrialized nations, the decrease in the ec/y ratio comes from the change in energy mix. That is, "The change from coal to petroleum and petroleum to primary electricity is associated with a general decline in the ec/y ratio". As the high income group countries improve the weight of using pollution free electrical energy, the decrease in the release of CO_2 is expected.

For the high income group countries, the improvement in the release of CO_2 and ec/y ratio confirms the causal relationship that economic growth leads energy consumption negatively. This discovery further provides the evidence that the $\Delta ec/\Delta y$ ratio (the required % increase in energy consumption resulting from a 1% increase in economic growth) changes from the energy crisis of 0.7% to the post crisis of 0.6%. It is manifest that, as the income increases in the high income group, energy use tends to diminish (the efficiency of energy use increases) and the release of pollution (CO_2) also tends to decrease. It is not surprising then that economic growth leads energy consumption growth negatively in the high income group countries.

Similarly, in the upper middle income group, there is a unidirectional causality running from economic growth positively to energy consumption. The release of CO_2 pollution after the energy crisis tends to increase and the efficiency of energy use tends to decrease. A 1% increase in income requires more than 1% of energy consumption. The release of CO_2 in this income group is the highest among the four

---

[13] For the survey literature of EKC, see Dinda (2004). There is an abundant literature on empirical studies of EKC such as Torras and Boyce (1998), Agras and Chapman (1999), and Dinda et al. (2000).

[14] All the energy-related data are collected from the Energy Balance CD published by the International Energy Agency (IEA). Besides the computation of means for the period between 1972 and 2002, means for both during and after energy crisis are also calculated to facilitate comparisons of energy use in all four different income groups.

groups and the ec/y ratio is the second highest (second only to the low income group) due to both high ind/y (industrial production/GDP, among the highest) and $\Delta ec/\Delta y$ (energy increase rate is greater than 1) ratios. Upper middle income group countries have relatively low production cost and are eager to raise the income level. They take advantage of low labor costs to encourage construction of factories from foreign investments and production of goods and services. Since standards are lax regarding environmental protection and related environmental regulations, relatively more pollution generating energy sources (e.g., coal and petroleum) are used. As a result, greater release of CO_2 and inefficient use of energy are expected. The economic reality in this income group seems to be in agreement with the causality from our panel VAR results that economic growth leads energy consumption positively.

For the lower middle income group, we discover that, like the upper middle income group, economic growth leads energy consumption positively as a uni-directional causality. Table 3 shows that the release of CO_2 is slightly lower than the high income group in the entire period from 1972 to 2002. A 1% increase in income requires slightly less than 1% (about 0.93%) in energy consumption, which is less than that of the upper middle income group but higher than that of the high income group. The release of CO_2 in the lower middle income group is only behind the upper middle income group after the energy crisis. The ind/y ratio is also behind that of the upper middle income group after the energy crisis. Like the upper middle income group, the same argument may apply to the lower middle income group. That is, with low production cost, this group of countries encourages capital inflow (factories in particular) to produce goods and services. It is one of the reasons that they produce relatively larger amounts of pollution (relatively high among these four groups). Compared to the upper income group, this income group releases lower amounts of CO_2 and has a higher efficiency of energy use. Our empirical results demonstrate the causality that, like the upper middle income group, economic growth leads energy consumption positively.

Finally, for the low income group countries, a 1% increase in economic growth requires more than a 1% increase in energy use. The efficiency of energy use (ec/y) is the highest and the industrial production ratio (ind/y), and the release of CO_2 are the lowest among these four groups. As such, our empirical results indicate no causal relationship between economic growth and energy consumption.

4. Policy Implications

The investigation of the causal relationship between energy consumption and income has important policy implications. When energy consumption leads income positively, it suggests that the benefit of energy use is greater than the externality cost of energy use. Conversely, if an increase in income brings about an increase in energy consumption, the externality of energy use (e.g., pollution) will set back economic growth. Under this circumstance, a conservation policy is necessary. The importance of these policy implications is evident by the size of the literature. Some focused on an individual country while others

concentrated on certain developing countries or developed countries. Because of insufficient observations in annual time series data, the power of statistical tests is suspect. On the other hand, when the panel data as a whole is used, the heterogeneity among countries will be neglected. In order to avoid the paucity of time series data and the "one size for all" homogeneity problem among countries, the data are grouped into four categories in those countries according to the World Bank definitions: low income group, lower middle income group, upper middle income group, and high income group. We employ the system GMM (GMM-SYS) model suggested by Blundell and Bond (1998) to estimate the correlation of the panel VAR taking into consideration the problem of correlation between the lagged dependent variable and residuals, and the near unit root of coefficients on lagged dependent variables. The result is interesting in terms of policy implication.

The policy implications derived from this study indicate that we need to take into consideration the degree of economic growth in each country when energy consumption policy is formulated. It is evident that global warming is mostly caused by the increase in CO_2 emission in the human consumption of fossil fuels. Our research also reveals that the countries with a greater weight of industrial production to GDP (the middle income group) tend to have a larger volume of CO_2 release. To those countries, a more conservative energy policy should be pursued. Cleaner energy sources should be used to replace fossil fuels. For the high income group countries, our empirical results indicate that energy consumption tends to decrease as GDP increases. Since global warming is becoming more serious, replacement of fossil fuels and more efficient energy use are needed to minimize the CO_2 emission. Those high income countries have greater resources and more advanced technology that enable them to do more to lessen global warming. Finally, in the low income countries, we find that energy consumption does not lead economic growth and hence substantial energy consumption is not likely to bring about significant economic growth. Instead, it will increase CO_2 emission. It is very important for those low income countries in implementing appropriate energy policy to promote economic growth. A one-size-for-all energy policy is not appropriate for it fails to implement correct policies for different income group countries.

5. Concluding remarks

We use the panel data of 82 countries from 1972 to 2002 and classify the data into income groups based on the World Bank definitions. In order to have uncorrelated residual series, we select the optimal lag in each income group from the panel VAR along with the GMM-SYS model. Using data for all countries as a whole, we discover that there is a bi-directional positive feedback relationship between economic growth and energy consumption. After the data are classified into four income groups, the causal relationship in each group is fairly different. For the low income group, there is no causal relationship between economic growth and energy consumption. For the middle income group countries (lower middle income group and upper middle income group), there is a uni-directional positive causal relationship running from economic growth to energy consumption. For the high income group countries, there is a negative uni-directional causal relationship running from economic growth to energy consumption. After we classify the data into four income groups, we do not find any uni-directional causal relationship running from energy consumption to economic growth, which we found when data are lumped into one group. It is apparent that the negative externality from the economic growth more than compensates the benefit from energy use. Therefore, a relatively stronger energy conservation policy should be pursued in all countries.

In order to further investigate the difference in causality in each income group between economic growth and energy consumption, more related data are used for an in-depth analysis. We found that the countries (the middle income groups) with a greater weight of industrial production to GDP (ind/GDP) tend to have a larger volume of CO_2 release. In addition, a 1% increase in economic growth requires closer to or greater than 1% of energy consumption. Those countries usually have the causality that economic growth leads energy consumption positively. As the income increases, the energy consumption will increase. Because of over-use in energy, there will be environmental pollution and the externality from resource use. After the energy crisis, the high income group countries tried to increase the efficiency of energy use and reduce industrial production share (ind/y), so as to reduce the release of CO_2. We discover that in the high income group, economic growth leads energy consumption negatively as a uni-directional causal relationship. Finally, in the low income group countries, the share of industrial production to GDP (ind/y) is low and the release of CO_2 is also low. Thus, there is no causal relationship between economic growth and energy consumption in the low income group countries. Our findings echo the concept of EKC in the literature. As income increases, pollution becomes a serious problem and as countries reach an even higher income level, the pollution begins to decline. Our findings indicate that, in the middle income group countries, economic growth leads energy consumption positively, which is disadvantageous to the environment. When the income is raised to the level of the high income group, economic growth leads energy consumption negatively as is shown in the high income group. It is manifest that, as the income is raised to the high income group level, those countries tend to reduce energy consumption in hopes to minimize the damage to the environment.

Appendix A

Table A1 – Sampled countries in ascending order based on 2000 GNI (82 countries)

Country name	GNI (2000 $)	Income group
Congo, Dem. Rep.	90	L
Nepal	230	L
Nigeria	260	L
Togo	320	L
Zambia	320	L
Ghana	330	L
Kenya	360	L
Bangladesh	390	L
Benin	390	L
Zimbabwe	440	L

Table A1 (continued)

Country name	GNI (2000 $)	Income group
India	450	L
Pakistan	480	L
Senegal	490	L
Haiti	500	L
Congo, Rep.	510	L
Cameroon	570	L
Indonesia	590	L
Cote d'Ivoire	690	L
Nicaragua	740	L
China	840	ML
Sri Lanka	850	ML
Honduras	860	ML
Syria	950	ML
Bolivia	1000	ML
Philippines	1030	ML
Morocco	1180	ML
Ecuador	1330	ML
Egypt, Arab Rep.	1490	ML
Paraguay	1510	ML
Algeria	1580	ML
Guatemala	1700	ML
Thailand	2010	ML
El Salvador	2020	ML
Colombia	2050	ML
Peru	2050	ML
Tunisia	2080	ML
Dominican Republic	2140	ML
Jamaica	2710	ML
Turkey	2980	ML
South Africa	3050	ML
Gabon	3120	ML
Malaysia	3390	MU
Brazil	3650	MU
Costa Rica	3820	MU
Panama	3870	MU
Venezuela	4100	MU
Hungary	4650	MU
Chile	4780	MU
Mexico	5110	MU
Trinidad and Tobago	5220	MU
Uruguay	6120	MU
Oman	6710	MU
Argentina	7490	MU
Saudi Arabia	8110	MU
Malta	9540	MU
Korea, Rep.	9790	MU
Portugal	10,930	H
Greece	11,290	H
New Zealand	13,700	H
Spain	14,790	H
Israel	17,060	H
Australia	20,090	H
Italy	20,160	H
Canada	21,820	H
Singapore	22,890	H
Ireland	23,030	H
France	23,990	H
Belgium	24,890	H
Finland	24,940	H
Germany	25,140	H
Netherlands	25,210	H
United Kingdom	25,410	H
Austria	25,700	H
Hong Kong, China	26,820	H
Sweden	28,650	H

Table A1 (continued)

Country name	GNI (2000 $)	Income group
Iceland	29,980	H
Denmark	31,460	H
United States	34,400	H
Japan	35,280	H
Norway	35,660	H
Switzerland	40,160	H
Luxembourg	43,550	H

Note: GNI = Gross National Income; L = Low Income (GNI < $826); ML = Lower Middle Income ($826 ≤ GNI ≤ $3255); MU = Upper Middle Income ($3256 ≤ GNI ≤ 0,065); and H = High Income (GNI > 0,065).

REFERENCES

Agras, J., Chapman, D., 1999. A dynamic approach to the Environmental Kuznets Curve hypothesis. Ecological Economics 28 (2), 267–277.

Akarca, A.T., Long, T.V., 1979. Energy and employment: a time series analysis of the causal relationship. Resources and Energy 2, 151–162.

Akarca, A.T., Long, T.V., 1980. On the relationship between energy and GDP: a re-examination. Journal of Energy Development 5, 326–331.

Alessie, R., Hochguertel, S., Van Soest, A., 2004. Ownership of stocks and mutual funds: a panel data analysis. Review of Economics and Statistics 86 (3), 783–796.

Altinay, G., Karagol, E., 2004. Structural break, unit root, and the causality between energy consumption and GDP In Turkey. Energy Economics 26, 985–994.

Arai, M., Kinnwall, M., Thoursie, P.S., 2004. Cyclical and Causal pattern of inflation and GDP growth. Applied Economics 36, 1705–1715.

Arellano, M., 2003. Panel Data Econometrics. Oxford University Press, New York.

Arellano, M., Bond, S., 1991. Some test of specification for panel data: Monte Carlo evidence and an application to employment equations. Review of Economic Studies 58, 277–297.

Arellano, M., Bover, O., 1995. Another look at the instrumental-variable estimation of error components models. Journal of Econometrics 68, 29–52.

Asafu-Adjaye, J., 2000. The relationship between energy consumption, energy prices and economic growth: time series evidence from Asian developing countries. Energy Economics 22, 615–625.

Baltagi, B.H., Bratberg, E., Holmas, T.H., 2005. A panel data study of physicians labor supply: the case of Norway. Health Economics 14, 1035–1045.

Blundell, R., Bond, S., 1998. Initial conditions and moment restrictions in dynamic panel data models. Journal of Econometrics 87 (1), 115–143.

Blundell, R.W., Bond, S.R., Windmeijer, F., 2000. Estimation in dynamic data models: improving on the performance of the standard GMM estimator. In: Baltagi, B. (Ed.), Nonstationary Panels, Panel Cointegration and Dynamic Panels. Advance in Econometrics, vol. 15. JAI Elsevier Science.

Bond, S.R., 2002. Dynamic panel data models: a guide to micro data methods and practice. Portuguese Economic Journal 1 (2), 141–162.

Bound, J., Jager, D.A., Baker, R.M., 1995. Problem with instrumental variable estimation when the correlation between the instruments and the endogenous explanatory variable is weak. Journal of the American Statistical Association 90, 443–450.

Cheng, B.S., Lai, T.W., 1997. An investigation of co-integration and causality between energy consumption and economic activity in Taiwan. Energy Economics 19, 435–444.

Cleveland, C.J., Kaufmann, R.K., Stern, D.I., 2000. Aggregation and the role of energy in the economy. Ecological Economics 32 (2), 301–317.

De Gregorio, Guidotti, P., 1995. Financial development and economic growth. World Development 23 (3), 434–448.

Dinda, S., 2004. Environmental Kuznets Curve hypothesis: a survey. Ecological Economics 49 (4), 431–455.

Dinda, S., Coondoo, D., Pal, M., 2000. Air Quality and economic growth: an empirical study. Ecological Economics 34 (3), 409–423.

Doornik, J.A., Arellano, M., Bond, S., (2006), "Panel Data Estimation using DPD for Ox," mimeo.

Engle, R., Granger, C.W.J., 1987. Cointegration and error correction: representation, estimation, and testing. Econometrica 55, 257–276.

Erol, U., Yu, E.H., 1987a. Time Series analysis of the causal relationship between U.S. energy and employment. Resources and Energy 9, 75–89.

Erol, U., Yu, E.S.H., 1987b. On the causal relationship between energy and income for industrialized countries. Journal of Energy and Development 13, 113–122.

Erol, U., Yu, E.S.H., 1989. Spectral analysis of the relationship between energy and income for industrialized countries. Resources and Energy 11, 395–412.

Falk, M., 2006. What drives business research and development (R&D) intensity across Organization for Economic Co-operation and Development (OECD) countries? Applied Economics 38, 533–547.

Feeny, S., Harris, M.N., Rogers, M., 2005. A dynamic panel analysis of the profitability of Australian tax entities. Empirical Economics 30, 209–233.

Felbermayr, G.J., 2005. Dynamic panel data evidence on the trade-income relation. Review of World Economics 141 (4), 583–611.

Fiorito, R., Kollintzas, T., 2004. Public goods, merit goods, and the relation between private and government consumption. European Economic Reviews 48 (6), 1367–1398.

Glasure, Y.U., Lee, A.R., 1997. Cointegration, error-correction, and the relationship between GDP and electricity: the case of South Korea and Singapore. Resource and Energy Economics 20, 17–25.

Granger, C.W.J., Newbold, P., 1974. Spurious regressions in econometrics. Journal of Econometrics 2, 111–120.

Grossman, G.M., Krueger, A., 1995. Economic growth and the environment. Quarterly Journal of Economics 10 (2), 353–377.

Holtz-Eakin, D., Newey, W., Rosen, H., 1998. Estimating vector autoregression with panel data. Econometrica 56, 1371–1385.

Hondroyiannis, G., Lolos, S., Papapetrou, E., 2002. Energy consumption and economic growth: assessing the evidence from Greece. Energy Economics 24, 319–336.

Hsiao, C., 1981. Autoregressive modeling and money income causality detection. Journal of Monetary Economics 7, 85–106.

Hwang, D.B.K., Gum, B., 1992. The causal relationship between energy and GDP: the case of Taiwan. Journal of Energy and Development 12, 219–226.

Jumbe, C., 2004. Cointegration and causality between electricity consumption and GDP: empirical evidence from Malawi. Energy Economics 26, 61–68.

Kraft, J., Kraft, A., 1978. On the relationship between energy and GNP. Journal of Energy and Development 3, 401–403.

Lee, C.C., 2005. Energy consumption and GDP in Developing countries: a cointegrated panel analysis. Energy Economics 27, 415–427.

Lee, C.C., 2006. The causality relationship between energy consumption and GDP in G-11 countries revisited. Energy Policy 34, 1086–1093.

Lee, C.C., Chang, C.P., 2007. "Energy Consumption and GDP revisited: a panel analysis of developed and developing countries". Energy Economics 29, 1206–1223.

Masih, A.M.M., Masih, R., 1996. Energy consumption, real income and temporal causality: results from a multi-country study based on cointegration and error-correction modeling techniques. Energy Economics 18, 165–183.

Masih, A.M.M., Masih, R., 1997. On the Temporal causal relationship between energy consumption, real income, and prices: some evidence from Asian-energy dependent NICs based on a multivariate cointegration/vector error-correction approach. Journal of Policy Modeling 19 (4), 417–440.

Nelson, C.R., Plosser, C.I., 1982. Trends and Random walks in macroeconomic time series. Journal of Monetary Economics 10, 139–162.

Oh, W., Lee, K., 2004a. Energy consumption and economic growth in Korea: testing the causality relation. Journal of Policy Modeling 26, 973–981.

Oh, W., Lee, K., 2004b. Causal relationship between energy consumption and GDP revisited: the case of Korea 1970–1999. Energy Economics 26, 51–74.

Paul, S., Bhattacharya, R.N., 2004. Causality between energy consumption and economic growth in India: a note on conflicting results. Energy Economics 26, 977–983.

Ram, R., 1986. Government size and economic growth: a new framework and some evidence from cross-section and time-series data. American Economic Review 76 (1), 191–203.

Soytas, U., Sari, R., 2003. Energy consumption and GDP: causality relationship in G-7 countries and emerging markets. Energy Economics 25, 33–37.

Stern, D.J., 2000. Multivariate cointegration analysis of the role of energy in the U.S. macroeconomy. Energy Economics 22, 267–283.

Toda, H.Y., Yamamoto, T., 1995. Statistical inference in vector autoregressions with possible integrated processes. Journal of Econometrics 66, 225–250.

Torras, M., Boyce, J.K., 1998. Income inequality and pollution: a reassement of the Environmental Kuznets Curve. Ecological Economics 25 (2), 147–160.

Yang, H.Y., 2000. A note on the causal relationship between energy and GDP in Taiwan. Energy Economics 22, 309–317.

Yao, S., 2006. On economic growth, FDI and exports in China. Applied Economics 38, 339–351.

Yildirim, J., Sezgin, S., Öcal, N., 2005. Military expenditure and economic growth in middle eastern countries: a dynamic panel data analysis. Defence and Peace Economics 16 (4), 283–295.

Yu, S.H., Choi, J.Y., 1985. The causal relationship between energy and GNP: an international comparison. Journal of Energy Development 10, 249–272.

Yu, E.S.H., Hwang, B.K., 1984. The relationship between energy and GNP: Further results. Energy Economics 6, 186–190.

Yu, S.H., Jin, J.C., 1992. Cointegration Tests of energy consumption, income and employment. Resources and Energy 14, 259–266.

Yu, E.S.H., Choi, P.C.Y., Choi, J.Y., 1988. The relationship between energy and employment: a re–examination. Energy Systems Policy 11, 287–295.

Economic Modelling 28 (2011) 2416–2423

Contents lists available at ScienceDirect

Economic Modelling

journal homepage: www.elsevier.com/locate/ecmod

Military expenditure and economic growth across different groups: A dynamic panel Granger-causality approach[☆]

Hsin-Chen Chang [a], Bwo-Nung Huang [b,*], Chin Wei Yang [c,d]

[a] *Department of Applied Economics, National Chung Hsing University, Taichung 402, Taiwan*
[b] *Department of Economics & Center for IADF, National Chung Cheng University, Chia-Yi 621, Taiwan*
[c] *Department of Economics, Clarion University of Pennsylvania, Clarion, PA 16214-1232, United States*
[d] *Department of Economics, National Chung Cheng University, Chia-Yi 621, Taiwan*

ARTICLE INFO

Article history:
Accepted 6 June 2011
Available online xxxx

JEL Classification:
H56
C22

Keywords:
Military expenditure
Economic growth
Dynamic panel data
GMM

ABSTRACT

Applying GMM (Arellano and Bond, 1991) to panel data of 90 countries spanning over 1992–2006, this paper explores possible relationships between military expenditure and economic growth. Based on the definitions of income levels by the World Bank – high, middle and low – our results indicate military spending leads negatively economic growth for the panels of low income countries with a marginally significance level of 10%. Of four different regional panels (Africa, Europe, the Middle East–South Asia and Pacific Rim), a negative but stronger (5% significance level) causal relationship from military expenditure to economic growth is found for the Europe and Middle East–South Asia regions.

© 2011 Elsevier B.V. All rights reserved.

1. Introduction

One of the prevalent views concerning arms race at the World Bank or the International Monetary Fund is the opportunity cost of military expenditure: slowdown in output and economic growth (Knight et al., 1996). The underpinnings of their claim lie in the crowding-out effect; that is, scarce resources are siphoned off from productive sectors to military development. Unless such resource transfers can result in profitable commercial applications in the future, we expect to see a negative causal relationship between military expenditure and economic growth for low income or developing economies, or in the region of heightened conflicts.

Literature on defense spending and economic growth dated back to the work by Benoit in 1973. Since then, there have been a plethora of studies written on the subject. Empirical results, nonetheless, have been rather disappointing: the relationship could be positive, negative or independent depending on length of sample data or methodology used. Many of the prior studies employed country-specific or cross-sectional data. Such an analysis is intrinsically static for it ignores the important

properties of time. As such the results may be biased or of limited use. With the recent advancements in panel data econometrics, researchers begin to classify panels based on regions (e.g., EU, the Middle East) or income levels. Note that crowding-out effects are particularly prevalent in developing counties. And some regions are historically war-prone such as the Middle East, South Asia and part of Africa. Consequently, we expect higher military expenditures for these economies, which may very well produce different causal relations. Many of the previous studies applied panel data approach to only one given region except for Knight et al. (1996) whose panel data comprised six regions. Well known in the literature, failure to use multiple-region framework can render conclusions to be of limited value. While the large-scale and multiple-region study by Knight et al. (1996) has the wider policy implications across different regions, it does not take dynamic elements into consideration and thus cannot really address the true dynamic causal relations between military spending and economic growth.

To address the country-specific and time-specific effects, and to allow for different economic developmental stages and varying endowment levels (regional effects), this paper applies a large panel data of 90 countries over 1992–2006 to re-explore the causal relationship between military spending and economic growth. Based on World Bank classifications, 90 countries are categorized into low-income, middle-income and high-income groups. At the same time, they are geographically partitioned into four regions: Africa, Europe, Middle East–South Asia and Pacific Rim (North, Central, South America and East Asia). In addition, we apply the

─────────

[☆] Financial support from the National Science Council (NSC96-2415-H-194-003-MY2) is gratefully acknowledged. We would like to thank two reviewers for detailed comments and suggestions. All remaining errors are our own.
 * Corresponding author.
 E-mail addresses: ecdbnh@ccu.edu.tw (B.-N. Huang), Yang@mail.clarion.edu (C.W. Yang).

dynamic panel data (DPD) model to analyze the true causal relations. A negative Granger causality is found in low-income group, Middle East–South Asia and Europe regions at 10%, 5% and 5% significance levels respectively. The remainder of this paper is organized as follows: Section 2 provides literature review. Section 3 describes data, model and the econometric methods used in the paper. Section 4 discusses the empirical results. A conclusion is given in Section 5.

2. Literature review

The literature on military spending and economic growth dated back to the seminal work by Benoit (1973) in which a positive relationship was found. Benoit's work inaugurated a vast array of studies in the hope of identifying a definitive pattern between the two variables (e.g., Atesoglu and Mueller, 1990; and Biswas, 1993; Faini et al., 1984; Ram, 1986; Smith, 1980). Research methods used in earlier literature largely employed unconditional correlation co-efficients (Benoit, 1973, 1978). Majority of later studies, on the other hand, relied on cross-sectional regressions (Antonakis, 1997; Biswas and Ram, 1986; Cohen et al., 1996; Deger, 1986; Faini et al., 1984; Grobar and Porter, 1989; Heo, 1997; and Lim, 1983). Cross-sectional regression lacks the time series effect. That is, it has little inferential power beyond the sample period studied. In addition, it has problem of heterogeneity and as such normally has low coefficient of determination. To circumvent these problems, our DPD model comprises 90 countries spanning over 1992–2006. To economize space, we limit the literature review of the military spending-economic growth relationship to multi-country models.

Frederiksen and Looney (1982) studied the relationship via a growth equation using, among other variables, military spending and investment to explain economic growth. Countries were divided into less developed countries (LDCs) with limited resources and those with rich resource endowments. The result of the cross-sectional study over the period of 1960–78 suggests that military spending is beneficial to economic growth for the LDCs with rich endowment. For the LDCs with limited resource, there existed no positive relationship between the two variables. Applying the Harrod-Domar growth model to 54 LDCs in the period of 1965–73, results by Lim (1983) pointed out large amount of military spending was harmful to economic growth. Making use of Feder's two-sector growth model (1983), Biswas and Ram (1986) investigated (i) the impact of private sector spending and defense expenditures on economic growth and (ii) externality of defense spending. Partitioning 58 LDCs into low income (17 countries) and lower-middle income (41 countries) groups and dividing sample period into 1960–1970 and 1970–1977 segments in order to consider potential structural break, Biswas and Ram (1986) identified a positive relationship for lower-middle income group over the 1960–1970 period.

Deger (1986) criticized the validity of non-defense growth rates used in Benoit's paper. In addition, Deger (1986) added a third variable, savings, into the military spending and economic growth model. That is, a system of three simultaneous equations with 50 LDCs over the 1965–73 period was estimated and the results showed (i) a positive relationship (coefficient of 0.2564) between military spending and economic growth, (ii) a negative relationship between military spending and savings (coefficient of −0.3939), and (iii) a negative relationship (not a positive one as claimed by Benoit) between military spending and economic growth after taking savings into consideration. Mintz and Stevenson (1995) attributed the divergent results to (i) lacking a consistent theory regarding military spending and economic growth, (ii) inappropriate research methodology and (iii) the failure to take into considerations the role of externality that military spending plays in the model. They criticize the limitation of cross-sectional regression, which at best gives comparative static results. As a result, they used a panel data model of 103 countries in which military spending positively leads economic growth only in less than 10% of the countries (Mintz and Stevenson, 1995).

To expand the scope of the study, Knight et al. (1996) incorporated 22 industrialized countries on top of 102 developing countries, which encompass six geographical regions: Asia, East Europe, Middle East, North Africa, Sub-Sahara and West Hemisphere. In Addition, the sample period was divided into 1975–1985 (Cold War era) and 1986–1990 (later part of Cold War) to account for a possible structural break. Different from prior studies, their empirical test was built upon the growth model by Solow (1956) and Swan (1956). The result identified a direct negative relationship between military spending and economic growth via investment and productivity. The important message was that reduced military spending gave rise to more peace dividend. On the other hand, built on Barro's growth model (1990) and using Levine and Renelt's cross-sectional data (119 countries over 1974–1989), Brumm (1997) found a significantly positive relationship between military/GDP ratio and economic growth.

To examine the existence of so-called "peace dividend" after the Cold War, Heo (1998) applied a nonlinear regression model to the longitudinal data consisting of 80 countries over 1961–1990. Not surprisingly, negative relationship between military spending and economic growth (or peace dividend) was found in two thirds of the countries. Based on the three-equation model (growth, savings and military spending), Galvin (2003) applied two-stage and three-stage least squares methods to the 1999 cross-sectional data of 64 countries. The result pointed to a negative relationship between the two variables. Regrouping the 64 countries into low-income and middle-lower income categories, Galvin found greater negative impact for the lower income group and the spin-off effect from its military spending was negligible.

Up to now, the literature of the topic in general dealt largely with large samples, world or longitudinal data. Some of the studies already take into consideration the regional segmentation or different resource endowments (Frederiksen and Looney, 1982; Galvin, 2003; Lim, 1983). Linden's application (1992) of Feder's two-sector model (1983) to the 13 countries from the Middle East suggested the existence of a negative relationship between military spending and economic growth but a positive relationship between size of government, oil price, capital formation and economic growth. In a similar vein, McNair et al. (1995) applied the three-sector (private, national defense and the public sector of non-national defense) Feder (1983) and Ram (1986) models to 10 NATO countries over the period of 1951–1988. Using the fixed effect and the random effect models of panel data along with testing the existence of a panel unit root and co-integration relation, they found a positive military spending-economic growth relationship from the supply side. However, by not considering demand side, the potential crowding-out effect was left out.

Dunne and Mohammed (1995) selected 13 relatively homogeneous Sub-Saharan countries (over 1967–1985) for cross-sectional and panel data regression analyses. Including strategic variables such as size of army in the time series (13 countries combined), they detected a negative relationship via accumulated human capital formation, investment allocation and international balance account. Likewise, Hassan et al. (2003) sampled 5 of 7 South Asian Regional Cooperation Council members (Bangladesh, India, Pakistan, Nepal and Sri Lanka) from 1985 to 1996 in their study. They showed military spending, human capital, domestic gross investment, foreign direct investment, information and communication technology and infrastructure significantly explained economic growth within the framework of the neoclassic growth model. In contrast, Kollias et al. (2004) first tested the existence of a unit root and co-integration for the 15 countries from EU. Results from the Granger causality test suggested that majority of the countries showed economic growth led military spending, but not vice versa.

In sum, the abovementioned studies did not seem to address the heterogeneity factors between countries. It seems that choice of data is more and more in favor of grouping countries into different regions

151

over time or using longitudinal data in place of homogeneous cross-sectional data. On the one hand, cross-sectional data models assuming the homogeneity for each member of the cross section have limited inferential power due to country-specific effects. In contrast, a time series model may address this problem; however, paucity of data (only annual data are available) renders unit root and co-integration tests biased due to small sample property. With the advent of the dynamic panel data (DPD) method using the longitudinal data, the DPD begins to play an important role in deciphering the military spending-economic growth relationship. Yildirim et al. (2005) was among the first to use the DPD approach to analyze such a relationship for the 13 countries (the Middle East and Turkey) over 1989–1999. Using the generalized method of moments (GMM) developed by Arellano and Bond (1991), they identified a positive relationship between military spending and economic growth. More recently, Kollias et al. (2007) applied the fixed effect model of panel data to 15 EU countries and found feedback relationships between the two variables.

3. Data, model and econometric methods

To improve on the limitations –failure to consider both country-specific and time series effects at the same time– that majority of the prior papers suffer, this study uses a panel data of 90 countries over the time period from 1992 to 2006. All the data are obtained from World Development Index (WDI) of the World Bank. We deliberately exclude the data before 1992 in which the Cold War prevailed to avoid a likely structural break. To account for the heterogeneity emanated from different economic developmental stages or regional characteristics, we divide countries (i) based on the World Bank classification, into 3 panels: high-income (32 countries), low income (20) and middle-income (38) groups, and (ii) based on geographic proximity into 4 panels: Africa (23 countries), Europe (22), the Middle East and South Asia (16), and Pacific Rim (26) regions.[1] It is to be pointed out that the vast majority of prior studies dealt with static or comparatives static models,[2] in which one assumes an a priori relationship: military spending causes economic growth.

As mentioned before, the vast majority of the prior studies applied cross-sectional data to perform static analysis: usually they are based on the frameworks of Feder (1983), Ram (1986, 1995) or the augmented Solow model. According to the survey article by Dunne et al. (2005), Feder and Ram models suffer severe theoretical and econometric problems and as such should be avoided. Instead, they suggested the growth model by Barro (1990). Furthermore, Aizenman and Glick (2003) introduced a threat variable along with an interaction variable (threat times military spending) into the Barro model in order to examine the economic growth-defense spending relationship when a nation is under external threat. They expect to see (i) threat without military expenditure reduces growth, (ii) military expenditure without threat would reduce growth as well, and (iii) military expenditure in the presence of sufficiently large threat increases economic growth.

A common thread of the above static models is the a priori assumption that military spending causes economic growth. To circumvent the problem, one may employ the dynamic panel data (DPD) model from which the Granger causality can be applied to investigate the defense spending-economic growth relationship. Note that the Granger causality test within the framework of vector autoregressive (VAR) model does not require any a priori theory, i.e., the choice of explanatory variables is data driven. Consequently variables such as threat and initial income may not be included as is the case in some static models.[3] In this paper, we include investment/GDP ratio (proxy for capital stock) and population growth (proxy for labor input) into the model.

To formulate the model, we use the following notations: g_{it} denotes the real GDP growth rate per capita of country i at time t; kme_{it} denotes military spending per capita of country i at time t; iy_{it} denotes the investment/GDP ratio of country i at time t, a proxy for capital good; and $gpop_{it}$ denotes the population growth of country i at time t. Thus, the military spending and economic growth can take the following form:

$$g_{it} = \alpha_1 g_{it-1} + \sum_{j=1}^{p} \beta_{1j} iy_{it-j} + \sum_{j=1}^{p} \beta_{2j} gpop_{it-j} \qquad (1)$$
$$+ \sum_{j=1}^{p} \beta_{3j} kme_{it-j} + \eta_i + \varepsilon_{it},$$

where ε_{it} is assumed to be i.i.d.; η_i denotes unobservable country-specific (non-time series) effect with $E(\eta_i) = \eta$ and $Var(\eta_i) = \sigma_\eta^2$. The lagged dependent variable in (1) renders traditional panel models (fixed or random effect) inappropriate [$E(g_{it-1}, \eta_i) \neq 0$]. To overcome this problem, one common way is to take the first difference of Eq. (1) to remove country-specific effect from the DPD model. That is, to rid $E(g_{it-1}, \eta_i) \neq 0$ problem by taking first difference leads readily to

$$\Delta g_{it} = \alpha_1 \Delta g_{it-1} + \sum_{j=1}^{p-1} \beta_{1j} \Delta iy_{it-j} + \sum_{j=1}^{p-1} \beta_{2j} \Delta gpop_{it-j} \qquad (1')$$
$$+ \sum_{j=1}^{p-1} \beta_{3j} \Delta kme_{it-j} + \Delta \varepsilon_{it},$$

where Δ is the operator for first difference. The OLS estimation on Eq. (1') can overcome the correlation problem between individual effects and explanatory variables. On the flip side, it creates another problem: correlation between lagged dependent variable and error term or $E(\Delta g_{it-1}, \Delta \varepsilon_{it}) \neq 0$. Arellano and Bond (1991) proposed to use lagged dependent variable (level value of g_{it-s} for s > 2) as instrument variable along with GMM to overcome the problem of $E(\Delta g_{it-1}, \Delta \varepsilon_{it}) \neq 0$. The reason lies in the possible correlation between the lagged dependent variable and the unobservable country specific effect. Thus we take fist difference of (1') ensure such correlation is rid of. If such a procedure fails to do the job, one should resort to the use of the lagged dependent variable with a higher lag as suggested by Arellano and Bond (1991).

Given that the maximum lag in $\Delta \varepsilon_{it} = \varepsilon_{it} - \varepsilon_{it-1}$, is 1 (or ε_{it-1}), using a lag of 2 or higher is a good choice for the instrument variable. In order to investigate whether military spending Granger-causes economic growth, we must estimate (1') before testing the null hypothesis of $H_0: \beta_{31} = \beta_{32} = ... = \beta_{3p-1} = 0$. Notice that we estimate the panel models for 90 countries, encompassing 3 income groups and 4 geopolitical regions over 1992–2006.

4. Analysis and discussion of the empirical results

One requirement for using the DPD model is that all the variables in Eq. (1) are of I (0). To examine such a property for the different panels, we opt for the unit root tests by Levin, Lin and Chu or LLC (2002) and Im, Pesaran and Shin or IPS (2003). The difference between the LLC and IPS

---

[1] The Arellano and Bond model (1991) applies nicely in the case of N > T where N denotes cross sectional units and T time series observations. Consequently, each sub-panel must have at least 15 observations (1992-2006). As for regional panels, we partition regions based on geographical characteristics. For instance the Middle East and South Asia are grouped together due to geographical proximity. We delineate 4 regional panels as shown in Fig. 1. Specific countries of each panel are listed in Appendix A.

[2] To the best of our knowledge, the paper by Yildirim et al. (2005) is the first using DPD model to investigate the military spending-economic growth relationship for the Middle East and Turkey.

[3] In addition, data pertaining to threat provided by Correlates of War (COW) are available only for the period of 1992-2001. Hence, we can not include it in the model.

Table 1
Panel unit root test results.

		g_{it}	iy_{it}	$gpop_{it}$	kme_{it}
90-country	LLC	-21.90^{***}	-6.32^{***}	-10.46^{***}	-13.91^{***}
	IPS	-18.17^{***}	-3.81^{***}	-5.92^{***}	-8.66^{***}
High-income	LLC	-14.21^{***}	-5.39^{***}	-3.90^{***}	-6.85^{***}
	IPS	-11.44^{***}	-3.12^{***}	-2.08^{**}	-4.51^{***}
Middle-income	LLC	-12.78^{***}	-3.10^{***}	-9.06^{***}	-8.26^{***}
	IPS	-11.13^{***}	-2.51^{***}	-7.50^{***}	-4.73^{***}
Low-income	LLC	-10.89^{***}	-2.43^{***}	-1.84^{**}	-9.90^{***}
	IPS	-8.74^{***}	-0.68	0.32	-6.13^{***}
Africa	LLC	-10.25^{***}	-4.73^{***}	-9.93^{***}	-11.53^{***}
	IPS	-8.99^{***}	-2.50^{***}	-8.78^{***}	-7.91^{***}
Europe	LLC	-11.59^{***}	-3.82^{***}	-2.68^{***}	-6.93^{***}
	IPS	-8.49^{***}	-3.45^{***}	-0.60	-3.81^{***}
Middle-East & South Asia	LLC	-11.94^{***}	-2.72^{***}	-3.83^{***}	-5.57^{***}
	IPS	-9.74^{***}	-2.37^{***}	-4.85^{***}	-3.95^{**}
Pacific-Rim	LLC	-9.91^{***}	-3.76^{***}	-7.48^{***}	-5.19^{***}
	IPS	-8.23^{***}	-1.47^{*}	n.a.	-1.77^{**}

Notes: LLC = Levin, Lin and Chu (2002) test statistic. IPS = Im, Pesaran and Shin (2003) test statistic. ***, **, * denote 1%, 5% and 10% significance levels respectively. g_{it} denotes real economic growth rate for the ith cross-sectional unit at time t. Similarly iy_{it}, $gpop_{it}$ and kme_{it} denote investment/GDP ratio, population growth and per capita real military expenditure for the ith cross-sectional unit at time t respectively.

Table 2
Estimated DPD results based on 3 income levels.

	All(90-country)	High-income	Middle-income	Low-income
α_1	0.005^2	-0.1099	-0.2081	-0.1128^{**}
	(0.04)	(-1.26)	(-1.47)	(-2.09)
β_{11}	0.1436^{***}	0.0360	0.1337^{*}	0.2309^{***}
	(3.18)	(0.54)	(1.80)	(3.90)
β_{21}	-0.2495	-0.3055^{*}	-1.1119^{***}	0.5734^{***}
	(-1.36)	(-1.65)	(-4.43)	(2.57)
β_{31}	-0.0002	-0.0005	-0.0041	-0.00236^{*}
	(-0.58)	(-0.75)	(-0.78)	(-1.78)
$N \times T$	90×14	32×14	38×14	20×14
m1	-1.84^{*}	-2.24^{**}	-1.71^{*}	-1.81^{*}
	$[0.07]$	$[0.03]$	$[0.09]$	$[0.07]$
m2	-0.14	-1.78^{*}	-1.53	-0.18
	$[0.89]$	$[0.08]$	$[0.13]$	$[0.85]$

Notes: Numbers in parentheses and brackets are t statistics and p values, respectively. *, **, *** denote significance level at 10%, 5% and 1% respectively. N = cross sectional unit; T = time series observations; g = per capita real economic growth; iy = investment/GDP; gpop = rate of population growth; kme = per capita real military expenditure. m1 and m2 denote test statistics of serial uncorrelated residuals of the first and second orders in the panel model. The empirical model is
$$g_{it} = \alpha_1 g_{it-1} + \sum_{j=1}^{p} \beta_{1j} iy_{it-j} + \sum_{j=1}^{p} \beta_{2j} gpop_{it-j} + \sum_{j=1}^{p} \beta_{3j} kme_{it-j} + \eta_i + \varepsilon_{it}.$$

models lies in that the LLC model assumes a common unit root process while the IPS model allows for individual unit root process.

As shown in Table 1, regardless of using LLC or IPS for 4 panels, all of the four variables g_{it}, iy_{it}, $gpop_{it}$ and kme_{it} obey the I (0) assumption. That is, we reject the null hypothesis H_0:variable is of I (1) in all cases. The results using STATA 10 for 4 variables under 3 panels for all 90 countries are reported in Table 2.[4]

If we lump all the 90 countries into one panel, military spending does not Granger-cause economic growth (Table 2): only iy_{it} (investment/GDP) Granger-causes economic growth. This is to say neither military spending nor labor input impacts economic growth. Instead capital stock is instrumental to economic growth, a phenomenon sometimes supported by the growth theory. However, if we partition per capita real income into high-, middle-, and low-income panels, different results emerge. In the high-income panel, only population growth Granger-causes (negatively) economic growth with 10% significance level. For the middle-income group, the Granger causality is observed in (i) iy_{it} (investment/GDP) with 10% significance level and in positive direction with respect to economic growth and (ii) $gpop_{it}$ (population growth) with $\alpha = 1\%$ significance level but in negative direction. Note that military spending does not impact economic growth in this income group. Finally, for the low income panel, both iy_{it} and $gpop_{it}$ are found to Granger-cause (positively) economic growth with $\alpha = 1\%$. In addition, kme_{it} (military spending per capita) Granger-causes negatively economic growth ($\alpha = 10\%$). It is to be pointed out factors that determine economic growth actually vary according to the stages of economic development (or income levels). Capital stock, commonly regarded as the engine for economic growth, plays a key role only in low- and middle-income groups. Technological advances rather than capital accumulation drive economic growth for high-income countries. Population growth however is found to Granger-cause (positively) economic growth for low-income countries but negatively for middle- and high-income countries with $\alpha = 1\%$ and 10% respectively. Surprisingly, military spending is found to impact economic growth

negatively ($\alpha = 10\%$) only in low-income countries. The crux of the matter is that capital stock and/or population growth impact economic growth in general. Defense spending plays a role only in low-income countries as was found in the study by Lim (1983) and Yang et al. (2011). However the negative causality runs counter to the result by Biswas and Ram (1986) in which military expenditure contributes positively to economic growth for lower-middle income group.

Though the Granger causality on military expenditure–economic growth is marginally significant (10%), it lends support to increasing evidence that developing countries are better off focusing on economic growth rather than producing guns and ammunition.

Also reported in Table 2 are mj (j = 1, 2) statistics for each panel. Here m1 and m2 denote test statistics of serial uncorrelated residuals of the first and second orders in the panel model.[5] The mj statistic is a standardized residual auto-covariance, which are asymptotically N (0, 1) under the null of no autocorrelation. According to Doornik et al. (2006), if the disturbance ε_t is not serially correlated, there should be evidence of significant and negative first order (j = 1) serial correlation in the difference form ($\hat{\varepsilon}_{it} - \hat{\varepsilon}_{it-1}$), and no evidence of second order (j = 2) serial correlation in the differenced residual. An examination of Table 2 indicates that all m1 statistics are significant and all m2 statistics are insignificant for each panel. This is to say, the VAR model with lag = 1 for all 3 panels and 90 countries is free of the serial correlation problem.[6]

Needless to say, military conflicts are more likely to cause war and certain regions are more prone to war perhaps due to history or geographical proximity. Note that cross-sectional units (N) must exceed time-series observations (T) in applying DPD model developed by Arellano and Bond (1991). As such we categorize all 90 countries into 4 geopolitical regions to satisfy N > T. Fig. 1 exhibits a panoramic view of the relative locations for the 4 geopolitical regions and Table 3 reports the estimation results.

An inspection of Table 3 reveals that population growth Granger-causes economic growth for the panel of 23 African countries. For the panel of 22 European countries neither capital stock nor population growth significantly Granger-causes economic growth. Instead, military spending is found to Granger-cause negatively economic

---

[4] We employ one-step estimator in our DPD model. Bond (2002) mentioned that "In fact, a lot of applied work using these GMM estimators has focused on results for the one-step estimator rather than the two-step estimator. This is partly because simulation studies have suggested very modest efficiency gains from using the two-step version, even in the presence of considerable heteroskedasticity (see Arellano and Bond, 1991; Blundell and Bond, 1998; Blundell et al., 2000), but more importantly because the dependence of the two-step matrix on estimated parameters makes the usual asymptotic distribution approximations less reliable for the two-step estimator (p.147)." For this reason, in our estimation, the robust one-step estimator is used.

[5] For details, see Arellano and Bond (1991).
[6] To overcome the heteroskedasticity problem in the one-step model, we employ robust-to-heteroskedasticity variance–covariance estimator and as such the Sargan test statistics cannot be presented.

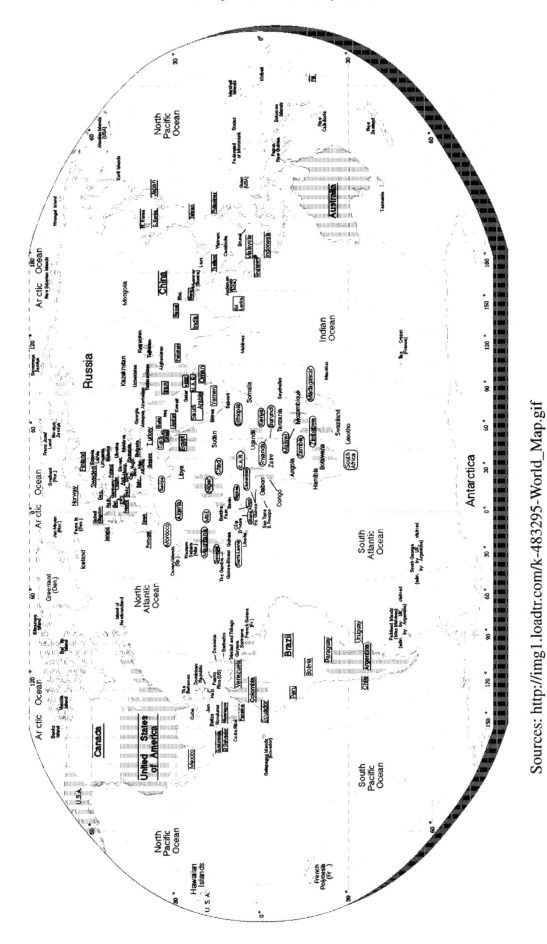

Sources: http://img1.loadtr.com/k-483295-World_Map.gif

Notes:
country/name : Africa, country/name : Europe, country/name : Middle East and South Asia, name : Pacific Rim.

Fig. 1. Relative geographical location and member countries in 4 geopolitical panels.

Table 3
Estimated DPD results based on 4 geopolitical regions.

	Africa	Europe	Middle East–South Asia	Pacific Rim
α_1	-0.5365^{***}	-0.0108	-0.1778^{**}	0.3916^{**}
	(-2.60)	(-0.05)	(2.17)	(2.34)
β_{11}	0.2254^{***}	-0.1691	-0.0258	-0.7004^{**}
	(3.96)	(-2.33)	(-0.95)	(-2.30)
β_{12}				0.7098^{***}
				(2.63)
β_{21}	0.5192^{**}	0.0326	-1.5202^{***}	0.6507
	(2.45)	(0.10)	(-5.27)	(1.17)
β_{22}				-1.6068^{**}
				(-2.43)
β_{31}	0.0060	-0.0030^{**}	-0.0006^{**}	0.0148
	(0.69)	(-2.40)	(-2.05)	(0.54)
β_{32}				-0.157
				(-0.57)
$N \times T$	23×14	22×14	16×14	29×14
m1	-1.87	-2.36	-3.44	-3.32
	$[0.06]$	$[0.02]$	$[0.00]$	$[0.00]$
m2	-0.11	0.08	-0.993	-1.63
	$[0.91]$	$[0.94]$	$[0.32]$	$[0.10]$
Per capita GDP growth (%)	-0.23	2.37	1.58	2.38
ME/GDP (%)	2.78	2.50	6.38	2.56

Note: Same as Table 2. ME = military expenditure; GDP = gross national product.

growth with $\alpha = 5\%$. Similarly, increased military expenditure Granger-causes negatively economic growth for the panel of the Middle East–South Asia with $\alpha = 5\%$. In addition, population growth is found to Granger-cause economic growth negatively in this panel ($\alpha = 1\%$). The optimum lag for the m1 and m2 statistics is 1 for the VAR model and for the 3 panels since all m1 statistics are significant (negative) and m2 statistics are insignificant (Table 3). Finally the Pacific Rim panel is defined to comprise entire America and East Asia. The optimum lag for the m1 and m2 statistics is found to be 2 in order to satisfy the serial independence condition by Doornik et al. (2006). In this region military spending does not Granger-cause economic growth, but population growth negatively impacts economic growth. Note that investment/GDP ratios (both lags 1 and 2) lead economic growth with conflicting signs, even though their cumulative impacts on economic growth are insignificant. In sum, only in Europe and Middle East–South Asia panels, do we find that military spending exerts negative impacts on growth of the economy! Attached to the bottom of Table 3 are per capita economic growth and military expenditure ratios (ME/GDP) for the four regions. A glimpse of Table 3 indicates that the ME/GDP ratio of the Middle East–South Asia (6.38%) was noticeably higher than the other three regions (averaged around 2.6%). However, per capita economic growth of this region (1.58%) exceeded only that of Africa. It is known that Middle East–South Asia region has been replete with military conflicts and clashes early in history. Since the Hebron Massacre of 1929, Israel–Arab–Iran conflicts include the Arab–Israel War (1948), the Six-Day War (1967), the Yom Kippur War (1973), Israel–Iraq War (1981), the Iran–Iraq War (1980–1988), the Gulf War (1991), the Iraq War (started 2003), and constant conflicts between Hamas militants and Israelis. In the South Asia area, there were Sino–Indian War of 1962, the Chola Incident (1967), the Sino–Indian Skirmish (1987), and the Liberation Tigers of Tamil over decades. The border dispute among China, India and Pakistan is far from resolved.

The frequency with which wars and conflicts occurred in the region speaks volumes about its high ME/GDP ratio. Political instability in Yemen, Pakistan, Iran, Iraq, India and Sri Lanka leads to the "crowding out effect" in macroeconomics: rising military expenditures in the countries bid away funds in the capital market. As a result, such spending drives up interest rates, which discourage investment in private sectors. Generally speaking, unless the

spillover effect is positive –technological advances from military spending lend themselves to profitable commercial applications-military spending is not likely to be more productive than private sector investment. Furthermore, administration and coordination costs of war drain scarce resources from private sectors. Aside from it, military conflicts scare away foreign investments and reduce international trade. This is to say, the crowding-out effect in these regions leads to economic slowdown.

On the one hand, the above result bears resemblance to that by Linden (1992) in which 13 Middle East countries comprise the sample space. On the other hand, our result runs counter to those (i) by Hassan et al. (2003), who employed 5 countries from South Asia and (ii) by Yildirim et al. (2005), who sampled Middle East countries and Turkey during 1989–1999. Their conclusion all point to the finding that military expenditure enhances economic growth in the Middle East countries and Turkey as a whole. One of the major results of this paper runs counter to that by Yildirim et al. (2005): increased military spending reduces, rather than enhances, economic growth in our model. There are three reasons that can explain the difference in the result. First, the definition of Middle East–South Asia region in our paper includes 10 of 13 Middle East countries used in their model along with the six countries from South Asia. Second, our sample period extends from 1992 to 2006 compared to 1989–1999 in their model. The justification is that we would like to avoid the structural break caused by the Cold War. Finally, the model used by Yildirim et al. (2005) is based on the Feder model (2003), which suffers serious theoretical and econometric problems according to Dunne et al. (2005). To avoid the pitfall, we perform our analysis using the Barro model (1990) as suggested by Dunne et al. (2005).

With respect to the military expenditure–economic Growth in Africa, Dunne and Mohammed (1995), using time series analysis, found a negative correlation for the 13 sub-Saharan countries. However, the correlation disappears when cross sectional or panel data, are used. Our analysis reinforces the insignificant relationship for the region. Kollias et al. (2007) applied panel data model to 15 EU countries and found feedback relationships while the 14-EU-country study by Mylonidis (2008) reveals a negative relationship. In addition, the negative relationship tends to intensify over time when 5-year average of cross-section regression is applied (Mylonidis, 2008).

The literature on military expenditure–economic growth relationship has come a long way but remains unsettled. Clearly, the result can vary depending on the definition of panels and sample periods. In this paper, except for the Pacific Rim panel that has eluded the literature thus far, the results of the other 3 panels are very much in line with those in the literature. In short, by partitioning sample space by income levels (economic development) or regional differences, more detailed results emerge: military expenditure Granger-causes economic growth negatively for some panels, the result easily masked if we lump all 90 countries into one group.

5. Conclusion

To tackle the problem of insufficient sample size that many of previous studies suffered, the recent trend is to use panel data analysis. Advances in panel data econometrics such as DPD enable researchers to investigate the Granger causality among variables. From the viewpoint of crowding-out effects, there is a kernel of truth in it especially for developing countries: resulting paucity in infrastructure and private sector investment can indeed hinder economic growth. In terms of geopolitical location, some region (e.g., Middle East & South Asia) is a tinderbox on the verge of frequent crises. As such, a noticeable portion of resources goes to ammunition production instead of factories with the ensuing economic slowdown.

To take into account the regional and income heterogeneity, this paper makes use of a 90-country panel data spanning over 1992–

2006. Before applying the DPD estimator by Arellano and Bond (1991), we divide the 90 countries in (i) 3 income panels: low income, middle income, and high income, and (ii) 4 geopolitical panels: Africa, Europe, Middle East–South Asia and Pacific Rim. In the 3 income panels, military spending is found to lead negatively economic growth in the low-income panel at 10% significant level. Of the 4 geopolitical panels, military spending Granger-causes negatively economic growth in the Middle East–South Asia and Europe regions. Well known throughout history, frequent conflicts between nations give rise to higher military expenditures, which more likely slowdowns economic growth. This intuition is supported by the result of our DPD panel model. One important conclusion is that regardless of using income or geopolitical panels, once the Granger-causality is identified, the sign of the causality points to the same direction: military expenditure indeed impedes economic growth.

Appendix Table A. Four geopolitical panels and their member countries

Africa (23)	Europe (22)	Middle East & South Asia (16)	Pacific Rim (29)
Country name	Country name	Country name	Country name
Algeria	Austria	Bahrain	El Salvador
Burundi	Belgium	Cyprus	Guatemala
Cameroon	Denmark	Egypt, Arab Rep.	Mexico
Central African Republic	Finland	Iran, Islamic Rep.	Nicaragua
Chad	France	Israel	Panama
Ethiopia	Germany	Jordan	China
Ghana	Greece	Oman	Indonesia
Kenya	Ireland	Saudi Arabia	Japan
Madagascar	Italy	Syrian Arab Republic	Korea, Rep.
Malawi	Luxembourg	United Arab Emirates	Malaysia
Mali	Netherlands	Yemen, Rep.	Philippines
Mauritania	Norway	Bangladesh	Singapore
Morocco	Portugal	India	Taiwan
Niger	Spain	Nepal	Thailand
Nigeria	Sweden	Pakistan	Canada
Rwanda	Switzerland	Sri Lanka	United States
Senegal	Turkey		Argentina
Sierra Leone	United Kingdom		Bolivia
South Africa	Bulgaria		Brazil
Togo	Hungary		Chile
Tunisia	Poland		Colombia
Zambia	Romania		Ecuador
Zimbabwe			Paraguay
			Peru
			Uruguay
			Venezuela, RB
			Australia
			New Zealand
			Fiji

Notes: Numerical values in parentheses denote numbers of member countries in each panel.

References

Aizenman, J., Glick, R., 2003. Military expenditure, threats, and growth. NBER Working Paper. #9618.

Antonakis, N., 1997. Defense spending and growth in Greece: a comment and further empirical evidence. Applied Economics Letters 4, 651–655.

Arellano, M., Bond, S., 1991. Some tests of specification for panel data: Monte Carlo evidence and an application to employment equation. The Review of Economic Studies 58 (2), 277–297.

Atesoglu, H.S., Mueller, M.J., 1990. Defense spending and economic growth. Defense Economics 2 (1), 19–27.

Barro, R.J., 1990. Government spending in a simple model of endogenous growth. Journal of Political Economy 98 (5), 103–126.

Benoit, E., 1973. Defense and Economic Growth in Developing Countries. Lexington Books.

Benoit, E., 1978. Growth and defense in developing countries. Economic Development and Cultural Change 26, 271–280.

Biswas, B., 1993. Defense spending and economic growth in developing countries. In: Payne, J.E., Sahu, A.P. (Eds.), Defense and Economic Growth. Boulder Westview Press, pp. 223–235.

Biswas, B., Ram, R., 1986. Military expenditures and economic growth in less developed countries: an augmented model and further evidence. Economic Development and Cultural Change 34, 361–372.

Blundell, R., Bond, S., 1998. Initial conditions and moment restrictions in dynamic panel data models. Journal of Econometrics 87 (1), 115–143.

Blundell, R.W., Bond, S.R., Windmeijer, F., 2000. Estimation in dynamic data models: improving on the performance of the standard GMM estimator. In: Baltagi, B. (Ed.), Non-stationary panels, panel co-integration and dynamic panels. : Advance in Econometrics, Vol. 15. JAI Elsevier Science.

Bond, S.R., 2002. Dynamic panel data models: a guide to micro data methods and practice. Portuguese Economic Journal 1 (2), 141–162.

Brumm, H., 1997. Military spending, government disarray, and economic growth: a cross-country empirical analysis. Journal of Macroeconomics 19 (4), 827–838.

Cohen, J.S., Minty, A., Stevernson, R., Ward, M.D., 1996. Defense expenditures and economic growth in Israel: the indirect link. Journal of Peace Research 33 (3), 341–352.

Deger, S., 1986. Economic development and defense expenditure. Economic Development and Cultural Change 35, 179–196.

Doornik, J. A., Arellano, M., and Bond, S., (2006), "Panel Data Estimation using DPD for Ox," mimeo.

Dunne, J.P., Mohammed, N.A.L., 1995. Military spending in Sub-Saharan Africa: some evidence for 1967–85. Journal of Peace Research 32 (3), 331–343.

Dunne, J.P., Smith, R.P., Willenbrockel, D., 2005. Models of military expenditures and growth: a critical review. Defence and Peace Economics 16 (6), 449–461.

Faini, R., Annez, P., Taylor, L., 1984. Defense spending, economic structure, and growth: evidence among countries and over time. Economic Development and Cultural Change 32 (3), 487–498.

Feder, G., 1983. On exports and economic growth. Journal of Economic Development 12, 59–73.

Frederiksen, P.C., Looney, R.E., 1982. Defense expenditures and economic growth in developing countries: some further empirical evidence. Journal of Economic Development 7, 113–126.

Galvin, H., 2003. The impact of defense spending on the economic growth of developing countries: a cross-section study. Defence and Peace Economics 14 (1), 51–59.

Grobar, L., Porter, R.C., 1989. Benoit revisited: defense spending and economic growth in LDCs. Journal of Conflict Resolution 33 (2), 318–345.

Hassan, M.K., Waheeduzzaman, M., Rahman, A., 2003. Defense expenditure and economic growth in the SAARC countries. Journal of Social Political and Economic Studies 28 (3), 275–293.

Heo, U.K., 1997. The political economy of defense spending in South Korea. Journal of Peace Research 34, 483–490.

Heo, U.K., 1998. Modeling the defense–growth relationship around the globe. Journal of Conflict Resolution 42 (5), 637–657.

Im, K.S., Pesaran, M.H., Shin, Y., 2003. Testing for unit roots in heterogeneous panels. Journal of Econometrics 115, 53–74.

Knight, M., Loayza, Norman, Villanueva, D., 1996. The peace dividend: military spending cuts and economic growth. IMF Staff Papers 43 (1), 1–37.

Kollias, C., Manolas, G., Paleologou, Suzanna-Maria, 2004. Defense expenditure and economic growth in the European Union: a causality analysis. Journal of Policy Modeling 26, 553–569.

Kollias, C., Mylonidis, N., Paleologou, S.M., 2007. A panel data analysis of the Nexus between defense spending and growth in the European Union. Defence and Peace Economics 18 (1), 75–85.

Levin, A., Lin, C.F., Chu, C., 2002. Unit root tests in panel data: asymptotic and finite-sample properties. Journal of Econometrics 108, 1–24.

Lim, D., 1983. Another look at growth and defense in less developed Countries. Economic Development and Cultural Change 31, 377–384.

Linden, M., 1992. Military expenditure, government size and economic growth in the Middle East in the period 1973–85. Journal of Peace Research 29, 265–270.

McNair, E., Murdoch, C., Sandler, T., 1995. Growth and defense: pooled estimates for the NATO Alliance, 1951–1988. Southern Economic Journal 61, 846–860.

Mintz, A., Stevenson, R.T., 1995. Defense expenditures, economic growth, and the "peace dividend: a longitudinal analysis of 103 countries. Journal of Conflict Resolution 39 (2), 283–305.

Mylonidis, N., 2008. Revisiting the nexus between military spending and growth in the European Union. Defence and Peace Economics 19 (4), 265–272.

Ram, R., 1986. Government size and economic growth: a new framework and some evidence from cross-section and time series data. The American Economic Review 76 (1), 191–203.

Ram, R., 1995. Defense expenditure and economic growth. In: Hartley, K., Sandler, T. (Eds.), Handbook of Defense Economics. Elsevier, Amsterdam, pp. 251–273.

Smith, R., 1980. Military expenditure and investment in OECD countries 1954–73. Journal of Comparative Economics 4, 19–32.

Solow, R.M., 1956. A contribution to the theory of economic growth. Quarterly Journal of Economics 50, 65–94.

Swan, T.W., 1956. Economic growth and capital accumulation. The Economic Record 32, 334–361.

Yang, Albert J.F., Trumbull, W.N., Yang, C.W., Huang, Bwo-Nung, 2011. On the relationship between military expenditure, threat, and economic growth-A nonlinear approach. Defence and Peace Economics 22 (4).

Yildirim, J., Sezgin, S., Öcal, N., 2005. Military expenditure and economic growth in Middle Eastern countries: a dynamic panel data analysis. Defence and Peace Economics 16 (4), 283–295.

Huang, Yang, and Hwang, International Journal of Applied Economics, 1(1), September 2004, 81-97

New Evidence on Demand for Cigarettes: A Panel Data Approach

Bwo-Nung Huang, Chin-wei Yang, and Ming-jeng Hwang*

National Chung Cheng University, Clarion University of Pennsylvania, and West Virginia University

Abstract. This paper estimates the demand for cigarettes using panel data – 42 states and Washington, D.C. – from 1961 to 2002. We first employ the panel unit root test before estimating the demand structure. We have found that (i) the price and income elasticities are approximately -0.41 and 0.06, (ii) the price elasticities of neighboring states is 0.09, (iii) decreasing tax elasticity gives rise to decreasing price elasticity, and smaller tax shares (real tax as percentage of real price) seem to be related to declining tax elasticity, (iv) overall antismoking campaigns have contributed to declining income elasticities across different income groups, and (v) the decline in income elasticity for dividend and transfer income recipients is the main cause for the decrease in overall income elasticity. It has interesting implications: cigarette consumption is a normal good to wage earners and transfer payment recipients, but an inferior good to the owners of stocks and the elderly population.

Keywords: Cigarette demand, demand elasticity, panel data.

JEL classification: L66, C33

Introduction

Cigarette demand estimation has been of paramount interest since the Surgeon General's warning (1964) on the potential causal relationship between cigarette smoking and smoking related diseases, lung cancer in particular. Prior studies focused on the estimates of price, income, and advertising elasticities using either time series (e.g., Fujii, 1980; Bishop and Yoo, 1985; Keeler *et al.* 1993) or cross-section (e.g., Lyon and Simon, 1968) data. As pointed out by Baltagi and Levin (1986) in their cigarette demand study, the pure cross-section model cannot effectively control for state specific (demographic) effects whereas the pure time-series analysis fails to effectively control for unobservable taste changes. However, an appropriate pooled technique can provide opportunities to accommodate for both shortcomings.

Baltagi and Levin (1986) pioneered panel data estimation by incorporating 47 states over the period between 1963 and 1980. Baltagi *et al.* (2000) most recently discussed the homogeneous and heterogeneous estimator problems in the cigarette demand estimation using panel data. They strongly endorse pooled models, which posit a set of homogeneous parameters over their heterogeneous counterparts based on the root mean square error criterion. In their panel estimation (1960-1990), Keeler *et al.* (2001) emphasized the biased elasticity estimates emanating from omitted variables in demand functions. Nelson (2003) investigated the effects of advertising bans on cigarette consumption.

The panel data approach undoubtedly provides heterogeneous information at the state level while ensuring more reliable estimates via large sample property. With the growing popularity of panel data applications, the time series properties such as unit root and cointegration have received increasingly more attention. If a unit root exists in the panel data, the result using the level (logarithmic) is suspect. To the best of our knowledge, analyses of demand for cigarettes have thus far ignored the panel unit root or panel cointegration properties with the exception of Nelson (2003) and Bask and Melkersson (2004). The purpose of this paper is to (1) estimate cigarette demand structure after applying the panel unit root technique; (2) disaggregate income into earning, dividend, transfer payment, and estimate respective income elasticities; (3) partition the cigarette price into pre-tax price and tax in demand estimation; and (4) estimate the time-varying price and income elasticities via using the rolling window approach in an attempt to offer explanations on decreasing elasticities.

It should be noted that different state cigarette taxes are a hotbed for "bootlegging" across states. Such a cross-border effect is significant and cannot be ignored (Coats, 1995). To this end, we include average neighboring state prices into the model to capture the substitution effect. Anti-smoking campaigns date back to 1965 (warning label), 1969 (advertisement bans on television and radio), 1972 (U.S. Surgeon General's warning about environmental hazards from tobacco smoking), 1979 (U. S. Surgeon General's report concluding nicotine was addictive), 1986 (release of U. S. Surgeon General's report on involuntary smoking), 1988 (smoking ban on all flights of two hours or less), 1990 (smoking ban on all flights), and 1992 (release of the U. S. Environmental Protection Agency report on environmental tobacco smoking). These events may well have contributed to the volatility of price and income elasticity of cigarette demand over time and as such, need to be considered in evaluating the price and income elasticities estimated from a pooling model.

Most recently, Sloan *et al.*, (2002) found an own price elasticity of -0.15 and income elasticity of 0.15 applying a two-stage least squares model over 1990-1998. Nelson (2003) employed a nonstationary panel model to improve on the Saffer-Chaloupka (2000) result. He found the price elasticity had increased from -0.4 to -0.621 with decreased income elasticity after the regime shift in 1985. Interestingly enough, Gallet (2003) estimated demand functions for the 45 states along with the conjectural variation coefficients and the Lerner indexes. While the majority of estimated average demand elasticity is less than unity, some are quite elastic: -2.78 (Louisiana), -2.73 (Alabama), and -1.84 (Kentucky). The estimated conjectural variation coefficients indicate an anti-competitive way of advertising, echoing findings by Tremblay and Tremblay (1995), which is in rough agreement with the sizes of the Lerner index. Kim and Seldon (2004) found the price elasticity of cigarette demand in South Korea to be -0.276 and a nearly zero income elasticity, which is comparable to -0.33 in Greece (Hondroyiannis and Papapetrou, 1997) and the average value of 22 OECD countries (Steward, 1993).

This paper employs the panel unit root technique for 42 states and Washington, D.C. before estimating cigarette demand. The null hypothesis of a unit root is easily rejected and as a result, the demand structure can be formulated in double logarithmic form. The next section introduces the model. The third section provides the data and panel unit root test result. The fourth section is the discussion of the empirical results, and the last section gives concluding remarks.

Research Model

According to Laughhunn and Lyon (1971), Hamilton (1972), Doron (1979), Baltagi and Levin (1986), and Baltagi, Griffin, and Xiong (2000), cigarette demand is formulated as below:

$$Q_{it} = f (P_{it}, Y_{it}, Pn_{it}, Z_{it}) \tag{1}$$

where Q_t is per capita sales of cigarettes by population in state i at time t; P_{it} is the retail price per pack; Y_{it} denotes per capita personal income; Pn_{it} represents the average cigarette price of neighboring states; and Z_{it} represents other explanatory variables[1] (*e.g.,* proportion of population ages 65 and older). All the variables are deflated by the consumer price index (CPI).

Two major strands of cigarette demand models dominate the literature: the partial adjustment model and the rational addiction model. The partial adjustment model emphasizes the habit formation phenomenon (past) while the rational addiction model argues that both of the past and future consumptions exert positive impacts on current consumption as developed by Becker and Murphy (1988). Consider the partial adjustment, or habit persistence model of cigarette demand:

$$\ln Q_{it} - \ln Q_{it-1} = \rho (\ln Q_{it}' - \ln Q_{it-1}) + \mu_{it} \tag{2}$$

in which $\ln Q_{it}^*$ represents the expected or desired consumption level as shown below:

$$\ln Q_{it}' = \alpha' + \beta_1' \ln P_{it} + \beta_2' \ln Y_{it} + \beta_3' \ln Pn_{it} + Z_{it}' \gamma' \tag{3}$$

Substituting (3) into (2) immediately yields

$$\ln Q_{it} = \alpha_1 + (1-\rho) \ln Q_{it-1} + \beta_1 \ln P_{it} + \beta_2 \ln Y_{it} + \beta_3 \ln Pn_{it} + Z_{it}' \gamma + \mu_{it} \tag{4}$$

Empirical estimates based on the partial adjustment model are generally oriented towards short run elasticities and speed of adjustment in the long run (Baltagi and Levin, 1986 and 1992; Baltagi *et al.*, 2000).

Alternatively, Becker and Murphy (1988) developed the rational addiction hypothesis assuming: (1) individuals make choices that span over several time periods and (2) addictive consumptions in different time periods are complementary. Consequently, past and future consumptions enter the demand structure under the rational addiction paradigm as:

$$\ln Q_{it} = \alpha_{1t}' + \alpha_1 \ln Q_{it-1} + \alpha_2 \ln Q_{it+1} + \beta_1 \ln P_{it} + \beta_2 \ln Y_{it} + \beta_3 Pn_{it} + Z_{it} \gamma + \mu_{it} \tag{5}$$

The rational addiction hypothesis is found to be consistent with cigarette demand by Chaloupka (1991), Becker *et al.* (1994), Labeaga (1993), Walters and Sloan (1995), and Escario and Molina (2001), Nelson (2003), and Sloan *et al.* (2002). However, it is deduced from long-term disaggregate individual behavior. Employing aggregate panel data (state), we estimate the demand structure of (4) based on the partial adjustment model.[2] It is to be pointed out the double-log functional form is employed for convenience. While incorporating interacting explanatory variables in a multiplicative form, its estimated elasticities are restricted to be constant. For example, flexible functional form – the Box-Cox transformation – was used by Chang and Hsing (1991) in estimating residential electricity with the transformation parameter of 0.23. In a separate study, however, Hsing and Chang (2000) found demand for real M2 to be linear with the transformation parameter of 1.01. The Box-Cox model can indeed yield more accurate elasticities for future policy implementation and remains an interesting future research topic.

Data Source and Panel Unit Root Result

The cigarette data used in this study are obtained from *The Tax Burden on Tobacco* published by the Tobacco Institute. Data for 42 states and Washington D.C. over the 1961–2002 sample period include per capita cigarette consumption measured by tax paid per capita sales in packs (Q_{it}), per capita personal income deflated by 1982 – 1984 dollars (Y_{it}), weighted average real price (cents) per pack of premium brands deflated in 1982 – 1984 dollars (P_{it}), and weighted average real tax (state and federal cigarette taxes) per pack of cigarettes deflated by 1982 – 1984 CPI (Tax_{it})[3].

To take the cross-border bootlegging effect into consideration, we include the cigarette price(s) of neighboring states into the demand equation. Baltagi and Levin (1986) used the minimum price of neighboring states to capture the cross-border effect whereas Keeler, *et al.* (1993) employed the average price of neighboring states as the proxy. Well-documented in time series studies, economic variables are fraught with the unit root problem as was first pointed out by Nelson and Plosser (1982).[4] A common pitfall is to apply a regression without checking the unit root property at level with the ensuing high R^2 and low Durbin-Watson statistic: the spurious regression termed by Granger and Newbold (1974).[5] The increasingly important role played by panel data in the time series analysis makes the panel unit root test a standard procedure. If the variables in equations (4) and (5) are not stationary, a panel cointegration technique may be needed to examine the cointegration relationship among the variables.

A primary purpose for using the panel unit root is to enhance the power of the test due to its large sample property. Even though such a property can also be obtained by increasing the length of the sample period, a long time period may well suffer from structural changes. On the other hand, the cross section property of panel data introduces heterogeneity and hence weakens the testing power. In the initial stage of the panel unit root models, heterogeneity is not allowed for (e.g., Levin and Lin, 1993; Breitung and Meyer, 1994; and Quah, 1994). A recent advance by Im, Pesaran, and Shin (IPS, 1997) takes cross-section heterogeneity into consideration. We briefly describe the IPS model:

Given y_{it} ($i = 1, \cdots, N$ and $t = 1, \cdots, T$), we generate $\tilde{y} = y_{it} - (\frac{1}{N}) \sum_{i=1}^{N} y_{it}$, and apply the

following model similar to the Augmented Dicky - Fuller tes t (ADF) on $\tilde{y}_{it}$

$$\Delta \tilde{y}_{it} = \alpha_i + \beta_i \tilde{y}_{it-1} + \sum_{j=1}^{w_i} \gamma_{ij} \Delta \tilde{y}_{it-j} + \varepsilon_{it} \tag{6}$$

We use $t_{ij}(w_i)$ to test the null of (6) $\hat{\beta}_i = 0$ and define the t - bar statistic from the following relation :

$$\bar{\Gamma_{\bar{t}}} = \sqrt{N}\,(\bar{t}_{NT} - a_{NT})/\sqrt{b_{NT}} \quad \sim \quad N(0,1)$$

$$\text{where} \quad \bar{t}_{NT} = \frac{1}{N}\sum_{i=1}^{N} t_{iT}(w_i),$$

$$a_{NT} = (\frac{1}{N})\sum_{i=1}^{N} E[\,t_{iT}(w_i)\,/\,\beta_i = 0\,], \text{and} \tag{7}$$

$$b_{NT} = \frac{1}{N}\sum_{i=1}^{N} V[\,t_{iT}(w_i)\,/\,\beta_i = 0\,].$$

Note that a_{NT} and b_{NT} in (7) need to be estimated before applying the IPS test. Table 2 of their paper provides the simulated values from the Monte Carlo experiments. The testing power is shown to be superior to that by Levin and Lin (1993). The t statistics from (6) for all states are then substituted into (7) to obtain the t–bar statistics. A comparison with a 5-variable critical value at $\alpha = 5\%$ is shown in Table 1.

A perusal of Table 1 indicates the panel variables are stationary. As such, we employ the pooling technique (fixed effect model) to estimate the demand structure of (4).

Discussion of the Empirical Results

The advantage of pooled estimators over heterogeneous estimators (individual state demand functions) is that individual regressions often yield unreliable and implausible coefficients. For example, long run price elasticity using OLS varies from 5.46 to −8.46 and from 5.23 to −7.46 using 2SLS (Baltagi et al., 2000) despite that Pesaran and Smith (1995) showed the average of these estimates (mean group or MG estimator) is consistent as N and T approach infinity. Besides, the MG estimator is sensitive to outliers (Pesaran et al., 1999, p. 629). Other heterogeneous estimators such as shrinkage methods by Maddala et al. (1994) produce equally unstable estimates (Baltagi et al., 2000, p. 122). This reduces the choice to the two remaining estimators: the dynamic fixed-effect (DFE) models in which parameter homogeneity prevails except intercepts and the pooled mean group (PMG) model where intercepts, short-run elasticities, and error variance can vary but the long-run price elasticity is constrained to be the same for all states. While it seems true that income elasticity of consumption should approach 1 in the long run for a country according to Pesaran et al. (2000), the price elasticity may not converge to a common value even in the long run for an addictive commodity. Such a non-convergent and volatile demand structure could be traced to the switching elasticity theorem by Greenhut et al. (1974). As the phenomenon of declining price elasticity at higher prices has been borne out in many empirical studies (e.g., Keeler et al., 2001), its role cannot be downplayed.

In the panel data estimation, one may choose between the random effect and fixed effect models. According to Judge et al. (1988), the difference in results is miniscule for a large T and a small N. In the case of a small T relative to N, the fixed-effect model is inefficient though consistent and, as such, a random effect model may be preferred. However, the correlation between characteristics pertaining to cross sections and explanatory variables may render estimates biased in the random effect model. For a relatively smaller T (as T = 42, N = 43 in our case) the worst in applying the fixed effect model is its inefficiency. Above all, 41 regions constitute nearly 85% of the entire U.S. and as such, treating parameters as fixed is a reasonable assumption.

Besides, Baltagi *et al.* (2000) and Freeman (2000) pointed out that the fixed effect model performs better in prediction among the pooling techniques. Consequently, we apply the fixed effect model to estimate (4). The choice of the fixed effect model is also confirmed by the Hausman specification test.[6] Since the cigarette price is determined by both demand and supply, an OLS estimate of (4) could very likely lead to a simultaneous equation bias. Notice that the lagged consumption (Q_{it-1}) is considered an endogenous variable for it is related to the error term. As such, we opt for the two-stage least square (2SLS) model with the instruments of real cigarette tax, elderly population proportion ages 65 years and older, lagged values of real cigarette tax, cigarette price, neighborhood price, per capita sale, and per capital income.[7] Results are shown in Table 2.

An inspection of Table 2 indicates a price, income, and substitution elasticity of -0.41, 0.06, and 0.3 respectively with expected signs and significant t values. The coefficient on the elderly population proportion appears to be statistically significant with the correct sign: as the elderly population proportion increases, cigarette consumption is expected to decrease. Note that the OLS estimates differ from that of the 2SLS which takes the endogeneity problem into consideration. Our 2SLS estimates (1954-1997) are very similar to that by Keeler *et al.* (2001) in which the omitted variable is considered for the sample period of 1960-1990 and panel estimate results by Yurekli and Zhang (2000). In fact, prior panel results are rather diverse. For example, the price and income elasticities by Baltagi *et al.* (2000), who used 1963-1992 panel data are in the range of -0.009 ~ -0.48 and -0.04 ~ 0.19. The elasticities are found to be between -0.47 ~ -0.61 and -0.0001 ~ 0.0002 by Yurekli and Zhang, who used 1970-1995 panel data. Clearly, the results vary with respect to sample lengths. As more evidence has surfaced against smoking, it is of interest to investigate the size of these elasticities. To this end, we make use of the rolling window approach in estimating the elasticities. That is, for a window length of 15 years, we first estimate the 1975 elasticities via the window period of 1961-1975. For the 1976 elasticities, we use the window of 1962-1976. Finally, the 2002 elasticities are estimated via the 1988-2002 windows. Both price and income elasticies, short run and long run, for the window length of 20 and 25 years are estimated and shown in Figure 1.

A perusal of Figure 1 suggests that short run price elasticity values are between -0.25 and -0.7 for all three windows. Roughly speaking, the longer the length of the window is, the less volatile the estimated price elasticity will be. Furthermore, it appears that the price elasticity is on the decline (absolute value). Moreover, volatility of the income elasticity is also in decline, most likely due to increasing alertness about harmful cigarette smoking. In sum, we have witnessed declining values for both elasticities as more smokers quit smoking; the remaining "hardcore" consumers represent the "trapped" price inelastic group. Beyond that, cigarette consumption tends to become an inferior good. Note that estimated time-varying elasticities seem to become stable once the sample length reaches 25 years. For instance, the rolling elasticity values are sensitive to the end period (1992 or 1995) if 15-year or 20-year sample data are used. Prior estimates on price elasticity, -0.2581 by Thursby and Thursby (2000) and -0.49 ~ -0.61 by Yurekli and Zhang (2000), vary noticeably once the sample length is not long enough. Based on our simulations, it seems a reliable estimate on elasticities entail the use of a longer sample length (*e.g.*, at least 25 years) in estimating cigarette demand.

Long-run price and income elasticities (-2.5 ~ -1.0) and (0.1 ~ 0.8) are similar as they largely mirror the short-run elasticities via the coefficient of lagged consumption (ρ). To dissect the price elasticity of cigarettes, we need to divide its price into pre-tax price PT_{it} and cigarette tax Tax_{it} deflated by CPI. The resulting demand can then be estimated based on the following formulation:

$$\ln Q_{it} = \alpha_1 + \rho \ln Q_{it-1} + \beta_{11} \ln PT_{it} + \beta_{12} \ln Tax_{it} + \beta_2 \ln Y_{it} + \beta_3 \ln Pn_{it} + \beta_4 T + \hat{a}_5 \text{ Age } 65_{it} + \mu_{it} \tag{8}$$

The time-varying pre-tax and tax elasticities of 20-year and 25-year moving windows estimated by the instrument variable approach are presented in Figure 2.

As with price elasticity of demand in which expenditure-income share is an important determinant, real tax / real price ratio must be considered in explaining demand price elasticity of cigarettes. Let us denote gross price and tax by P and t. Hence, the net price PT is

$$PT = P - t. \tag{9}$$

A few algebraic manipulations lead readily to

$$\varepsilon_p = \frac{\varepsilon_{PT} \cdot \varepsilon_t}{(1 - \alpha)\varepsilon_t + \alpha \varepsilon_{PT}}, \tag{10}$$

where ε_p and ε_{PT} denote price and pre-tax elasticity, ε_t is tax elasticity; and α is tax share shown as a percentage to price or t/P. In general, tax elasticity decreased in absolute value when tax share α decreased until about 1992 (Figure 2 (a)). In contrast, cigarette consumption was relatively more tax elastic when the tax share was higher in most of the 1970's. With a greater tax share, cigarette consumption is more sensitive to the tax price. In terms of the denominator of equation (10), for each increase in α, the decrease in $(1-\alpha)\varepsilon_t$ is greater than an increase in $\alpha \varepsilon_{PT}$: tax elasticity is generally greater than pre-tax price elasticity in absolute value (Figures 2a and 2b), i.e., $|\varepsilon_t| > |\varepsilon_{PT}|$ for a stable numerator. The smaller denominator in (10) signals a relatively price elastic cigarette demand around 1974 ~ 1975 (Figure 1). On the other hand, a declining tax share (α) gave rise to a greater denominator in (10). Hence, the price elasticity became much more inelastic around 1992 when the tax share bottomed out to about 25% of the price. That is, the cigarette taxes were relatively small during the period.

Note that equation (10) is derived in deterministic form for small change whereas the time-varying rolling-window elasticties are statistical estimates with random variations. Equation (10) provides simulated elasticity that generally fits estimated elasticity, especially when real price changes are moderate, i.e., -0.025 (simulated value based on (10)) versus −0.029 (rolling window value) for 1995; -0.04 versus 0.04 for 1996; and −0.036 versus 0.044 for 1997.

Income elasticity peaked to 0.24 around 1980 and then gradually plummeted to 0.02 in 1995 (Figure 1). The declining trend can be better understood by disaggregating income (Y_{it}) into earning $(EARN_{it})$, dividend (Dvd_{it}), and transfer payment (Trp_{it}) components.[8] The three income elasticities within the 20-year (Figure 3a) and 25-year (Figure 3c) rolling window frameworks along with the corresponding income shares (Figures 3b and 3d) are shown in Figure 3.[9] An examination of Figure 3a and Figure 3c indicates that both earning elasticity and transfer income elasticity were on the rise before 1992. On the other hand, the dividend income elasticity after 1992 was in sharp decline (25-year window) as was the transfer income elasticity after 1990. Generally speaking, decreasing income elasticity after 1992 (Figure 1b) can be attributed to its three components. Dividend income elasticity had the largest decline, followed by transfer income elasticity and earning elasticity (Figure 3). After 2000, the earning elasticity was actually on a mild rise, due to a recession as cigarette consumption may well be considered counter-cyclical. In short, decreases in both dividend income and transfer income elasticities account to a large degree for the declining income elasticity in the U.S.

As indicated by Keeler *et al.* (1993), cigarette consumption is related to the education level. Generally speaking, a better educated group is more accessible and acceptable to the information of severe health problems from smoking. Beyond that, there appears to be a positive relationship between income and education level. Dividend, as well as interest income, can be viewed as a proxy for the income of that group with more education. In this light, it becomes evident that cigarette smoking is an inferior good for the higher income group. Furthermore, the information about smoking from the Surgeon General in 1986, 1988, 1990, and 1992 has also discouraged smoking to some extent, even for wage income and transfer payment groups. The declining income elasticity can be attributed to (i) increasing income share for dividend recipients and decreasing share for wage earners, (ii) decreasing income elasticities from a sequence of warnings about hazardous health especially in the 1990's, (iii) negative income elasticity for a dividend recipient group, and (iv) declining transfer income elasticity with the principal recipients of the elderly and the poor, who cannot afford more cigarette consumption in the presence of rising prices. This negative relation between the elderly population and cigarette consumption can also be detected in Table 2.

Concluding Remarks

Undoubtedly, the Surgeon General's report on harmful effects caused by cigarette smoking has affected its demand structure. A tax policy could be effective for forward-looking consumers in reducing consumption. Based on the Markov Switching model, Coppejans and Sieg (2002) estimated about 31% of smokers in Los Angeles believe the future price will rise by at least 10%. The relative tax – which went up close to 50% in the beginning but plummeted since the 1970's – is on the rise again. As litigation against big tobacco companies continues to favor the plaintiff, the ability to raise tobacco price plays a critical role. That is, only a very inelastic demand in the future can essentially provide a reliable source for the astronomical amount of settlement costs. This paper makes use of the panel unit root technique recently developed by Im *et al.* (1997) to test the spurious regression problem. Considerable amount of data – 42 states and Washington, D.C. spanning from 1961 to 2002 – are assembled via the fixed effect model to yield the price, income, and substitution elasticities of cigarettes of –0.41, 0.06, and 0.30, respectively. Three interesting results emerge from the panel estimates of the cigarette demand structure. First, this paper using 15-year, 20-year, or 25-year rolling window approaches finds decreasing price (absolute value) and income elasticities. In addition, we show that demand price elasticity is dependent largely on tax share and tax elasticity. The decreasing tax shares contribute to smaller tax elasticity (absolute value), which in turn gives rise to a more inelastic demand. Given the rising state budget deficits, cigarette taxes will increase as will the cigarette price owing to the enormous size of settlement costs imposed on the tobacco industry.

Last, the declining income elasticity can be better analyzed by breaking income down into three categories. The time varying elasticities of individual groups indicates (i) anti-smoking campaigns have contributed to declining income elasticity; (ii) increasing income shares for dividend recipients and decreasing income share for wage earners generally account for decreasing income elasticity because the former are more likely to quit or reduce cigarette smoking; (iii) for dividend recipients, cigarette consumption has recently become an inferior good; and (iv) decreasing transfer income elasticity. Therefore, future cigarette consumption will critically hinge on tax share, income share, price, and income elasticities of remaining hard-core smokers.

Footnotes

* The authors would like to gratefully thank the Editor and a referee for their insightful comments and suggestions.

1. Variables such as beverage, alcohol and bubble gum consumption could have been included in Z_{it}, if available.

2. The partial adjustment model still remains the main framework in studies by Keeler *et al.* (2001), Yurekli and Zhang (2000), and Thursby and Thursby (2000) since the estimated discount rates in the rational addiction model are not feasible or negative.

3. Forty-two states are included in the study owing to the fact that neighboring state prices, and/or elderly population proportion data are not available. The missing eight states are Alaska (AK), Colorado (CO), Hawaii (HI), Maine (ME), North Carolina (NC), New Hamshire (NH), Oregon (OR), and Washington (WA).

4. ADF statistics of average income, average cigarette price, and consumption for 42 states and Washington, D.C. are −1:42, -0.89, and −0.92, respectively with the corresponding critical value of −3.52, -3.52, and −2.93, respectively: an indication of the existence of an unit root without using panel data.

5. The spurious regression leads to estimation bias due to non-standard distribution, hence the correlation estimation may well be biased.

6. We employ the Hausman (1978) specification test with the χ^2 statistic (d. f. = 5) of 39.96 to reject the null hypothesis. Hence, we do not use the random effect model.

7. One lag period is used.

8. Official definitions of the three income groups are due to Current Population Survey and the U.S. Census Bureau.

9. Figures 3b and 3d contain the same information since income shares are not estimated under either window length.

References

Baltagi, B. H. and D. Levin. 1986. "Estimating Dynamic Demand for Cigarettes Using Panel Data: The Effects of Bootlegging, Taxation and Advertising Reconsidered," *The Review of Economics and Statistics,* 68(1), 148-155.

Baltagi, B. H. and D. Levin. 1992. "Cigarette Taxation: Raising Revenues and Reducing Consumption," *Structural Change and Economic Dynamics,* 3(2), 321-335.

Baltagi, B. H., J. M. Griffm and W. W. Xiong. 2000. "To Pool or Not to Pool: Homogeneous versus Heterogeneous Estimators Applied to Cigarette Demand," *Review of Economics and Statistics,* 82(1), 117-126.

Bask, M. and M. Melkersson. 2004. "Rationally Addicted to Drinking and Smoking?" *Applied Economics,* 36(4), 373-381

Becker, Gary, and Kevin Murphy. 1988. "A Theory of Rational Addiction," *Journal of Political Economy,* 96(4), 675-700.

Becker, G. S., M. Grossman, and K. Murphy. 1994. "An Empirical Analysis of Cigarette Addiction," *American Economic Review,* 84(3), 396-418.

Bishop, J. A. and J. H. Yoo. 1985. "Health Care, Excise Taxes and Advertising Ban on the Cigarette Demand and Supply," *Southern Economic Journal,* 52(2), 402-411.

Breitung, J, and W. Meyer. 1994. "Testing for Unit Roots in Panel Data: Are Wages on Different Bargaining Levels Cointegrated," *Applied Economics,* 26(4), 353-361.

Chaloupka, F. 1991. "Rational Addictive Behavior and Cigarette Smoking," *Journal of Political Economy,* 99(4), 722-742.

Chang, H. S. and Y. Hsing. 1991. "The Demand for Residential Electricity: New Evidence on Time-varying Elasticities," *Applied Economics,* 23(7), 1251-1256.

Coats, R. Morris. 1995. "A Note on Estimating Cross - Border Effects of State Cigarette Taxes," *National Tax Journal,* 48(4), 573-584.

Coppejans, M and H. Sieg. 2002. "Price Uncertainty, Tax Policy, and Addiction: Evidence and Implications," NBER Working Paper Series. # 9073, 1-36.

Doron, G. 1979. *The Smoking Paradox: Public Regulation in the Cigarette Industry,* Cambridge, MA: Abt Books.

Escario, J. J. and J. A. Molina. 2001. "Testing for the Rational Addiction Hypothesis in Spanish Tobacco Consumption," *Applied Economic Letter,* 8(4), 211-215.

Freeman, D. G. 2000. "Alternative Panel Estimates of Alcohol Demand, Taxation, and the Business Cycle," *Southern Economic Journal,* 67(2), 325-344.

Fujii, E. T. 1980. "The Demand for Cigarettes: Further Empirical Evidence and Its Implications for Public Policy," *Applied Economics,* 12(4), 479-489.

Gallet, C. A. 2003. "Advertising and Restrictions in the Cigarette Industry: Evidence of State-by-State Variation," *Contemporary Economic Policy,* 21(3), 338-348.

Granger, C. W. J. and P. Newbold. 1974. "Spurious Regressions in Economics," *Journal of Econometrics,* 2(2), 111-120.

Greenhut, M. L., M. J. Hwang, and H. Ohta. 1974. "Price Discrimination by Regulated Motor Carriers: Comment," *American Economic Review,* 64(4), 780-784.

Hamilton, J. L. 1972. "The Demand for Cigarettes: Advertising, the Health Care, and the Cigarette Advertising Ban," *The Review of Economics and Statistics,* 54(4), 441-411.

Hausman, J. A. 1978. "Specification Tests in Econometrics," *Econometrica,* 46(6), 1251-1272.

Hondroyiannis, G. and E. Papapetrou. 1997. "Cigarette Consumption in Greece: Empirical Evidence from Cointegration Analysis,"*Applied Economics Letters,* 4(9), 571-574.

Hsing, Y. and H. S. Chang. 2000. "Testing the Portfolio Theory of Money Demand in the United States," *Economia Internazionale/International Economics,* 56(1), 13-22.

Im, K., H. Pesaran, and Y. Shin. 1997. ""Testing for Unit Roots in Heterogenous Panels," Manuscript, Department of Applied Economics, University of Cambridge.

Judge, G. G., R C. Hill, W. Griffiths, H. Lutkepohl, and T. C. Lee. 1988. *Introduction to the Theory and Practice of Econometrics,* 2nd ed. New York, NY: John Wiley & Sons.

Keeler, T. E., T.-W. Hu, P. G. Barnett, and W. G. Manning. 1993. "Taxation, Regulation, and Addiction: A Demand Function for Cigarette Based on Time Series Evidence," *Journal* of *Health Economics,* 12(1), 1-18.

Keeler, T. E., T.-W. Hu, W. G. Manning, and H.-Y. Sung. 2001. "State Tobacco Taxation, Education, and Smoking: Controlling for the Effects of Omitted Variables," *National Tax Journal,* 54(1), 83-102.

Kim, S. J. and B. J. Seldon. 2004. "The Demand for Cigarettes in the Republic of Korea and Implications for Government Policy to Lower Cigarette Consumption," *Contemporary Economic Policy,* 22(2), 299-308.

Labeaga, J. M. 1993. "Individual Behavior and Tobacco Consumption: A Panel Data Approach," *Health Economics,* 2(2), 103-112.

Laughhunn, D. J. and H. L. Lyon. 1971. "The Feasibility of Tax-induced Price Increases as a Deterrent to Cigarette Consumption," *Journal of Business Administration,* 3(1), 27-35.

Levin, A. and C.-F. Lin. 1993. "Unit Root Tests in Panel Data: New Results," UCSC Working Paper 93-56.

Lewit, E. M. and D. Coate. 1982. "The Potential for Using Excise Taxes to Reduce Smoking," *Journal of Health Economics,* 1(2), 121-145.

Lyon, H. L. and J. L. Simon. 1968. "Price Elasticity of Demand for Cigarettes in the United States," *American Journal of Agricultural Economics,* 50(4), 881-893.

Maddala, G. S., V. K. Srivastava, and H. Li. 1994. "Shrinkage Estimators for the Estimation of Short-run and Long-run Parameters from Panel Data Models," Working Paper, Ohio State University.

Mundlak, Y. 1978. "On the Pooling of Time Series and Cross Section Data," *Econometrics,* 46(1), 69-85.

Nelson, J. P. 2003. "Cigarette Demand, Structural Change and Advertising Bans: International Evidence, 1970-1995," *Contribution to Economic Analysis and policy,* 2(1), 1-27.

Nelson, C. R., and C. I. Plosser. 1982. "Trends and Random Walks in Macroeconomic Time Series: Some Evidence and Implications," *Journal of Monetary Economics,* 10(2), 139-162.

Pesaran, M. H. and R. Smith. 1995. "Estimating Long-run Relationship from Dynamic Heterogeneous Panels," *Journal of Econometrics,* 68(1), 79-113.

Pesaran, M. Hashem, Yongcheol Shin, and Ron J. Smith. 1999. "Pooled Mean Group Estimation of Dynamic Heterogeneous Panels," *Journal of the American Statistical Association,* 94(446), 621-634.

Pesaran, M. H., Y. Shin, and R. P. Smith. 2000. "Structural Analysis of Vector Error Correction Models with Exogenous I(1) Variables," *Journal of Econometrics,* 97(2), 293-343.

Quah, D. 1994. "Exploiting Cross Section Variation for Unit Root Inference in Dynamic Data," *Economics Letters,* 44(1-2), 9-19.

Saffer, H. and F. Chaloupka. 2000. "The Effect of Tobacco Advertising Bans on Tobacco Consumption," *Journal of Health Economics,* 19(6), 1117-1137.

Sloan, F. A., V. K. Smith, and D. H Taylor, Jr. 2002. "Information, Addiction, and 'Bad Choices': Lessons from a Century of Cigarettes," *Economics Letters,* 77(2), 147-155.

Steward, M. J. 1993. "The Effect on Tobacco Consumption of Advertising Bans in OECD Countries," *International Journal of Advertising,* 12(2), 155-180.

Thursby, J. G. and M. C. Thursby. 2000. "Interstate Cigarette Bootlegging: Extent, Revenue Losses, and Effects of Federal Intervention," *National Tax Journal,* 53 (1), 59-77.

Tremblay, C. H. and V. J. Tremblay. 1995. "The Impact of Cigarette Advertising on Consumer Surplus, Profit, and Social Welfare," *Contemporary Economic Policy,* 13(1), 113-124.

Waters, T. M. and F. A. Sloan. 1995. "Why Do People Drink? Tests of the Rational Addiction Model," *Applied Economics,* 27(8), 727-736.

Yurekli, A. A. and P. Zhang. 2000. "The Impact of Clean Indoor – Air Laws and Cigarette Smuggling on Demand for Cigarettes: An Empirical Model," *Health Economics,* 9(2), 159-170.

Table 1. Results of the Panel Unit Root Test

Variables	$\overline{\Gamma}_\iota$
Q_{it}	-2.82*
P_{it}	-2.11*
Pn_{it}	-2.31*
Y_{it}	-4.25*

Note: The 5% critical value, for a_{NT} = 1.470, b_{NT} = 0.801 (with one lag and time trend) is −1.96. The null hypothesis is: the variable has a unit root. * = 1% significance level. ** = 5% significance level.

Table 2. Panel Estimates of the Cigarette Demand

	$\ln Q_{it}$	$\ln P_{it}$	$\ln Y_{it}$	$\ln Pn_{it}$	T	$Age65_{it}$	$\overline{R}^2$
OLS	0.8874 (48.76)	-0.2114* (-7.75)	0.0421 (1.38)	0.1322* (5.37)	-0.0016** (-2.26)	-0.0032*** (-1.67)	0.98
TSLS	0.8490 (42.31)	-0.4067* (-8.81)	0.0579*** (1.64)	0.3036* (7.62)	-0.0019** (-2.31)	-0.0046** (-2.34)	0.98

Note: Numbers in parentheses are t statistics; * = 1% significance level. The estimation is based on the fixed-effect model. Standard errors are hetero-scedasticity-consistent estimator by White. Q_{it} = tax-paid per capita sales for state i and time period t (subscripts); P_{it} = price per pack deflated by CPI; Y_{it} = personal income deflated by CPI; Pn_{it} = average price of adjacent states deflated by CPI; and T = time trend (1961= 1). Age 65 = proportion of population ages 65 years and older in the ith state.

Figure 1. Time Varying Elasticities of Cigarette Demand

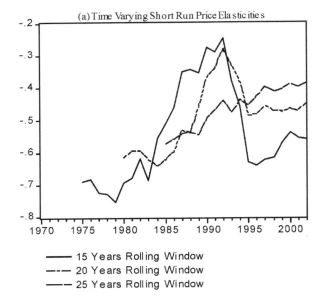

(a) Time Varying Short Run Price Elasticities

——— 15 Years Rolling Window
--- 20 Years Rolling Window
—-— 25 Years Rolling Window

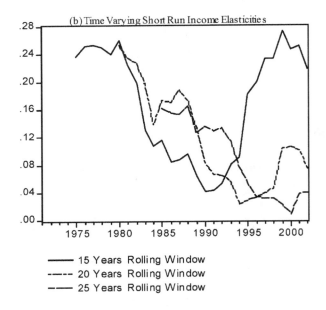

(b) Time Varying Short Run Income Elasticities

——— 15 Years Rolling Window
--- 20 Years Rolling Window
—-— 25 Years Rolling Window

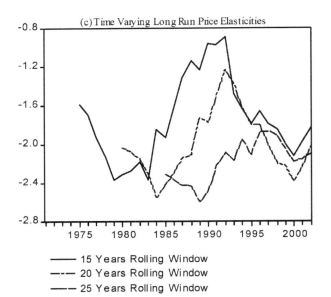

(c) Time Varying Long Run Price Elasticities

——— 15 Years Rolling Window
--- 20 Years Rolling Window
—-— 25 Years Rolling Window

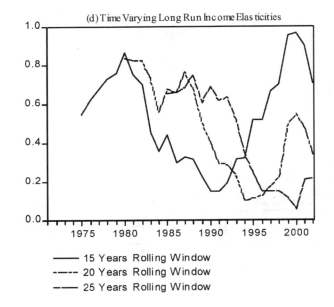

(d) Time Varying Long Run Income Elasticities

——— 15 Years Rolling Window
--- 20 Years Rolling Window
—-— 25 Years Rolling Window

Figure 2. Tax and Pre-Tax Elasticities and Tax Share(20-year rolling window)

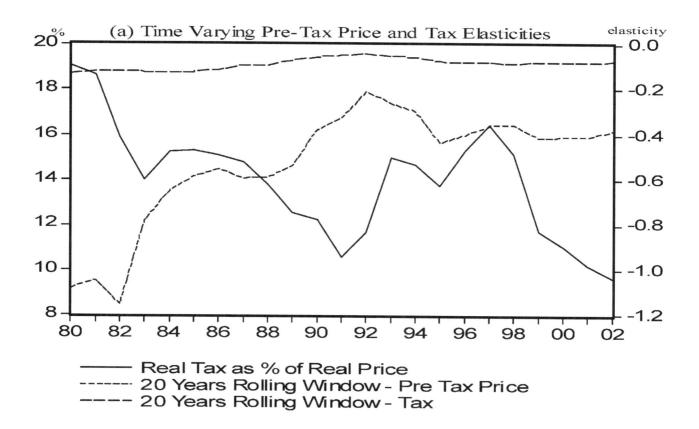

Tax and Pre-Tax Elasticities and Tax Share(25-year rolling window)

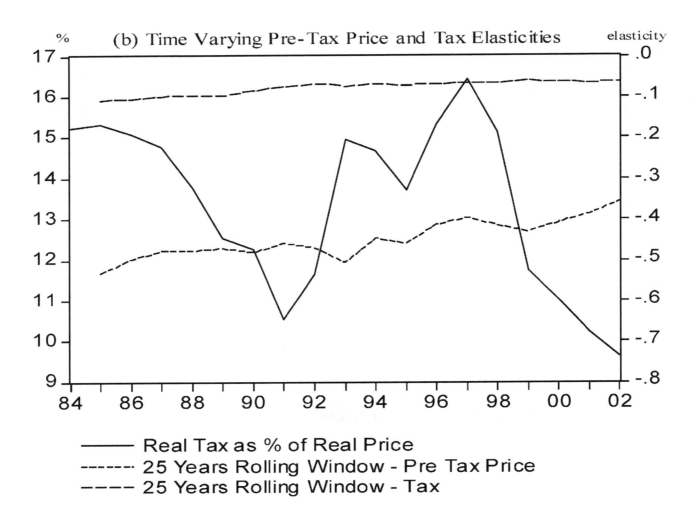

(b) Time Varying Pre-Tax Price and Tax Elasticities

——— Real Tax as % of Real Price
------- 25 Years Rolling Window - Pre Tax Price
– – – 25 Years Rolling Window - Tax

Figure 3. Component Income Elasticities and Income Shares

20-year rolling window

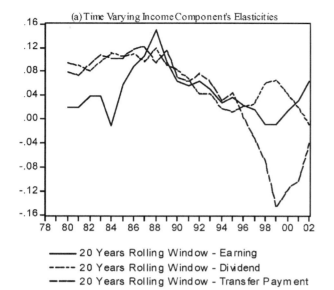

(a) Time Varying Income Component's Elasticities

——— 20 Years Rolling Window - Earning
----- 20 Years Rolling Window - Dividend
—-— 20 Years Rolling Window - Transfer Payment

25-year rolling window

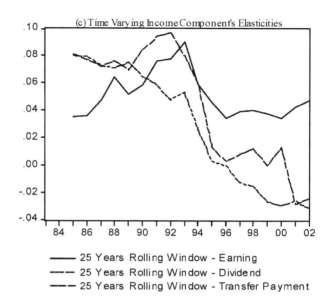

(c) Time Varying Income Component's Elasticities

——— 25 Years Rolling Window - Earning
----- 25 Years Rolling Window - Dividend
—-— 25 Years Rolling Window - Transfer Payment

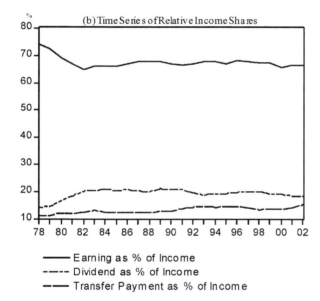

(b) Time Series of Relative Income Shares

——— Earning as % of Income
----- Dividend as % of Income
—-— Transfer Payment as % of Income

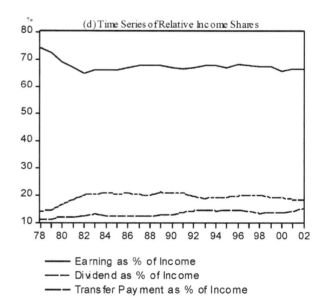

(d) Time Series of Relative Income Shares

——— Earning as % of Income
—-— Dividend as % of Income
—-— Transfer Payment as % of Income

175

III. Granger Causality Models with Thresholds

Threshold regression dates back to the pioneering work by Tong (1978, 1983). It has a very wide array of applications in Biostatistics and Economics. For instance, dosage applied before reaching some threshold value may have no or minimum effect. But after the threshold amount of dosage, it may very well possess desirable effects. In economic theory, for example, it is well known that consumption pattern alters noticeably from and after certain threshold income. The problem with the Chow test is: it requires a priori knowledge of break points and hence the model is not data driven. In the case when there are more than one break point, small sample property prevails to the extent that hampers reasonable statistical estimation.

A threshold point c determined by data (not by a priori known event) divides sample space into two or more regimes. Each one has a distinctively different response level, and this makes it a nonlinear model. The problem of the threshold regression lies in its discontinuity and as such the inference on the threshold point may be suspect. In addition, its asymptotic distribution of the threshold estimator is typically nonstandard. Yu (*Econometric Theory*, vol.30, pp. 676-714, 2014) advocated a bootstrapping approach. However, technics of bootstrapping are not generally regarded as an antidote. Caner and Hansen (2001) developed an asymptotic theory of inference for two-region threshold autoregressive model with an autoregressive unit root (see *Econometrica*, 2000). In this case, they considered both the nonlinear threshold effect and non-stationarity problem. It was found that there existed a strong threshold effect embedded in the monthly unemployment rate. The threshold effect came from short-run dynamics, not in the unit root. As a result, care must be exercised that threshold effect can be confounded in the guise of non-stationarity.

The paper "Demand for Cigarettes Revisited: An Application of the Threshold Regression Model" utilized a four-regime panel model (dynamic fixed effect) to estimate cigarette demand in the US. The results include: (1) income elasticity is positively significant for income less than $8568, but it became insignificant for income exceeding $18196. It signals that low income consumers increased cigarette consumption as long as their income increases. (2) price elasticity reached its largest value for income threshold exceeding $18196 suggesting kicking the habit was found only in the high income group. (3) inelastic demand for the threshold income range between $11296 and $18196 reflects the difficulty in kicking the smoking habit.

The procedure of the Hansen model (1999) was used to search for optimum threshold value c* (minimum sum of squared error from the threshold autoregressive model). If the null hypothesis that linear regression is sufficient is rejected, one may opt for a threshold regression with lagged dependent variable. The price elasticity in this case was estimated to be around -0.249.

Using 82 countries over the period of 1971-2002, Hwang et al. (2008) applied the nonlinear threshold model by Hansen (1997) to investigate the energy policies in the sample. This paper "Does Energy Consumption Bolster Economic Growth: An Application of the Nonlinear Model" addressed such relationships between energy consumption and economic growth. The model has near optimum power with threshold value determined by data. Within the framework of panel Granger causality model with thresholds, it was found that 48 of the 82 countries has their thresholds in terms of the 4 variables below the optimal values, indicating aggressive energy policy was taken to combat pollution. On the other hand, the remaining 34 of the 82 countries have their threshold values exceeding their optimal values signaling that policy of energy conservation ought to be taken.

Cointegration and threshold regression can be combined to study futures market in Finance. The paper "The Dynamics of a Nonlinear Relationship between Crude Oil Spot and Futures Prices: A Multivariate Threshold Regression Approach" did just that. Daily data from January of 1986 to April of 2007 were collected and divided into 3 sub-periods. The first two were characterized by the fact that spot price was in general greater than the futures price. In contrast, Period 3 spanning from September 11, 2001 to April 30, 2007 had futures price greater than spot price. If the absolute value of the basis (futures price minus spot price) exceeds threshold value, it implies there is a chance for arbitrage. In this case, one of the error correction coefficients is expected to be significant, indicating that an adjustment is in the making toward the equilibrium.

To facilitate the hypothesis testing, the ADF unit root, Engle-Granger two step cointegration, the AIC as well as the cd statistics by Tsai (1998) were employed. Cointegration relationship was identified between logarithmic spot and futures prices. And at least one causal relationship was found between crude oil spot and futures prices. It is good to see the threshold model we presented predicts better than the linear model in terms of Diebold and Mariano's criterion (1995).

Military spending has a huge potential impact on economies since arms race has become a norm across the world. Does military spending give rise to economic growth? The paper of "On the Relationship between Expenditure, Threat, and Economic Growth: A Nonlinear Approach " addresses this problem via using the threshold regression model . An F statistics of two residual variances (linear versus threshold models) was calculated to test the null hypothesis that the linear model is sufficient.

We rejected the null and some interesting causal relationships were found. For the 23 countries whose income is less than \$475.93, military expenditure Granger-caused negatively the economic growth. This is a significant finding: for developing countries, it doesn't pay to get into arms race with their neighbors. These were countries when threat was heightened, economic growth would truly suffer significantly. The remaining 69 countries whose income exceeded \$475.93, military spending led economic growth but only insignificantly. It is little wonder that it pays for a developed economy such as the US to increase military spending. On the other hand, it is unwise for a third world country to purchase relatively obsolete weaponry: it puts a dent on economic growth on the one hand; it contributes scarce funds to the already rich economies on the other hand.

A multivariate threshold model was proposed by Bwo-Nung Huang (2008, *Energy Journal*) to study economy's tolerance and length of response to a positive oil shock. More often than not, we are talking about the impact of oil price increase. Three results surfaced from the analysis. (1) As an economy matures, its level of energy use in industry and transportation is expected to be lower. As a result, the threshold of tolerance is higher when a shock of positive oil price is administered into the system. (2) a lower ratio of energy use and/or the lower import ratio is, the longer the delay. (3) As an economy becomes more advanced, the length of response time from a positive oil price shock remains longer. The paper deals with unusual topics which actually sheds a new light on the topic.

Arms race across the Taiwan Straits has been an important episode on the stage of international politics. What are the real economic consequences and why? We had a unique opportunity to join General Lai, an expert on the topic. Our paper "Defense Spending and Economic Growth across the Taiwan Straits: A Threshold Regression Model "clearly illustrated the consequences of such a catch-up game. Data of military spending and economic growth from 1953-2000 were employed along with the Phillips-Perron unit root test, Johansen (1988) Johansen and Juselius (1990) cointegration tests before conducting the Granger causality.

Using rival's spending as threshold variable, we obtained the following results. First, China's defense spending Granger-caused that of Taiwan. Second, there is a two way feedback relationship between defense spending and economic growth. Third, China's national defense spending Granger-caused her own economic growth. The results indicate China's defense spending is the cause of the arms race. And China's defense spending led to economic growth while that of Taiwan did not really have a clearly recognizable pattern.

The paper "Tourism Development and Economic Growth: A Nonlinear

Approach" was the first paper that applied threshold regression to the tourism-economic growth nexus (Po and Huang 2008). Employing data on 88 countries over 1995-2005, the results indicated segmentation of the sample into three different regimes is actually appropriate based on the threshold variable: the degree of tourism specialization or receipts from international tourism as a percentage of GDP. In regime I where the tourism specialization is less than 4.0488% (57 countries) and regime III where tourism specialization is greater than 4.7337% (23 countries).

In both regime I and III, there was a significant and positive Granger causality relationship between tourism specialization and economic growth. In this light, tourism being an important source for the economic growth ought to use as a primary tool to boost economic growth. In regime II (8 countries), however, such a causal relationship is lacking. An in-depth analysis indicated that, in these countries, the ratio of the value added of the service industry to GDP was among the lowest as was the forest-covered size as percentage of total area of the country. It suggested that these issues ought to be improved upon for countries of regime II in the near future.

Agricultural Economics 34 (2006) 81–86

Demand for cigarettes revisited: an application of the threshold regression model

Bwo-Nung Huang [a,b,*], Chin-Wei Yang[c]

[a]Department of Economics and Center for IADF, National Chung-Cheng University, Chia-Yi, Taiwan 621
[b]Department of Applied Economics, National Chia-Yi University, Chia-Yi, Taiwan 600
[c]Department of Economics, Clarion University of Pennsylvania, Clarion, PA 16214-1232, USA

Received 4 June 2004; received in revised form 19 November 2004; accepted 24 February 2005

Abstract

Recent estimates of the income elasticity of cigarette demand have pointed to a disturbing result: a nearly zero or sometimes negative income elasticity. In order to explore the nonlinearity embedded in the cigarette demand structure, we employ a four-regime panel model (dynamic fixed effect) to estimate the cigarette demand function in the United States. The results indicate that income elasticity is (i) positively significant for the income level less than 8,568 US\$, (ii) positive but statistically insignificant for the income greater than 18,196 US\$, and (iii) negatively significant for the income range between 8,568 and 18,196 US\$. In addition, we find that the price elasticity assumes the greatest absolute value for the income level in excess of 18,196 US\$, but becomes most inelastic for the income level between 11,129 and 18,196 US\$.

JEL classification: D12

Keywords: Threshold regression model; Multi-regime; Cigarettes demand; Elasticity

1. Introduction

The issue of cigarette demand has long been an important topic and has become especially salient in the wake of the recent mammoth settlements between state governments and cigarette companies. In 1989, the Surgeon General of the United States concluded that smoking was the chief single avoidable cause of death in American society and the most important public health issue of our time (U.S. Department of Health and Human Services, 1989). Not surprisingly, many policies over the next two decades have targeted cigarette consumption. Clearly, an accurately estimated demand structure would contribute greatly to the creation of effective policy. Nonetheless, prior estimates on price and income elasticities present rather divergent results: using the U.S. panel data (1963–80), Baltagi and Levin (1986) estimated the price elasticity to be between −0.215 and −0.225 and income elasticity to be between −0.002 and 0.004 (statistically insignificant). Using panel data for the years 1972–1990, Thursby and Thursby (2000) found the price and income elasticities to be −0.2581 and 0.1089, respectively. Similarly, Coats (1995) obtained price and income

elasticities in the ranges −0.005 and −0.016, and 0.0083 and 0.078, respectively. Yurekli and Zhang's (2000) results suggest that the elasticity values are between −0.48 and −0.62, and −0.0001 and −0.0002, respectively, while Baltagi et al. (2000) report values between −0.21 and −0.50, and −0.04 and 0.14, respectively. Furthermore, Keeler et al. (2001) estimated price and income elasticities to lie between −0.22 and −0.29, and 0.23 and 0.28, respectively. Recently, Huang et al. (2004), using panel data for the years 1961–2002, found the price and income elasticities estimates to be −0.41 and 0.06, respectively. In sum, the demand estimates of the U.S. cigarette market have not been consistent. Two trends, however, seem to stand out. First, declining income elasticity suggests that cigarette consumption has become less dependent on income. According to some analyses, cigarettes have become inferior goods, although these results are statistically insignificant. Second, the price elasticity is found to vary according to income level.

The purpose of this article is to estimate the demand for cigarettes in the United States by applying Tong's (1990) threshold regression method to a dynamic fixed effects (DFE) demand model and data for 47 U.S. states from 1963 to 1997. The organization of the article is as follows: Section 2 discusses the data and research methodology used in this analysis. Section 3 provides empirical results and Section 4 concludes.

*Corresponding author. Tel.: +88 652720411 ; fax: +88 52720816.
E-mail address: ecdbnh@ccu.edu.tw (B.-N. Huang).

2. Source of data and empirical methodology

Our panel data consist of 47 states and the District of Columbia from 1963 to 1997 with a total of 1,645 observations obtained from the Tax Burden on Tobacco compiled by the Tobacco Institute.[1] Per capita consumption, Q_{it}, is measured by tax-paid sales in packs. Per capita personal income, Y_{it}, is in 1982–1984 dollars, and P_{it} is the real price including taxes per pack. Following Baltagi and Levin (1986) and Baltagi et al. (2000), we estimate the demand function based on the partial adjustment model:[2]

$$\ln Q_{it} = a_{0i} + \beta_1 \ln Q_{it-1} + \beta_2 \ln P_{it} + \beta_3 \ln Y_{it} + \beta_4 X_{it} + \varepsilon_{it}, \tag{1}$$

where ε_{it} is the white noise, $i = 1, 2, \ldots, 47$, and X_{it} denotes other explanatory variables such as prices of substitutes or complements, and education level.[3]

Generally speaking, a demand function can be derived from the utility maximization paradigm (e.g., the Stone Geary utility function or other separable preference functions). For empirical estimates to be consistent with theoretical results, budget restrictions and homogeneity are imposed: $e_{i1} + e_{i2} + \cdots + e_{in} + e_{iy} \equiv 0$ and $\sum_I e_{iy} s_i \equiv 1$, where e_{i1} is the cross price elasticity between commodity i and 1; e_{iy} is the income elasticity for the ith commodity; and s_i is the expenditure share of commodity i. The former can be readily tested if prices of related goods are included in the demand function for commodity i. For a system of demand functions, symmetry and homogeneity can also be imposed (Silberberg, 1990). However, the validity of the restriction test relies on the Monte Carlo method in which an exact finite sample distribution is not known (Bera et al., 1981). Furthermore, the prices of related goods in the cigarette demand function are not available for the entire panel data set of 47 states and 35 years. Consequently, we do not test homogeneity. Given that demand for cigarette takes its form directly from the rational addiction model developed by Becker and Murphy (1988), whose theoretical implications have been extensively studied, we, like most prior studies, estimate the cigarette demand based on the Eq. (1): the partial adjustment model or a restricted version of the rational addiction model. As is typically done with panel data, Eq. (1) can be estimated

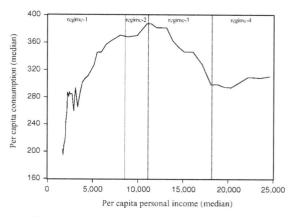

Note: The demarcation points are derived from the four-regime model based on income levels.

Fig. 1. Nonlinear relationship between per capita consumption (median) and per capita personal income (median).

via the fixed effect model as 47 states comprise nearly the entire cross section.[4]

Before estimating the demand function Eq. (1), it is important to examine the relationship between income (median) and cigarette consumption (median).[5] As is evident from Fig. 1, a positive relationship prevails for incomes below 11,000 US\$. However, this relationship disappears and even turns negative for incomes greater than 11,000 US\$ and less than 18,000 US\$. Finally, the relationship between cigarette consumption and income become slightly positive for income levels greater than 18,000 US\$. Clearly, cigarette demand seems to have changed along with income, and this provides ample justification for employing a nonlinear specification. The nonlinearity between income and cigarette consumption renders the conventional one-regime regression model suspect, especially as more and more income elasticity estimates turn out to be negative or insignificant. Fig. 1 lends support to recent empirical findings that speak to the phenomenon of declining income elasticity across income levels.

An appropriate nonlinear model is the threshold autoregressive (TAR) model developed by Tong (1990). Within the framework of the two-regime model, our cigarette demand function can be formulated as

$$q_{it} = (\alpha_{10i} + \beta_{11} q_{it-1} + \beta_{12} p_{it} + \beta_{13} y_{it})(1 - I[z_{it-d} > c]) + (\alpha_{20i} + \beta_{21} q_{it-1} + \beta_{22} p_{it} + \beta_{23} y_{it}) I[z_{it-d} > c] + \varepsilon_t, \tag{2}$$

where z_{it-d} is the threshold variable; d denotes delay or lag length; c represents the threshold value; and $I[A]$ is the index variable which equals one for A (greater than the threshold

---

[1] Colorado, North Carolina, Oregon, and Virginia are excluded from the study in order to maintain a balanced panel.

[2] Even though the rational addition model developed by Becker and Murphy (1988) is more general by including future cigarette consumption, the econometrics of the model is problematic. For instance the implied discount rate is −225% estimated by Becker et al.(1994) and is 40% by Baltagi and Griffin (2001). Both rates are clearly implausible.

[3] Panel data in demeaned forms are often used in nonlinear estimation and as such qualitative indices are not germane to the estimation of cigarette demand. The paucity of the panel data on other explanatory variables makes full estimation extremely difficult. Beyond that, the purpose of this article is to investigate the potential changes in both price and income elasticities under different regimes using the threshold variable, which is determined within the model. We restrict our model to a limited number of explanatory variables.

[4] We employ the Hausman (1978) specification test and obtain a χ^2 statistic (df = 3) of 42.87 to reject the null hypothesis of random effects.

[5] Median income and median consumption are calculated across 47 states.

value) and zero otherwise. The q_{it}, p_{it}, and y_{it} denote the natural logarithms of Q_{it}, P_{it}, and Y_{it}, respectively. Equation (2) indicates that the demand structure of cigarette is represented by $\mathbf{B}_1' = \{\alpha_{10i}, \beta_{11}, \beta_{12}, \beta_{13}\}'$ when the value of the threshold variable is less than or equal to c, but by $\mathbf{B}_2' = \{\alpha_{20i}, \beta_{21}, \beta_{22}, \beta_{23}\}'$ when the value is greater than c. As z_t needs to be a predetermined variable, a common practice is to choose lagged values of some variable, z_{t-d}, in which d denotes delay. To ascertain whether z_{t-d} is an appropriate threshold variable, we perform a test linearity versus nonlinearity.

TAR models are usually applied to time series or cross-section data. The panel TAR model, however, is often "demeaned" first (within model) to remove α_i as suggested by Hansen (1999):[6]

$$q_{it}^* = (\beta_{11}q_{it-1}^* + \beta_{12}p_{it}^* + \beta_{13}y_{it}^*)(1 - I[z_{it-d} > c])$$
$$+ (\beta_{21}q_{it-1}^* + \beta_{22}p_{it}^* + \beta_{23}y_{it}^*)I(z_{it-d} > c) + \varepsilon_t^*, \quad (3)$$

where $q_{it}^* = q_{it} - \bar{q}_i$; $p_{it}^* = p_{it} - \bar{p}_i$; $y_{it}^* = y_{it} - \bar{y}_i$; $\varepsilon_{it}^* = \varepsilon_{it} - \bar{\varepsilon}_i$; $\bar{q}_i = (1/T)\sum_{t=1}^{T} q_{it}$; $\bar{p}_i = (1/T)\sum_{t=1}^{T} p_{it}$; $\bar{y}_i = (1/T)\sum_{t=1}^{T} y_{it}$; and $\bar{\varepsilon}_i = (1/T)\sum_{t=1}^{T} \varepsilon_{it}$. Recall that lower case variables are in natural logarithm. In estimating Eq. (3), we arrange the value of the threshold variable (z_{it-d}) in ascending order to search for the optimum threshold value c^* via a grid search or

$$c^* = \arg\min S_1(c), \quad (4)$$

where $S_1(c)$ denotes sum of squared errors (SSE) from Eq. (3) under different c values.[7] Before estimating Eq. (3), it is necessary to examine the existence of the nonlinear relationship or Eq. (3) in terms of z_{it-d}. The null hypothesis ($H_0 : \mathbf{B}_1 = \mathbf{B}_2$) is that a traditional linear model is sufficient. Only after the null is rejected, can we estimate TAR model or Eq. (3). In this case, we employ the likelihood ratio to test the nonlinear relationship:

$$LR_j(c^*) = \left(\frac{S_j - S_{j+1}}{\hat{\sigma}_{j+1}^2}\right), \quad (5)$$

where S_j and S_{j+1} are the SSE under $H_0 : \mathbf{B}_1 = \mathbf{B}_2$, and $H_1 : \mathbf{B}_1 \neq \mathbf{B}_2$, respectively, $\hat{\sigma}_{j+1}^2$ denotes the residual variance under H_1. We let $j = 1$ when the one-regime is tested against the two-regime model; $j = 2$ when the two-regime model is tested against the three-regime model; and so forth. Strictly speaking, the statistic in Eq. (5) does not have an exact χ^2 distribution as c is not identified under H_0 (Hansen, 1999). As a result, critical values may be simulated by a bootstrap method.

---

[6] See Hansen (1999) for details.

[7] The data are arranged in ascending order for estimation. Due to the degrees of freedom problem, the smallest and largest 5% or 10% data are removed. For the remaining sample space bounded by threshold values $[\underline{c}, \bar{c}]$, n grids are partitioned to estimate n Eq. (3) so that the optimum threshold value c^* satisfies Eq. (4).

Table 1
LR statistics and critical values of the static demand model: bootstrap method (10,000 replications)

H_0	LR statistic	90% Critical values	95% Critical values	99% Critical values
Linear relation	196.35	23.47	27.76	36.83

Note: The LR statistic is calculated based on Eq. (5).

In the case of $j = 3$, if $LR_3(c^*)$ exceeds the critical values derived from the bootstrap method, we may apply a four-regime model as below.

$$q_{it}^* = (\beta_{11}q_{it-1}^* + \beta_{12}p_{it}^* + \beta_{13}y_{it}^*) \cdot I(z_{it-d} \leq c_1)$$
$$+ (\beta_{21}q_{it-1}^* + \beta_{22}p_{it}^* + \beta_{23}y_{it}^*) \cdot I(c_1 < z_{it-d} \leq c_2)$$
$$+ (\beta_{31}q_{it-1}^* + \beta_{32}p_{it}^* + \beta_{33}y_{it}^*) \cdot I(c_2 < z_{it-d} \leq c_3)$$
$$+ (\beta_{41}q_{it-1}^* + \beta_{42}p_{it}^* + \beta_{43}y_{it}^*) \cdot I(z_{it-d} > c_3) + \varepsilon_t, \quad (6)$$

in which c_1, c_2, and c_3 represent three different threshold values. The number of regimes used in our analysis depends on the LR statistic.

3. Discussion of empirical results

We first estimate the single-regime cigarette demand function. Since cigarette price is determined by both demand and supply, an OLS estimate of Eqs. (1), (3), or (6) could suffer from simultaneous equation bias. Note that the lagged consumption (Q_{it-1}), typically considered a predetermined variable in demand estimations, can actually be correlated with error terms. This is especially the case under rational addiction or in partial adjustment models in which past consumption is a key factor in habit formation. For these reason, we opt for the two-stage least square (2SLS) approach with the instruments of cigarette tax, lagged values of cigarette price, and per capita sales.[8] Since Hansen's methodology is typically used in nondynamic fixed effect models, we first exclude q_{it-1}^* from the demand function. Prior to estimating the equation we need to examine, using the previous year's income as the threshold variable, whether the nonlinear effect indeed exists via the likelihood ratio of Eq. (5).[9] The likelihood ratio along with values from the bootstrap method (10,000 repetitions) are reported in Table 1.[10]

Inspection of Table 1 reveals that the null hypothesis of a linear demand function can easily be rejected as the calculated likelihood ratio far exceeds the 99% critical value. Table 2 reports the estimation results of the TAR model using the previous year's income as the threshold variable.

---

[8] The instrument variables include the following p_{it-1}, p_{it-2}, q_{it-2}, q_{it-3}, Tax_{it}, Tax_{it-1}. However, the result from using only p_{it-1}, q_{it-2}, and Tax_{it} are similar.

[9] Fig. 1 indicates that cigarette consumptions have varied with per capita income as suggested by the theory of demand. Thus, the previous year's income is a good candidate for the threshold variable.

[10] The procedures for panel data bootstrap are described by Hansen (1999, p. 351). We used Eviews 4.0 for model estimation.

Table 2
Estimates of one- and two-regime demand functions

	p_{it}^*	y_{it}^*	$\bar{R}^2$	SSE	Threshold value	Sample size
One-regime model	−0.970 (−51.29)	0.005 (0.349)	0.676	14.868		1,645
Two-regime model			0.710	13.278		
Regime I	−0.882 (−43.856)	0.036 (2.598)			$y_{it-1} \leq 15,847$	1,478
Regime II	−1.070 (−23.456)	−0.318 (−6.687)			$y_{it-1} > 15,847$	167

Note: Values in parenthesis are *t*-statistics. SSE = sum of squared errors.

The single-regime (static) demand function provides an estimated price elasticity of nearly one (absolute value) and an income elasticity of 0.005 (insignificant). Obviously, while the price elasticity is relatively high when we drop lagged consumption q_{it-1}, the income elasticity remains consistent with prior estimates. The results of the two-regime nonlinear model shown in the lower half of Table 2 suggest that the threshold income level is 15,847 US$. At or below this threshold income level, the income elasticity is found to be significant at 0.036. However, above the threshold value of 15,847 US$ the income elasticity turns significantly negative (−0.318). There seems to be a clear message from the two-regime demand structure: cigarette smoking is becoming an inferior good for consumers in higher income states who are more responsive to warnings of harmful effects from smoking.[11] Without the income threshold, the income elasticities in the different regimes get lumped together and become in aggregate insignificant, a result that has likely affected many prior studies.

The above demand function has thus far ignored a key characteristic in addictive behavior: the persistence of past consumption. A majority of prior studies have included lagged consumption in DFE models (Baltagi and Levin, 1986; Baltagi et al., 2000; Keeler et al. 1993, 2001). According to Baltagi et al. (2000) and Freeman (2000), DFE models outperform other pooling techniques in terms of forecast accuracy. In addition, Hansen's approach can be readily applied to the DFE model via Eq. (3).[12] Thus, we employ the conventional partial adjustment model with lagged consumption to capture the addictive behavior component of cigarette demand. As in our above analysis, we first examine the existence of a two-regime (1 threshold) model. We then continue to test the $i + 1$-regime against the i-regime model until the H_0 cannot be rejected. The likelihood ratio statistics and corresponding *P*-values generated by the bootstrap method are reported in Table 3.

Examination of Table 3 indicates that we cannot reject the null hypothesis of four regimes in favor of five. That is, we feel justified in using four different regimes based on income in our model. Table 4 reports the results of the DEF model for one and four regimes.

Table 3
LR statistics of the DFE model

Hypotheses	LR statistic
Linear vs. two-regime model	19.17 (0.0000)
Two-regime vs. three-regime model	12.23 (0.0000)
Three-regime vs. four-regime model	6.68 (0.0413)
Four-regime vs. five-regime model	4.60 (0.1254)

Note: P-values in parentheses are calculated using the bootstrap method with 10,000 replications.

The estimated price and income elasticities are −0.249 and −0.013 (insignificant) using the one-regime model. The value of the price elasticity is in agreement with prior studies (Baltagi et al., 2000; Keeler et al., 2001). The size of the income elasticity is comparable to that in Baltagi et al. (2000) and Yurekli and Zhang (2000). The estimated income elasticity is, however, different from that in Coats (1995) and in Thursby and Thursby (2000), in which the income elasticity was found to be significantly positive. It seems reasonable to believe the cigarette consumption has undergone some structural change as illustrated in Fig. 1. Failure to take regime-shift phenomenon into consideration would render biased estimates. Since the four-regime model is supported relative to its three-regime and two-regime counterpart, we focus on its estimates as shown at the bottom of Table 4.

The threshold income levels of the four-regime model are (i) below 8,568 US$, (ii) between 8,568 and 11,129 US$, (iii) between 11,129 and 18,196 US$, and (iv) greater than 18,196 US$. For regime I (less than 8,568 US$), the estimated price and income elasticities are −0.369 and 0.074, both statistically significant.[13] For regime II, the corresponding estimates are −0.416 and −0.073, both statistically significant. For regime III, the corresponding estimates are both significant as well at −0.198 and −0.072. For regime IV, they are −0.522 and 0.022 with only the price elasticity being statistically significant. Three results stand out: first, the income elasticity turns negative when per capita income exceeds 8,568. Second, the price elasticity (absolute value) is the greatest for the highest income group (regime IV). Third, the income elasticity becomes positive albeit statistically insignificant in regime IV. This result can be interpreted as a transitional phase for income

---

[11] The threshold income level corresponds to the sample period roughly between 1987 and 1988.

[12] Applying Hansen's approach to the fixed effect model may entail changes in the underlying probability distribution. However, we derive critical values via the bootstrap method to circumvent this problem.

---

[13] The Newey–West variance–covariance matrix is used to correct for possible heteroskedasticity and autocorrelation problems.

Table 4
Estimates of the one- and four-regime DFE models

	q_{it-1}^*	p_{it}^*	y_{it}^*	σ	SSE	Threshold value	Sample size
One-regime model	0.821 (42.419)	−0.249 (−12.536)	−0.013 (−1.553)	0.0579	5.508		1,645
Four-regime model				0.0561	5.142		
Regime I	0.790 (14.438)	−0.369 (−3.386)	0.074 (3.912)			$y_{it-1} \leq 8,568$	176
Regime II	0.595 (10.677)	−0.416 (−7.680)	−0.073 (−2.942)			$8,568 < y_{it-1} \leq 11,129$	448
Regime III	0.826 (24.006)	−0.198 (−6.480)	−0.072 (−3.617)			$11,129 < y_{it-1} \leq 18,196$	989
Regime IV	0.763 (8.434)	−0.522 (−3.628)	0.022 (0.372)			$y_{it-1} > 18,196$	32

Note: Numbers in parentheses are *t*-statistics. σ = standard error of regression; SSE = sum of squared residuals.

elasticity between positively significant values estimated earlier by Thursby and Thursby (2000) and negative (significant or insignificant) values by Yurekli and Zhang (2000). For some states, cigarette is becoming an inferior good. Care must be exercised because homotheticity of the utility function implies unitary income elasticity for all goods in the long run. Given the growing evidence of declining income elasticity in studies, it is of great importance to find out whether the preference function is nonhomothetic, which gives rise to dwindling income elasticity (Silberberg, 1990, p. 341). Or more importantly, cigarettes may have gradually turned into an inferior good for higher income consumers.

As shown in Fig. 1, the relationship between median cigarette sales and median income has been changing throughout the sample period. Steep and positive slopes are witnessed in regime I whereas a positive but relatively flat relationship prevails in regime II. The slope turns negative in regime III before it becomes slightly positive in regime IV. Aside from the narrow range of regime II, our statistical estimates are very much in line with the picture in Fig. 1. However, the different income levels that define the four regimes do not exactly correspond to the four regimes based on different years. This is because state income levels vary considerably in the sample period. Roughly speaking, regime II, III, and IV start from 1983, 1988, and 1997, respectively, if minimum income is used. The starting years for the regimes are 1975, 1976, and 1982 when maximum income is used and become 1979, 1982, and 1991 when median income is used. The majority of the states have entered regime IV by 1991.

Price elasticity of demand for cigarette assumes (i) the greatest absolute value for the state with income 18,196 US\$ or greater and (ii) the smallest value for the income level between 11,129 and 18,196 US\$. This suggests that the price increase is most effective in states with median income 18,196 US\$ or greater, followed by the states with income levels between 8,568 and 11,129 US\$, less than 8,568, and between 11,129 and 18,196 US\$.

Is it necessarily true that the multi-regime demand function outperforms the single-regime model in terms of pretest sampling bias? To provide an answer, we perform the out-of-sample test for both single-regime and multi-regime models. First, we estimate the models using 1963–1994 panel data based on Eqs. (1) and (6), respectively. Second, we calculate the predicted values of cigarette consumption for both the single-

Table 5
Comparison of out-of-sample forecast results

	Mean ($\hat{\mu}_i$) (*t*-statistic)	Significance level for rejecting $H_0: \mu_i = 0$
$f_{1t} = u1_t^2 - u4_t^2$	0.00076 (1.83)	6.93%
$f_{2t} = |u1_t| - |u4_t|$	0.00568 (1.93)	5.61%

regime and four-regime models using the 1995–1997 data set, denoted as $qf1_{it}$ and $qf4_{it}$, respectively. Following the generalized testing method for comparing two forecast results developed by Diebold and Mariano (1995), we compute the forecast errors: $u1_t = q_{it} - qf1_{it}$ and $u4_t = q_{it} - qf4_{it}$, where q_{it} is the actual cigarette consumption and $qf1_{it}$ and $qf4_{it}$ are predicted values from both models. The discrepancy in forecast can then be calculated as

$$f_{1t} = u1_t^2 - u4_t^2, \tag{7}$$

and

$$f_{2t} = |u1_t| - |u4_t|. \tag{8}$$

Given $\mu_i \equiv Ef_{it}(i = 1, 2)$, a failure to reject $H_0: \mu_i = 0 (i = 1, 2)$ suggests that the two models have similar predictive power. In the case of $\mu_i > 0$, the four-regime model is deemed to be superior in forecasting cigarette sales, and *vice versa* for $\mu_i < 0$. In particular, Diebold and Mariano (1995) prove the following property:

$$\sqrt{P}(f_i - \mu_i) \xrightarrow{a} N(0, S_{i,ff}), \tag{9}$$

in which $S_{i,ff} \equiv \sum_{-\infty}^{\infty} \Gamma_{i,j}$, $\Gamma_{i,j} \equiv E(f_{i,t} - \mu_i)(f_{i,t-j} - \mu_i)$ and P is a forecast period. For actual estimation, one can use the Newey-West estimator to approximate $\hat{S}_{i,ff} \equiv \sum_{-\infty}^{\infty} \hat{\Gamma}_{i,j}$.

The forecast results consisting of 141 estimated sales in three years are reported in Table 5.[14]

Examination of Table 5 reveals that the out-of-sample test results based on Eq. (9) are in favor of the four-regime model irrespective of whether Eqs. (7) or (8) is used. That is, the forecast errors of the single-regime model exceed those of the four-regime model at a significance level between 6% and 7%.

[14] To conserve space, we do not list in-sample estimates using the 1963–1994 data set for either model. However, the in-sample estimates are similar to those in Table 4.

This evidence lends further support to the use of a multi-regime demand function for the U.S. cigarette market.

4. Conclusion

Thirty-five years have elapsed since the first cigarette demand estimation in 1968. With the substantial amount of data that is available nowadays, panel estimations can readily be used to provide more accurate results. Recently, as estimates of negative income elasticity have begun to surface, cigarettes have been viewed as an inferior good. Prior studies using linear one-regime models suggest mixed results because the data are aggregated. This article applies a multi-regime model to 47 states using data from 1963 to 1997. Within the framework of the DFE model, the threshold regression by Hansen (1999) is used to estimate cigarette demand. First, we examine whether there is a nonlinear relationship between different regimes via the likelihood ratio test. *P*-values are calculated from a bootstrap method and the null hypothesis is easily rejected at the 1% significance level. Similarly, the alternative demand function of four regimes is supported at the 5% significance level. As can be detected from Fig. 1, there exists a nonlinear relationship between personal income and cigarette consumption. Hence, we use the previous year's income as the threshold variable to partition the data set into different regimes in which the elasticities can take on distinct values. Evidently, cigarettes become an inferior good as income rises.

The results from the four-regime model suggest that the income elasticity is negative when per capita income is greater than 8,568 US$, but turns positive though insignificant for incomes above 18,196 US$. In the income range between 8,568 and 18,196 US$ the income elasticity is significantly negative. As for the price elasticity, a nonlinear relationship prevailed: price elasticity has the greatest value for incomes above 18,196 US$, followed by the income ranges 8,568–11,129 US$, below 8,568 US$, and 11,129–18,196 US$.

The nonlinear relationship manifested in the four-regime model is strikingly different from the results of prior studies. In addition, based on the out-of-sample forecast results, the four-regime model has a significantly smaller forecast error. This suggests that the use of a threshold regression may be necessary in situations of changing demand structure. However, linear restrictions on homogeneity for a demand function and restrictions on symmetry for a system of demand functions could probably sharpen estimates even further. This is an interesting topic for future research.

Acknowledgments

We are grateful to the two reviewers whose suggestions have improved the quality of the article. With usual caveats in order, all errors are authors' only.

References

Baltagi, B. H., Griffin, J. M., 2001. The econometrics of rational addiction: the case of cigarettes. J. Bus. Econ. Stat. 19(4), 449–454.

Baltagi, B. H., Levin, D., 1986. Estimating dynamic demand for cigarettes using panel data: the effects of bootlegging, taxation and advertising reconsidered. Rev. Econ. Stat. 68(1), 148–55.

Baltagi, B. H., Xiong, W., Griffin, J. M., 2000. To pool or not to pool: homogeneous versus heterogeneous estimators applied to cigarette demand. Rev. Econ. Stat. 82(1), 117–126.

Becker, G. S., Murphy, K. M., 1988. A theory of rational addiction. J. Polit. Econ. 96, 675–700.

Becker, G. S., Grossman, M., Murphy, K. M., 1994. An empirical analyses of cigarette addiction. Am. Econ. Rev. 84, 396–418.

Bera, A., Byron, R. P., Jarque, C. M., 1981. Further evidence on asymptotic tests for homogeneity and symmetry in large demand system. Econ. Lett. 8, 101–105.

Coats, R. M., 1995. A note on estimating cross—border effects of state cigarette taxes. Natl. Tax J. 48(4), 573–584.

Diebold, F. X., Mariano, R. S., 1995. Comparing predictive accuracy. J. Bus. Econ. Stat. 13(3), 253–263.

Freeman, D. G., 2000. Alternative panel estimates of alcohol demand, taxation, and the business cycle. South. Econ. J. 67(2), 325–344.

Hansen, E. B., 1999. Threshold effects in non-dynamic panels: estimation, testing and inference. J. Econom. 93(2), 345–368.

Hausman, J. A., 1978. Specification tests in econometrics. Econometrica 46, 1251–1272.

Huang, B. N., Yang, C. W., Hwang, M. J., 2004. New evidence on demand for cigarettes: a panel data approach. Int. J. Appl. Econ. 1(1), 81–97.

Keeler, T. E., Hu, T. W., Barnett, P. G., Mannning, W. G., 1993. Taxation, regulation and addiction: a demand function for cigarette based on time-series evidence. J. Health Econ. 12, 1–18.

Keeler, T. E., Hu, T. W., Manning, W. G., Sung, H., 2001. State tobacco taxation, education, and smoking: controlling for the effects of omitted variables. Natl. Tax J. 54(1), 83–102.

Silberberg, E., 1990. The Structure of Economics: A Mathematica Analysis, 2nd ed. Mc Graw-Hill Publishing Company, New York.

Thursby, J. G., Thursby, M. C., 2000. Interstate cigarette bootlegging: extent, revenue losses, and effects of federal intervention. Natl. Tax J. 53(1), 59–77.

Tong, H., 1990. Non-linear Time Series: A Dynamical Systems Approach. Oxford University Press, Oxford.

U.S. Department of Health and Human Services, 1989. Reducing the health consequence of smoking 25 years of progress. A Report of the Surgeon General 1989.

Yurekli, A. A., Zhang, P., 2000. The impact of clean indoor-air laws and cigarette smuggling on demand for cigarette: an empirical model. Health Econ. 9, 159–170.

Volume 36   Issue 2   February 2008   ISSN 0301-4215

ENERGY POLICY

Energy Policy 36 (2008) 755–767

www.elsevier.com/locate/enpol

Does more energy consumption bolster economic growth? An application of the nonlinear threshold regression model[✩]

Bwo-Nung Huang[a,b,*], M.J. Hwang[c], C.W. Yang[d]

[a]*Department of Economics & Center for IADF, National Chung Cheng University, Chia-Yi 621, Taiwan*
[b]*Department of Economics, West Virginia University, West Virginia, USA*
[c]*Department of Economics, College of Business and Economics, West Virginia University, West Virginia, USA*
[d]*Department of Economics, Clarion University of Pennsylvania, Pennsylvania, USA*

Received 31 July 2007; accepted 26 October 2007

Abstract

This paper separates data extending from 1971 to 2002 into the energy crisis period (1971–1980) and the post-energy crisis period (1981–2000) for 82 countries. The cross-sectional data (yearly averages) in these two periods are used to investigate the nonlinear relationships between energy consumption growth and economic growth when threshold variables are used. If threshold variables are higher than certain optimal threshold levels, there is either no significant relationship or else a significant negative relationship between energy consumption and economic growth. However, when these threshold variables are lower than certain optimal levels, there is a significant positive relationship between the two. In 48 out of the 82 countries studied, none of the four threshold variables is found to be higher than the optimal levels. It is inferred that these 48 countries should adopt a more aggressive energy policy. As for the other 34 countries, at least one threshold variable is higher than the optimal threshold level and thus these countries should adopt energy policies with varying degrees of conservation based on the number of threshold variables that are higher than the optimal threshold levels.
© 2007 Elsevier Ltd. All rights reserved.

Keywords: Threshold regression; Energy consumption; CO_2 emissions

1. Introduction

On February 2, 2007, a report from the Intergovernmental Panel on Climate Change (IPCC) indicated that global atmospheric concentrations of carbon dioxide (CO_2), methane, and nitrous oxide have increased markedly as a result of human activities since 1750 and now far exceed pre-industrial levels determined from ice cores spanning many thousands of years. The increase in CO_2 has mainly originated from the use of fossil fuels and partly from the changes in land use. The emissions of CO_2 and other gases from burning fossil fuels and other processes trap heat from the sun in the atmosphere, much like a greenhouse. Global warming, of course, raises the temperature in the air and in the ocean, and raises the sea level as well. Without properly controlling for the phenomenon, serious consequences on the eco-system and the extinction of many creatures will be unavoidable. The IPCC report places the blame for global warming mainly on the release of CO_2 from the use of fossil fuels.

Fossil fuels are a major source of energy for industrial production, residential consumption, and transportation.[1] It is obvious that energy use plays an important role in our daily lives and economic activities. Since the fossil fuels are relatively cheap and readily available, part of the benefit

---

[✩] Financial support from the National Science Council (NSC94-2415-H-194-002) is gratefully acknowledged. We would like to thank the editor and two anonymous referees for their helpful comments and suggestions. All errors are clearly our own.

*Corresponding author at: Department of Economics & Center for IADF, National Chung Cheng University, Chia-Yi 621, Taiwan. Tel.: +886 52720411x34101; fax: +886 52720816.

E-mail address: ecdbnh@ccu.edu.tw (B.-N. Huang).

0301-4215/$ - see front matter © 2007 Elsevier Ltd. All rights reserved.
doi:10.1016/j.enpol.2007.10.023

---

[1] Fossil fuels account for about 90% of energy use in the world today. Petroleum accounts for the largest share at 40%, followed by coal at 24% and natural gas at 22%. Fossil fuels are currently among the most economically available sources of power for both residential and commercial uses.

from a greater use of fossil fuels is the increase in production and living standards. These benefits, however, must be offset, to some extent, by a negative externality that arises from an increase in global warming and environmental pollution.

Whether or not the increased economic benefits from energy consumption will outweigh the negative externality depends on the empirical evidence of a positive causal relationship between energy consumption and economic growth. If the empirical evidence indicates that energy consumption (in terms of the growth rate) Granger-causes economic growth, it is suggested that a more aggressive energy policy should be followed. On the other hand, if economic growth Granger-causes energy consumption or there is no causal relationship between energy consumption growth and economic growth, the policy-maker should implement a more conservative energy policy since energy consumption may not bring about economic growth but may increase CO_2 emissions into the atmosphere and accelerate global warming.[2]

Granger causality has been widely used in the literature in analyzing the relationship between energy consumption and economic growth. Kraft and Kraft (1978) discovered that economic growth leads energy consumption. Employing the same US data, however, Stern (2000) found that energy consumption leads economic growth. Therefore, two conflicting empirical results for the same country using the same data were found. Furthermore, Akarca and Long (1980), Yu and Jin (1992), and Soytas and Sari (2003) discovered that there is no causal relationship between energy consumption and economic growth. In addition, Lee (2005) showed that a feedback relationship existed between the two.[3] Such inconsistent results were also found in emerging countries such as Taiwan and South Korea. The main reason for the inconsistencies may involve using different periods in time series data, obtaining an insufficient sample, or failing to take into account the nonlinearity due to certain country-specific factors. In addition, using the Granger-causality approach to investigate the causal relationship between energy consumption and economic growth in the previous literature led to three possible problems: (a) whether or not the yearly data were sufficient to represent the long-term relationship between the two; (b) the inability of the yearly data to eliminate the problems of short-term fluctuations due to business cycles and structural change; and (c) the failure to delineate countries with special features in terms of different causal relationships.

Since the relationship between energy consumption and economic growth is inherently a long-term one, a biased estimate may be the result of an insufficiently large sample size in the time series, the existence of structural changes, or short-term economic fluctuations. Another reason for the inconsistent empirical results may emanate from the omission of specific characteristics in certain countries affecting the relationship between energy consumption and economic growth (Soytas and Sari, 2006). For example, high CO_2 emission countries may be characterized by an overuse of energy or a lack of regulations to enforce proper energy consumption. As a result, the environmental damage from energy use may outweigh the benefits from economic growth. To tackle the insufficient sample size problem, many researchers have used the panel data approach. Lee and Chang (2007a) separated the data used into 18 developing countries and 22 developed countries and employed a dynamic panel data (DPD) approach to test the causal relationship between energy consumption and economic growth. He discovered that economic growth leads energy consumption growth in the developing countries, while in the developed countries there is a feedback relationship between the two. Huang et al. (2008) used panel data for 82 countries and grouped the data into four categories based on the income levels defined by the World Bank: low-income group, lower middle-income group, upper middle-income group, and high-income group. They employed the DPD approach to investigate Granger causality. They discovered that (a) in the lower and upper middle-income groups, economic growth leads energy consumption positively; (b) in the high-income group, economic growth leads energy consumption negatively; and (c) in the low-income group, no causal relationship exists between energy consumption and economic growth. Their conclusion—a reduction in energy use in the high-income group—was consistent with the policy of decreasing the use of fossil fuels (e.g., extending the summer daylight time and the more extensive use of fluorescent lighting).

By grouping on the basis of the difference in the degree of economic development in their research, Lee and Chang (2007a) and Huang et al. (2008), in effect, assume a nonlinear relationship between energy consumption and GDP. Moon and Sonn (1996) employed an endogenous growth model to infer that the economic growth rate rises initially with productive energy expenditure but subsequently declines. In other words, there is an inverse U-shaped nonlinear relationship between energy consumption and economic growth as was evidenced by their empirical results from the yearly data extending from 1968 to 1989 in Korea. Lee and Chang (2007b) used the level of total energy consumption as a threshold variable to investigate the existence of a nonlinear relationship under the one-sector and two-sector growth models. The empirical result from the 1955–2003 annual data in Taiwan indicates that there is an inverse U-shaped relationship between energy consumption growth and economic growth. That is, the relationship between energy consumption and economic growth indicated above is nonlinear and the traditional linear model is no longer appropriate.

---

[2]With fossil fuels as a major source of energy consumption, they may promote economic growth, pollute the environment, and also deplete the available resources. Therefore, there is a nonlinear relationship between energy consumption and GDP growth.

[3]For the empirical literature on the causal relationship between energy consumption and economic growth, refer to Stern and Cleveland (2003).

Although the DPD approach may overcome the short-coming of insufficient sample size, the problem of biased estimates remains unresolved. The biasness may arise from the inadequate representation of a long-run relationship, the inability to allow for structural change, and the presence of short-term fluctuations from business cycles when using annual data. Last, but not least, it can emanate from failure to segment the data into different panel groups based on a country's characteristics. In order to tackle these problems, many researchers employ average values over time for each cross-sectional unit.[4] The advantage of this is that the cross-sectional data are separated into different regimes based on certain threshold values in order to investigate different relationships between energy consumption and economic growth in each regime. This approach helps us calculate the conditions for the existence of different relationships between the growth of energy consumption and economic growth.

It is widely recognized that global warming will have a great impact on the welfare of future generations.[5] If we can identify the conditions under which a significant relationship exists or does not exist between energy consumption and economic growth, more relevant energy policies can be formed. If energy consumption cannot bring about economic growth in a country, the best policy may be to reduce energy consumption with a view of avoiding a negative impact on the environment. The major contribution of this research is to apply 82-country cross-sectional data to investigate the relationship between energy consumption and GDP growth based on four energy-related threshold variables. These four variables include CO_2 emissions (CO_2), the efficiency of energy use represented by the total primary energy supply to production per \$1 of GDP (tpes/Y), the ratio of industrial energy consumption to total energy consumption (ind/tfc), and per capita energy consumption (ec). Among those four variables, the ec variable was used in the past by Moon and Sonn (1996) and Lee and Chang (2007b). As far as we know, this paper is the first to use the other three threshold variables.

Although the cross-sectional data analysis has advantages as indicated above, it also has certain disadvantages. To test the causal relationship using cross-sectional data, the potential relationship between consumption and economic growth is assumed a priori in the growth equation. In other words, the energy consumption growth variable is assumed to be an explanatory variable for explaining economic growth as was indicated in prior research.

This paper uses relevant data covering the period from 1972 to 2002 for 82 countries. Since the energy crisis led to structural change, the data are separated into two periods: the energy crisis period (1972–1980), or Period-I, and the post-energy crisis period (1981–2002), or Period-II. First, we take the yearly averages in these two periods to establish the 82-country cross-sectional data set. These variables are then tested statistically and grouped into different regimes based on certain energy-related features (called threshold variables, q_i). Under different regimes, the threshold regression model is employed to investigate the relationships between energy consumption and economic growth with different q_i's. In the nonlinear test, the candidates for the threshold variables in our regression model are CO_2 emissions (CO_2), the efficiency of energy use (tpes/yr), the ratio of industrial energy consumption to total energy consumption (ind/tfc), and per capita energy consumption (ec). From the different regimes classified by these threshold variables, we discover that the relationship between energy consumption and economic growth is rather different. Furthermore, irrespective of whether a linear or nonlinear structure is used, the relationship between energy consumption and economic growth also differs between the energy crisis period and the post-energy crisis period.

The major finding of this paper is that it distinguishes countries with strong relationships between energy consumption and economic growth from countries without such relationships. We also discover that, in Period-II, 34 of the 82 countries exhibit a positive relationship between energy consumption and economic growth, regardless of which threshold variable is used. As such, these countries should adopt a more aggressive energy policy. For the remaining 48 countries, no such relationship can be found between the two using either of the threshold variables. These 48 countries should, to different extents, adopt conservative energy policies depending on the number of threshold variables through which no such a relationship is found.

This paper is organized as follows. Following Section 1, which discusses the motivation behind the paper and provides a review of the related literature, Section 2 introduces the econometric models and data. Section 3 then analyzes and discusses the empirical results. The final section, Section 4, provides the concluding remarks and the policy implications derived from this research.

2. Model specification, econometric method, and data

In reference to the often-used growth models in the literature, for instance that of Barro (1990),[6] let the *i*th

---

[4]For the literature on the use of cross-sectional data using annual averages, refer to Khan and Senhadji (2001) on the relationship between inflation and economic growth and see Levine and Zervos (1998) on the relationship between banking and economic growth.

[5]On April 7, 2007, in Brussels, IPCC published a report on the possible adverse effects of global warming, ranging from increased flooding, hunger, drought, and disease, to the extinction of many species.

[6]Refer to Barro (1990) for the growth-related models, Levine and Zervos (1998) for the relationship between monetary development and economic growth, Aizman and Glick (2003) for the relationship between defense spending and economic growth, and Khan and Senhadji (2001) for the relationship between inflation and economic growth.

189

country's yearly average economic growth be Δy_i, the yearly average per capita energy consumption growth be Δec_i, and other explanatory economic growth variables be referred to as x_i. In the literature, the x_i include inflation π_i, the proxy variable for capital stock (gross fixed capital formation as a percentage of GDP, I/Y), the share of government expenditure in GDP (gc_i), and the growth of labor (Δlf_i). The output growth equation, which includes energy consumption, is shown below:[7]

$$\Delta y_i = \alpha_0 lyo_i + \alpha_1 \Delta ec_i + \beta x_i + \varepsilon_i, \qquad (1)$$

where Δ is the first difference operator, y_i is the logarithm of per capita real GDP for the ith country; lyo_i represents the logarithm of initial income for the ith country, and ec_i denotes the log of energy consumption (kilogram of oil equivalent per capita). The purpose of this paper is to apply the threshold regression model to investigate the difference in the relationship between energy consumption and economic growth using certain threshold variables. The threshold regression model originated from the threshold autoregressive model in the time series model developed by Tong (1978) and Tong and Lim (1980). By denoting a threshold variable as q_i and an optimal threshold value as c_1^*, the two-regime threshold equation can be expressed as

$$\Delta y_i = (\alpha_{10} lyo_i + \alpha_{11} \Delta ec_i + \beta_{11} \pi_i + \beta_{12}(I/Y)_i + \beta_{13} gc_i + \beta_{14} \Delta lf_i)I(q_i \leqslant c_1^*) + (\alpha_{20} lyo_i + \alpha_{21} \Delta ec_i + \beta_{21} \pi_i + \beta_{22}(I/Y)_i + \beta_{23} gc_i + \beta_{24} \Delta lf_i)I(q_i > c_1^*) + \varepsilon_i, \qquad (2)$$

where ε_i is assumed to be i.i.d. and to follow the white noise process. $I(.)$ is an indicator function. If the relationship in (.) is present, then $I(.)$ is 1, otherwise, $I(.)$ is 0. Eq. (2) is a simple two-regime model delineated by the value of a threshold variable. It indicates that the relationship between the explanatory variables and economic growth is represented by $B_1 = (\alpha_{10}, \alpha_{11}, \beta_{11}, \beta_{12}, \beta_{13}, \beta_{14})$ when q_i (the threshold variable) is less than or equal to a threshold value c_1^* (regime 1), but by $B_2 = (\alpha_{20}, \alpha_{21}, \beta_{21}, \beta_{22}, \beta_{23}, \beta_{24})$ when q_i is greater than the threshold value c_1^* (regime 2). Before estimating Eq. (2), it is necessary to examine the existence of the nonlinear relationship. Only after the null hypothesis ($H_0 : B_1 = B_2$) or linear relationship is rejected can we estimate Eq. (2). Given the fact that the threshold value (c) is typically unknown, the traditional F-test is not appropriate for testing the null hypothesis. Under the assumption that the error term is i.i.d., Hansen (1997) suggests employing a test method derived from Davies (1977, 1987) and Andrews and Ploberger (1994), namely, a test with near-optimum power against an alternative

distance from the null hypothesis. It takes the form of a standard F statistic:

$$F = n\left(\frac{\tilde{\sigma}_n^2 - \hat{\sigma}_n^2}{\hat{\sigma}_n^2}\right), \qquad (3)$$

in which $\tilde{\sigma}_n^2$ is the residual variance of Eq. (1) and $\hat{\sigma}_n^2$ is the residual variance of Eq. (2) under c_1^*. However, the asymptotic distribution of Eq. (3) does not follow an F distribution since c is unknown. Hansen (1996) circumvents the problem by using a bootstrap method to produce an asymptotic distribution for testing.[8]

Once Eq. (3) is employed, we may be able to reject the null hypothesis. That is, there is a nonlinear relationship for the data where q_i is used as a threshold variable as shown in Eq. (2). First, the data need to be arranged in ascending order in q_i in order to estimate Eq. (2). To accommodate the degrees of freedom problem, the smallest and largest 15% of observations are removed. The remaining sample space bounded by threshold values $[\underline{c}, \overline{c}]$ is partitioned into n grids. For each $c_i(i = 1, \ldots, n)$, we obtain a $\hat{\sigma}(c_i)$. Next, we select the smallest $\hat{\sigma}(c_i)$ to be the optimal threshold value c_1^* via Eq. (2):

$$c_1^* = \arg\min\hat{\sigma}(c_i)^2, \quad i = 1, \ldots, n,$$
$$c_i \in [\underline{c}, \overline{c}], \qquad (4)$$

where $\hat{\sigma}(c_i)^2$ denotes the residual variance from Eq. (2) under the threshold value c_i. If the data exhibit a second threshold level (c_2^*), the three-regime growth model is shown as

$$\Delta y_i = (\alpha_{10} lyo_i + \alpha_{11} \Delta ec_i + \beta_{11} \pi_i + \beta_{12}(I/Y)_i + \beta_{13} gc_i + \beta_{14} \Delta lf_i) \times I(q_i \leqslant c_1^*) + (\alpha_{20} lyo_i + \alpha_{21} \Delta ec_i + \beta_{21} \pi_i + \beta_{22}(I/Y)_i + \beta_{23} gc_i + \beta_{24} \Delta lf_i)I(c_1^* < q_i \leqslant c_2^*) + (\alpha_{30} lyo_i + \alpha_{31} \Delta ec_i + \beta_{31} \pi_i + \beta_{32}(I/Y)_i + \beta_{33} gc_i + \beta_{34} \Delta lf_i)I(q_i > c_2^*) + \varepsilon_i. \qquad (5)$$

To test the existence of the three regimes versus the two regimes, the null hypothesis is set up as H_0: two-regime versus three-regime. The test statistic is similar to the F-statistic in Eq. (3) with the exception that $\tilde{\sigma}_n^2$ is the residual variance of Eq. (2) and $\hat{\sigma}_n^2$ is the residual variance of Eq. (5). Since the threshold level is unknown and the asymptotic F-statistic cannot be derived from the χ^2 distribution, a bootstrap procedure is needed to test the hypothesis.

Since the energy consumption of fossil fuels can bring about economic growth or a negative externality, which gives rise to environmental pollution and global warming, the selection of threshold variables should be based on energy-related variables. These variables include (1) CO_2 emissions (CO_2, kg per 2000 PPP $ of GDP), (2) energy efficiency represented by the ratio of total primary energy supply to production per $1 of GDP ($tpes/Y$), (3) the ratio of industrial energy consumption to total energy

---

[7]There is a difference between Eq. (1) and the model derived from the neo-classical production function (Ghali and El-Sakka, 2004). The neo-classical production function does not include initial income lyo_i and its capital stock uses a flow rate rather than a stock concept (proxy variable I/Y). There is also a difference between Eq. (1) and the two-sector model (Lee and Chang, 2007b). The two-sector model does not include the initial income either. The energy consumption growth under the two-sector model is $(ec/Y)\Delta ec$, while Eq. (1) uses Δec.

[8]See Hansen (1997, p. 6) for the bootstrap procedure.

consumption (ind/tfc), and (4) per capita energy consumption (ec, kg of oil equivalent).

Moon and Sonn (1966) employed the concept of an endogenous growth model and treated energy use as an important factor in a production function. They further assumed that a hypothetical social planner sets optimal energy intensity and derives the steady-state growth rate g as[9]

$$g = -\frac{1}{\sigma}[\rho - (1 - \beta\tau)A^{1/\alpha}\tau^{(1-\alpha)/\alpha}], \qquad (6)$$

where g is output growth rate, α is a positive constant that reflects the level of technology, A is the capital share in the Cobb–Douglas production function, ρ is the rate of time preference, σ is the coefficient of relative risk aversion, β is the exogenously given energy price in the world energy market, and τ is the energy intensity defined as a ratio of energy input to output.

Different energy intensities (i.e., different values for τ) have two effects on the growth rate in Eq. (6). An increase in τ reduces g, but an increase in τ raises the marginal product of K (capital), which raises g. By following different τ, the size and sign of g will also be different. Basically, before τ is raised to a certain level (the most optimal level), g may increase. When the energy use has surpassed this optimal level, g begins to decrease and may even become negative. Such theory shows that there is a nonlinear relationship between energy and production growth. By using per capita energy consumption (ec) as a threshold variable, it is expected that when ec is smaller than the optimal threshold level, there is a positive significant relationship between energy consumption and output growth. Conversely, when ec is greater than the optimal threshold level, there is an insignificant or negative relationship between the two.

The overuse of energy will not only bring about a loss of efficiency but will also lead to an increase in CO_2 emission. Thus, both CO_2 emission and tpes/Y (representing the efficiency of energy use) may become threshold variables delineating a nonlinear relationship between energy consumption and output growth. Further, the idea for the use of ind/tfc is similar to the use of a combination of ec and CO_2. The threshold variable ind/tfc represents the share of energy use in the industrial sector to total energy consumption, and the industrial production is the source of production and pollution in a country. If ec and CO_2 are used as the threshold variables of the nonlinear relationship between energy consumption and output growth, ind/tfc can also be viewed as a threshold variable. We elaborate more on the selection of these threshold variables.

If the CO_2 emissions are higher in a particular country, it is likely that the major source of energy may be the cheaper and dirtier fossil fuels in that country. The use of relatively cheaper fossil fuels may increase production and hence pollution as well. If the pollution is higher than a certain

threshold level, the benefit may be offset to a degree by the cost of the pollution resulting from the energy used. It is, therefore, expected that if pollution is higher than a certain threshold level, there will be no significant relationship between energy consumption and economic growth (or a negative correlation). Since we use CO_2 emissions to represent the extent of the pollution, it is also expected that, if CO_2 is used as a threshold variable, the growth equation will exhibit a nonlinear relationship. That is, when the CO_2 emissions in a particular country exceed a certain threshold level, the growth of energy consumption will not benefit economic growth as much based on the concept of green GDP. A similar concept can be applied to the selection of another threshold variable, namely, ind/tfc. If an industrial sector in a country uses more energy but without much regulation, the pollution from the industrial sector will be expected to be higher. The greater the ind/tfc ratio in a country, the relatively more serious is the pollution problem in that country. Therefore, if the ind/tfc threshold variable is higher than a certain threshold level, energy consumption will do little good to economic growth.

The use of energy efficiency as represented by the ratio of total primary energy supply to production per $1 of GDP (tpes/Y) can be used as a threshold variable. When energy is used more efficiently, it is expected that energy consumption will be conducive to economic growth. As the energy use becomes increasingly more inefficient, the consumption of energy may not contribute to economic growth. If the level of tpes/Y is higher than a certain threshold value, there will be no significant relationship between energy consumption and economic growth.

CO_2 emission is the most important variable among the four threshold variables of the nonlinear relationship between energy consumption growth and economic growth. Energy consumption may bring a positive and a negative externality. For the negative externality, the CO_2 emission from an excessive use of fossil fuel is a major source in generating the problem of global warming. The other three threshold variables (ind/tfc, tpes, and ec) are all energy-use-related variables. The excessive energy use or the lack of efficiency in energy use may imply that a negative externality (increase in CO_2 emission) may be greater than its positive externality. Hence, the policy-makers in their decision as to whether to increase energy consumption need to consider the various consequences when CO_2 is used as a threshold variable. In other words, countries, with CO_2 emission higher than our estimated optimal threshold level, should reduce energy consumption because more energy consumption is no longer conducive to output increase. It may even aggravate global warming as a result of greater and unnecessary increase in CO_2 emission.

The purpose of this paper is to investigate whether there exists a nonlinear long-term relationship between energy consumption and economic growth. We have collected the annual data for 82 countries covering the period from 1972

---

[9]For detailed derivation, see Moon and Sonn (1996, pp. 191–193).

to 2002. The data set includes per capita energy consumption (ec), per capita real GDP (y), inflation (π), a capital stock proxy variable (I/Y), the labor force (lf), CO_2 emissions (CO_2), energy efficiency (tpes/Y), and the ratio of industrial energy consumption to total energy consumption (ind/tfc). The energy data are obtained from Energy Balance published by the Energy Information Agency.[10] To investigate the difference in the nonlinear relationship between that during the energy crisis (1972–1980 as Period-I) and that during the post-energy crisis (1981–2002 as Period-II), two sets of cross-sectional data encompassing yearly averages are used based on Eq. (1). We use Eq. (2) to decipher the relationship under the two-regime model and Eq. (5) to decipher that under the three-regime model.

3. Empirical results and discussion

Before applying the threshold regression, it is necessary to test the existence of the nonlinear relationship in terms of the threshold variables (i.e., CO_2, tpes/Y, ind/tfc, and ec) during these two periods for the 82 countries. Table 1 displays the nonlinear test results.

Based on Table 1, when CO_2 is used as a threshold variable, we cannot reject the linear assumption for the energy crisis period (Period-I). However, for the post-energy crisis period (Period-II), we reject the two-regime model at the 5% level and there is a three-regime relationship. That is, for Period-II, the 82 countries are grouped into three regimes for analysis based on the threshold levels of $CO_2 = 0.17703$ and 0.69478. When tpes/Y is used as the threshold variable, we cannot reject the linear assumption in Period-I. For Period-II, the 82 countries need to be separated into two groups for the estimation. When ind/tfc is used as the threshold variable, the 82 countries need to be separated into two groups in both periods. Finally, in the case of the ec threshold variable, the two-regime threshold regression is needed for the estimation in both periods. The nonlinear test results indicate that, when CO_2 and tpes/Y are used as the threshold variables, there is no nonlinear relationship between energy consumption and economic growth in Period-I. The other results strongly support the use of the threshold regression to investigate the relationship between energy consumption and economic growth. The estimated results using the two-regime and three-regime models in both periods are shown in Table 2.

Table 2 displays both the linear and nonlinear results estimated using four different threshold variables in both periods. From column (1) or the linear model, it can be seen that during the energy crisis period (Period-I), there is no significant relationship between energy consumption and economic growth, but there is a significant positive relationship between the two in the post-energy crisis period (Period-II). This implies that an increase in energy

consumption is conducive to economic growth in Period-II. For these two periods, both the linear and nonlinear models yield different relationships between energy consumption and economic growth. If we do not take structural change into consideration, the empirical results will certainly be biased. The results of the linear estimation indicate that (1) inflation is negatively related to economic growth in both periods; (2) I/Y is positively related to economic growth in both periods; (3) there is no significant labor–economic growth relationship during Period-I, but a significant negative relationship during Period-II; and (4) as for the relationship between initial income and economic growth, there is no significant association between the two during Period-I, but there is a negative relationship during Period-II. The estimated results of other explanatory variables are consistent with previous estimated results in the literature. It is thus clear that our chosen explanatory variables are adequate in describing economic growth.[11] From the size of the standard error in the lower part of column (1), the model in Period-II fits the data better than the model in Period-I. Finally, according to the F test (White) results, there is no heteroscedasticity problem with the linear model irrespective of Period-I or Period-II.[12]

With respect to the nonlinear estimation during Period-I when CO_2 and tpes/Y are used as the threshold variables, our test results indicate that we cannot reject the null hypothesis of linear assumption. Thus, the Period-I columns of these two variables under nonlinear estimation (Table 2) are left empty. The CO_2 threshold variable in Period-II has a three-regime relationship. If CO_2 emissions variable (column (2)) is less than 0.17703 (12 countries, regime 1), a 1% increase in energy consumption growth may contribute to a 1.2700% increase in economic growth. If the CO_2 emissions variable is greater than 0.17703 and less than 0.69478 (regime 2, 56 countries), a 1% increase in energy consumption growth is expected to contribute only to a 0.4951% increase in economic growth. When the CO_2 emissions variable exceeds 0.69478 (14 countries, regime 3), energy consumption growth may not contribute anything to economic growth. The estimated results of the three-regime model indicate that, if the use of energy produces only a slightly negative externality (CO_2), energy consumption is conducive to economic growth. Conversely, if the use of energy gives rise to too much negative externality, that negative externality will to a considerable degree offset the benefit derived from the use of energy. The standard error (0.0100) obtained from the three-regime nonlinear model is significantly smaller than that derived from the linear model (0.0116). All of the nonlinear models in both periods are free of the heteroscedasticity problem. If the linear model is used for estimation, there is a positive

---

[10]The names of the 82 countries are displayed in Table A1 in the Appendix A.

[11]Since the purpose of this paper is to investigate the relationship between energy consumption and economic growth, the rest of the discussion will concentrate on the empirical results between the two rather than the estimated results of other control variables.

[12]In all estimations, we have used the White heteroscedasticity-consistent standard error.

Table 1
Test results of the nonlinear threshold model

Threshold variables H_0	CO_2		tpes/Y		ind/tfc		ec	
	Period-I	Period-II	Period-I	Period-II	Period-I	Period-II	Period-I	Period-II
1 vs. 2	1.99 (0.12)	3.73 (0.02)	2.49 (0.09)	7.59 (0.00)	4.10 (0.01)	7.59 (0.00)	5.06 (0.00)	5.12 (0.00)
2 vs. 3		3.13 (0.04)		1.98 (0.12)	0.87 (0.50)	2.43 (0.07)	1.70 (0.16)	1.18 (0.36)
2-Regime (c1)	0.62603	0.69478	0.11990	0.25564	0.22877	0.40133	477.0290	4367.6940
3-Regime (c1, c2)		0.17703		0.12867	0.22877	0.23651	592.75658	782.62276
		0.69478		0.25564	0.37957	0.40383	1036.11708	4367.6940

Note: Period-I: the energy crisis period from 1972 to 1980. Period-II: post-energy crisis period from 1981 to 2002; H_0: 1 vs. 2 or linear vs. 2-regime, H_0: 2 vs. 3 or 2-regime vs. 3-regime; c1, c2, respectively, represent first and second threshold levels; number inside ( ) denotes *p*-value derived based on 1000 replications of bootstrap; CO_2 denotes CO_2 emission (kg per 2000 PPP $ of GDP); tpes/Y = total primary energy supply/GDP (adjusted for purchasing power parity using the base year of 2000); ind/tfc = total industry sector energy consumption/total final energy consumption; ec = per capita energy consumption (kg of oil equivalent).

relationship between energy consumption and economic growth during Period-II. However, these linear results cannot indicate the impact of energy consumption in different countries with different degrees of CO_2 emissions. As was indicated above, the estimated results of the three-regime model do support the use of a nonlinear model in that they better describe the relationship between energy consumption and economic growth.

If the energy efficiency threshold variable (tpes/Y) is used, our test results do support the use of the two-regime threshold regression in Period-II and the results are displayed in the third column of Table 2. When tpes/Y is smaller than 0.25564 (in 61 countries), there is a positive relationship between energy consumption and economic growth. When tpes/Y is larger than 0.25564 (in 21 countries), there is a negative but insignificant relationship between the two. Most countries began to pay more attention to the efficiency of energy use in the post-energy crisis period. It is expected that, for those countries with the least energy efficiency, an increase in energy consumption growth may not contribute to economic growth at all.

The use of ind/tfc as a threshold variable is similar to that of CO_2. Since industrial production is the major source of CO_2 emissions, when the share of industrial energy use increases, the more CO_2 emissions there will be. It is expected that there will be no relationship between energy consumption and economic growth. As is indicated in Period-II of column (4), when ind/tfc is smaller than 0.40133 (in 68 countries) there is a positive relationship between the two. When the ind/tfc threshold value exceeds 0.40133, there exists no significant relationship between the two. Conversely, in Period-I, when ind/tfc is smaller than 0.22877 (in 60 countries), there is no relationship between the two. When ind/tfc is higher than 0.22877 (in 22 countries), there is a positive relationship between the two.

Finally, with per capita energy consumption (ec) as a threshold variable, the nonlinear test results indicate that there is a two-regime nonlinear relationship in both periods. In Period-I, when ec is less than 477.0290 (in 61 countries), there is no relationship between energy consumption and economic growth. As the ec level exceeds

477.0290, there exists a significant positive relationship between the two. In Period-II, when ec is less than 4367.6940 (in 69 countries), there is a significant positive relationship between the two. With the ec level greater than 4367.6940 (in 13 countries), there is a significant negative relationship between the two. Such results show that, during the energy crisis, low energy-use countries may not enjoy the benefits to economic growth from energy use. During the post-energy crisis period, overuse of energy may have a negative impact on economic growth. Lee and Chang (2007b) applied Taiwanese time series data and failed to discover a relationship between energy consumption and economic growth when energy consumption exceeds a certain threshold level. Our research indicates that too much energy consumption may actually reduce economic growth during the post-energy crisis period. Such a result is consistent with the prediction shown by the theoretical model of Moon and Sonn (1996). As the energy consumption intensity becomes greater than a certain level, output growth may become smaller or may even become negative.

To summarize the above empirical results, the relationship between energy consumption and economic growth differs between the energy crisis period and the post-energy crisis period. Following the energy crisis, oil prices went up and people realized that energy was becoming a scarce resource. The use of fossil fuels as a major source of energy thus might have a negative impact on the environment. Our empirical results indicate that the overuse of energy is not conducive to economic growth and as such may even adversely affect economic growth. The efficient use of energy and maintaining CO_2 emissions below a certain threshold level will both contribute to economic growth. Otherwise, the overuse of energy that gives rise to a negative externality from environmental pollution may outweigh the benefits from economic growth. In the presence of such an offsetting effect, there is no significant positive (or even negative) relationship between energy consumption and economic growth.

The advantage of using the cross-sectional nonlinear model is that we can identify countries under certain

Table 2
Estimated results of the linear and nonlinear threshold regressions

Independent	Linear (1) Period-I	Period-II	Nonlinear (2) CO$_2$ Period-I	Period-II	Nonlinear (3) tpes/Y Period-I	Period-II	Nonlinear (4) ind/tfc Period-I	Period-II	Nonlinear (5) ec Period-I	Period-II
R1										
lyo$_{i,1}$	−0.0005	−0.0016*		−0.0026		0.0001	0.0025	0.0001	−0.0009	−0.0016
Δec$_{i,1}$	0.1467	0.4390***		1.2700***		0.6346***	0.1243	0.5256***	0.0357	0.5397***
π$_{i,1}$	−0.0174	−0.0142**		−0.0232***		−0.0066	−0.1030	−0.0219***	−0.0948	−0.0147**
(I/Y)$_{i,1}$	0.0012*	0.0016***		0.0015***		0.0007**	0.0019**	0.0009***	0.0021***	0.0015***
Δlf$_{i,1}$	−0.0611	−0.6677***		−0.4413		−0.4744***	−1.0398	−0.5189***	−0.3533	−0.5318***
R2										
lyo$_{i,2}$				−0.0003		−0.0027**	−0.0002	−0.0059***	−0.0005	−0.0062***
Δec$_{i,2}$				0.4951***		−0.0159	0.4462***	−0.0024	0.4767***	−0.9697***
π$_{i,2}$				−0.0071*		−0.0229***	−0.0067	0.0055	−0.0062	0.1770***
(I/Y)$_{i,2}$				0.0009***		0.0027***	0.0008**	0.0033***	0.0009***	0.0041***
Δlf$_{i,2}$				−0.4273***		−1.2144***	0.0050	−0.9197***	0.0376	−0.7290***
R3										
lyo$_{i,3}$				−0.0022						
Δec$_{i,3}$				0.3130						
π$_{i,3}$				−0.0013						
(I/Y)$_{i,3}$				0.0025***						
Δlf$_{i,3}$				−1.4842***						
n_1	82	82		12		61	60	68	21	69
n_2				56		21	22	14	61	13
n_3				14						
s.e.	0.0207	0.0116		0.0100		0.0097	0.0185	0.0097	0.0180	0.0103
White F	1.25 [0.28]	0.66 [0.75]		1.09 [0.39]		0.85 [0.59]	1.00 [0.46]	0.70 [0.74]	0.86 [0.59]	1.50 [0.14]

Note: lyo = initial income; Δec = energy consumption growth; π = inflation; (I/Y) = gross fixed capital formation/GDP; gc = government consumption/GDP; Δlf = growth of labor force; the dependent variable is per capita real GDP growth; *,**,and ***represent, respectively, 10%, 5%, and 1% significance levels; values inside the ( ) are *t*-values; values inside the [ ] are *p*-values; s.e. = standard error of estimates; n_1, n_2, and n_3, respectively, represent sample sizes under regimes 1, 2, and 3 (under linear model, n_1 represents total sample size). White F = F statistics by White's heteroscedasticity test. R1 = regime 1; R2 = regime 2; R3 = regime 3.

B.-N. Huang et al. / Energy Policy 36 (2008) 755–767

conditions in view of the different relationships between energy consumption and economic growth in the long term. If there is a positive relationship between energy consumption and economic growth in a country, then that country should adopt a more aggressive energy policy. Conversely, if there is no (or a negative) relationship between the two, then that country should adopt a more conservative energy policy to avoid environmental pollution and wasted energy. To address more specifically which countries should adopt which policies, this paper provides a more in-depth empirical analysis as follows.

Four threshold variables (CO_2, tpes/Y, ind/tfc, and ec) are used as the standards of demarcation.[13] Based on our nonlinear empirical results, when these four variables in a country exceed certain levels, there is no significant positive relationship between energy consumption and economic growth in that country. Since energy consumption cannot bring about economic growth, that country ought to adopt a more conservative energy policy to avoid the pollution resulting from excessive energy consumption. Let the dummy variable for a country be equal to 1 if one of these four variables exceeds the optimal threshold levels, otherwise the dummy variable is equal to 0. In other words, there is no significant relationship (or a negative relationship) if a dummy variable for a country is equal to 1. We may add dummy variables for all four variables (the sum should be between 0 and 4). If the sum is 0, the threshold variables in that country are all lower than the optimal levels. If the sum is greater than 0, at least one threshold variable in that country is greater than the optimal threshold levels. Hence, at least in one case, there exists no positive relationship between energy consumption and economic growth. This suggests that such a country should adopt a more conservative energy policy. In its extreme case, all four dummy variables may equal 1 (i.e., the sum is 4). In other words, the volume of CO_2 emissions is too high, energy efficiency is too low, the share of industrial energy consumption is too high, and overall energy consumption is too high. As a result, there is not a single significant positive relationship (or even the existence of a negative relationship) between energy consumption and economic growth and hence a more aggressive conservative policy should be pursued to reduce energy consumption and CO_2 emissions. For such a country, an increase in energy consumption is unable to help increase economic growth. Table 3 displays four dummy variables and their sum for all 82 countries during the post-energy crisis period.

As indicated in Table 3, 48 out of the 82 countries have all four threshold variables lower than the optimal threshold levels. This means that those 48 countries may use energy consumption to promote economic growth. The remaining 34 countries have at least one threshold variable

higher than the estimated optimal levels. Among them, 17 countries have one threshold variable higher than the estimated optimal level. In six out of those 17 countries, the average per capita income exceeds $10,000. Of the six countries with higher per capita income, four countries (Belgium, the Netherlands, Sweden, and Norway) have a high ec and two countries (Denmark and Japan) have a high ind/tfc. Among the rest of the 11 countries with per capita incomes lower than $10,000, five countries (Nepal, DR Congo, Benin, Kenya, and Zimbabwe) have a high tpes/Y, five countries (Egypt, Turkey, Brazil, Portugal, and Greece) have a high ind/tfc, and one country (Gabon) has high CO_2 emissions. A total of 17 countries with the exception of Gabon still have CO_2 emissions lower than the optimal threshold level. When tpes/Y, ind/tfc and ec are adequately controlled, energy consumption is still conducive to economic growth for these 17 countries.

Of the 82 countries, eight countries have two threshold variables that are higher than the estimated optimal levels. Of these, three countries (Australia, Canada, and Iceland) have average per capita income in excess of $10,000. Each of the three high-income countries with two variables that exceed the optimal levels has a high ec with Australia having high CO_2 emissions and the other two countries having high tpes/Y. All five countries with per capita income lower than $10,000 have higher tpes/Y. For the other threshold variables, Zambia has a high ind/tfc and the other four countries have high CO_2 emissions. Since these eight countries have two variables higher than the optimal levels and most countries have high CO_2 emissions, they should adopt a more conservative energy policy to reduce ec, ind/tfc, and CO_2 emissions and thereby increase tpes/Y.

A total of seven countries (China, South Africa, Venezuela, Singapore, Finland, the US, and Luxembourg) from among the 82 countries have three threshold variables higher than the estimated optimal levels. Of these, the four countries (Singapore, Finland, the US, and Luxembourg) with per capita income in excess of $10,000 all have ec and tpes/Y values that are higher than the optimal levels. In terms of the other threshold variables, Finland has a high ind/tfc and the other three countries have CO_2 emissions that exceed the optimal level. In the case of the other three countries with per capita incomes lower than $10,000, all have values of CO_2 emissions, tpes/Y, and ind/tfc that are higher than the optimal levels. Only two countries (Trinidad and Tobago, and Saudi Arabia) have all four threshold variables higher than the optimal level. Of the nine countries with three or more threshold variables higher than the optimal levels, most countries (with the exception of Finland) have CO_2 emissions higher than the optimal level. All nine countries have a higher tpes/Y, and six countries have higher values of ec and ind/tfc. For these nine countries, a more aggressive conservative energy policy should be pursued to reduce ec and CO_2 emissions, and to increase tpes/Y.

---

[13]The purpose in delineating Period-II from Period-I in order to decipher the nonlinear relationship is that we are now facing the post-energy crisis period.

Table 3
Summary of significant threshold variables during the post-energy crisis period in each country

Country name	CO$_2$	tpes/Y	ind/tfc	ec	sum	%coal	%pet	%elec	kgdp	kgdpid
Ghana	0	0	0	0	0	0.00	22.69	8.13	235.87	3
Bangladesh	0	0	0	0	0	1.25	16.66	4.50	263.92	4
India	0	0	0	0	0	26.71	36.02	11.71	297.56	5
Togo	0	0	0	0	0	0.00	19.39	3.00	313.27	6
Pakistan	0	0	0	0	0	6.11	36.78	10.48	408.38	11
Senegal	0	0	0	0	0	0.00	71.48	7.41	435.13	13
Indonesia	0	0	0	0	0	1.61	39.26	3.83	548.95	14
Sri Lanka	0	0	0	0	0	0.02	27.80	4.64	575.06	15
Cameroon	0	0	0	0	0	0.00	54.93	12.53	629.45	17
Haiti	0	0	0	0	0	0.67	17.22	1.66	672.40	18
Cote d'Ivoire	0	0	0	0	0	0.00	61.01	10.39	803.82	19
Honduras	0	0	0	0	0	0.44	32.90	6.55	893.98	20
Philippines	0	0	0	0	0	3.50	49.05	11.54	910.48	21
Bolivia	0	0	0	0	0	0.58	53.30	7.59	975.83	23
Morocco	0	0	0	0	0	5.44	74.80	12.78	1008.92	25
Nicaragua	0	0	0	0	0	0.00	31.60	5.67	1032.72	26
Thailand	0	0	0	0	0	4.93	54.11	11.62	1254.10	28
Ecuador	0	0	0	0	0	0.00	75.29	8.31	1278.34	29
Paraguay	0	0	0	0	0	0.00	25.95	6.68	1394.52	30
Tunisia	0	0	0	0	0	1.18	72.64	12.34	1481.19	31
Guatemala	0	0	0	0	0	0.02	29.14	4.56	1554.72	32
Dominican	0	0	0	0	0	0.01	56.24	9.25	1648.39	33
Colombia	0	0	0	0	0	8.19	45.93	11.55	1724.20	34
Algeria	0	0	0	0	0	1.47	61.23	8.40	1738.23	35
El Salvador	0	0	0	0	0	0.00	36.93	8.17	1898.25	36
Peru	0	0	0	0	0	2.43	58.41	11.36	2058.33	37
Malaysia	0	0	0	0	0	2.16	68.77	13.50	2433.00	39
Chile	0	0	0	0	0	5.89	50.82	13.32	3146.75	42
Costa Rica	0	0	0	0	0	0.06	54.52	16.85	3308.85	44
Panama	0	0	0	0	0	1.52	54.48	14.92	3349.94	45
Hungary	0	0	0	0	0	10.74	33.55	12.98	3770.81	46
Mexico	0	0	0	0	0	1.89	63.31	10.52	4890.34	48
Uruguay	0	0	0	0	0	0.07	57.69	17.55	4927.59	49
Malta	0	0	0	0	0	0.00	75.04	24.99	5755.91	52
Korea, Rep.	0	0	0	0	0	16.88	66.28	13.10	5881.11	53
Oman	0	0	0	0	0	0.00	72.57	15.49	6703.81	54
Argentina	0	0	0	0	0	0.77	51.49	11.94	6972.96	55
Spain	0	0	0	0	0	4.78	66.46	17.60	10,235.52	58
New Zealand	0	0	0	0	0	8.84	45.02	23.80	11,667.06	60
Ireland	0	0	0	0	0	15.61	57.56	13.82	13,350.26	62
Israel	0	0	0	0	0	0.24	73.25	22.12	14,010.89	63
Italy	0	0	0	0	0	2.80	54.80	15.85	14,582.52	64
Hong Kong	0	0	0	0	0	0.05	68.01	27.06	15,934.40	66
France	0	0	0	0	0	4.46	54.22	17.34	17,596.07	69
Germany	0	0	0	0	0	11.35	49.48	15.77	17,769.31	71
Austria	0	0	0	0	0	7.32	46.60	17.34	17,937.93	72
UK	0	0	0	0	0	6.81	46.29	16.17	18,360.01	74
Switzerland	0	0	0	0	0	1.43	65.38	20.21	30,115.76	82
Nepal	0	1	0	0	1	1.23	5.91	0.93	180.10	1
Congo, DR	0	1	0	0	1	8.99	38.35	14.57	211.71	2
Benin	0	1	0	0	1	0.00	60.85	7.30	320.13	7
Kenya	0	1	0	0	1	2.06	59.51	6.91	346.64	8
Zimbabwe	0	1	0	0	1	31.61	23.70	18.42	578.33	16
Egypt	0	0	1	0	1	1.74	68.74	13.95	1102.41	27
Turkey	0	0	1	0	1	16.78	49.53	10.54	2304.31	38
Brazil	0	0	1	0	1	3.64	48.29	15.44	3073.62	41
Gabon	1	0	0	0	1	0.00	60.54	10.16	4438.85	47
Portugal	0	0	1	0	1	3.21	67.55	15.62	7287.50	56
Greece	0	0	1	0	1	6.50	70.87	16.95	8610.25	57
Belgium	0	0	0	1	1	9.01	53.08	14.64	17,055.86	67
Netherlands	0	0	0	1	1	1.75	39.45	12.30	17,661.30	70
Sweden	0	0	0	1	1	2.57	44.26	29.81	21,267.88	75
Denmark	0	0	1	0	1	2.53	56.42	16.64	23,629.51	77

Table 3 (*continued*)

Country name	CO_2	tpes/Y	ind/tfc	ec	sum	%coal	%pet	%elec	kgdp	kgdpid
Norway	0	0	0	1	1	4.70	44.07	45.48	26,170.13	78
Japan	0	0	1	0	1	7.57	63.66	21.66	28,736.60	81
Nigeria	1	1	0	0	2	0.31	56.86	4.28	348.33	9
Zambia	0	1	1	0	2	11.07	30.43	27.03	429.46	12
Syrian	1	1	0	0	2	0.02	82.88	8.26	969.68	22
Congo, Rep.	1	1	0	0	2	0.00	66.00	6.84	1007.28	24
Jamaica	1	1	0	0	2	1.16	78.98	14.37	2785.02	40
Australia	1	0	0	1	2	7.11	52.63	18.88	15,710.12	65
Canada	0	1	0	1	2	2.16	43.83	21.42	18,120.79	73
Iceland	0	1	0	1	2	3.46	38.27	21.33	23,082.41	76
China	1	1	1	0	3	56.64	18.07	8.92	384.19	10
South Africa	1	1	1	0	3	31.12	30.00	22.23	3177.87	43
Venezuela, RB	1	1	1	0	3	0.45	53.91	13.67	5394.71	50
Singapore	1	1	0	1	3	0.00	81.14	17.80	12,977.41	61
Finland	0	1	1	1	3	6.05	41.18	22.74	17,398.14	68
United States	1	1	0	1	3	3.19	52.82	17.18	26,220.37	79
Luxembourg	1	1	0	1	3	16.91	55.23	12.82	27,250.41	80
Trinidad and Tobago	1	1	1	1	4	0.00	20.22	7.21	5598.67	51
Saudi Arabia	1	1	1	1	4	0.00	65.69	10.49	11,318.52	59

Note: 1 represents the case in which values of variables are greater than the optimum value, while 0 represents the case where values of variables are less than its optimum level; sum denotes the summation of dummy variables indicated under the column of CO_2, tpes/Y, ind/tfc, and ec with a value of 1; %coal, %pet, and %elec represent, respectively, the percentages of coal, petroleum, and electricity consumption to total energy consumption; kgdp = the yearly average of per capita real GDP (1981–2002); kgdpid = the ranking of per capita GDP in 82 countries (e.g., kgdpid = 1 denotes the lowest ranked country in terms of the 1981–2002 yearly average of per capita GDP among 82 countries, and kgdpid = 82 denotes that of the highest ranked country).

Table 3 also shows the ratios of coal (% coal), petroleum (% pet), and electricity (% elec) to total energy consumption and per capita income (kgdp) average (1981–2002) data. The correlation coefficients between the variables and numbers of threshold variables that are higher than the estimated optimal levels are 0.24 for coal, −0.03 for petroleum, 0.14 for electricity, and 0.25 for per capita income. The greater the use of coal (a correlation coefficient of 0.24) and the higher the per capita income (a correlation coefficient of 0.25), the greater the tendency to have a higher number of threshold variables that exceed the estimated optimal levels.

4. Conclusions and policy implications

With weather patterns becoming more erratic, there is increasing evidence to suggest that CO_2 emissions from the use of fossil fuels is the main source of global warming. The major purpose of energy use is to increase economic growth. However, the use of energy also has a negative impact on global warming, environmental pollution, and resource exhaustion. Whether or not the use of energy can bolster economic growth has been a focus of attention for academics and policy-makers alike. Unfortunately, there have not been consistent findings in the literature based on the results of the Granger-causality tests. One important reason for this involves the use of time series data. Problems of small sample size, structural change, and/or short-term variations due to the business cycle may render the time series estimates invalid. Biased results may also ensue. To overcome the above-mentioned pitfalls as a result of using time series data, this paper separates the

data covering the period from 1971 to 2002 into the energy crisis period (1971–1980) and the post-energy crisis period (1981–2000) for 82 countries in order to accommodate the problem of structural change. The cross-sectional data for yearly averages in these two periods are used to smooth out short-term variations.

To identify the conditions under which economic growth is related to energy consumption, we selected CO_2 emissions (CO_2), energy efficiency (tpes/Y), the ratio of industrial energy consumption to total energy consumption (ind/tfc), and per capita energy consumption (ec) as threshold variables. Our nonlinear test results support the fact that these variables be used as threshold variables because the energy consumption–economic growth equation clearly exhibits a nonlinear relationship. Moreover, the estimated results from these two periods indicate that the relationship between energy consumption and economic growth differs between the energy crisis period and the post-energy crisis period. During the post-energy crisis period, the nonlinear results indicate that, when the CO_2, tpes/Y, ind/tfc, and ec threshold variables are higher than their optimal threshold values, there is no significant positive relationship (or even a presence of negative relationship) between energy consumption and economic growth. Conversely, there exists a significant positive relationship between the two if these threshold values are below the estimated optimal values. We also find that, among the 82 countries, 48 countries have no single threshold variable higher than the estimated optimal values. Therefore, these 48 countries may increase their use of energy consumption to promote economic growth. In addition, 17 countries have one threshold variable that is higher than the optimal value (only one country has CO_2 emissions higher than the optimal value) and these countries may still use energy consumption to help promote economic growth if CO_2 emissions are appropriately controlled. There are eight countries with two threshold variables higher than the estimated optimal values (of which five have a common threshold variable: CO_2 emissions). These eight countries should adopt a more conservative energy policy compared with the 17-country group with only one threshold variable. Finally, nine countries have at least three threshold variables higher than the estimated optimal values. Among them, eight countries have CO_2 emissions higher than the optimal value. For this nine-country group, the negative externality outweighs the benefits brought about by the energy consumption and as such the energy policy adopted should be to reduce energy consumption and CO_2 emissions, and to increase energy efficiency.

The policy implications derived from this study need to be clarified. In order to design a sound energy consumption policy, each country needs to take into consideration information on the CO_2 emission (CO_2), the efficiency of energy use (tpes/Y), the share of industrial energy consumption to total energy consumption (ind/tfc), and per capita energy consumption (ec). It is evident that the CO_2 emission from the use of fossil fuels is a major source of global warming. When CO_2 emission is higher than a certain threshold level, energy consumption can no longer bring about significant output increase because of the nonlinear relationship between energy consumption and economic growth. The CO_2 emission in 14 out of 82 countries is higher than the threshold level estimated from our threshold regression. For such high CO_2 emission countries, the use of fossil fuels needs to be reduced. Other energy policies need to be implemented such as increasing the efficiency of energy use, more stringent regulations of CO_2 emission, and/or the replacement of fossil-type fuels by cleaner energy sources. For 20 countries with at least one threshold variable higher than the estimated optimal threshold levels, appropriate reduction of energy use, the use of cleaner energy sources, and more efficient use of energy to avoid unnecessary CO_2 emission could be a good energy policy to follow. Lastly, for the 48 countries without a threshold variable higher than the estimated level, though energy consumption will bring about output growth, more efficient use of energy and the use of cleaner energy sources are encouraged to combat global warming.

Though the use of cross-sectional data is able to identify the possible existence of a correlation between energy consumption growth and economic growth via certain threshold variables, it is nevertheless unable to test the Granger causality between the two. Future analysis on the nonlinear Granger causality between energy consumption growth and economic growth across various economies over the long run requires the use of a dynamic panel data (DPD) model developed by Arellano and Bond (1991). However, one of the problems in applying the DPD model

is how to segment the data into different regimes via threshold variable(s) in order to decipher the relationship in detail.[14]

Appendix A

See Table A1 for details.

Table A1
Names of 82 countries

Algeria	Greece	Pakistan
Argentina	Guatemala	Panama
Australia	Haiti	Paraguay
Austria	Honduras	Peru
Bangladesh	Hong Kong	Philippines
Belgium	Hungary	Portugal
Benin	Iceland	Saudi Arabia
Bolivia	India	Senegal
Brazil	Indonesia	Singapore
Cameroon	Ireland	South Africa
Canada	Israel	Spain
Chile	Italy	Sri Lanka
China	Jamaica	Sweden
Colombia	Japan	Switzerland
Congo, DR	Kenya	Syrian Arab Republic
Congo, Rep.	Korea, Rep.	Thailand
Costa Rica	Luxembourg	Togo
Cote d'Ivoire	Malaysia	Trinidad and Tobago
Denmark	Malta	Tunisia
Dominican Republic	Mexico	Turkey
Ecuador	Morocco	UK
Egypt, Arab Rep.	Nepal	US
El Salvador	Netherlands	Uruguay
Finland	New Zealand	Venezuela, RB
France	Nicaragua	Zambia
Gabon	Nigeria	Zimbabwe
Germany	Norway	
Ghana	Oman	

References

Aizman, J., Glick, R., 2003. Military expenditure, threats and growth. NBER Working Paper no. 9618.

Akarca, A.T., Long, T.V., 1980. On the relationship between energy and GDP: a re-examination. Journal of Energy Development 5, 326–331.

Andrews, D.W.K., Ploberger, W., 1994. Optimal tests when a nuisance parameter is present only under the alternative. Econometrica 62 (6), 1383–1414.

Arellano, M., Bond, S.R., 1991. Some tests of specification for panel data: Monte Carlo evidence and an application to employment equations. Review of Economic Studies 58, 277–297.

Barro, R.J., 1990. Government spending in a simple model of endogenous growth. Journal of Political Economy 98 (5), 103–126.

Davies, R.B., 1977. Hypothesis testing when a nuisance parameter is present only under the alternative. Biometrika 64, 247–254.

Davies, R.B., 1987. Hypothesis testing when a nuisance parameter is present only under the alternative. Biometrika 74, 33–43.

---

[14]Hansen (1999) provided a method of estimating and testing the panel data threshold effect. However, this method is useful only in the non-dynamic panels of fixed effect. As far as we know, the technique to handle dynamic panel threshold effect is yet to be developed in the literature.

Ghali, K.H., El-Sakka, M.I.T., 2004. Energy use and output growth in Canada: a multivariate cointegration analysis. Energy Economics 26, 225–238.

Hansen, B., 1996. Inference when a nuisance parameter is not identified under the null hypothesis. Econometrica 64, 413–430.

Hansen, B.E., 1997. Inference in TAR models. Studies in Non-linear Dynamics and Econometrics 2 (1), 1–14.

Hansen, B.E., 1999. Threshold effects in non-dynamic panels: estimation, testing, and inference. Journal of Econometrics 93, 345–368.

Huang, B.N., Huang, M.J., Yang, C.W., 2008. Causal relationship between energy consumption and GDP growth revisited: a dynamic panel data approach. Ecological Economics, forthcoming.

Khan, M.S., Senhadji, A.S., 2001. Threshold effects in the relationship between inflation and growth. IMF Staff Papers 48 (1), 1–21.

Kraft, J., Kraft, A., 1978. On the relationship between energy and GNP. Journal of Energy and Development 3, 401–403.

Lee, C.C., 2005. Energy consumption and GDP in developing countries: a cointegrated panel analysis. Energy Economics 27, 415–427.

Lee, C.C., Chang, C.P., 2007a. Energy consumption and GDP revisited: a panel analysis of developed and developing countries. Energy Economics 29 (6), 1206–1223.

Lee, C.C., Chang, C.P., 2007b. The impact of energy consumption on economic growth: evidence from linear and nonlinear models in Taiwan. Energy 32, 2282–2294.

Levine, R., Zervos, S., 1998. Stock markets, banks, and economic growth. American Economic Review 88 (3), 537–558.

Moon, Y.S., Sonn, Y.H., 1996. Productive energy consumption and economic growth: an endogenous growth model and its empirical application. Resource and Energy Economics 18, 189–200.

Soytas, U., Sari, R., 2003. Energy consumption and GDP: causality relationship in G-7 countries and emerging markets. Energy Economics 25, 33–37.

Soytas, U., Sari, R., 2006. Can China contribute more to the fight against global warming? Journal of Policy Modeling 28, 837–846.

Stern, D.I., 2000. Multivariate cointegration analysis of the role of energy in the US macroeconomy. Energy Economics 22, 267–283.

Stern, D.I., Cleveland, C.J., 2003. Energy and economic growth. Rensselaer Working Papers in Economics, 0410.

Tong, H., 1978. On a threshold model. In: Chen, C.H. (Ed.), Pattern Recognition and Signal Processing. Sijthoff & Noordhoff, Amsterdam, pp. 101–141.

Tong, H., Lim, K.S., 1980. Threshold autoregressions, limit cycles, and data. Journal of the Royal Statistical Society B 42, 245–292 (with discussion).

Yu, S.H., Jin, J.C., 1992. Cointegration tests of energy consumption, income and employment. Resources and Energy Economics 14, 259–266.

199

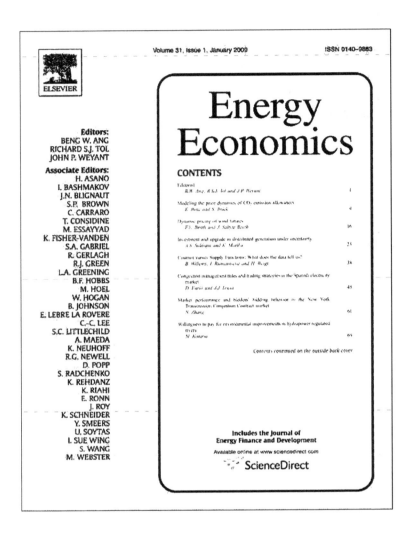

Volume 31, Issue 1, January 2009   ISSN 0140-9883

Energy Economics

CONTENTS

Contents continued on the outside back cover

Includes the Journal of
Energy Finance and Development

Available online at www.sciencedirect.com
ScienceDirect

200

Energy Economics 31 (2009) 91–98

Contents lists available at ScienceDirect

Energy Economics

journal homepage: www.elsevier.com/locate/eneco

The dynamics of a nonlinear relationship between crude oil spot and futures prices: A multivariate threshold regression approach

Bwo-Nung Huang [a,b,*], C.W. Yang [c], M.J. Hwang [d]

[a] Department of Economics & Center for IADF, National Chung Cheng University, Chiayi 621, Taiwan
[b] Department of Economics, West Virginia University, Morgantown, WV 26505-6025, United States
[c] Department of Economics, Clarion University of Pennsylvania, Clarion, PA 16214-1232, United States
[d] Department of Economics, College of Business and Economics, West Virginia University, P. O. Box 6025, Morgantown, WV 26505-6025, United States

ARTICLE INFO

Article history:
Received 2 November 2007
Received in revised form 20 July 2008
Accepted 6 August 2008
Available online 13 August 2008

JEL classification:
Q43
G13
C22

Keywords:
Price discovery
Regime
Multivariate threshold
Backwardation
Contango

ABSTRACT

This paper segments daily data from January of 1986 to April of 2007 into three periods based on certain important events. Both periods I and II indicate that the spot prices in general are higher than futures prices as was well-known in the literature. Only period-III (2001/9/11–2007/4/30) displays a reverse phenomenon: futures prices, in general, exceed spot prices. When the absolute value of a basis (futures-spot) is greater than the threshold value in the arbitrage area (regime 1 and 3), at least one of the error correction coefficients, representing adjustment towards equilibrium, is statistically significant. That is, there exists a tendency in the oil market in which prices move toward equilibrium. With respect to the short-run dynamic interaction between spot price change (Δs_t) and futures price change (Δf_t), our results indicate that when the spot price is higher than futures price, and the basis is less than certain threshold value (regime 3), there exists at least one causal relationship between Δs_t and Δf_t. Conversely, when the futures price is higher than spot price and the basis is higher than certain threshold value (regime 1), there exists at least one causal relationship between Δs_t and Δf_t. Finally, we use the method suggested by Diebold and Mariano [Diebold, Francis X., Mariano, Roberto S., 1995. Comparing predictive accuracy. Journal of Business and Economic Statistics 13 (3), 253–263] to compare the predictive power between the linear and nonlinear models. Our empirical results indicate that the in-sample prediction of the nonlinear model is clearly superior to that of the linear model.

© 2008 Elsevier B.V. All rights reserved.

1. Introduction

The major function of futures markets is to hedge or reduce either the producer's or consumer's risk. Because futures prices may reflect future trend of spot prices, it has an important function of price discovery.[1] The literature on the function of price discovery spreads ink on primarily financial products such as the stock index (Antoniou et al., 1998; Huang, 2002; Illueca and Lafuente, 2003). Past research on futures commodities and spot commodities focused mainly on agriculture products (see Koontz et al., 1990; Oellermann et al., 1989; Schroeder and Goodwin, 1991) and oil products (Schwarz and Szakmary, 1994; Foster, 1996; Silvapulle and Moosa, 1999; Moosa, 2002). It is to be pointed out that majority of the research on futures and spot prices of crude oil followed the model by Garbade and Silber (1983): an error correction model

(henceforth ECM), or a bivariate vector error correction model (henceforth VECM). Based on the lagged spot prices, futures prices, and error terms, their models help identify the lead–lag relationship between spot prices and futures prices, and hence, the mechanism of price discovery: the transmission of information into price formation. According to Silvapulle and Mossa (1999), there exists a nonlinear relationship between the spot and futures prices (rather than a linear relationship frequently discussed in the past literature). Employing the nonlinear causality model by Baek and Brock (1992) to reevaluate the lead–lag relationship between spot and futures prices, they found that a two-way feedback relationship, rather than a one-way causality, exists where futures prices always lead spot prices. The nonlinear relationship can be attributed to the difference in transaction costs, the role of noise traders, and the microstructure effects of markets (Abhyankar, 1996).

It is worth investigating whether a dynamic relationship between the spot and futures prices implies a certain nonlinear relationship. When futures prices are greater than spot prices by a certain amount (upper limit of a threshold band),[2] investors will start the arbitrage by selling futures commodities and buying spot commodities. Likewise, investors will find arbitrage profitable when spot prices are greater

---

* Corresponding author. Department of Economics & Center for IADF, National Chung Cheng University, Chiayi 621, Taiwan.
E-mail address: ecdbnh@ccu.edu.tw (B.-N. Huang).

[1] Under the unbiasedness hypothesis, futures prices reflect future trend of spot prices. In fact, there are strong arguments and substantial empirical evidence that do not support this hypothesis. This would be the case particularly if we consider the cost-of-carry relationship as a determinant of futures prices, which implies that spot and futures prices are related contemporaneously. In this case, the futures price has nothing to do with the spot price that is to prevail in the future. We appreciate greatly one referee for this valuable comment.

[2] A basis (b_t) is defined as the futures price (F_t) minus the spot price (S_t). In general, the size of the threshold band is determined mainly from relevant transaction costs.

0140-9883/$ – see front matter © 2008 Elsevier B.V. All rights reserved.
doi:10.1016/j.eneco.2008.08.002

than futures prices to a certain degree (lower limit of a threshold band) by selling spot commodities and buying futures commodities. When the difference between futures and spot prices (basis) is within that threshold band, there will be no arbitrage as the basis (difference between futures and spot prices) is smaller than the transaction costs. As a result, a possibility exists for a nonlinear dynamic relationship between the spot and futures prices. In sum, there are three regimes: Regime 1 is where the basis (positive) is greater than the upper limit of a threshold band; Regime 3 is where the basis (negative) is smaller than the lower limit of the threshold band; and, of course, Regime 2 is where the basis is between the upper and lower limits of the threshold band. The basis in the first and third regimes may affect futures prices, or spot prices, or both, while the basis in the second regime has no real effect on futures or spot prices. If the ECM is employed to investigate the dynamic relationship between spot and futures prices, it is expected that the coefficient on the error correction term is statistically significant in regimes 1 and 3 indicating adjustments toward an equilibrium as the basis of spot and futures prices is either higher or lower than a certain threshold level (transaction cost). However, under regime 2, the error correction coefficient is expected to be insignificant as the basis does not surpass any threshold bounds.

Past studies on stock index emphasize the dynamics regarding arbitrage from changes in futures and spot prices (Dwyer et al., 1996; Brooks and Garrett, 2002). Results from their studies can only tell the existence of an arbitrage, but not on how investors profit from the arbitrage (taking the long or short position on futures commodities) and as such, the short-term dynamic relationship between futures and spot prices is left unexplained. A limitation in their estimation is that they did not use a bivariate VECM model in which a basis with a lag of 1 period (b_{t-1}) should have been used to partition the whole sample into three different regimes. Instead, the entire sample was estimated using a univariate autoregressive (AR) model. Therefore in this paper, we employ the multivariate threshold autoregressive (henceforth MVTAR) model to investigate the short-term dynamics and the arbitrage behavior between futures and spot prices of crude oil using b_{t-1} as threshold variable.[3]

The price of crude oil is affected not only by the change in demand for and supply of crude oil but also by changes in politics in the world (e.g., war). Since crude oil price can be affected by structural change due in part to external shocks, the data under study from January, 1986 to April, 2007 are separated into three periods: structural breaks are based on Gulf War of 1990 and 9/11 in 2001 (including Iraq War) to investigate the arbitrage behavior and the dynamics of nonlinear relationship between spot and futures prices in the crude oil market. The data indicate that what happened in the crude oil market where spot prices are higher than futures prices seems to have no longer existed after 9/11, 2001. By using the nonlinear bivariate threshold autoregressive model and using a basis with a lag of 1 period as the threshold variable, our empirical results show that, regardless of what period used, when the basis is within the upper and lower limits of the threshold band (outside the arbitrage regime), no statistically significant terms on error correction coefficient and/or short-term interaction between spot price change (Δs_t) and futures price change (Δf_t) can be found.[4] On the other hand, when the absolute value of the basis is greater than a certain threshold value (regime 1 or 3), at least in one regime we can find statistically significant error correction coefficients in the spot commodities market. Further, when spot prices in general are higher than futures prices (defined as period 1), the Granger causality between Δs_t and Δf_t starts to

manifest in the presence of a negative basis whose value are less than the optimal threshold value (regime 3). When the frequencies of negative and positive bases are roughly the same (defined as period 2), there exists significant Granger causality in both regimes 1 and 3. When futures price is greater than spot price (period 3), the Granger causality between Δs_t and Δf_t appears only in regime 1.

Following this section, Section 2 introduces the theoretical and empirical models. Section 3 describes the data and reports the results from unit root, cointegration and nonlinear tests. Section 4 analyzes and discusses empirical results. Section 5 compares the in-sample prediction between the linear and nonlinear models. Section 6 gives concluding remarks.

2. Model specification

Our model is formulated on the basis of the Cost-of-Carry hypothesis, which is also known as the Theory of Storage and was first formalized by Kaldor (1939) and Working (1948, 1949). It is based on the traditional view that a commodity futures price is the current commodity spot price net of the costs of storage (interest charges, insurance and warehousing rent) and a convenience yield. This model is based on an arbitrage argument that the futures price is equal to the spot price plus the carrying costs. In essence, the futures price must be high enough to offset the storage costs incurred while an arbitrageur awaits delivery. If the futures price is too low, the arbitrageur holding the commodity in inventory could sell on the spot market and buy the futures contract to avoid incurring the carrying costs until the maturity of the futures contract.

To illustrate a simple arbitrage strategy, consider an economy with no transaction costs and with two time periods, t_0 and t_1. Individuals wishing to take a speculative position in a commodity essentially have two choices:

1. buy a futures contract;
2. buy the spot commodity and store it.

In either case, consider a fully leveraged position. Let $F_{1|0}$ be the futures price at t_0 for delivery at t_1, and $\tilde{S}_1$, be the unknown future spot price at time t_1. If a futures contract is purchased at t_0, the net cash flow at the expiration of the contract is given by

$$-F_{1|0} + \tilde{S}_1 = \text{Cashflow (futures)}.$$

If, on the other hand, the commodity is purchased and stored for speculation on the future spot price, the net cash flow at t_1 is given by:

$$-S_0(1 + r_{1|0}) + \tilde{S}_1 - w_{1|0} = \text{Cashflow (storage)},$$

where S_0 is the spot price at t_0, $r_{1|0}$ is the risk-free rate, and $w_{1|0}$ is the storage cost over the contract period. For futures contracts to be held in equilibrium, it is necessary that

$$\text{Cashflow (futures)} \geq \text{Cashflow (storage)}.$$

The relationship between the cash flow from the two alternative strategies implies that

$$F_{1|0} \leq S_0(1 + r_{1|0}) + w_{1|0}. \tag{1}$$

In a two-period economy, Eq. (1) should hold. For a multi-period economy, however, the price of a futures contract maturing more than a single period ahead may not necessarily be equivalent to the stored commodity. Kaldor (1939) and Working (1948, 1949) used the concept of convenience yield derived from inventories of the commodity held in storage to explain the disparity between these two prices.

Unlike costs of storage, the convenience yield can be likened to a liquidity premium, usually being described as the convenience of holding inventories, as many commodities are inputs in the production process, or as the convenience of having inventory to meet unexpected demand. When the motive for holding money is referred to, the

---

[3] Since arbitrageurs respond very quickly to incentives for pure profits, arbitrage activity exists when the basis exceeds the threshold level of transactions costs. As threshold variable is a predetermined variable, with $d=1$ in the daily data, we should be able to capture the arbitrage behavior in the crude oil market.

[4] Even though Dwyer et al. (1996) divided the data into three regimes to separately investigate the dynamic relation using VECM, they applied a two-step approach using a single-basis equation in estimating upper and lower threshold levels to locate three regimes. They did not use a bivariate model to simultaneously estimate the upper and lower limits of a threshold band.

convenience motive is usually distinguished from the speculative motive. However, the convenience motive for holding commodity stocks would only apply strictly to those stocks held in order to avoid the nuisance and costs of frequent deliveries for processing, and frequent revisions of the production schedule to meet increased sales. That is, the convenience yield may arise from situations such as increased flexibility in production runs, meeting unexpected demand and maintaining a given level of output at a lower cost.

In equilibrium, the Cost-of-Carry hypothesis implies that the return from purchasing a commodity at t and selling it for delivery at $t+k$, $F_{t+k|t}-S_t$, equals the interest foregone during storage (financing or interest costs), $S_t r_{t+k|t}$, plus the marginal warehousing cost, net of the marginal convenience yield. Therefore, in a multi-period world, Eq. (1) becomes

$$F_{t+k|t} = S_t(1 + r_{t+k|t}) + w_{t+k|t} - c_{t+k|t}, \quad (2)$$

where $w_{t+k|t}$ is the storage costs and $c_{t+k|t}$ is the convenience yield over k periods.

Kaldor (1939) and Working (1948, 1949) hypothesized the marginal convenience yield to be a decreasing function of the level of inventories. The higher the inventory level, the smaller will be the convenience yield of an additional unit. In the case where the level of inventory is sufficiently high, the convenience yield may be near or equal to zero, resulting in the phenomenon of contango, that is, the futures price is above the spot price. Conversely, when stocks of the commodity are extremely low, the marginal convenience yield may actually exceed marginal storage costs, resulting in the phenomenon of backwardation, that is, the spot price is above the futures price.[5]

Defining basis b_t as $F_{t+k|t}-S_t$ and using Eq. (2), we have

$$b_t - S_t r_{t+k|t} = w_{t+k|t} - c_{t+k|t} = tc, \quad (3)$$

where $tc = w_{t+k|t} - c_{t+k|t}$ denotes total cost. Under the Cost-of-Carry framework, if the basis minus interest is still larger (or smaller) than the threshold band, an arbitrage is profitable. Consequently, a long arbitrage position (positive basis) is justified as

$$b_t - S_t r_{t+k|t} > tc. \quad (4)$$

Conversely, a short arbitrage position (negative basis) may be preferred if

$$b_t - S_t r_{t+k|t} < -tc. \quad (5)$$

Since the investor's decision on arbitrage is based on b_{t-1} (basis of 1 period before), not b_t, we substitute b_{t-1} for b_t in estimating Eqs. (4) and (5). It is manifest from the above that the driving force behind the arbitrage is the basis. Past literature did take the basis into consideration via traditional VAR or VECM model to evaluate the interaction between futures and spot commodities. Such a linear model fails to explain how different bases affect the interaction between and the volatility of futures and spot prices. To circumvent the problem, this paper applies the following three-regime VECM using b_{t-1} as a threshold variable to investigate the interaction between futures and spot prices.

$$\Delta f_t = \alpha_i^f ecm_{t-1} + \beta_i^f(L)\Delta f_{t-1} + \gamma_i^f(L)\Delta s_{t-1} + \varepsilon_t^f. \quad (6)$$
$$\Delta s_t = \alpha_i^s ecm_{t-1} + \beta_i^s(L)\Delta f_{t-1} + \gamma_i^s(L)\Delta s_{t-1} + \varepsilon_t^s$$

Let $F_{t+1|t}=F_t$ and define $b_t=F_t-S_t$, we have $f_t=\ln(F_t)$ and $s_t=\ln(S_t)$. $\Delta = 1-L$, where L is the lag operator. $ecm_{t-1}=f_{t-1}-s_{t-1}$. If $b_{t-1}>tc$, then $i=1$; if $-tc\leq b_{t-1}\leq tc$, then $i=2$; and if $b_{t-1}<-tc$, then $i=3$.

In Eq. (6), we need to estimate the upper and lower limits of the threshold band $\pm tc$,[6] along with the coefficients, α_i, β_i, and γ_i. From

the bivariate threshold model in Eq. (6), we derive three different regimes by the size of b_{t-1}. If $i=1$, the arbitrageur will sell futures commodities and buy spot commodities and we expect α_i^f to be negative and statistically significant, α_i^s to be positive and statistically significant, and/or both. On the other hand, if $i=3$, the arbitrageur will sell spot commodities and buy futures commodities and again, we expect a significant α_i^s (positive), significant α_i^f (negative), and/or both.[7] As to the dynamic relationship between Δf_t and Δs_t in these three regimes (i.e., $i=1$, 2 and 3), we shall explore the potential Granger causality ($\rightarrow$) of $\Delta f_t \rightarrow \Delta s_t$ or $\Delta s_t \rightarrow \Delta f_t$ and feedback ($\leftrightarrow$) of $\Delta f_t \leftrightarrow \Delta s_t$ via empirical estimations.[8]

Eq. (6) better explains arbitrage behavior despite the difficulty in estimation (e.g., simultaneous estimation of $\pm tc$). Dwyer et al. (1996) and Brook and Garrett (2002) simply applied the linear autoregressive (AR) model from the cointegration vector to find the threshold levels. Based on these threshold levels, the data are grouped into different regimes for the two-step estimation using the different VECM. As such, they were unable to analyze simultaneously both the interaction between Δf_t and Δs_t under different regimes ($i=1,2,3$) and the relationship between α_i^f and α_i^s. In order to determine the threshold values for arbitrage endogenously within the VECM framework, this paper uses the MVTAR proposed by Tsay (1998) to estimate the system of equations defined in Eq. (6).[9]

3. Description of data and the results from unit root, cointegration and nonlinear test

The WTI crude oil data of 5563 daily spot oil prices (S_t) and futures prices (F_t) from January 2, 1986 to April 30, 2007 are used (published by the Energy Information Administration, www.eia.doe.gov).[10] By taking the logarithm of both variables, spot prices and futures prices are denoted respectively as s_t ($=\ln S_t$) and f_t ($=\ln F_t$). Fig. 1 exhibits trends of spot prices and basis during this period.

A perusal of Fig. 1 reveals characteristics of spot prices. There are two important events (the Gulf War of 1990–91 and 9/11 of 2001) resulting in a greater increase in oil price changes. At the same time, if we look at the bases (futures price minus spot price) for the whole period, the number of negative bases (spot price is greater than futures price) exceeds the number of positive bases. Table 1 presents descriptive statistics of the bases for the entire period and three separate periods.

From Table 1, for the 22 years from 1986 to 2007, negative basis shows up 51.32% of the time. By separating the whole data period into three periods based on the two important events, the percentages of negative basis are respectively 55.42% for period-1 (01/02/1986–02/28/1991), 51.32% for period-2 (03/31/1991–08/31/2001), and 46.48% for period-3 (09/01/2001–04/30/2007). In the crude oil market, spot prices, most of the time, are higher than futures prices. This particular result of negative basis is termed "backwardation" as coined by Litzenberger and Rabinowitz (1995).[11] As pointed out by Pindyck (2001), when the marginal convenience yield (net of storage cost) is large, the spot price will exceed the futures price. When the inventory of crude oil is low (e.g., as a result of outside shocks), the marginal convenience yield will be

[5] For the discussion of the cost–of-carry model, please see Chow et al. (2000, pp. 217–217).
[6] The threshold values ($\pm tc$) incorporate both implicit (e.g., convenience yield) and explicit (e.g., storage) transaction costs.

[7] If $i=1$ and the basis > 0, the futures price may drop and the spot price may increase, and we expect that $\alpha_1^f>0$, or $\alpha_1^s<0$, or both. With $i=3$ and the basis <0, the futures price may increase and the spot price may decrease, and we expect that $\alpha_3^f<0$, or $\alpha_3^s>0$, or both. Such relations indicated by α^f and α^s reflect the potential for arbitrage by investors.
[8] The feedback relationship between Δf_t and Δs_t suggests a simultaneous reaction of futures and spot markets to new information. Silvapulle and Mossa (1999) used a nonlinear causality model to test the interaction relationship between futures and spot commodities, and discovered that there was a feedback relationship between them.
[9] For the statistical test and the estimation process of MVTAR, see Tsay (1998, pp. 1186–1196).
[10] The nearby futures series are constructed from the daily closing prices on futures contracts one month prior to the expiration month.
[11] Backwardation in future contracts was first called "normal backwardation" by John. M. Keynes in his 1930 book, A Treatise on Money, the Theory of Normal Backwardation Applies to Commodity Futures.

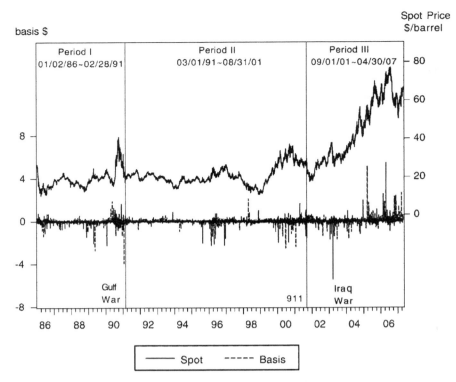

Fig. 1. Time series plot of WTI spot prices and bases.

high; therefore, from Eq. (3), one should be able to witness that spot price is higher than futures price. Pindyck (2001) indicated that marginal convenience yield is not directly observable, and as such, can be determined from the spread between futures and spot prices (basis). The basis is defined as the difference of futures price minus spot price. A large negative value of a basis, in fact, reflects a large marginal convenience yield (Pindyck, 2001). In other words, as the volatility of oil price from outside shocks increases, the demand for crude oil will rise and the inventory is expected to decrease. This would lead to an increase in the marginal convenience yield or a larger negative basis indicating the spot price is far greater than the futures price. Fig. 2 displays the relationship between the volatility of crude oil spot prices derived from the conditional variance of the GARCH(1,1) model and the basis.

As indicated in Fig. 2, when the volatility of crude oil spot prices is high, the basis tends to be negative (reflect a backwardation relationship) with a large absolute value. This is in consistence with the result of Pindyck (2001). Note that, as the crude oil price rose to a higher level after 2004, the volatility remained relatively low and the basis appeared to be positive (the futures price is higher than spot price). It is not of a backwardation relationship mentioned earlier. As indicated in the basic statistics of Table 1 during period-III, number of negative

bases shows up at less than 50% of the time. It is manifest that the relationship between the spot and futures prices during period-III is no longer a backwardation process. Are these changes during period-III going to affect the short-run relationship and arbitrage behavior between the spot prices and futures prices in the crude oil market? Are the short-run dynamic relationship and arbitrage behavior between the spot prices and futures prices going to be the same between the periods of high volatility or low volatility of crude oil prices? These are part of questions we try to find answers to.

Before applying the Tsay model, we need to ascertain that our model is a bivariate VECM. In other words, it is necessary to first test the cointegration relationship between the logarithm of futures price (f_t) and logarithm of spot price (s_t). Table 2 lists the test results of unit root and cointegration for these two variables.

It is clear from Table 2 that there is a cointegration relationship between the spot and futures prices with a $(1,-1)$ relationship. The Engle and Granger (1987) residual based cointegration test results do support the use of the bivariate error correction model to investigate the short-run dynamics and arbitrage relationship between the spot prices and futures prices under different threshold levels.

According to Tsay (1998), before estimating Eq. (6), it is necessary to test the existence of the nonlinear relation using $_tb_{t-1}$ as a threshold variable in the data. Tsay (1998) proposed a $C(d)$ statistic to examine the existence of the nonlinear threshold relationship. Note that the $C(d)$ statistic based on the arranged regression follows a asymptotically chi-square distribution with $k(pk+1)$ degree in which p is the lag length of the VAR model and k is the number of variable. Some statistical problems related to the testing of the nonlinear threshold model warrant further discussion. The multivariate threshold model proposed by Tsay (1998) did not take into consideration the cointegration relationship because it gives rise to the complicated statistical testing problem of linearity vs. nonlinearity and cointegration vs. non-cointegration. At the present time, the two-step test approach adopted by Balke and Fomby (1997) is fairly accepted. That is, under the linear assumption, we test for the existence of a cointegration relation first. If there is a cointegration relationship, the pre-specified cointegration vector needs to be included

Table 1
Descriptive statistics of bases

	Period-I	Period-II	Period-III	All
Mean	−0.0357	−0.0223	0.0375	−0.0097
Max	1.89	2.10	5.45	5.45
Min	−3.92	−2.53	−5.45	−5.45
S.D.	0.3029	0.2483	0.4633	0.3221
Pos.	0.4458	0.4808	0.5352	0.4868
Neg.	0.5542	0.5192	0.4648	0.5132
N	1346	2741	1476	5563

Note: Mean = arithmetic mean; Max = the largest value in the period; Min = the lowest value in the period; S.D. = standard deviation; pos. = percentage of positive bases in the period; neg. = percentage of negative bases in the period; N = sample size, where period-I = 01/02/1986–02/28/1991, period-II = 03/31/1991–08/31/2001, and period-III = 09/01/2001–04/30/2007.

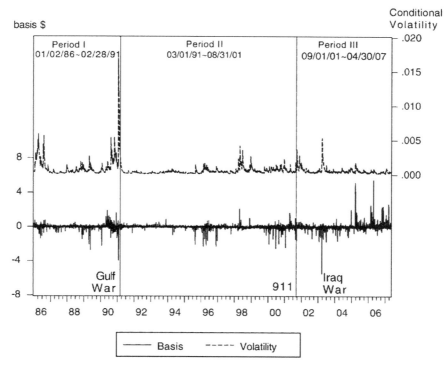

Fig. 2. Time series plot of conditional volatility and bases.

in the model to test whether it is a linear or nonlinear relationship (Lo and Zivot, 2001; Seo, 2006). Our test essentially follows the two-step approach.[12] By applying the AIC criterion to Eq. (6), we find the optimal lag for $p = 10$ irrespective of using the whole period, period-I, period-II, or period-III. Table 3 reports the $C(d)$ statistics using $d = 1$.

It is clear from Table 3 that we reject the null hypothesis of the linear model for all periods. The optimal threshold band tc* is estimated given the assumption of symmetrical upper and lower limits of the threshold band (±tc). In other words, the transaction cost is assumed to be the same irrespective of taking the long or short position.[13] We then proceed to use the grid-search algorithm to estimate the best threshold value. In order to estimate the optimal threshold level tc* for a given p and d, we arrange b_{t-1} in ascending order to search for tc* based on the criterion proposed by Weise (1999).[14]

$$\arg \min_{tc \in TC} \log \det |\Sigma| = \pm tc^*, \tag{7}$$

where $\Sigma = E(\varepsilon \; \varepsilon')$, $\varepsilon = [\varepsilon_t^f \; \varepsilon_t^s]$ represents the variance and covariance matrix from Eq. (6).

4. Analysis of empirical results

The estimated causal relationships among Δs_t, Δf_t, coefficients of α_i^f and α_i^s in three regimes, and the original linear model indicated in Eq. (5) are reported in Table 4.[15]

[12] Theoretically, the test involves four categories: (1) linear without a cointegration relation; (2) nonlinear without a cointegration relation; (3) linear with a cointegration relation; and (4) nonlinear with a cointegration relation. So far, the literature does not seem to have a good test in the case of the nonlinear with or without a cointegration relation as it involves a different distributional problem (Seo, 2006).

[13] Dwyer et al. (1996) employed parametric and nonparametric approaches and were unable to reject the hypothesis of symmetrical transaction cost.

[14] The extent of a search target depends on the number of data points, the number of variables, and the lagged periods (p and d).

[15] To save space, both the AIC results and the Ljung-Box Q statistics regarding the existence of possible serial correlation in residuals are not provided. They are available upon request.

Table 4 displays the estimated results of the linear and nonlinear models (three regimes) in all three periods as well as that for the whole period. The last column of Table 4 shows the optimal threshold value of ±0.0015 for all periods (periods I, II, and III as well as the whole period). Using the error correction model, we find that the futures price change (Δf_t) is found to lead spot price change (Δs_t) in the linear model when the whole period is considered. This is consistent with price discovery indicated in the past literature. With respect to arbitrage, when futures prices are higher than spot prices, the arbitrageurs would sell futures contracts and buy spot contracts. As a result, spot prices will be higher and futures prices tend to be lower. Therefore, the relationship between the bases and spot price changes is positive and significant ($\alpha^s > 0$). At the same time, the relationship is negatively significant ($\alpha^f < 0$) between the bases and the futures prices. For the whole period, the signs of estimated

Table 2
Unit root and cointegration test

Period	ADF unit root test		Cointegration test	Cointegration vector
	f_t	s_t	Engle–Granger two-step	ecm_{t-1}
All	−3.04	−3.09	−10.94 [0.00]	$f_{t-1} - 1.000s_{t-1}$
Period-I	−2.46	−2.38	−21.89 [0.00]	$f_{t-1} - 0.999s_{t-1}$
Period-II	−1.92	−1.91	−27.20 [0.00]	$f_{t-1} - 1.000s_{t-1}$
Period-III	−3.09	−3.00	−19.61 [0.00]	$f_{t-1} - 1.000s_{t-1}$

Note: ADF = Augmented Dickey–Fuller Test; cointegration test for H_0: $r = 0$, r represents the number of cointegration vector; values inside [.] are p values; f_t and s_t represent respectively the futures and spot prices after taking the logarithm.

Table 3
The $C(d)$ statistics (Tsay, 1998)

Period	$d = 1$
I	22.41 [0.00]
II	37.90 [0.00]
III	38.67 [0.00]
All	98.34 [0.00]

Note: values in [.] are p values. $C(d)$ statistic follows χ^2 distribution.

205

Table 4
Estimated results of three-regime and linear models

Period	Regime (i)	α_i^f	α_i^s	$H_0: \Delta s_{t-j} \nrightarrow \Delta f_t$	$H_0: \Delta f_{t-j} \nrightarrow \Delta s_t$	n_i	|c|
I	1	−0.09 (−0.41)	0.27 (1.30)	7.43 [a] [0.68]	11.33 [0.33]	420 {31.46%}	0.015
	2	−4.01 (−1.06)	−2.39 (−0.63)	2.38 [0.99]	6.73 [0.75]	257 {9.25%}	
	3	−0.11 (−0.58)	0.42** (2.13)	20.88 [0.02]	24.22 [0.01]	658 {49.29%}	
	Linear	−0.12 (−0.98)	0.41*** (3.30)	8.90 [0.54]	16.82 [0.08]	1335	
II	1	−0.22 (−1.50)	0.43* (2.73)	15.50 [0.12]	25.53 [0.00]	893 {32.71%}	0.015
	2	1.12 (0.57)	1.14 (0.60)	0.42 [0.78]	8.43 [0.59]	648 {23.74%}	
	3	−0.08 (−0.71)	0.49*** (4.22)	20.13 [0.03]	33.23 [0.00]	1189 {43.55%}	
	Linear	−0.22*** (−2.71)	0.39*** (4.69)	23.11 [0.01]	43.97 [0.00]	2730	
III	1	0.03 (0.18)	0.42** (2.23)	18.26 [0.05]	19.64 [0.03]	643 {43.89%}	0.015
	2	12.87* [1.95]	10.04 [1.64]	7.36 [0.69]	11.56 [0.32]	240 {16.38%}	
	3	−0.20 (−0.93)	0.53** (2.31)	13.02 [0.22]	17.32 [0.07]	582 {39.73%}	
	Linear	−0.06 (−0.51)	0.49*** (3.82)	15.01 [0.13]	26.37 [0.00]	1465	
All	1	−0.03 (−0.26)	0.43*** (4.15)	11.01 [0.36]	15.39 [0.12]	1962 {35.34%}	0.015
	2	0.85 (0.50)	0.98 (0.59)	4.31 [0.93]	10.95 [0.36]	1148 {20.68%}	
	3	−0.09 (−1.02)	0.48*** (5.24)	18.68 [0.04]	15.60 [0.11]	2442 {43.98%}	
	Linear	−0.13** [−2.26]	0.42*** (7.01)	12.24 [0.27]	19.86 [0.03]	5552	

Note: values in ( ) are t statistics; α_i^f and α_i^s are estimated coefficients; [a] indicates χ^2 statistics; numbers in [ ] are p values; $\nrightarrow$ denotes "does not Granger-cause"; values inside { } are share (%) of sample size in a regime to the total sample size in the period; *, **, *** indicate 10%, 5%, and 1% significance levels, respectively. c denotes optimal threshold values. n_i is sample size for regime i.

coefficients from the linear model are all as expected. However, by using the whole period, we are unable to point out under what circumstances, when an arbitrageur will enter or will not enter the market. Further, an arbitrageur is not sure if there exists the phenomenon of price discovery between Δf_t and Δs_t in an arbitrage zone. A nonlinear model is needed to help identify the arbitrage zone. If the basis is greater than the threshold value, it is under regime I. And if the basis is smaller than the threshold value it is under regime III. Regime II is the area where arbitrage is not profitable (or out of the arbitrage zone). By understanding the process of basis adjustment and the dynamic interaction between Δf_t and Δs_t, investors and policy makers alike will be able to predict more accurately the future trend of crude oil price.

From Table 4 (whole periods), under regime II where arbitrage is not profitable, the coefficients for α^f and α^s are not statistically significant and there is no Granger causality between Δf_t and Δs_t. However, in the arbitrage zone, at least one of α^f and α^s is significant and there is a Granger causality between Δf_t and Δs_t. Under regime I (where futures price is noticeably greater than spot price), spot prices tend to rise ($\alpha_1^s > 0$) as a response to a large basis and thus there exists no Granger causality relationship between Δf_t and Δs_t. Under regime 3 (where spot price is greater than futures price), there is a tendency for the spot prices to drop ($\alpha^s > 0$ since basis <0, and, therefore, $\Delta s_t < 0$) and a change in spot price (Δs_t) is found to lead a change in futures price (Δf_t). This result is consistent with the backwardation relationship in the crude oil market indicated by Pindyck (2001) in the literature. Most of the time, spot prices are greater than futures prices in the crude oil market and spot price plays an important role in the process of price discovery as evidenced by the estimated results of the arbitrage behavior for the entire period. Finally, in the column of n_i in Table 4, the shares of sample size for regimes 1, 2, and 3 relative to the total sample size (whole period) are respectively 35.34%, 20.68%, and 43.98%. Little wonder that Regime 3 where spot prices exceed futures prices has the highest percentage in terms of the frequency.

The volatility of crude oil prices may lead to different dynamic relationships in term of the spot and futures prices. To investigate these relationships, the data of entire period are grouped into three periods based on certain important events. Afterwards, the MVTAR model is employed to investigate the dynamic and arbitrage relationships in terms of the spot and futures prices in the crude oil market under different regimes.

Period-I includes the Gulf War of 1990. According to Fig. 2, period-I has the most volatility among the three periods and the negative basis shows up 55% of the time, a typical backwardation relationship. The linearly estimated results indicate that the arbitrage behavior appears in the spot market because α^s is positive and statistically significant. How-

ever, there is no significant Granger causality relationship between Δf_t and Δs_t. If we separate the data into three regimes based on the optimal threshold values, the significant arbitrage appears only in regime 3. That is, spot price is higher than futures price (negative basis) and arbitrage happens in the spot market ($\alpha_3^s > 0$). Again, it is a backwardation relationship. Evident from the period-I section of Table 4, the dynamic relationship between Δf_t and Δs_t (a feedback relationship) occurs only in regime 3.

As far as the volatility of oil prices is concerned, period-II is relatively stable. During this period, negative basis shows up approximately 52% of the time, again, a backwardation relationship as in period-I. But, the number of negative bases is closer to that of the positive bases when compared to period-I. The linear results in period-II indicate that α^f and α^s are statistically significant and signs are as expected. The Granger causality test indicates a feedback relationship between Δf_t and Δs_t. The nonlinear results explain better the arbitrage and short-run dynamic relationship in terms of spot market and futures market when basis is either positive or negative. If the basis is negative and its absolute value is higher than transaction cost (regime 3), the arbitrage occurs in the spot market (because $\alpha_3^s > 0$ and significant). Within regime 3, there is a feedback relationship between Δf_t and Δs_t. As such, the change in futures prices can be used to predict the change in spot prices and vice versa. When the basis is positive and higher than transaction cost (regime 1), the arbitrage takes place still in the spot market ($\alpha_1^s > 0$ and significant) but the short-run dynamic relationship in terms of Δf_t and Δs_t is different in regime 3: the change in futures price leads the change in spot price in regime 1. Under the non-arbitrage area (regime 2), there is no arbitrage (α_2^s and α_2^f) nor short-run relationship between Δf_t and Δs_t.

Finally, the period-III started on 9/11 of 2001. The major volatility of oil price changes happened right after 9/11 attack on New York World Trade Center. Though the oil price spiked tremendously after 2004, the conditional volatility remains relatively stable. The relationship between the spot and futures prices in this period is different from that of period-I and period-II. During period-III, the positive basis shows up approximately 54% of the time, more frequently than does negative basis. That is, the futures price is greater than the spot price (a contango relationship), which differs from the result of past literature. Are such different results going to create different arbitrage behavior and different impact on short-run interaction? The estimated results from period-III of Table 4 may shed a light on this question. Results from the linear model indicate the arbitrage happens in the spot market ($\alpha^s > 0$ and significant) and Δf_t leads Δs_t. These results are not much different from those in other periods. If, however, we separate the data into arbitrage and non-arbitrage zones based on the basis, we discover in period-III that the arbitrage happens in both

Table 5
Comparison of in-sample fit results

Period	Variable	Mean (t statistic)	Significant level for rejecting $\mu=0$
I	Δf_t	-6.30×10^{-5} (−2.51)	0.01
	Δs_t	-6.61×10^{-5} (−2.31)	0.02
II	Δf_t	-1.08×10^{-5} (−4.27)	0.00
	Δs_t	-8.73×10^{-6} (−3.30)	0.00
III	Δf_t	-2.36×10^{-5} (−3.27)	0.00
	Δs_t	-2.21×10^{-5} (−2.61)	0.01
All	Δf_t	-1.17×10^{-5} (−2.95)	0.00
	Δs_t	-1.13×10^{-5} (−2.63)	0.01

Note: Based on the comparison of predictability by Diebold and Mariano (1995), we test H_0: $\mu=0$, where μ is defined as the average of $d_t = u_{1t}^2 - u_{2t}^2$, u_{1t}^2 represents the square of residuals from the nonlinear model, and u_{2t}^2 represents the square of residuals from the linear model.

regimes 1 and 3 and signs are as expected ($\alpha_1^s > 0$, $\alpha_3^s > 0$ and significant). However, the dynamic relationship between Δf_t and Δs_t appears only in regime 1 when the basis is positive and higher than the transaction cost. When the futures price is greater than spot price by a threshold value, there exists a significant interaction: a feedback relationship in regime 1 between Δf_t and Δs_t.

From our three-regime and three period model, we summarize results as follows: (i) whether it is a backwardation or a contango market, there exists no arbitrage relationship in the non-arbitrage zone and there is no short-run interaction between futures price change and spot price change; (ii) the arbitrage activity in the crude oil market happens in the spot market as evidenced by the adjustment coefficient α^3 (statistically significant); (iii) in a backwardation market (the spot price is greater than futures price), the short-run dynamic relationship between futures price change and spot price change appears to happen more often in regime 3 when the basis is negative and its absolute value is greater than transaction cost, while, in a contango market, the short-run dynamic relationship happens in regime 1 where the basis is positive and its absolute value exceeds transaction cost.

5. Comparison of in-sample predictive power

Irrespective of using linear or nonlinear model, there exists an arbitrage and short-run interaction between the spot price change and futures price change. Is there a better reason to use the nonlinear model to investigate the relationship between the spot and futures prices in the crude oil market? Theoretically speaking, the arbitrage happens when the basis is greater than certain threshold value (transaction cost). Statistically, however, we need to go one step further to demonstrate that the nonlinear model can describe better the dynamic relationship between the spot and futures prices. In other words, we need to test whether the in-sample predictability for the nonlinear model is better than that of the linear model. Therefore, we follow the generalized testing method by Diebold and Mariano (1995) in comparing the in-sample prediction on different models.[16] Let the prediction be based upon a function $\hat{y}_t(\beta)$, such that the time $t=R,\dots, T$ prediction of y_t is $\hat{y}_t=\hat{y}_t(\hat{\beta})$. If we let P denotes the number of prediction errors $\hat{u}=y_t-\hat{y}_t$ and $R+P=T$. When two predictive models exit an addition index $i=1,2$ will be used to distinguish between each of the two models. Let u_{1t} denote the

prediction errors of the nonlinear model or from Eq. (6) and u_{2t} the prediction errors of the linear model. The discrepancy in in-sample prediction can then be calculated as:

$$d_t = u_{1t}^2 - u_{2t}^2. \tag{8}$$

Given that $\mu \equiv Ed_t$, where E is expectations operator, a failure to reject the null hypothesis of $\mu=0$ suggests that the two models have similar in-sample prediction power. In the case of $\mu<0$, nonlinear model seems to be superior in explaining price changes and vice versa for $\mu>0$. Let the autocovariance be defined as $\gamma_j = Ed_t d_{t-j}$ and let $\hat{S}_{dd}$ denote a consistent estimate of $S_{dd} = \gamma_0 + 2\sum_{j=1}^{\infty} \gamma_j$, the long run covariance is associated with the covariance stationary sequence d_t. Diebold and Mariano (1995) prove that the statistics

$$\sqrt{P}\sum_{t=R}^{R+P} d_t \Big/ \sqrt{\hat{S}_{dd}} \xrightarrow{a} N(0,1). \tag{9}$$

Hence, normal table can be used to test the null of equal predictive power. In actual estimation, one can use the Newey–West estimator $\hat{S}_{dd} = \hat{\gamma}_0 + \sum_{v=1}^{q}[1-v/q+1](\hat{\gamma}_v + \hat{\gamma}_{-v})$ to approximate S_{dd}. Table 5 compares the in-sample predictability between the linear and nonlinear models for the whole period and three separate periods.

From the t test and p value of Table 5, the in-sample prediction errors from the nonlinear MVTAR model are all smaller than those of the linear model, irrespective for the whole period or three different periods. The significance levels are all less than 1% with the exception of Δs_t in period-1 at the 2% level. Such results indicate that the nonlinear MVTAR model can better understand the dynamic relation between the spot and futures prices in the crude oil market than can the linear model.

6. Conclusion

Literature abounds regarding the interaction between futures and spot prices. The research on the crude oil futures and spot markets, however, is somewhat limited. The investigation into the interaction between the two in terms of the arbitrage activity is scanty. To further understand the dynamics between the crude futures and spot prices, it is necessary to divide the data into three regimes based on the size of the bases, which are used as a guideline for arbitrage by investors.

WTI crude oil futures and spot price data (daily) are used in the multivariate threshold regression model proposed by Tsay (1998). Using the basis (a lag of 1) as a threshold variable, futures and spot prices are analyzed in the VECM to explain investors' behavior in the arbitrage regimes (upper and lower regimes) and inside the non-arbitrage regime (middle regime). We then investigate the dynamic interaction between the futures and spot prices within the three regimes. Within the non-arbitrage zone (regime 2), both error correction coefficients, representing adjustment towards equilibrium, and the dynamic relation based on the Granger causality between Δs_t and Δf_t are all statistically insignificant.

By separating almost 22 years of data into three periods based on certain important events, both periods I and II indicate that the spot prices in general are higher than futures prices as is well-known in the literature. Only period-III (2001/9/11–2007/4/30) displays that futures prices, in general, are higher than spot prices. When the absolute value of a basis is greater than the threshold value in the arbitrage area, at least one of the error correction coefficients, representing adjustment towards equilibrium, is statistically significant. When the spot price is higher than futures price, significant interaction is found when the basis is found to be negative and is less than certain threshold value (in regime 3). Conversely, when futures price exceeds the spot price, significant interaction exists only when the basis is found to be positive and is higher than certain threshold value (in regime 1). Finally, we use the in-sample prediction criterion by Diebold and Mariano (1995) to compare the predictive power between the linear and nonlinear models. Our empirical results indicate that the in-sample

---

[16] We are not using the out-of-sample forecasting model. As was indicated by Ashley and Patterson (2006), *The results from out-of-sample forecast can be idiosyncratic to the particular model validation period chosen unless the hand-out sample is lengthy, in which case an insufficient number of observations may remain for model specification and estimation. (In particular, one might expect that an adequate post-sample forecast period for evaluating a state-switching model would need to be sufficiently long to include a number of state switches.)* In general, the out-of-sample forecasting effectiveness is more appropriate for the selection of model choice. For the evaluation of the MVTAR model, we need a longer out-of-sample period so that more different observations can be included. Otherwise, a biased estimation could result due to small sample size. We, therefore, use the in-sample approach to compare the predictability of the linear and nonlinear model.

207

predictability of the nonlinear model is clearly superior to that of the linear model.

Acknowledgments

The valuable comments from the editor and referees are greatly appreciated. Any errors are our own.

References

Abhyankar, A., 1996. Does the stock index futures market tend to lead the crash? New evidence from the FTSE 100 stock index futures market. Working Paper No 96-01. Department of Accounting and Finance, University of Stirling.

Antoniou, A., Holmes, P., Priestley, R., 1998. The effects of stock index futures trading on stock index volatility: an analysis of the asymmetric response of volatility to news. Journal of Futures Markets 18, 151–166.

Ashely, R.A., Patterson, D.M., 2006. Evaluating the effectiveness of state-switching time series models for U.S. real output. Journal of Business & Economic Statistics 24 (3), 266–277.

Baek, E., Brock, W., (1992). A general test for nonlinear Granger causality: bivariate model. Working paper, Iowa State University and University of Wisconsin, Madison.

Balke, N., Fomby, T., 1997. Threshold cointegration. International Economic Review 38, 627–645.

Brooks, G., Garrett, I., 2002. Can we explain the dynamics of the UK FTSE 100 stock and stock index futures markets? Applied Financial Economics 12 (1), 25–31.

Chow, Ying-Foon, McAleer, M., Sequeira, J.M., 2000. Pricing of forward and futures contracts. Journal of Economic Surveys 14 (2), 215–253.

Diebold, Francis X., Mariano, Roberto S., 1995. Comparing predictive accuracy. Journal of Business and Economic Statistics 13 (3), 253–263.

Dwyer Jr., G.P., Locke, P., Yu, W., 1996. Index arbitrage and nonlinear dynamics between the S&P 500 futures and cash. The Review of Financial Studies 9 (1), 301–332.

Engle, R., Granger, C.W.J., 1987. Co-integration and error correction: representation, estimation and testing. Econometrica (55), 251–276.

Foster, Andrew J., 1996. Price discovery in oil markets: a time varying analysis of the 1990–91 Gulf conflict. Energy Economics 18, 231–246.

Garbade, K.D., Silber, W.L., 1983. Price movement and price discovery in futures and cash markets. Review of Economics and Statistics 65, 289–297.

Huang, Y.C., 2002. Trading activity in stock index futures markets: evidence of emerging markets. Journal of Futures Markets 22 (10), 983–1003.

Illueca, M., Lafuente, J.A., 2003. The effect of spot and futures trading on stock index market volatility: a nonparametric approach. The Journal of Futures Markets 23 (9), 841–858.

Kaldor, N., 1939. Speculation and economic stability. Review of Economic Studies 7, 1–27.

Koontz, S., Garcia, R., Hudson, M.A., 1990. Dominant-satellite relationship between live cattle cash and futures markets. The Journal of Futures Markets 10, 123–136.

Litzenberger, R.H., Rabinowitz, N., 1995. Backwardation in oil futures markets: theory and empirical evidence. The Journal of Finance 50 (5), 1517–1545.

Lo, M., Zivot, E., 2001. Threshold cointegration and nonlinear adjustment to the law of one price. Macroeconomic Dynamics 5, 533–576.

Moosa, I.A., 2002. Price discovery and risk transfer in the crude oil futures market: some structural time series evidence. Economic Notes by Banca Monte dei Paschi di Siena SpA 31, 155–165.

Oellermann, C.M., Brorsen, B.W., Farris, P.L., 1989. Price discovery for feeder cattle. The Journal of Futures Markets 9, 113–121.

Pindyck, R.S., 2001. The dynamics of commodity spot and futures markets: a primer. Energy Journal 22 (3), 1–29.

Seo, M., 2006. Bootstrap testing for the null of no cointegration in a threshold vector error correction model. Journal of Econometrics 134, 129–150.

Schroeder, T.C., Goodwin, B.K., 1991. Price discovery and cointegration for live hogs. The Journal of Futures Markets 11, 685–696.

Schwarz, T.V., Szakmary, A.C., 1994. Price discovery in petroleum markets: arbitrage, cointegration, and the time interval of analysis. The Journal of Futures Markets 14, 147–167.

Silvapulle, P., Moosa, I.A., 1999. The relationship between the spot and futures prices: evidence from the crude oil market. Journal of Futures Markets 19, 175–193.

Tsay, R.S., 1998. Testing and modeling multivariate threshold models. Journal of the American Statistical Association 93 (443), 1188–1202.

Weise, G.L., 1999. The asymmetric effects of monetary policy: a nonlinear vector autoregression approach. Journal of Money, Credit, and Banking 31 (1), 85–108.

Working, H., 1948. Theory of the inverse carrying charge in futures markets. Journal of Farm Economics 30, 1–28.

Working, H., 1949. The theory of the price of storage. American Economic Review 39, 1254–1262.

Defence and Peace Economics, 2011
Vol. 22(4), August, pp. 449–457

ON THE RELATIONSHIP BETWEEN MILITARY EXPENDITURE, THREAT, AND ECONOMIC GROWTH: A NONLINEAR APPROACH

ALBERT J.F. YANG[a], WILLIAM.N. TRUMBULL[b], CHIN WEI YANG[c] and BWO-NUNG HUANG[d],*

[a]*Department of Marketing/Distribution Management, National Kaohsiung First University of Science & Technology, Taiwan;* [b]*Department of Economics, College of Business and Economics, West Virginia University, PO Box 6025, Morgantown, WV 26506 – 6025, USA;* [c]*Department of Economics, Clarion University of Pennsylvani Clarion, PA 16214-1232, USA & Department of Economics, National Chung Cheng University, Chia-Yi 621, Taiwan;* [d]*Department of Economics & Center for IADF, National Chung Cheng University, Chia-Yi 621, Taiwan*

(Received 11 January 2010; in final form 23 April 2010)

The main objective of the paper is to decipher the military expenditure–economic growth relationship, taking the level of economic development (income) into consideration. Our findings suggest the following: (i) military expenditure has a significantly negative relationship to economic growth for the 23 countries with initial incomes (threshold variable) less than or equal to \$475.93; (ii) when the threat level is heightened, economic growth (23 countries) is expected to decrease. However, military expenditure in the presence of sufficiently large threats increases growth; (iii) for the remaining 69 countries whose initial incomes (real GDP per capita in 1992 price) exceed \$475.93, no significant relationship exists whether the threat variable is taken into consideration or not.

Keywords: Military expenditure; Economic growth; Threat; Threshold

JEL Codes: H56; C22

1. INTRODUCTION

The pioneering work by Benoit (1973) ushered its way into exploring the relationship between military expenditure and economic growth. Despite a great number of ensuing studies, the results are inconclusive, perhaps due to using different countries, sample periods and/or methodologies. That is, the relationship could be positive, negative or insignificant depending on the data and/or methodology used. Only a few studies focus on such relationships while holding the level of economic development (income) constant. For developed countries, it is not unusual to see that increased military expenditure gives rise to higher economic growth. Would the same result hold for low income countries?

Quite a few humanitarian organizations raise the issue that low-income countries are more willing to spend their already scare resources on weapons of defense or destruction rather than

*Corresponding author. Email: ecdbnh@ccu.edu.tw

ISSN 1024-2694 print: ISSN 1476-8267 online © 2011 Taylor & Francis
DOI: 10.1080/10242694.2010.497723

on child welfare, education, infrastructure and so forth, which would otherwise be conducive to economic growth. Therefore, it is of great importance to investigate whether the relationship between military expenditure and economic growth depends on the stage (level) of economic development.

Employing average cross-sectional data over 10 years (1992–2003) for 92 countries and treating initial incomes, of 1992, as the threshold variable, we include the threat variable as suggested by Aizenman and Glick (2003, 2006) into the growth model (Barro and Sala-i-Martin, 1995). The main objective of the paper is to decipher the military expenditure–economic growth relationship taking the level of economic development (income) into consideration. Our findings point to the following phenomena: (i) military expenditure has a significantly negative relationship with economic growth for the 23 countries with initial incomes (threshold variable) less than or equal to $475.93; (ii) when the threat level is heightened, economic growth for the 23 countries is expected to decrease. However, military expenditure in the presence of sufficiently large threats increases growth; (iii) for the remaining 69 countries whose initial incomes exceed $475.93, these exists no significant relationship between military expenditure and economic growth whether the threat variable is taken into consideration or not. That is, economic growth is determined by inflation rate, investment, education, but not by military expenditure in these countries. The result is in agreement with that by Aizenman and Glick (2003, 2006): the relationship between threat and economic growth is observed only in low income countries. This result supports the concerns of the humanitarian organizations.

The structure for the remainder of the paper is as follows. Section II discusses the literature. Section III provides the analytical model, econometric methods, and data. Section IV analyzes and discusses the empirical results. The final section presents concluding remarks.

2. LITERATURE REVIEW

The literature on the relationship between military spending and economic growth dates back to the early 1970s. Benoit's (1973) seminal work, pointing out that military spending can positively impact the growth of developing countries, ushered in well-publicized debates (e.g., Smith, 1980; Faini *et al.*, 1984; Ram, 1986; Atesoglu and Muller, 1990; Biswas, 1993). Early empirical studies focus to a great extent on simple unconditional correlation coefficients (Benoit, 1973, 1978). Later studies rely more on cross-sectional multiple regression models to examine the effects of military spending on economic growth (Lim, 1983; Faini *et al.*, 1984; Biswas and Ram, 1986; Deger, 1986; Grobar and Porter, 1989; Cohen *et al.*, 1996; Antonakis, 1997; and Heo, 1998).[1]

Well known in the literature, correlation does not imply causality and thus the Granger causality becomes the major tool in analyzing the military expenditure–economic growth relationship (Joerding, 1986; Chowdhury, 1991; LaCivita and Frederiksen, 1991; Khilji and Mahmood, 1997; Chang *et al.*, 2000; and Kollias *et al.* 2004). Unfortunately, no conclusive statement can be reached regarding the relationship in terms of the Granger causality. As was pointed out in a survey article by Dunne *et al.* (2005), it is difficult to obtain any robust empirical regularity depending on the choice of sample periods, target countries, and/or econometric methods. Nevertheless, if one must summarize the reported empirical findings, then, as Dunne *et al.* (2005) observed, there tends to be an insignificant or negative impact of such expenditures on economic growth in the case of developing countries and a comparatively stronger negative effect in developed countries (Kollias *et al.*, 2007: 75).

---

[1] Defense spending and military spending is used interchangeably in this paper.

To tackle the country-specific problem in terms of degrees of economic development or in terms of military expenditure/GDP ratio, researchers start to employ some nonlinear or panel approaches. For instance Cuaresma and Reitschuler (2003) found a negative relationship between military expenditure and economic growth for the countries whose military expenditure/GDP (threshold variable) is less than or equal to 3.25%. The relationship becomes insignificant for the countries whose threshold ratio exceeds 3.25%. The results by Kollias *et al.* (2007), who applied a panel approach for the 15 countries in the European Union, indicated that there existed feedback relationships between the two variables.

In addition to military expenditure, researchers generally added explanatory variables such as investment/GDP ratio, human capital (education) into the model. However, a hostile threat emanating from external intimidations is also important in exploring the military expenditure–GDP relationship. As such Aizenman & Glick (2003, 2006) first proposed to include a threat variable in the hope of explaining the negative relationship. Applying average values of annual data (1970–1998) over 90 countries, they found a significantly negative relationship between military expenditure and GDP but an insignificant (negative) relationship between economic growth and the threat variable. In contrast, the results by Araujo Junior and Shikrda (2008) indicate the opposite relationship if the conventional OLS is applied.

3. MODEL SPECIFICATION, ECONOMETRIC METHOD, AND DATA

Adding the threat variable suggested by Aizenman and Glick (2003, 2006) to the Barro and Sala-i-Martin (1995) type of growth equation yields the following linear equation:

$$g_i = \alpha_0 y0_i + \alpha_1 m_i + \alpha_2 threat_i + \alpha_3 m_i \cdot threat_i + \beta X_i + \varepsilon_i \qquad (1)$$

where g_i is the economic growth of country i; $y0_i$ is the initial income of 1992 adjusted for inflation (real GDP per capita in 2000 US dollars). m_i denotes per capita real military expenditure (deflated by GDP deflator 2000=100) of country i; $threat_i$ represents degree of external threat.[2] X is a vector of commonly used explanatory variables in modeling economic growth, such as the inflation rate (π_i), capital stock proxy $(I/Y)_i$, where I_i is investment and Y denotes GDP; edu_i represents level of education or secondary school enrollment ratio (%). As is often seen in the growth model, we expect (i) a positive relationship between (I/Y), education and economic growth, and (ii) a negative relationship between inflation π_i and economic growth. In the presence of the conditional convergence the estimated coefficient (α_0) of $y0_i$ is most likely to be significantly negative. The crucial relationship between m_i, $threat_i$ and m_i $threat_i$ could be very subtle. Intuitively, economic growth normally heads for a decline or slowdown when national sovereignty is under threat at the same time as military spending remains unchanged. The scene has been played out dynasty after dynasty in Chinese history. From the theoretical viewpoint, military expenditure is very likely to crowd out spending from private sectors when no real threat exists. The crowding-out effect was particularly prevalent in developing countries (e.g., Taiwan before 1970). Increased military spending in response to real threat(s) more often than not increases economic growth since such an increase may well stimulate private industries that complement and/or collaborate in joint projects. This is consistent with Aizenman and Glick (2003), where threats without expenditure for military security reduce growth; military

---

[2] This variable was constructed from data on the Militarized Interstate Dispute collected by the Correlates of War Project (COW) at the University of Michigan (http://www.correlatesofwar.org/). Threat is defined as the number of years a country was at war with each of its adversaries during the period of 1992 to 2002 summed over the set of its adversaries. Refer to Aizenman and Glick (2003: 3) for the information on how to construct this variable and source of the data.

expenditure without threats reduces growth; and military expenditure in the presence of suffi-ciently large threats increases growth. Consequently, we expect a negative relationship between m_i and g_i ($\alpha_i < 0$), as well as between *threat*$_i$ and g_i ($\alpha_2 < 0$). However, we expect a positive relationship between g_i and the interaction term m_i *threat*$_i$. Limited by the availability of the threat variable, the sample period of all variables except education is from 1992 through 2003. The paucity of education data makes its sample span rather short: from 1999 to 2003. Average values over the sample period are taken except for $y0_i$, which by definition assumes 1992 values. All the data of the 92 countries except Taiwan are taken from the WDI database by World Bank. The Taiwan data are obtained from Directorate-General of Budget, Account-ing and Statistics, Executive Yuan and Minister of Education of Taiwan.

Given that the purpose of this research is to investigate the relationship between military expenditure and economic growth for a country with different income levels (initial per capita real income for each country), we employ the threshold autoregressive regression (TAR) model by Tong (1978) and Tong and Lim (1980). As mentioned before, the nature of defense spend-ing–economic growth may be structurally different depending on different levels of income. The magnitude of the crowding-out effect for low-income countries may very well exceed that of high income countries due to resources constraints. In order to explore potentially varying relationships between military expenditure and economic growth at different income levels, we use $y0_i$ as a threshold variable. Hence, the two-regime TAR model is shown below:

$$g_i = (\alpha_{10} y0_i + \alpha_{11} m_i + \alpha_{12} threat_i + \alpha_{13} m_i \cdot threat_i + \beta_{11} X_i) \cdot I(y0_i \leq c^*)$$
$$+ (\alpha_{20} y0_i + \alpha_{21} m_i + \alpha_{22} threat_i + \alpha_{23} m_i \cdot threat_i + \beta_{21} X_i) \cdot I(y0_i > c^*) + \varepsilon_i \qquad (2)$$

where ε_i is assumed to follow an i.i.d process. I(.) is an indicator function in which, if the relation in (.) is present, I(.) is 1; otherwise, I(.) is 0 and $i=1, 2,\ldots,92$. Equation (2) is a simple two-regime model. It indicates that the relationship between explanatory variables (including military expenditure) and economic growth is represented by $(\alpha_{10}, \alpha_{11}, \alpha_{12}, \alpha_{13}, \beta_{11}) = B_1$ when $y0_i$ (threshold variable) is less than or equal to a threshold value c^* (regime 1), but by $(\alpha_{20}, \alpha_{21}, \alpha_{22}, \alpha_{23}, \beta_{21}) = B_2$ when $y0_i$ is greater than a threshold value c^* (regime 2). Before estimating equation (2), it is necessary to examine the existence of the nonlinear relationship. Only after the null hypothesis ($H_0 : B_1 = B_2$) is rejected, can we estimate the TAR model or equation (2). Given the fact that the threshold value (c) is unknown while testing equation (2), the traditional F-test is not appropriate for testing the null hypothesis. Under the assumption that the error term is i.i.d., Hansen (1997) suggests employing a test method derived from Davies (1977, 1987) and Andrew and Ploberger (1994), a test with near-optimal power against an alternative hypothesis distant from the null. This F test can be shown as

$$F = n \left(\frac{\tilde{\sigma}_n^2 - \hat{\sigma}_n^2}{\hat{\sigma}_n^2} \right) \qquad (3)$$

where $\tilde{\sigma}_n^2$ is the residual variance of equation (1), and $\hat{\sigma}_n^2$ is the residual variance of equation (2) given c^*. However, the asymptotic distribution of equation (3) does not follow a χ^2 distri-bution since c is unknown. Hansen (1996) suggests a procedure to overcome the problem by using a bootstrap method to obtain an asymptotic distribution for testing.[3]

---

[3] See Hansen (1997: 6) for the bootstrap procedure.

Once equation (3) is employed, we may be able to reject the null hypothesis. That is, there is a nonlinear relationship in the data where $y0_i$ is used as a threshold variable as shown in equation (2). First, the data need to be arranged in ascending order in $y0_i$ in order to estimate equation (2). To accommodate the degrees of freedom problem, the smallest and largest 15% of observations are removed.[4] The remaining sample space bounded by threshold values $[\underline{c}, \overline{c}]$ is partitioned into n grids. For each c_j ($j = 1, \ldots, n$), we can obtain a $\hat{\sigma}(c_j)$. Next, we select the smallest $\hat{\sigma}(c_j)$ to be the optimal threshold value c^* via equation (2) or

$$c^* = \arg \min \hat{\sigma}(c_j) \quad j = 1, \cdots, n$$
$$c_j \in [\underline{c}, \overline{c}] \tag{4}$$

where $\hat{\sigma}(c_j)$ is the residual variance from equation (2) using the threshold value c_j.

Table I reports names of 92 countries along with their respective per capita real income of 1992 as initial income levels.

An examination of Table I immediately reveals a garden variety of initial incomes from the lowest per capita GDP of Ethiopia (US\$93.54) to the highest per capita GDP of Luxembourg (US\$34,951.45) in 1992. It is to be noted that less-developed, developing, and developed countries are all included in this research.

4. ANALYSIS AND DISCUSSION OF RESULTS

Prior to estimating equation (2) we need to examine, using $y0_i$ as a threshold variable, the existence of the nonlinear relationship. The results of the nonlinearity test indicate that the relationship between military expenditure growth and economic growth is nonlinear. Table II presents nonlinearity test results, estimated threshold values, and the results of estimation in linear (OLS) as well as in nonlinear models.

Estimated results using OLS are reported in column 1 of Table II. A close inspection indicates readily that (i) military expenditure is negatively but somewhat significantly related to economic growth, (ii) threat variable and economic growth are positively related at $\alpha = 5\%$, (iii) the interactive term ($m \cdot threat$) and economic growth are negatively but insignificantly related, (iv) initial income and economic growth are negatively but insignificantly related indicating a lack of conditional convergence during the sample period, (v) a significant and negative relationship exists between inflation rate and economic growth and (vi) proxy for capital stock variable (I/Y) and economic growth are positively and significantly related as are education and economic growth. The signs of the estimated coefficient are largely consistent with conventional wisdom except for the relationship between threat and economic growth according to Aizenman and Glick (2003, 2006), which Araujo Junior and Shikida (2008) ascribe it to the problem of endogeneity when using linear OLS. It is our opinion that the unusual result of the Aizenman and Glick (2003, 2006) could have been avoided by taking nonlinearity into consideration.

In terms of nonlinear estimation, the estimated optimal threshold value is \$475.93 (in 1992 dollars) below which there are 23 countries. This is regime 1 or R1 whose estimated results

---

[4] The grid-searching procedure for the optimal c^* requires upper and lower 15% observations be truncated for sample estimation. However, for a large sample only the upper and lower 10% of observations need to be removed. The a priori assumption is that optimal c will not occur in the truncated region.

TABLE I Initial Incomes (y0) of 92 Countries in Ascending Order

Country name	y0	Nation name	y0	country name	y0
Ethiopia	$93.54	Syria	$1,058.92	Portugal	$8,973.39
Malawi	$127.55	Morocco	$1,092.19	Greece	$9,102.41
Burundi	$155.34	Egypt	$1,195.16	Saudi Arabia	$9,707.55
Niger	$160.13	Ecuador	$1,325.53	Bahrain	$10,433.06
Nepal	$185.94	Paraguay	$1,407.62	Taiwan	$10,589.00
Mali	$191.61	Iran	$1,468.76	Cyprus	$10,992.93
Chad	$201.60	Bulgaria	$1,490.89	New Zealand	$11,253.76
Sierra Leone	$205.10	Guatemala	$1,504.26	Spain	$11,673.51
Ghana	$218.01	Romania	$1,532.66	Ireland	$14,342.66
Central African Rep	$238.27	Tunisia	$1,614.51	Israel	$15,009.79
Madagascar	$242.56	Peru	$1,624.69	Singapore	$15,658.00
Togo	$244.91	Thailand	$1,657.62	Australia	$16,657.42
Rwanda	$287.70	Jordan	$1,659.36	Italy	$16,886.46
Bangladesh	$293.71	El Salvador	$1,754.93	Kuwait	$17,000.00
Sudan	$297.96	Algeria	$1,756.65	Finland	$17,937.84
India	$324.19	Fiji	$1,824.49	Canada	$18,541.52
Zambia	$335.82	Colombia	$1,938.90	United Kingdom	$19,318.54
Nigeria	$376.72	Turkey	$2,569.11	Belgium	$19,339.20
Senegal	$395.24	Malaysia	$2,881.58	France	$19,592.95
Mauritania	$397.14	South Africa	$2,929.03	Netherlands	$19,870.90
Kenya	$425.43	Poland	$2,938.28	Austria	$20,529.89
Yemen Arab Rep	$457.83	Brazil	$3,285.51	Germany	$20,566.24
China	$475.93	Panama	$3,342.94	Sweden	$22,507.17
Pakistan	$496.42	Chile	$3,588.69	United Arab Emirates	$24,077.45
Zimbabwe	$580.98	Hungary	$3,656.42	Denmark	$24,753.58
Sri Lanka	$615.55	Mexico	$5,168.60	United States	$28,365.70
Cameroon	$665.51	Venezuela	$5,366.58	Norway	$28,915.70
Nicaragua	$679.01	Uruguay	$5,448.18	Switzerland	$31,988.40
Indonesia	$691.61	Libya	$5,500.00	Japan	$34,535.46
Philippines	$876.64	Argentina	$6,861.00	Luxembourg	$34,951.45
Bolivia	$888.61	Oman	$7,356.99		

Note: y0=1992 real GDP per capita (in 2000 US dollars).

are presented in Column 2 of Table II. A perusal of it points to the observations that (i) there is a significantly negative relationship between military expenditure and economic growth for countries in R1; (ii) much in agreement with Aizenman and Glick (2006), threat variable and economic growth also exhibits a negative and significant relationship; (iii) the interaction term and economic growth show a significantly positive relationship indicating that as the threat level increases at a given level of military spending, economic growth is expected to increase; (iv) initial income has a negative but insignificant relationship with respect to economic growth; (v) consistent with the literature review, there exists a positive and significant relationship between the proxy for capital stock variable I/Y and economic growth; and (vi) inflation rate, as well as education, plays no significant role in economic growth. The remaining 69 countries with initial incomes greater than $475.93 constitute regime 2 or R2. In R2, initial income, inflation, proxy for capital stock, and education variables are significant in explaining economic growth. Military expenditure is found to be negatively related to economic growth albeit insignificantly. The threat variable and its interactive term are found to be statistically insignificant, indicating that external threat, military expenditure and economic growth are

TABLE II   Estimated Results for the Linear and Non-linear Models of 92 Countries

	(1) OLS	(2) Threshold regression	
		R1	R2
$y0_i$	−0.0000005	−0.00003	−0.0000003
	(−1.62)	(−1.12)	(−0.98)
π_i	−0.0004**	−0.0003	−0.0003***
	(−2.72)	(−1.17)	(−2.62)
$(I/Y)_i$	0.0532***	0.1079***	0.0462**
	(2.58)	(2.49)	(2.07)
edu_i	0.0002***	0.0002	0.0002***
	(3.35)	(0.63)	(2.66)
m_i	−0.000013*	−0.0019***	−0.000012
	(−1.65)	(−3.90)	(−1.48)
$threat_i$	0.00016**	−0.0005**	0.000042
	(1.97)	(−1.95)	(1.13)
$m_i \cdot threat_i$	−0.00000008	−0.00013***	−0.000000002
	(−1.45)	(5.46)	(−0.07)
N	92	23	69
$\bar{R}^2$	0.3324	0.5096	
s. e.	0.0143	0.0123	
White-F	2.27	0.78	
	[0.01]	[0.68]	
TAR-F		3.52	
		[0.04]	
c_1		$475.93	

Note: g_i(economic growth) is the dependent variable; the numbers inside the ( ) and [ ] are t statistics and p-values, respectively, and *, ** and *** respectively represent the 10%, 5%, and 1% significance levels. OLS indicates the linear model. R1 indicates regime 1 (initial income smaller than or equal to c_1). R2 denotes regime 2 (initial income greater than the c_1) s.e. = standard error of residual. c_1 is the optimal threshold value of the two-regime model; $\bar{R}^2$= adjusted R^2; $y0_i$= initial income (real GDP per capita in 1992 price); π_i= inflation rate; $(I/Y)_i$= gross fixed capital formation/GDP; m_i = per capita real military expenditure; $threat_i$ = threat variable; White-F= heteroskadasticity-consistent F statistic for the model including the interactive term; TAR-F=threshold autoregressive F statistic with the null hypothesis H_0: linear model versus nonlinear model.

insignificantly related in high-income countries or R2. The findings of our nonlinear model echo the concerns by some humanitarian organizations: military expenditure can certainly crowd out investment in education, infrastructure, or child welfare in low-income countries, as shown in regime 1. Hence, external threat can be a real hindrance to its economic growth. Not surprisingly, education is not an important predictor for economic growth in R1 but plays an important role in R2.

For comparison purposes, we present in Table III descriptive statistics of the variables for countries in both R1 (23 countries) and R2 (69 countries).

A close inspection indicates that economic growth (g) in R1 is not only lower than that in R2 but also possesses a greater standard deviation. Even though there exists a significant difference in income, the I/Y (capital stock) ratio is relatively close. This may explain that I/Y is a significant predictor in the growth model. It is to be pointed out that there is a huge difference in education variable (secondary school enrollment ratio) between R1 and R2: 28.18% versus 85.28%. It explains why education is certainly conducive to economic growth in R2 but not in R1. The per capita military expenditure in R2 far exceeds that in R1, 299.69 versus 10.35 US dollars but military expenditure/GDP ratio (or mey) is comparatively close: 2.73% versus 2.94%. For developing countries to spend nearly 3% on the military sector, it is almost certain to generate a significant crowding-out effect. Little wonder it constitutes a hindrance to economic growth.

TABLE III   Descriptive Statistics of Countries in Two Regimes

		mey (%)	m ($)	y0($)	g (%)	I/Y (%)	edu (%)
R1	mean	2.7251	10.3504	321.88	0.9435	17.11	28.1777
	S.D.	1.9920	12.2760	153.05	2.3940	5.86	17.1629
R2	mean	2.9405	299.6873	10061.66	1.7556	20.76	85.2839
	S.D.	2.6121	368.6791	9652.54	1.3574	4.31	23.9641

Note: R1 = regime 1 that consists of 23 countries with initial income less than or equal to \$475.93; R2 = regime 2 that consists of 69 countries with initial income greater than \$475.93; mean = average value; S.D. = standard deviation; mey = military expenditure /GDP; m = per capita real military expenditure (in 2000 US dollars); $y0$=1992 per capita real income (in 2000 US dollars); g = growth rate (%). (I/Y)= investment/GDP ratio (%); edu = secondary school enrollment ratio (% gross).

5. CONCLUDING REMARKS

In order to investigate the relationship between economic growth and military expenditure under different scenarios, we employ a threshold regression using average values over the 1992–2003 period for 92 countries. Included in the model are a proxy for capital stock, human capital, inflation and initial income. Furthermore, we add a threat variable and its interactive terms to the model as suggested by Aizenman and Glick (2003, 2006). The results indicate that there exists a significantly negative military expenditure–economic growth relationship for the group of 23 countries with initial income less than or equal to \$475.93 or R1. Beyond that, we find a significantly negative nexus between the threat variable and economic growth in R1. In addition, there exists a positive relationship between economic growth and the interaction term of military spending and threat variable. The results from our nonlinear model in R1 are consistent with the conjecture by Aizenman and Glick (2003). Failure to take the nonlinearity into consideration could lead to a biased estimate, as was the case in our OLS model. While education is conducive to economic growth in R2, our paper echoes the concerns of some humanitarian organization that low-income countries are more willing to dish out already scare resources on weapons of defense rather than on child welfare, education, infrastructure and so forth, which could be otherwise beneficial to economic growth.

Despite the findings from prior studies – a positive relationship between economic growth and military spending – a prevailing view held by the World Bank or IMF has it that increased military spending carries an opportunity cost: it slows down output and economic growth (Knight *et al.*, 1996). The theoretical underpinning is that when scarce resources are used in preparing war, they are siphoned off from more productive sectors of their economies: unless increased military spending (generally in developed countries) leads to advanced research and development, which will translate into profitable commercial activities in the future, we expect a negative military expenditure–economic growth relationship especially in low-income countries. Our empirical results using threshold regression support this view.

ACKNOWLEDGEMENTS

The valuable comments from the editor and two referees are greatly appreciated. Any errors are our own. Financial support from the National Science Council (NSC96-2415-H-194-003-MY2) is gratefully acknowledged. A preliminary draft was presented at the 83rd WEA International Conference in Hawaii, 2008.

References

Aizenman, J. and Glick, R. (2003) Military expenditure, threats, and growth. *NBER Working Paper* 9618.

Aizenman, J. and Glick, R. (2006) Military expenditure, threats, and growth. *Journal of International Trade & Economic Development* 15(2) 129–155.

Andrews, D.W.K. and Ploberger, W. (1994) Optimal tests when a nuisance parameter is present only under the alternative. *Econometrica* 62(6) 1383–1414.

Antonakis, N. (1997) Defence spending and growth in Greece: a comment and further empirical evidence. *Applied Economics Letters* 4 651–655.

Araujo Junior, A. F. and Shikida, C. D. (2008) Military expenditures, external threats and economic growth. *Economics Bulletin* 15(16) 1–7.

Atesoglu, H.S. and Muller, M. (1990) Defence spending and economic growth. *Defence Economics* 2 19–27.

Barro, R.J. and Sala-i-Martin, X. (1995) *Economic Growth*. Cambridge: MIT Press.

Benoit, E. (1973) *Defense and Economic Growth in Developing Countries*. Lexington: Lexington Books.

Benoit, E. (1978) Growth and defense in developing countries. *Economic Development and Structure Change* 26 271–280.

Biswas, B. (1993) Defense spending and economic growth in developing countries. In the *Defense and Economic Growth*, edited by J. E. Payne and A.P. Sahu. Boulder: Westview Press, 223–235.

Biswas, B. and Ram, R. (1986) Military spending and economic growth in less developed countries: an augmented model and furtits evidence. *Economic Development and Cultural Change* 34(2) 361–372.

Chang, T., Fang, W., Wen, L.F. and Liu, C. (2000) Defense spending, economic growth and temporal causality: evidence form Taiwan and mainland China, 1952–1995. *Applied Economics* 33(10) 1289–1299.

Chowdhury, A. (1991) A causal analysis of defence spending and economic growth. *Journal of Conflict Resolution* 35 80–97.

Cohen, J.S., Minty, A., Stevernson, R. and Ward, M.D. (1996) Defense expenditures and economic growth in Israel: the indirect link. *Journal of Peace Research* 33(3) 341–352.

Cuaresma, J.C. and Reitschuler, G. (2003) A non-linear defence–growth nexus? Evidence from the US economy. *Defence and Peace Economics* 15(1) 71–82.

Davies, R.B. (1977) Hypothesis testing when a nuisance parameter is present only under the alternative. *Biometrika* 64 247–254.

Davies, R.B. (1987) Hypothesis testing when a nuisance parameter is present only under the alternative. *Biometrika* 74 33–43.

Deger, S. (1986) *Military Expenditures in Third World Countries*. London: Routledge and Kegan Paul.

Dunne, J.P., Smith, R.P. and Willenbrockel, D. (2005) Models of military expenditures and growth: a critical review. *Defence and Peace Economics* 16(6) 449–461.

Faini, R., Annez, P. and Taylor L. (1984) Defense spending, economic structure, and growth: evidence among countries and over time. *Economic Development and Cultural Change* 32(3) 487–498.

Grobar, L. and Porter, R.C. (1989) Benoit revisited: defense spending and economic growth in LDCs. *Journal of Conflict Resolution* 33(2) 318–345.

Hansen, B. (1996) Inference when a nuisance parameter is not identified under the null hypothesis. *Econometrica* 64 413–430.

Hansen, B.E. (1997) Inference in TAR models. *Studies in Non-linear Dynamics and Econometrics* 2(1) 1–14.

Heo, U.K. (1998) Modeling the defense-growth relationship around the globe. *Journal of Conflict Resolution* 42(5) 637–657.

Joerding, W. (1986) Economic growth and defense spending: Granger causality. *Journal of Development Economics* 21 35–40.

Khilji, N.M. and Mahmood, A. (1997) Military expenditures and economic growth in Pakistan. *The Pakistan Development Review* 36(4II) 791–808.

Knight, M., Loayza, Norman and Villanueva, D. (1996) The peace dividend: military spending cuts and economic growth. *IMF Staff Papers* 43(1) 1–37.

Kollias, C., Manolas, G. and Paleologou, S.M. (2004) Defence expenditure and economic growth in the European Union: a causality analysis. *Journal of Policy Modeling* 26 553–569.

Kollias, C., Mylonidis, N. and Paleologou, S.M. (2007) A panel data analysis of the nexus between defence spending and growth in the European Union. *Defence and Peace Economics* 18(1) 75–85.

LaCivita, C.J. and Frederiksen, P.C. (1991) Defense spending and economic growth- an alternative approach to the causality issue. *Journal of Development Economics* 35(1) 117–126.

Lim, D. (1983) Another look at growth and defense in less developed countries. *Economic Development and Structure Change* 31 377–384.

Ram, R. (1986) Government size and economic growth: a new framework and some evidence from cross-section and time series data. *American Economic Review* 76(1) 191–203.

Smith, R. (1980) Military expenditure and investment in OECD countries 1954–73. *Journal of Comparative Economics* 4 19–32.

Tong, H. (1978) On a threshold model. In *Pattern Recognition and Signal Processing*, edited by C.H. Chen. Amsterdam: Sijthoff & Noordhoff, 101–141.

Tong, H. and Lim, K.S. (1980) Threshold autoregressions, limit cycles, and data. *Journal of the Royal Statistical Society* B 42 245–292 (with discussion).

Factors Affecting an Economy's Tolerance and Delay of Response to the Impact of a Positive Oil Price Shock

Bwo-Nung Huang*

This paper applies a multivariate threshold model to estimate a country's threshold level of economic tolerance (c) and delay of response (d) to the impact of a positive price change and its shock. Regression analysis is employed to investigate the factors affecting c and d. We find: (1) as a country becomes more advanced in economic development and acquires a lower ratio of energy use in its industry and transportation sectors, the threshold of tolerance is greater as evidenced by the positive impact of an oil price change and its shock; (2) if a country has a lower ratio of energy use in the industry sector, a lower energy import ratio and is more advanced in economic development, it will have a longer delay; and (3) as an economy becomes more advanced, the length of the response time from the impact of the shock of an oil price change will be longer.

1. INTRODUCTION

Since the global energy crises of the 1970s and 1980s and their effects on the world economy, the impact of an oil price change and its shock on economic activities have been a focus of research over the past three decades. Following the 1992 Gulf War, oil prices remained relatively stable and discussions on the subject were not as frequent. More recently, because of the increased demand for oil from developing countries, China and India in particular, and the uncertainty in the Middle East fueled by the tensions arising because of Iran's alleged nuclear ambitions, the oil price began to rise again in early 2004. By March 13, 2008, the oil price had spiked to a historical high of $110.21 per barrel (WTI spot). With

The Energy Journal, Vol. 29, No. 4. Copyright ©2008 by the IAEE. All rights reserved.

* Department of Economics, West Virginia University and Department of Economics and Center for IADF, National Chung Cheng University, Chia-yi 621, Taiwan. Email: ecdbnh@ccu.edu.tw.

Financial support from the National Science Council (NSC94-2415-H-194-002) is gratefully acknowledged. The valuable comments from the editor and four referees are greatly appreciated. Any errors are my own. This paper was completed during my visit to West Virginia University during the 2006 school year. In particular, I would like to thank the Department of Economics at WVU for the many resources it provided and Professor Ming-Jeng Hwang for his helpful suggestions.

oil prices increasing, economists may refocus their attention on the issue of an oil price change and its impact on economic activities. While there are related studies on the use of the nonlinear model to investigate the impact of an oil price change on the economy, they do not take into consideration the difference in tolerance and the speed of adjustment due to an oil price change resulting from the differences in economic development, energy dependence, and the efficiency of energy use in each country.[1]

Is there a degree of economic tolerance which shields the economy from the impact of an oil price change or its shock?[2] If there is, what are the factors affecting the degree of economic tolerance? The major contribution of our paper is that it uses Tsay's MVTAR (Multivariate Vector Threshold Autoregressive) model to estimate the threshold value (c) regarding the impact of an oil price change or its shock, and the length of response (d) of the threshold variable in each of the 21 countries.[3] We then employ the multiple regression model to test and identify the relationship between energy variables and c or d in each country. The monthly data used are the real stock price ($rstkp_t$), industrial production (y_t), the real oil price ($roilp_t$), and the interest rate (r_t). We find: (1) as a country becomes more advanced in economic development (real per capita GDP is used to denote the degree of economic development) and acquires a lower ratio of energy use in its industry and transportation sectors, the threshold of tolerance is greater as evidence by the positive impact of an oil price change and its shock; (2) if a country has a lower ratio of energy use in the industry sector, a lower energy import ratio and is more advanced in economic development, it will have a longer delay in terms of its economic response from the positive impact of an oil price change; and (3) as an economy becomes more advanced, the length of response time from the impact of the shock of an oil price change will be longer. Such discoveries will help policy planners in different countries design their own energy policy based on the degree of economic tolerance and the speed of response from the impact of an oil price shock.

This paper is organized as follows. Section I states the motivation for this research. Section II reviews the literature. Section III introduces the empirical models and related procedures. Section IV describes the data, and tests for the unit root and cointegration. In Section V, the MVTAR model is used to estimate the

---

1. The important literature on using the nonlinear model to investigate the impact of an oil price change on the macroeconomy includes Mork (1989), Lee et al. (1995), Hamilton (1996, 2003) and Jiménez-Rodriguez and Sánchez (2005). By separating a price increase from a price decrease or a positive price change (positive shock) from a negative price change (negative shock), they then explore the impact of both a positive shock and a negative shock on aggregate economic activities. However, no consideration is given as to the differences in response to the oil price shocks based on the differences in economic development, energy dependability, and the efficiency of energy use in each country.

2. Following Lee et al. (1995), standardized residuals estimated from the GARCH model of oil price changes are used to represent oil price shocks where the oil price change is defined as percentage changes of the oil price.

3. We define the value of c as a measure of the "threshold of tolerance", with a larger value of c indicating that it requires a larger oil price increase to have an effect, while the value of d measures the "length of delay" of the economic response, with a larger value of d indicating a longer delay.

threshold values for each country. Section VI analyzes the estimated results based on threshold levels (c) and the delay (d) in each country. Section VII concludes the analysis.

2. LITERATURE REVIEW

Prior research on the relationship between an oil price change and economic activities can be traced back to the pioneering work of Hamilton (1983),[4] who was the first to indicate that oil price increases partly accounted for every U.S. recession after World War II (except for one in 1960). Since then, using alternative estimation procedures on new data such as the U.S., the U.K. and Japan, researchers have extended and largely reinforced Hamilton's results (Gisser and Goodwin, 1986; Burbidge and Harrison, 1983).

Gradually, attention has shifted to the relationship between an oil price change and GDP. Gilbert and Mork (1986) and Mork (1989) first reasoned that the impact of an oil price increase or decrease on production may be nonlinear. An oil price change may increase the cost of resource allocation regardless of the direction of the price change. When the oil price is decreasing, the decrease in the cost of production and the increase in the cost of resource allocation often offset each other. With the increase in the oil price, these two forces reinforce each other and the corresponding impact on GDP can be significant. Hence, the oil price change should have an asymmetrical effect on the economy. By separating oil price changes into price increases and price decreases, Mork (1989) found that there exists an asymmetrical relationship between an oil price change and real output. That is, there is a negative relationship between an oil price increase and GDP. Lee et al. (1995) and Hamilton (1996) investigated the impact of an oil price change on aggregate economic activities after first going through certain nonlinear transformations of oil price data. They discovered that, when the oil price is raised, there is a negative relationship between the oil price and real activity in the U.S. economy. On the other hand, when the oil price falls, there is an insignificant relationship between the two. Cuñado and Pérez de Gracĭa (2003) investigated the asymmetric relationship between an oil price change and real activity in 14 European countries by following three indexes often used in the literature. They discovered that, if the problem of asymmetry is not considered, seven out of the fourteen countries will indicate that an oil price change does not lead (negatively) industrial production. When a positive oil price change is considered, only three countries do not indicate that an oil price change leads (negatively) industrial production. By using the same indexes with the multivariate VAR model to study the relationship between an oil price change and aggregate economic activities in eight OECD countries, Jiménez-Rodriguez and Sánchez (2005) found that the impact of an oil price increase on economic growth is far greater than the impact of an oil price decrease. In addition, they discovered that an oil price rise in five oil

---

4. Brown and Yücel (2002) provide an excellent related survey on the relationship between an oil price change and aggregate economic activity.

importing countries (excluding Japan) has a negative impact on economic growth. In two oil exporting countries out of the eight OECD countries, an oil price increase was found to have a different impact. In the U.K., an oil price rise was found to have a negative impact on GDP, but Norway was found to have a positive impact between the two.

A few economists have claimed in the literature that an economic recession does not originate from an oil price increase, but from the government's tight monetary policy. Bernanke, Gertler, and Watson (1997, BGW) employed the VAR model, using different policy options, to show that the decrease in GDP from the oil price increases of 1973, 1979-1980, and 1990 was caused by monetary policy, and not by the oil price increase itself. The results of BGW (1997), of course, gave rise to further discussions. Subsequently, Hamilton and Herrera (2004, H&H) employed the same data as BGW and the same estimating procedures, but used two different assumptions and came to a different conclusion. H&H first used a lag of 12 periods, rather than seven periods as used by BGW. Second, BGW assumed that the Federal Reserve's increasing of the federal funds rate by 900 (basis points) was deemed impractical. As a result, H&H came to support the negative correlation between the oil price increase and GDP.

Balke, Brown, and Yücel (2002) used a model similar to the BGW one by adding certain variables and allowed for a nonlinear impact relationship. They found that oil price changes may affect GDP even without any change in monetary policy. In addition, an asymmetrical relationship was found between an oil price change and GDP or a short-term interest rate. Hooker (1999) added an interest rate variable in a multivariate VAR model to investigate the role of monetary policy in the correlation between an oil price change and GDP growth. He indicated that the impact of an oil price change on GDP is significantly reduced. He therefore concluded that an oil price increase indirectly affects GDP through monetary policy. Tatom (1988, 1993) found that the asymmetric response of economic activity to oil shocks disappears when the monetary policy or changes in the misery index (which combines unemployment and inflation rates) are taken into account. On the other hand, Brown and Yücel (2002) also showed that the Federal Reserve's response to oil price shocks is not the cause of the asymmetry. They found that the asymmetry does not go away when both the federal funds rate is held constant or the federal funds rate and expectations of the federal funds rate are held constant. Hence, monetary policy does not appear to be the sole cause of asymmetry on the real side.

Besides discussing the relationship between an oil price change and GDP, some economists have emphasized the relationship between an oil price change and other macro-economic variables. Davis and Haltiwanger (2001) used plant-level census quarterly data to investigate the relationship between an oil price change and job creation or job loss. Keane and Prasad (1996) used the individual data of the U.S. National Longitudinal Survey of Young Men to investigate the relationship between an oil price change and the real wage rate. They found that an oil price increase causes the overall wage rate to decline but the wage rate of skilled workers to increase. Carruth, Hooker, and Oswald (1998) employed the error correction model (ECM) to test the relationship between an oil price change and unemployment. They found that the oil price has more of an impact on unemployment than the interest rate.

The research on the impact of an oil price change on the capital market started relatively late compared to its impact on the labor market and product market. Jones and Kaul (1996) were the first to test the impact of an oil price change on the stock market using 1947-1991 data and discovered that the oil price did have an impact on the overall stock return. However, Huang et al. (1996) employed 1979-1990 daily future oil price data only to find no evidence of the impact of an oil price change on the overall stock return. Subsequently, Sadorsky (1999) used 1947-1996 data on U.S. monthly interest rates, the oil price, industrial production, and stock returns to construct a VAR model to investigate the impact of an oil price change on the overall stock return and industrial production. He found that the oil price change or its shock does have an impact on the real stock returns. In particular after 1986, the oil price, rather than the interest rate, better explains the predicted error of the stock return.

Besides the variable for an oil price change, some researchers have used the variable of its volatility (uncertainty) to investigate its impact on certain macro-economic variables. Both Sadorsky (1999) and Kaul and Seyhum (1990) used the volatility of oil price changes to test its relationship with the stock returns. The literature on the subject is generally based on U.S. data. However, some researchers used data from different countries. For example, Mork et al. (1994) used data for seven OECD countries; Papapetrou (2001) used data for Greece; Cuñado and Pérez de Gracïa (2003) used data from 14 European countries; de Miguel, Manzano, and Martin-Moreno (2003) used Spanish data; Jiménez-Rodriguez and Sánchez (2005) used data for eight OECD countries; and Kilian (2008) employed G7 data. Some studies not only investigated the relationship between the oil price and aggregate economic activity, but also looked at the extent of an economic response from an oil price change (e.g., Mork et al., 1994; Cuñado and Pérez de Gracïa, 2003; Jiménez-Rodriguez and Sánchez, 2005 and Kilian, 2008).

There is an abundant literature on the nonlinear impact of an oil price shock on economic activity. However, due to the differences in the degree of economic development, energy dependence, and the efficiency of energy use, the economic tolerance and the speed of economic response in each country as a result of the impact of a positive oil price change and its shock are expected to be different. One important issue lacking in the literature is that we are unable to calculate appropriate threshold values for the economic tolerance in relation to an oil price shock in each country. Huang et al. (2005) employed the multivariate threshold autoregressive model (MVTAR) of Tsay (1998) to find the threshold value of an oil price change and its shock in each country. They found: (1) the most appropriate threshold value seems to vary according to how an economy depends on imported oil and its attitude towards adopting energy-saving technology; (2) an oil price change or its shock has a limited impact on the economy if the change is below the threshold levels; (3) if the change is above the threshold levels, it appears that the change in the oil price explains the macroeconomic variables better than the shock caused by the oil price; and (4) if the change is above the threshold levels, a change in the oil price or its shock explains the variation in GDP growth better than the real interest rate. Huang et al. (2005) have also taken into consideration the economic tolerance of the impact of an oil price change and its shock based on different natural endowments in each country. However, only three countries (the U.S., Canada, and Japan) are included in the study and no statistical tests are used to verify its validity. More countries and more refined statistical techniques are needed to test the relationship between the economic tolerance emanating from the impact of an oil price change (or its shock) and the degree of dependence on crude oil.

3. EMPIRICAL MODELS AND RELATED PROCEDURES

Due to the differences in the degree of economic development, energy dependence, and the efficiency of energy use, the economic tolerance and the speed of economic response in each country from the impact of a positive oil price change and its shock are expected to be different. It is necessary to use the MVTAR model to first estimate the "threshold of tolerance (c)" and the "length of delay (d)" in each country. Then, we search for energy related or growth related variables in an attempt to explain how an oil price change and its shock have different impacts on the threshold of tolerance and length of delay in each country.

The origin of the MVTAR model of Tsay (1998) dates back to the univariate threshold autoregressive model (TAR) of Tong (1978). The objective of a TAR model is to divide data into different regimes through a threshold variable q_{t-d}, where d represents the delay periods of the threshold variable. At the beginning, the threshold variable uses the delay (d) period of the dependent variable; hence, the univariate TAR model is also referred to as SETAR (self-exciting TAR). The two-regime model or SETAR(1) can be expressed as

$$y_t = (\phi_{0,1} + \phi_{1,1}y_{t-1})(1 - I[q_{t-1} > c]) + (\phi_{0,2} + \phi_{1,2}y_{t-1})\,I[q_{t-1} > c] + \varepsilon_t, \quad (1)$$

where ε_t epresents an iid white-noise process and $q_{t-1} = y_{t-1}$. $I[.]$ is an index function, which equals 1 or $I[.] = 1$ if the relation in the brackets holds. $I[.]$ equals zero otherwise. Equation (1) can be viewed as a multivariate threshold VAR(1) if y_t is a vector with q_{t-1} a scalar based on one of the elements of y_t. If there is a cointegration relationship among the variables, the multivariate threshold error correction (MVTEC) model can be expressed as

$$y_t = (c_1 + \delta_1 z_{t-1} + \sum_{i=1}^{p} \phi_{i,1}y_{t-i})(1 - I[q_{t-d} > c]) +$$
$$(c_2 + \delta_2 z_{t-1} + \sum_{i=1}^{p} \phi_{i,2}y_{t-i})\,I(q_{t-d} > c]) + \varepsilon_t, \quad (2)$$

where $E(\varepsilon) = 0$, $E(\varepsilon\varepsilon') = \Sigma$, y_t is a (4x1) vector, and z_{t-1} is an error correction term.

If there is no cointegration among the variables ($\delta_1 = \delta_2 = 0$), equation (2) becomes the MVTAR model. In order to estimate the "threshold of tolerance (c)" and "length of delay (d)" from an oil price change or shock in each country, the estimation of the nonlinear relationship in (2) must include the restriction that the column of $\phi_{i,1}$ corresponding to the effect of oil prices is identically zero for $i=1, ..., p$.[5]

If only the oil price variable and production variable are used to investigate the impact of an oil price change on economic activity, misleading results may appear as an oil price change may affect other economic variables such as stock prices and interest rates. Since the capital market plays an increasingly important role in economic activity, it is necessary to include the stock price and interest rate variables to study the impact of an oil price change on aggregate economic activities as was already indicated in the literature (Jones and Kaul, 1996; Sadorsky, 1999 or Huang et al., 2005). Thus, the study of the response of aggregate economic activities from an oil price change should include in the vector y_t items such as the relationships among an oil price change (Δlroilp$_t$), an interest rate change (Δr_t), a real stock price change (Δlrstkp$_t$), and a production change (Δly$_t$). If the study seeks to investigate the impact of an oil price shock on aggregate economic activities, the standardized residual (v_t) is used for the shock.

Prior to the estimation of equation (2), we need to test if the nonlinear relation indeed exists. That is, the null hypothesis: $H_0: \phi_{i,1} = \phi_{i,2}$, $i = 1,2..., p$, $c_1 = c_2$, and $\delta_1 = \delta_2$ should be tested. The C(d) statistic based on the arranged regression (Tsay, 1998) can be used to test the H_0: linear relation vs. H_A: nonlinear relation. Note that C(d) is a random variable that follows an asymptotically chi-square distribution with k(pk+1) degrees of freedom in which p is the lag length of the MVTEC or MVTAR model and k is the number of variables in y_t.

In order to estimate the optimal threshold level c^* for a given p and d, we arrange q_{t-d} in ascending order to search for c^* based on the following criterion (Weise, 1999):

$$(c^*, d) = \arg\min_i \det \left| \hat{\Sigma}(c_i) \right| , \qquad (3)$$

where $\det \left| \hat{\Sigma} \right|$ is the determinant value of the variance and covariance matrix of equation (2).[6]

Equation (2) is a MVTEC model with the presence of a cointegration relationship. Sometimes, if the data are not in long-run equilibrium (i.e., $\delta_1 = \delta_2 = 0$), equation (2) becomes a MVTAR model. In order to apply our data to either the MVTEC or the MVTAR model, we need to test the four variables for a unit root. These four variables are the log real oil price (lroilp$_t$), interest rate (r_t), log

---

5. The suggestion made by one of the referees for the restriction is greatly appreciated.

6. See Appendix A for the detail on how to calculate the C(d) statistic and how to find the optimal threshold level c^*.

real stock price (lrstkp$_t$), and log industrial production index (ly$_t$). Of these four variables, if at least two variables are in the form of I(1), tests of the cointegration relationship are needed to determine the use of either the MVTEC or MVTAR model.

To investigate the relationship between an oil price shock and other economic variables, the GARCH(1,1) – AR(p) model (Lee, Ni and Ratti, 1995) is used to estimate the standardized residual (v$_t$) as the proxy for an oil price shock and the related threshold levels of the oil price shock. The GARCH model of the real oil price change ($\Delta lroilp_t$) is defined as

$$\Delta lroilp_t = \beta_0 + \sum_{i=1}^{p} \Delta lroilp_t + \varepsilon_t \qquad \varepsilon_t \mid I_{t-1} \sim N(0, h_t) \qquad (4)$$

$$h_t = \alpha_0 + \alpha_1 e_{t-1}^2 + \alpha_2 h_{t-1},$$

where I_{t-1} represents the information set at time t-1. The oil price shock $\hat{v}_t$ is estimated through the standardized residual error ($\varepsilon_{t-1} / \sqrt{h_{t-1}}$) of equation (4).

4. DATA SOURCES AND RESULTS OF TESTS FOR UNIT ROOTS AND COINTEGRATION

The monthly data for 21 countries are used in this research. The oil price data (oilp) are obtained from the average crude price in the commodity price section of the IFS CD-ROM provided by International Monetary Fund. With the exception of Taiwan, the data for the other three variables are also obtained from the IFS CD-ROM. The money market rate (line 60b) is used as a proxy for the interest rate. The share price (line 62) is used to reflect the stock price and industrial production (line 66) is used to approximate output. The data for Taiwan are taken from AREMOS. Logarithmic transformations of industrial production, the real stock price, and real oil prices are all taken before conducting the analysis. The real oil price and real stock price are deflated by the base year 2000 consumer price index (cpi). The length of the available data in each country is different. Among these 21 countries, 6 countries (Austria, France, Japan, the Netherlands, South Africa and the U.S.) have the longest data (1970.1~2005.4 for 424 data points). Portugal has the shortest data (1988.1~2005.4 for 208 data points). The data for all 21 countries are available up to 2005.4 except for India (only up to 1998.5). [7] Table 1 displays the data periods for all 21 countries.

Since the VAR or VEC (vector error correction) model is used to estimate the nonlinear relation, and the delay periods cannot be too short for the statistical analysis, at least 200 data points are needed for a delay of 12 periods as suggested by Hamilton and Herrera (2004). The availability of data for the industrial produc-

---

7. In order to collect all the necessary monthly data with the required length of time for the stock return, industrial production, the real oil price, and the interest rate, only 21 countries are found to meet these requirements in the International Financial Statistics (IFS) database. Many countries lack stock index data or industrial production index data, or the length of the stock index data is too short.

Table 1. Sample Countries and Data Periods

Country	Abbreviation	Data period	Number of Observations (T)
Austria	AT	1970:01~2005:04	424
Canada	CA	1975:01~2005:04	364
Denmark	DK	1974:01~2005:04	376
Finland	FI	1977:12~2005:04	329
France	FR	1970:01~2005:04	424
India	IN	1970:01~1998:05	341
Israel	IL	1984:06~2005:03	250
Italy	IT	1971:01~2005:04	412
Japan	JP	1970:01~2005:04	424
Korea	KR	1978:01~2005:04	328
Malaysia	MY	1980:01~2005:04	304
Mexico	MX	1984:01~2005:04	256
Netherlands	NL	1970:01~2005:04	424
Norway	NO	1971:08~2005:04	405
Philippines	PH	1981:01~2005:04	292
Portugal	PT	1988:01~2005:04	208
South Africa	ZA	1970:01~2005:04	424
Spain	ES	1974:01~2005:04	376
United Kingdom	GB	1972:01~2005:04	400
USA	US	1970:01~2005:04	424
Taiwan	TW	1981:01~2005:04	292

Note: ISO-3166 list of two-letter abbreviations for country code is used.

tion index, the interest rate, and stock price are also needed in choosing countries. After taking all these into consideration, only 21 countries are selected. Before the MVTEC or MVTAR model is formally employed in the statistical analysis, all the variables need to be tested for unit roots. Since the proxy for the oil price shock is estimated from the standardized residuals of equation (4), it needs to be I(0) and, as such, no unit root test is necessary. We use the DF-GLS of Elliott et al. (1996) and the MZ_d of Ng and Perron (2001) to test for the existence of a unit root for the time series data in each country. The two unit root test results for all 21 countries are displayed in Appendix Table 1.

According to Appendix Table 1, among the 21 countries, 13 countries exhibit I(1) for all four variables, while eight countries exhibit I(1) for three variables. Overall, for eight countries with I(0) for at least one level variable (variable in its level form), the interest rate variable is not I(1) in such countries. Since we are investigating the impact of an oil price change (Model I) and the impact of an oil price shock (Model II) on the interest rate, stock price, and industrial production, all the I(1) variables (lroilp$_t$, r$_t$, lrstkp$_t$, ly$_t$ in Model I and r$_t$, lrstkp$_t$, ly$_t$ in Model II) need to be examined regarding the existence of a cointegration relation. We apply the trace method and maximum eigenvalue proposed by Johansen (1988) to test for the existence of a cointegration relation for these I(1) variables. Appendix Table 2 displays Johansen's cointegration test results with the model types used.

The third column of Appendix Table 2 lists the estimated trace statistics; the fourth column consists of the maximum eigenvalues and related statistics; the fifth column lists the variables of I(1) for which cointegration tests are needed; and the sixth column reports the model selected to estimate threshold levels according to the cointegration test results. Appendix Table 2-1 shows estimated results from Model I (lroilp$_t$, r$_t$, lrstkp$_t$, ly$_t$) and Appendix Table 2-2 displays estimated results from Model II (r$_t$, lrstkp$_t$, ly$_t$). When the trace statistics and maximum eigenvalues reject H$_0$ at less than the 5% significance level, there exist cointegration relations among variables. As shown in Appendix Table 2-1 under Model I, Israel, Mexico, Norway, the U.K. and Taiwan provide evidence of a cointegration relation and the VEC model is used for these five countries. The remaining 16 countries do not exhibit a cointegration relation and the VAR model is sufficient for the statistical analysis. Under Model II, Israel, Korea, Mexico, Spain, the U.K., and Taiwan show the existence of a cointegration relation and as such the VEC model is used, while the VAR model is used in the other 15 countries for the purposes of the analysis.

5. ESTIMATING THE THRESHOLD LEVELS AND THE DELAY OF THRESHOLD VARIABLES

For countries with the presence of a cointegration relation as indicated in Appendix Table 2, we need to use the MVTEC model to analyze the data:

$$
y_t = (\sum_{j=1}^{12} \alpha_{1j} d_{tj} + \delta_1 z_{t-1} + \sum_{i=1}^{p} \phi_{i,1} y_{t-i})(1 - I[q_{t-d} > c]) +
$$
$$
(\sum_{j=1}^{12} \alpha_{2j} d_{tj} + \delta_2 z_{t-1} + \sum_{i=1}^{p} \phi_{i,2} y_{t-i}) I[q_{t-d} > c] + \varepsilon_t, \tag{5}
$$

where y_t is (Δlroilp$_t$, Δr$_t$, Δlrstkp$_t$, Δly$_t$) or (v$_t$, Δr$_t$, Δlrstkp$_t$, Δly$_t$). $d_{tj} = 1$ if observation t is characterized by month j and is zero otherwise. z_{t-1} denotes the cointegration vector among the I(1) variables, q_{t-d} denotes the threshold variable with a delay of d periods, c is the threshold level, and ε_t is assumed to follow an iid with N(0,1) distribution. For countries without a cointegration relation, the following MVTAR model is used:

$$
y_t = (\sum_{j=1}^{12} \alpha_{1j} d_{tj} + \sum_{i=1}^{p} \phi_{i,1} y_{t-i})(1 - I[q_{t-d} > c]) +
$$
$$
(\sum_{j=1}^{12} \alpha_{2j} d_{tj} + \sum_{i=1}^{p} \phi_{i,2} y_{t-i}) I[q_{t-d} > c] + \varepsilon_t^*, \tag{6}
$$

where y_t is ($\Delta lroilp_t$, Δr_t, $\Delta lrstkp_t$, Δly_t) or (v_t, Δr_t, $\Delta lrstkp_t$, Δly_t). Before equation (5) or (6) is used for the estimation, the C(d) test proposed by Tsay (1998) for the testing of nonlinearity needs to be performed first. Note that C(d) is a random variable that follows an asymptotically chi-square distribution with k(pk+1) degrees of freedom in which p is the lag length of the VAR model and k is the number of variables in y_t. We set p at 12 periods (one year) so that seasonal and dynamic relations can be detected by the data.[8] With p = 12, the upper bound of the delay d for the threshold variable is also set at 12 periods. Since our focus is on the impact of an oil price change (percentage) and its shock (size) on economic activities, the threshold variable (q_{t-d}) is chosen as $\Delta lroilp_{t-d}$ for Model I and v_{t-d} for Model II, respectively. Appendix Table 3-1 (Model I) and 3-2 (Model II) display the test results of the C(d) statistics.

Some statistical problems related to the testing of the nonlinear threshold model warrant further discussion. The multivariate threshold model proposed by Tsay (1998) did not take into consideration the cointegration relation because it gives rise to the complicated statistical testing problem of linearity vs. nonlinearity and cointegration vs. non-cointegration. At the present time, the two-step test approach adopted by Balke and Fomby (1997) is widely used. That is, under the linear assumption, we test for the existence of a cointegration relation first. If there is a cointegration relation, the pre-specified cointegration vector needs to be included in the model to test whether it is a linear or nonlinear relation (Lo and Zivot, 2001; Seo, 2006). Our test follows the two-step approach.[9]

Based on Appendix Tables 3-1 and 3-2, every country, under some delay of d periods, indicates the existence of the nonlinear relation as the impact of an oil price change or its shock reaches certain threshold levels. Using the estimation method proposed by Tsay (1998), the estimated results of a nonlinear relation based on the delay selected by the C(d) test statistics in each country under Models I and II are displayed in Tables 2-1 and 2-2. Since we assume that the threshold variable is lower than a certain level, an oil price change or shock will not bring about changes in the interest rate, stock price or production variables. Thus, Table 2 only displays Granger causality test results when the threshold variable is higher than a certain level for both the linear model and the nonlinear model.

8. We have used the AIC or SIC to select the optimal lag for each country. However, the estimated results from these criteria are often too short to handle the dynamic relation between an oil price change and aggregate economic activities well. By following the suggestion of Hamilton (1994) with monthly data, it is a good idea to include at least 12 lags in the regression (p. 583). In addition, Hamilton and Herrera (2004) also indicated that the impact of an oil price change on production may have a delay of three or four quarters. Therefore, we select a lag of p = 12 to estimate the MVTAR or MVTEC model.

9. Theoretically, the test involves four categories: (1) linear without a cointegration relation; (2) nonlinear without a cointegration relation; (3) linear with a cointegration relation; and (4) nonlinear with a cointegration relation. So far, the literature does not seem to have a good test in the case of the nonlinear with or without a cointegration relation as it involves a different distributional problem (Seo, 2006).

Table 2-1. Estimated Results of Nonlinear Multivariate Threshold Model and Directions of Causality Tests (Model I: $\Delta lroilp$, Δr, $\Delta lrstkp$, Δly)

	M	d	c	n_2	Linear			R2		
					$\Delta lroilp \nrightarrow \Delta r$	$\Delta lroilp \nrightarrow \Delta lrstkp$	$\Delta lroilp \nrightarrow \Delta ly$	$\Delta lroilp \nrightarrow \Delta r$	$\Delta lroilp \nrightarrow \Delta lrstkp$	$\Delta lroilp \nrightarrow \Delta ly$
AT	A	9	5.0001	79	9.37	15.23	23.72**	9.42	9.18	28.58***
CA	A	3	2.1460	117	8.66	18.33	9.62	8.62	22.75**	11.65
DK	A	8	4.4314	80	19.94*	13.03	7.99	46.43***	22.96**	6.80
FI	A	2	0.5710	137	9.56	20.75**	12.56	17.46	26.30***	11.61
FR	A	9	4.2763	87	12.34	14.89	10.52	4.21	15.54	32.98***
IN	A	1	0.2587	120	18.67*	9.23	18.49*	18.84*	20.15*	24.05**
IL	E	6	0.0747	108	28.02***	20.23*	25.78**	37.15***	17.27	24.84**
IT	A	2	4.5631	77	30.64***	22.93**	23.14**	23.55**	18.36*	10.61
JP	A	2	2.5405	124	15.10	10.97	7.61	23.66**	28.01*	11.21
KR	A	5	1.0570	120	9.87	16.40	10.83	13.75	25.79***	17.87
MY	A	4	0.3367	136	17.17	3.63	31.15***	18.50*	9.38	33.40***
MX	E	3	0.4835	98	12.19	33.48***	12.41	12.51	34.15***	12.90
NL	A	1	0.9370	155	9.59	24.58**	15.76	24.75**	11.82	29.04***
NO	E	3	2.3615	122	16.64	24.16**	9.07	11.46	30.87***	4.71
PH	A	1	0.3248	121	11.98	3.03	15.66	21.79**	3.29	4.98
PT	A	1	0.1920	95	11.64	14.63	7.81	24.89**	12.40	18.33*
ZA	A	1	3.4484	97	9.09	10.03	8.66	10.20	21.55**	7.40
ES	A	2	0.7209	144	6.35	24.68**	14.48	15.87	13.98	21.53**
GB	E	9	0.3435	153	24.63**	30.05***	45.53***	30.07***	38.45***	41.35***
US	A	11	2.3829	120	23.53**	31.81***	27.73***	22.56**	21.51**	21.73**
TW	E	2	1.5473	110	15.05	7.94	24.13**	7.54	10.29	28.47***

Note: d is the delay of threshold variable; c is the optimal upper bound threshold level; n_2 represent the sample sizes of regime 2 (R2); $\nrightarrow$ denotes "does not Granger cause"; and *, **and *** represent 10%, 5%, and 1% significance levels, respectively. M=A denotes VAR model while M= E denotes VECM model.

From the estimated results of Model I in Table 2-1, the optimal threshold levels c are as follows: the highest level is Austria at 5.0001%, the next highest is Italy at 4.5631%, and the lowest level is Israel at 0.0747%. The next lowest is Portugal at 0.1920% and India is at 0.2587%. The threshold levels c for all other countries are in between. For the delay d of the threshold variables, the shortest period is found for India, the Netherlands, the Philippines, Portugal, and South Africa for 1 period and the longest is the U.S. for 11 periods. For some countries, the impact of a positive oil price change on production and the stock price is rapid (one month), while for other countries the impact is slow (almost one year). The other part of Table 2-1 investigates the Granger causality of the impact of a positive oil price change on the interest rate, production, and stock price. As expected, there is no causal relationship in many countries under the linear model. When

Table 2-2. Estimated Results of Nonlinear Multivariate Threshold Model and Directions of Causality Tests (Model II: v, Δr, $\Delta lroilp$, Δly)

	M	d	c	n_1	Linear $v \nrightarrow \Delta r$	Linear $v \nrightarrow \Delta lrstkp$	Linear $v \nrightarrow \Delta ly$	R2 $v \nrightarrow \Delta r$	R2 $v \nrightarrow \Delta lrstkp$	R2 $v \nrightarrow \Delta ly$
AT	A	9	0.0313	182	8.71	14.35	20.41*	7.87	18.45*	20.71**
CA	A	1	0.5494	91	12.56	14.13	12.03	12.11	26.02***	16.05
DK	A	3	0.3968	119	22.20**	12.78	10.10	12.00	21.13**	15.75
FI	A	7	0.4063	106	4.72	15.32	17.70	14.48	20.51*	30.09***
FR	A	3	0.4625	102	23.17**	11.46	6.18	14.07	27.61***	9.45
IN	A	3	0.0574	136	16.54	6.66	14.34	15.20	12.94	23.92**
IL	E	3	0.0198	111	20.89*	20.45*	18.58	7.56	26.77***	12.23
IT	A	10	0.0455	186	31.64***	21.72**	21.58**	35.72***	26.84***	17.15
JP	A	12	0.3820	112	11.05	11.69	4.19	26.92***	7.33	5.79
KR	E	3	0.0472	146	24.81**	12.71	13.71	12.09	19.18	28.47***
MY	A	1	0.1327	125	16.02	5.57	41.90***	6.41	11.87	29.07***
MX	E	1	0.0189	111	14.26	28.58***	12.25	16.99	26.08**	13.12
NL	A	10	0.8608	62	16.76	23.62**	12.92	3.45	12.08	97.78***
NO	A	7	0.0217	177	11.88	19.78*	2.45	18.64*	24.06**	9.63
PH	A	3	0.4311	90	15.00	4.96	15.49	9.80	23.73**	8.45
PT	A	6	0.0067	93	18.23	13.77	6.73	9.65	11.38	23.36**
ZA	A	4	0.0637	187	12.26	8.95	9.50	14.90	27.73***	5.54
ES	E	2	0.0191	171	11.54	32.83***	26.91**	10.12	28.86***	28.62***
GB	E	12	0.0066	187	17.56	27.24**	40.28***	19.09	22.23**	31.77***
US	A	11	0.2317	147	18.69*	27.92***	21.45**	18.54*	25.47***	24.92**
TW	E	5	0.0057	136	20.94*	6.78	25.14**	17.85	10.56	26.80***

Note: d is the delay of threshold variable; c is the optimal upper bound threshold level; n_2 represent the sample sizes of regime 2 (R2); $\nrightarrow$ denotes "does not Granger cause"; and *, **and *** represent 10%, 5%, and 1% significance levels, respectively. M=A denotes VAR model while M=E denotes VECM model.

the data are above the threshold level (regime 2), certain causal relationships appear. In the linear case, and for nine out of 21 countries, the causality test does not find any linkage between an oil price change and the interest rate, or stock price, or production. Under regime 2, all countries display at least one causal relation: the impact of an oil price change leads the interest rate, stock price, or production. Nine countries show that the oil price change leads the interest rate change (four countries in the linear case); nine countries indicate that the oil price change leads the stock price change (eight countries in the linear case); and ten countries suggest that the oil price change leads industrial production (seven countries in the linear case). It is clear that the economic impacts of an oil price change are nearly unavoidable when the oil price change exceeds a certain threshold. For some countries, the economic impact of an oil price change is immediate, but it may take six months or more for some other countries. For some countries, the oil price change needs to exceed the threshold level of 4% or 5% to exert any economic impact. Yet, for other countries, it may take less than 1% or 2% of an oil price change to have an economic impact. What are the factors affecting the differences in the threshold level and the speed of response in different countries? These discussions are at the center of our contribution and are also applicable to the economic impact emanating from the shock of an oil price change.

Table 2-2 displays the economic impact of the shock of an oil price change. The fastest response is found in Canada, Malaysia and Mexico with one period, while the U.K. and Japan are the slowest with 12 periods. Some economies can sustain the greater shock of an oil price change and some cannot. The greatest degree of sustainability is in the Netherlands at 0.8608. The next is in Canada at 0.5494 followed by France at 0.4625. The lowest sustainability of the economy to the shock is found in the U.K. at 0.0066. The next lowest is in Portugal at 0.0067. The Granger causality test demonstrates that ten out of 21 countries under the linear case exhibit no causal relationship. However, if the data are classified as higher than the threshold value (regime 2), the impact of the shock of an oil price change on the interest rate change, stock price change, or industrial production change is much greater. For 12 countries, the shock of an oil price change is found to lead the stock price change under regime 2, but only impacts six countries in the linear case. The shock of an oil price change leads industrial production change in 11 countries under regime 2, but only in six countries in the linear case. The relationship between the shock of an oil price change and interest rate change indicates that an oil price change leads the interest rate in four countries under the linear model. Yet, under the nonlinear model (regime 2), the shock of an oil price change leads the interest rate change in only two countries. Similarly, we need to know whether the economic impact of the shock of an oil price change is immediate or takes more than six months, and whether the tolerance of the impact is different among countries.

Our results show that, overall, the statistical results from the impact of an oil price change are stronger than those from the impact of the shock of an oil price change. These findings are consistent with the findings of a three-country study by Huang et al. (2005). Furthermore, the previous literature has been lacking in terms of taking into consideration the differences in economic endowments and energy conditions among countries; therefore, earlier studies find little relationship between an oil price change or the shock of an oil price change and its impact on economic variables. With the threshold regression, we are able to identify the economic impact of a positive oil price change and its shock on the speed and tolerance of the impact in each country.

6. ELEMENTS AFFECTING THE THRESHOLD LEVEL AND THE DELAY PERIODS OF THRESHOLD VARIABLES

The delay (d) of the threshold variable (q_{t-d}) reflects the speed of response based on the economic impact of a positive oil price change and its shock. The value of the threshold level (c) reflects the tolerance of the impact. Due in part to the difference in economic development, energy dependence, and the efficiency of energy use, the speed of economic response (d) and the degree of tolerance (c) from an oil price change and its shock are expected to be different in each country. Tables 2-1 and 2-2 display our estimates of the speed of response (delay periods d) and the degree of tolerance (or threshold level c) as a consequence of the impact of a positive oil price change and its shock. The next step is to investigate the factors affecting the speed and the tolerance of the impact. The possible potential factors affecting the speed and tolerance of the impact of an oil price change and its shock are the degree of economic development, oil imports, and the proportion of the use of energy in the industry and/or transportation sector to the total consumption of energy. Those factors could all be used to partially explain the impact of a positive oil price change and its shock on the speed of response (d) and the tolerance level (c) in each country.[10]

We collect the data from the Energy Balance published by the International Energy Agency (IEA). Those data include: (1) the percentage of energy imports to total final energy consumption (imp/tfc); (2) the percentage of energy use in industry to total energy consumption (ind/tfc); (3) the percentage of energy use in transportation to total energy consumption (tra/tfc); and (4) the log of real per capita GDP (lkgdp) to represent the degree of economic development.

Because the values of c and d in Table 2 are obtained from cross-sectional data consisting of 21 countries, the variables used to interpret c and d need to be cross-sectional as well. Most of our data start in the 1970s with some starting in the 1980s; hence, we need to calculate the yearly average of the four above-mentioned variables from 1971-2003 and 1981-2003 to calculate their cross-sectional values. The reasons for using the periods 1971-2003 and 1981-2003 are: (1) the availability of most of the data for the 21 countries studied in those two periods; and (2) to study whether there is a change in imports of energy and final energy use. Table 3 shows the means and standard deviations of the related energy data in those two periods.

In terms of the percentage of oil imports to total energy consumption (imp/tfc), the 1981- 2003 averages in most countries exhibit a downward trend (9 countries indicate an upward trend). Among the 21 countries, Denmark (since 1996), Canada, the U.K., Mexico, Malaysia, and Norway are all oil exporting

---

10. Other variables such as different monetary policy responses, may well impact both the magnitude and the length of the effect emanated from an oil price change (Bernanke et al., 1997; Hamilton and Herrera, 2004). In addition, Kilian (2007) categorizes source of the shock into demand and supply driven. While these variables are of great value in the analysis, they are not easily accessible and quantifiable.

Table 3.   Basic Statistics of Related Energy Data During 1971-2003 and 1981-2003

	imp/tfc -71	imp/tfc -81	tra/tfc -71	tra/tfc -81	ind/tfc -71	ind/tfc -81	kgdp -71	kgdp -81
AT	0.886	0.892	0.241	0.244	0.294	0.278	18329.94	20218.06
	(0.045)	(0.048)	(0.015)	(0.017)	(0.028)	(0.011)	(3818.89)	(2860.66)
CA	0.314	0.287	0.274	0.276	0.354	0.358	18364.89	19802.54
	(0.079)	(0.066)	(0.011)	(0.009)	(0.010)	(0.008)	(3009.36)	(2335.49)
DK	1.248	1.166	0.276	0.297	0.195	0.189	23829.15	25563.23
	(0.179)	(0.143)	(0.038)	(0.021)	(0.013)	(0.009)	(3670.35)	(2979.00)
FI	0.955	0.946	0.168	0.179	0.405	0.429	17605.87	19413.59
	(0.065)	(0.060)	(0.019)	(0.009)	(0.049)	(0.037)	(3546.73)	(2548.34)
FR	0.988	0.934	0.258	0.282	0.305	0.286	17301.14	18752.13
	(0.094)	(0.043)	(0.042)	(0.023)	(0.033)	(0.016)	(2882.70)	(2089.78)
IN	0.278	0.259	0.213	0.181	0.393	0.376	304.02	343.96
	(0.057)	(0.054)	(0.083)	(0.080)	(0.084)	(0.095)	(94.13)	(85.69)
IL	1.519	1.734	0.377	0.359	0.257	0.242	14110.50	15363.66
	(0.518)	(0.125)	(0.038)	(0.030)	(0.049)	(0.051)	(2550.21)	(1972.14)
IT	1.282	1.240	0.278	0.304	0.364	0.336	14707.32	16215.32
	(0.085)	(0.036)	(0.045)	(0.021)	(0.047)	(0.018)	(2876.60)	(1927.70)
JP	1.289	1.264	0.241	0.258	0.441	0.403	29025.01	32530.81
	(0.053)	(0.033)	(0.030)	(0.016)	(0.067)	(0.032)	(6749.16)	(4739.12)
KR	1.202	1.322	0.200	0.219	0.407	0.404	6127.57	7608.81
	(0.260)	(0.219)	(0.042)	(0.034)	(0.031)	(0.036)	(3259.24)	(2788.25)
MY	0.736	0.604	0.361	0.369	0.398	0.398	2480.83	2916.33
	(0.304)	(0.117)	(0.024)	(0.022)	(0.020)	(0.019)	(934.78)	(771.17)
MX	0.098	0.111	0.353	0.362	0.385	0.385	5017.76	5269.06
	(0.077)	(0.083)	(0.030)	(0.031)	(0.030)	(0.036)	(529.05)	(333.13)
NL	1.898	2.016	0.198	0.214	0.361	0.350	17830.93	19217.39
	(0.257)	(0.193)	(0.033)	(0.025)	(0.029)	(0.025)	(3182.83)	(2789.29)
NO	0.450	0.325	0.216	0.223	0.416	0.400	26537.28	30129.89
	(0.203)	(0.049)	(0.018)	(0.016)	(0.037)	(0.031)	(7258.46)	(5465.38)

Note: imp/tfc = the ratio of energy import to total energy consumption; tra/tfc = the ratio of energy use in transportation sector to total energy consumption; ind/tfc = the ratio of energy use in industry sector to total energy consumption; 71 = 71-03 yearly average; and 81 = 81-03 yearly average. kgdp= per capita real GDP (2000 USD). Numbers inside ( ) are standard deviation.

Table 3.   Basic Statistics of Related Energy Data During 1971-2003 and 1981-2003 (continued)

	imp/tfc -71	imp/tfc -81	tra/tfc -71	tra/tfc -81	ind/tfc -71	ind/tfc -81	kgdp -71	kgdp -81
PH	0.855	0.849	0.220	0.232	0.291	0.283	914.99	934.19
	(0.069)	(0.081)	(0.076)	(0.087)	(0.026)	(0.027)	(74.94)	(63.83)
PT	1.213	1.236	0.309	0.309	0.421	0.417	7378.58	8254.91
	(0.082)	(0.077)	(0.016)	(0.019)	(0.029)	(0.033)	(1912.21)	(1610.51)
ZA	0.313	0.300	0.229	0.224	0.481	0.480	3110.92	3048.78
	(0.070)	(0.080)	(0.019)	(0.015)	(0.047)	(0.050)	(207.76)	(204.41)
ES	1.198	1.205	0.348	0.364	0.390	0.361	10399.46	11375.86
	(0.036)	(0.034)	(0.033)	(0.023)	(0.054)	(0.036)	(2288.29)	(2048.30)
GB	0.603	0.530	0.278	0.303	0.303	0.270	18591.30	20330.48
	(0.156)	(0.062)	(0.045)	(0.027)	(0.056)	(0.022)	(3797.38)	(3198.49)
US	0.334	0.355	0.361	0.377	0.286	0.272	26471.13	28942.87
	(0.084)	(0.085)	(0.030)	(0.021)	(0.029)	(0.023)	(5136.59)	(4054.37)
TW	1.287	1.369	0.221	0.237	0.550	0.538	7138.59	8891.22
	(0.201)	(0.169)	(0.036)	(0.031)	(0.035)	(0.031)	(3771.57)	(3148.52)

countries. The remaining 15 countries are oil importing countries. When a country is dependent on oil imports, this country will be more sensitive in response to an oil price change and the degree of tolerance is expected to be smaller. If these two are related, we expect a negative relationship between the imp/tfc ratio and c or d.

With regard to energy consumption in the industry and transportation sectors in each country, we do not see a noticeable change before and after the energy crisis, but the relative importance of the industry and transportation sectors to energy demand in these 21 countries is different. In some countries, such as Finland, Japan, South Korea, Norway, Portugal, South Africa, and Taiwan, energy consumption in industry is the largest sector in terms of energy use and the percentage of energy use in industry to total energy consumption (ind/tfc) is over 40%. However, in Israel, Malaysia, Mexico, Spain and the U.S., the energy use in the transportation sector (tra/tfc) accounts for over one-third of total energy consumption.

The literature on the relationship between energy consumption and economic growth such as Akarca and Long (1979) indicates that energy consumption can bring about economic growth. If that is the case, GDP is directly related to industry output and indirectly related to transportation, and the impact of an oil price change and its shock for countries with a high percentage of energy consumption (ind/tfc or ind/tfc+tra/tfc) should be greater than countries with a low percentage of energy consumption. If these two are related, we expect a negative relationship between energy consumption (ind/tfc or ind/tfc+tra/tfc) and c or d.

Finally, in a more developed country, the economic impact of an oil price change (or its shock) can be more easily adjusted through economic policy and technological advances. Hence, we expect a positive relationship between the economic development variable and c or d. The log of real per capita GDP (deflated by the 2000 consumer price index for average values of 1971-2003 and 1981-2003) is used to represent different degrees of economic development.

In order to explore how these variables explain c and d, the regression model is used in the statistical analysis. Since c is a random variable, a multiple regression is adequate. However, d is a multiple-response type variable and the size of d indicates the response periods; an ordered multiple-choice model needs to be used (e.g., ordered probit or logit). The multiple regression of c can be shown as:

$$c_{\Delta lroilp} \text{ (or } c_v) = \alpha_0 + \alpha' X_i + \varepsilon_i, \tag{7}$$

where X_i represents explanatory variables (imp/tfc, ind/tfc+tra/tfc, lkgdp). α_0 is a constant term and ε_i follows an iid distribution. $c_{\Delta lroilp}$ and c_v represent tolerance for the threshold level from an oil price change and from the shock of an oil price change, respectively. Since we use cross-sectional data for 21 countries, we may run into the problem of heteroskedasticity. The heteroskedasticity-consistent covariance model is employed to correct the data for heteroskedasticity. Since d is an ordered variable, the ordered multiple-choice model is established as:

$$d^*_{\Delta lroilp} \text{ (or } d^*_v) = \beta' X_i + \varepsilon_i, \tag{8}$$

where * denotes an unobserved latent variable; X_i denotes the explanatory variables (imp/tfc, ind/tfc, lkgdp); and ε_i is assumed to follow an iid process. $d_{\Delta lroilp}$ and d_v represent the speed of response from an oil price change and from the shock of an oil price change, respectively. If the cumulative probability density function is assumed to follow a normal distribution, we have an ordered probit model. If ε_i is a logistic type distribution, we have an ordered logit model. Table 4 displays the estimated results of $c_{\Delta lroilp}$ and c_v from the multiple regression model, and the estimated results of $d_{\Delta lroilp}$ and d_v from the ordered variable model with a normal distribution.[11]

From the column for $c_{\Delta lroilp}$ in Table 4, we observe a significant correlation between $c_{\Delta lroilp}$ and two variables (ind/tfc+tra/tfc and lkgdp). Both the ind/tfc+tra/tfc and lkgdp variables are statistically significant at less than 5% and 1%,

---

11. Since we cannot reject the null hypothesis that residuals are normally distributed, Table 4 only reports the estimated results from the probit model. The logit model is also used in our estimation. Both sets of results are very much alike. To save space, only the estimated results from the probit model are displayed in Table 4. Interested readers may request the results from the logit model.

Table 4. The Estimated Results from the Threshold Level of the Multiple Regression Model and the Delay of Threshold Variables from the Ordered Dependent Variable Model

Indep. \ dep	c_{Airmip}		d_{Airmip} (probit)		c_v		d_v (probit)	
	71-03	81-03	71-03	81-03	71-03	81-03	71-03	81-03
lkgdp	0.6268*** (3.31)	0.6527*** (3.23)	0.7763*** (2.86)	0.7769*** (2.87)	0.0606** (2.26)	0.0581** (2.12)	0.4400* (1.96)	0.4208** (2.02)
ind/tfc+ tra/tfc	-5.3468** (-2.23)	-5.7096** (-2.20)			-0.7043** (-2.38)	-0.6566** (-2.07)		
ind/tfc			-8.5145** (-2.40)	-8.7384*** (-2.49)			1.8236 (0.76)	0.6003 (0.25)
imp/tfc	-0.5675 (-0.91)	-0.6595 (-1.16)	-1.1551** (-2.03)	-1.0802*** (-2.49)	0.0988 (0.51)	0.0837 (0.60)	0.3225 (0.60)	0.2596 (0.56)
$\bar{R}^2$	0.20	0.21			0.14	0.10		
White F test	0.84 [0.56]	0.89 [0.53]			2.21 [0.11]	2.80 [0.06]		
J-B test	1.00 [0.61]	0.92 [0.63]	2.03 [0.36]	2.18 [0.34]	1.21 [0.55]	1.23 [0.54]	0.67 [0.72]	0.76 [0.68]
Pseudo R^2			0.21	0.22			0.06	0.05

Note: numbers inside ( ) are t statistics; numbers inside [ ] are p-value; *, **, *** represent 10%, 5%, and 1% significance levels, respectively; dep. and indep. represent the dependent and independent variables, respectively; probit represent the ordered dependent variable model based on normal error distributions ε_t; J-B = Jarque-Bera normality test; and White F test = White heteroskedasticity test. lkgdp= log of per capita GDP.

respectively. The estimated signs are as expected. In other words, the countries with a higher level of economic development tend to have a higher level of tolerance in responding to a positive oil price change or shock. If the industry and transportation sectors consume more energy in a country, the ability to tolerate a positive oil price change or shock will be lower. Irrespective of whether the countries are oil importing or oil exporting countries, we do not find the relation to be significant for the imp/tfc variable to interpret $c_{\Delta lroilp}$. In the c_v column, we observe that there is a negative significant relationship (lower than 5%) between ind/tfc+tra/tfc and c_v, and a positive significance relationship (lower than 5%) between c_v and lkgdp. Overall, both variables explain about 20% and 14% of the variation in $c_{\Delta lroilp}$ and c_v, respectively. The test results for heteroskedasticity using the White test indicate that our model is free of heteroskedasticity. Furthermore, the Jarque-Bera (J-B) statistic indicates that the residual follows a normal distribution. The explanatory power of the model using 1971-2003 data is about the same as that of the model using 1981-2003 data.

For the interpretation of the delay period (d), we use lkgdp, ind/tfc and imp/tfc variables to explain $d_{\Delta lroilp}$ and the pseudo R^2 of 0.21 (based on the probit model) is much higher than the 0.05 obtained from d_v. When the economic development is more advanced in a country with lower energy consumption in the industrial sector and lower energy imports, the response period (d) from a positive oil price change should be longer. By using the same variables to explain d_v, only lkgdp can significantly explain the variation in d_v. The other two variables (ind/tfc and imp/tfc) cannot significantly explain the variation in d_v. Overall, lkgdp explains approximately 5% of the variation in d_v. The J-B test results indicate that both the residuals of $d_{\Delta lroilp}$ or d_v cannot reject the null hypothesis of normality. It is clear that the use of the probit model to estimate the delay period is appropriate.

From the above regression estimation, economic development (per capita real GDP) is the most important factor affecting the threshold of tolerance and response time emanating from the impact of a positive price change and its shock (proxied by the standardized residuals). In other words, the more a country's economic growth advances, the better the country can withstand the impact of an oil price change and its shock. The percentage of energy use in the industrial sector and oil imports as a percentage of total consumption can be used to explain the delay in the response to an oil price change. The percentage of energy consumption in both the industry and transportation sectors can be used to explain the threshold of tolerance from an oil price change and its shock.

7. CONCLUDING REMARKS

Subsequent to the 9/11 terrorist attack in 2001, the Dow Jones Industrial Average finally broke the 12,000 mark on October 19, 2006. The economic impact of a recent oil price hike was not as severe as that of the previous energy crises. The recent price hike did not seem to harm the U.S. economy as much as was expected. For some countries, however, the impact was more severe. It is clear that the impact of a positive oil price change on the threshold of tolerance and the speed of response is different for each country. This paper first employs a nonlinear model to estimate threshold level (c) and the delay of response (d) from the impact of an oil price change and its shock. We also employ energy use and economic growth-related variables to explain different possible values of c and d in each country.

In general, the length of the response to the economic impact of an oil price change is about 3 ~ 4 quarters (Hamilton and Herrera, 2004). Hence, a delay of 12 periods in monthly data is used in four-variable threshold VAR or threshold VEC analysis. Since the required sample period cannot be too short, only 21 countries with at least 200 data points in each country are available to collect the necessary data. Using the delay (d) of an oil price change (Δlroilp$_{t-d}$) and the shock of an oil price change (v_{t-d}) as threshold variables, the nonlinear statistical test in each country supports the existence of the nonlinear relation. We, therefore, can estimate without much difficulty the threshold level (c) and the delay of threshold variable (d) from the nonlinear model.

The purpose of this paper is to identify evidence affecting different values of c and d for each country. We find possible explanatory variables from energy balance data published by the IEA. A nonlinear model is used to estimate the threshold level (c) and the delay of threshold variable (d) in each country. A multiple regression model is then employed to test and identify the relationship between energy variables and c or d in each country. We find: (1) as a country becomes more advanced in economic development (real per capita GDP is used to denote the degree of economic development) and acquires a lower ratio of energy use in its industry and transportation sectors, the threshold of tolerance is greater as evidence by the positive impact of an oil price change and its shock; (2) if a country has a lower ratio of energy use in the industry sector, a lower energy import ratio and is more advanced in economic development, it will have a longer delay in terms of its economic response from the positive impact of an oil price change; and (3) as an economy becomes more advanced, the length of the response time from the impact of the shock of an oil price change will be longer. The results from the regression analysis suggest that the economic development variable is the most important variable affecting the threshold of tolerance ($c_{\Delta lroilp}$ and c_v) and the speed of response ($d_{\Delta lroilp}$ and d_v) emanating from the impact of a positive price change and its shock.

Although we have found possible factors to explain the threshold level and the speed of adjustment from the impact of positive oil price shock, this research is not without limitations. The potential shortcomings of this research are as follows: (1) With the regression results based on 21 countries, the estimation results might not be as reliable. (2) With such a sample size, it might be a bit of a stretch to rely on asymptotic theory to justify the use of t-statistics in conducting inference. (3) There is a host of possible omitted variables that may affecting the "tolerance level" and the delay of the effect. Some possibilities pertain to monetary policy and the type of shocks, but there are other possibilities such as degree of openness of the economy, fiscal and exchange rate policy, etc. (see for instance Bohi, 1991) These omitted variables may be included in future analyses to test the robustness of the result.

REFERENCES

Akarca, A. T. and T. V. Long (1979). "Energy and Employment: A Time Series Analysis of the Causal Relationship." *Resources and Energy* 2: 151-162.

Balke, N. S., S. P. A. Brown, and M. K. Yücel (2002). "Oil Price Shocks and the U.S. Economy: Where Does the Asymmetry Originate?" *Energy Journal* 23(3): 27-52.

Balke, N. and T. Fomby (1997). "Threshold Cointegration." *International Economic Review* 38: 627-645.

Bernanke, B. S., M. Gertler, and M. Watson (1997). "Systematic Monetary Policy and the Effects of Oil Price Shocks." *Brookings Papers on Economic Activity* 1: 91-157.

Bohi, D. R. (1991). "On the Macroeconomic Effects of Energy Price Shocks." *Resources and Energy* 13: 145-162.

Brown, S. and M. K. Yücel (2002). "Energy Prices and Aggregate Economic Activity: An Interpretive Survey." *Quarterly Review of Economics and Finance* 42: 193-208.

Burbidge, J. and A. Harrison (1984). "Testing for the Effects of Oil Price Rises Using Vector Autoregression." *International Economic Review* 25: 459-484.

Carruth, A. A., M. A. Hooker, and A. J. Oswald (1998). "Unemployment Equilibria and Input Prices: Theory and Evidence from the United States." *Review of Economics and Statistics* 80(4): 621-28.

Cuñado, J. and Fernando Pérez de Gracĭa, (2003). "Do Oil Price Shocks Matter? Evidence from Some European Countries." *Energy Economics* 25: 137-154.

Davis, S. J. and J. Haltiwanger (2001). "Sectoral Job Creation and Destruction Response to Oil Price Changes." *Journal of Monetary Economics* 48: 465-512.

de Miguel, C., B. Manzano, and J. M. Martin-Moreno (2003). "Oil Price Shocks and Aggregate Fluctuations." *Energy Journal* 24(2): 47-61.

Elliott, G., T. J. Rothenberg and J. H. Stock (1996). "Efficient Test for an Autoregressive Unit Root." *Econometrica* 64: 813-836.

Gilbert, R. J. and K. A. Mork (1986). "Efficient Pricing During Oil Supply Disruptions." *The Energy Journal* 7(2): 51-68.

Gisser, M. and T. H. Goodwin (1986). "Crude Oil and the Macroeconomy: Tests of Some Popular Notions." *Journal of Money, Credit and Banking* 18: 95-103.

Hamilton, J. D. (1983). "Oil and the Macroeconomy Since World War II." *Journal of Political Economy* 91: 228-248.

Hamilton, J. D. (1994). *Time Series Analysis.* New Jersey: Princeton University Press.

Hamilton, J. D. (1996). "This is What Happened to the Oil Price-Macroeconomy Relationship?" *Journal of Monetary Economics* 38(2): 215-220.

Hamilton, J. D. (2003) "What is an Oil Shock?" *Journal of Econometrics* 113: 363–98.

Hamilton, J. D. and A. M. Herrera (2004). "Oil Shocks and Aggregate Macroeconomic Behavior: The Role of Monetary Policy: A Comment." *Journal of Money, Credit and Banking* 36(2): 265-286.

Hooker, M. A. (1999). "Oil and the Macroeconomy Revisited." Mimeo, *Federal Reserve Board*, Washington, D.C., August.

Huang, B. N., M. J. Hwang, and H. P. Peng (2005). "The Asymmetry of the Impact of Oil Price Shocks on Economic Activities: An Application of the Multivariate Threshold Model." *Energy Economics* 27: 455-76.

Huang, R. D., R. W. Masulis, and H. R. Stoll (1996). "Energy Shocks and Financial Markets." *Journal of Futures Markets* 16(1): 1-27.

Jiménez-Rodriguez, R. and M. Sánchez (2005). "Oil Price Shocks and Real GDP Growth: Empirical Evidence for Some OECD Countries." *Applied Economics* 37: 201-228.

Johansen, S. (1988). "Statistical and Hypothesis Testing of Cointegration Vectors." *Journal of Economic Dynamics and Control* 12: 231-254.

Jones, C. M. and G.. Kaul (1996). "Oil and the Stock Market." *Journal of Finance* 51: 463-491.

Kaul, G. and H. N. Seyhun (1990). "Relative Price Variability, Real Shocks, and the Stock Market." *Journal of Finance* 45: 479-496.

Keane, M. P. and E. S. Prasad (1996). "The Employment and Wage Effects of Oil Price Changes: A Sectional Analysis." *Review of Economics and Statistics* 78: 389-399.

Kilian, Lutz (2008). "A Comparison of the E ects of Exogenous Oil Supply Shocks on Output and Inflation in the G7 Countries." *Journal of the European Economic Association* March: forthcoming.

Kilian, Lutz (2007). "Not All Oil Price Shocks Are Alike: Disentangling Demand and Supply Shocks in the Crude Oil Market." Accessed at http://www-personal.umich.edu/~lkilian/ aer111507r1.pdf

Lee, K., S. Ni, and R. A. Ratti (1995). "Oil Shocks and the Macroeconomy: The Role of Price Variability." *Energy Journal* 16(4): 39-56.

Lo, M. and E. Zivot (2001). "Threshold Cointegration and Nonlinear Adjustment to the Law of One Price." *Macroeconomic Dynamics* 5: 533-576.

Mork, K. A. (1989). "Oil and the Macroeconomy when Prices Go Up and Down: An Extension of Hamilton's Results." *Journal of Political Economy* 97: 740-744.

Mork, K., O. Olsen, and H. Mysen (1994). "Macroeconomic Responses to Oil Price Increases and Decreases in Seven OECD Countries." *Energy Journal* 15: 19-35.

Ng, S. and P. Perron (2001). "Lag Length Selection and the Construction of Unit Root Tests with Good Size and Power." *Econometrica* 69: 1519-1554.

Papapetrou, E. (2001). "Oil Price Shocks, Stock Markets, Economic Activity and Employment in Greece." *Energy Economics* 23: 511-532.

Sadorsky, P. (1999). "Oil Price Shocks and Stock Market Activity." *Energy Economics* 21: 449-469.

Seo, M. (2006). "Bootstrap Testing for the Null of No Cointegration in a Threshold Vector Error Correction Model." *Journal of Econometrics* 134: 129-50.

Tatom, J. A. (1988). "Are the Macroeconomic Effects of Oil Price Changes Symmetric?" *Carnegie-Rochester Conference Series on Public Policy* 28: 325–368.

Tatom, J. A. (1993). "Are There Useful Lessons from the 1990–91 Oil Price Shock?" *Energy Journal* 14 (4): 129 –150.

Tong, H. (1978). "On a Threshold Model." in C. D. Chen (ed.), *Pattern Recognition and Signal Processing*, Amsterdam: Sijthoff and Noordhoff, 101-41.

Tsay, R. S. (1998). "Testing and Modeling Multivariate Threshold Models." *Journal of the American Statistical Association* 93: 1188-1202.

Weise, C. L. (1999). "The Asymmetric Effects of Monetary Policy: A Nonlinear Vector Autoregression Approach." *Journal of Money, Credit, and Banking* 31(1): 85-108.

APPENDIX A

The Definition of C(d) Statistic:

The C(d) statistic is calculated as follows. First, the linear multivariate model can be denoted in matrix form as shown below:

$$y'_t = X'_t \Phi + \varepsilon_t, \quad t = 1,.....,n \tag{A-1}$$

where h=max(p, d), $X_t = (1, y'_{t-1}, y'_{t-2},..., y'_{t-p})'$ is a (pk+1)-dimensional regressor, and Φ denotes the parameter matrix. The data are arranged in the order of small to large according to the values of the threshold variable q_{t-d}. The arranged regression becomes:

$$y'_{t(i)+d} = X'_{t(i)+d} \Phi + \varepsilon_{t(i)+d}, \quad i = 1,....,n - h \tag{A-2}$$

where t(i) is the time index of $q_{(i)}$. Tsay (1998) used the recursive least squares approach to estimate (A-2) and, at the same time, predictive residuals were obtained. If the threshold relation does not exist, the predictive residuals approach white noise. Consequently, the predictive residuals are uncorrelated with the regressor $X_{t(i)+d}$. On the other hand, if y_t follows a threshold model, then the predictive residuals are no longer white noise, because the least squares estimator is biased. In this case, the predictive residuals are correlated with the regressor $X_{t(i)+d}$.

Let Φ_m be the least squares estimate of Φ of equation (A-2) with i=1, ..., m; i.e., the estimate of the arranged regression using data points associated with the m smallest values of q_{t-d}. Let

$$\hat{e}_{t(m+1)+d} = y_{t(m+1)+d} - \hat{\Phi}'_m X_{t(m+1)+d} \tag{A-3}$$

and

$$\hat{\eta}_{t(m+1)+d} = \hat{e}_{t(m+1)+d} / [1 + X'_{t(m+1)+d} V_m X_{t(m+1)+d}]^{1/2}, \tag{A-4}$$

where $V_m = [\Sigma_{i=1}^{m} X_{t(i)+d} X'_{t(i)+d}]^{-1}$ is the predictive residual and the standardized predictive residual of regression (A-2). Next, consider the regression

$$\hat{\eta}_{t(l)+d} = X'_{t(l)+d} \Psi + w'_{t(l)+d}, \qquad l = m_0 + 1,...,n - h, \tag{A-5}$$

where m_0 denotes the starting point of the recursive least squares estimation. The problem of interest is then to test the hypothesis $H_0: \Psi = 0$ versus the alternative $H_a: \Psi \neq 0$ in regression (A-5). The C(d) statistic is therefore defined as:

$$C(d) = [n - h - m_0 - (kp + 1)] \times \{\ln[\det(S_0)] - \ln[\det(S_1)]\} \tag{A-6}$$

where the delay d signifies that the test depends on the threshold variable q_{t-d}, det(A) denotes the determinant of the matrix A, and

$$S_0 = \frac{1}{n - h - m_0} \sum_{l=m_0+1}^{n-h} \hat{\eta}_{t(l)+d} \hat{\eta}'_{t(l)+d}$$

and

$$S_1 = \frac{1}{n - h - m_0} \sum_{l=m_0+1}^{n-h} \hat{w}_{t(l)+d} \hat{w}'_{t(l)+d},$$

where $\hat{w}_t$ is the least squares residual of regression (A-5). For more details, see Tsay (1998).

Finding the Optimal Threshold Level:

The data are arranged in ascending order based on the threshold variable q_{t-d}. Depending on the abundance of the data, 15% to 25% of the smallest and largest values are discarded. Let the largest value be $\bar{c}$ or the 85th percentile (if 15% of the larger data are discarded) or the 75th percentile (if 25% are discarded). Let the smallest value c be the 15th percentile (if 15% of the smaller data are discarded) or the 25th percentile (if 25% of the values are discarded). Then, a grid search pro-

Appendix Table 1. Results of Unit Root Tests

	AT(T=424)				CA(T=364)			
	Level		Difference		Level		Difference	
	DF-GLS	MZ_α	DF-GLS	MZ_α	DF-GLS	MZ_α	DF-GLS	MZ_α
r	-1.89	-8.69**	-2.54**	-146.65***	-2.15**	-11.93**	-22.61***	-72.90***
lroilp	-1.42	-4.03	-16.70***	-202.29***	-2.31	-9.06	-14.36***	-127.57***
lrstkp	-1.20	-3.77	-15.96***	-198.24***	-2.54	-16.08*	-16.48***	-160.40***
ly	-1.59	-5.04	-2.06**	-89.51***	-2.51	-13.61	-8.87***	-295.98***
	DK(T=376)				FI(T=329)			
	Level		Difference		Level		Difference	
	DF-GLS	MZ_α	DF-GLS	MZ_α	DF-GLS	MZ_α	DF-GLS	MZ_α
r	-5.14***	-75.09***	-5.21***	-70.07***	-1.89	-7.48	-12.24***	-135.40***
lroilp	-2.33	-9.55	-14.71***	-134.86***	-2.05	-7.24	-12.35***	-115.97***
lrstkp	-1.54	-9.08	-17.78***	-287.44***	-2.29	-8.44	-12.65***	-131.18***
ly	-2.95**	-177.74***	-6.83***	-8.38**	-1.72	-13.54	-19.15***	-72.78***
	FR(T=424)				IN(T=341)			
	Level		Difference		Level		Difference	
	DF-GLS	MZ_α	DF-GLS	MZ_α	DF-GLS	MZ_α	DF-GLS	MZ_α
r	-1.31	-4.09	-4.44***	-111.79***	-2.72***	-42.29***	-15.43***	- 63.85***
lroilp	-1.50	-4.12	-16.77***	-206.68***	-0.81	-1.51	-14.47***	- 161.70***
lrstkp	-1.21	-3.05	-5.63***	-185.78***	-1.36	-3.43	-11.92***	- 111.05***
ly	-1.36	-14.95*	-2.45**	-65.80***	-1.59	-12.33	-3.19***	- 42.63***

continued

Appendix Table 1. Results of Unit Root Tests (continued)

	IL(T=250)				IT(T=412)			
	Level		Difference		Level		Difference	
	DF-GLS	MZ_α	DF-GLS	MZ_α	DF-GLS	MZ_α	DF-GLS	MZ_α
r	-0.89	-1.31	-2.35**	-8.83**	-1.46	-4.11	-4.89***	-86.64***
lroilp	-0.22	0.30	-2.36**	-41.10***	-1.46	-4.18	-2.27**	-43.67***
lrstkp	-4.35***	-23.90***	-7.42***	-77.62***	-1.26	-3.52	-6.54***	-179.45***
ly	-1.61	-11.78	-2.00**	-34.71***	-1.85	-8.85	-4.59***	-83.73***

	JP(T=424)				KR(T=328)			
	Level		Difference		Level		Difference	
	DF-GLS	MZ_α	DF-GLS	MZ_α	DF-GLS	MZ_α	DF-GLS	MZ_α
r	-0.98	-2.28	-3.42***	-191.98***	-1.00	-2.91	-3.49***	-81.16***
lroilp	-1.60	-4.43	-17.00***	-210.10***	-2.33	-9.43	-13.21***	-108.87***
lrstkp	-1.44	-4.36	-13.49***	-165.15***	-1.70	-5.42	-12.57***	-140.86***
ly	-1.87	-3.57	-2.08**	-297.71***	-1.45	-7.90	-21.81***	-202.23***

	MY(T=304)				MX(T=256)			
	Level		Difference		Level		Difference	
	DF-GLS	MZ_α	DF-GLS	MZ_α	DF-GLS	MZ_α	DF-GLS	MZ_α
r	-3.02***	-26.45***	-4.08***	-93.91***	-3.12**	-13.61	-11.59***	-32.02***
lroilp	-1.79	-3.90	-12.71***	-114.60***	-0.23	0.39	-10.30***	-101.44***
lrstkp	-2.86***	-10.73	-2.81***	-43.71***	-1.71	-4.81	-3.48***	-61.84***
ly	-2.13	-15.13*	-3.02***	-94.79***	-2.17	-10.16	-1.98**	-44.05***

continued

cedure targets the middle 70% or 50% of the rank-ordered data set to estimate the determinant of the variance and the covariance matrix det $|\Sigma|$ according to different threshold levels c_i. Finally, equation (3) is used to find the optimal threshold value c^*. Since we are interested only in the positive (price rise) impact of an oil

Appendix Table 1. Results of Unit Root Tests (continued)

	NL(T=424)				NO(T=405)			
	Level		Difference		Level		Difference	
	DF-GLS	MZ_α	DF-GLS	MZ_α	DF-GLS	MZ_α	DF-GLS	MZ_α
r	-2.97***	-18.09***	-22.78***	-208.71***	-2.02**	-8.32**	-18.26***	-245.00***
lroilp	-1.49	-4.04	-16.63***	-196.21***	-1.46	-4.28	-16.22***	-192.73***
lrstkp	-1.19	-1.81	-4.70***	-94.96***	-1.17	-3.10	-7.78***	-84.72***
ly	-1.42	-14.59*	-2.82***	-51.31***	-0.75	-3.21	-15.56***	-666.96***
	PH(T=292)				PT(T=208)			
	Level		Difference		Level		Difference	
	DF-GLS	MZ_α	DF-GLS	MZ_α	DF-GLS	MZ_α	DF-GLS	MZ_α
r	-4.41***	-35.50***	-14.17***	-822.94***	-1.13	-21.37***	-3.11***	-25.50***
lroilp	-1.08	-3.72	-10.51***	-116.05***	-2.18	-11.59	-9.22***	-85.35***
lrstkp	-1.68	-5.58	-16.68***	-144.56***	-1.15	-2.56	-1.64**	-7.47**
ly	-1.30	-3.69	-16.88***	-144.99***	-1.94	-7.34	-21.30***	-47.61***
	ZA(T=424)				ES(T=376)			
	Level		Difference		Level		Difference	
	DF-GLS	MZ_α	DF-GLS	MZ_α	DF-GLS	MZ_α	DF-GLS	MZ_α
r	-1.59	-6.71	-20.13***	-210.59***	-4.47***	-33.55***	-11.80***	-15.35***
lroilp	-1.20	-2.70	-16.79***	-203.32***	-1.94	-6.79	-13.29***	-227.17***
lrstkp	-2.68***	-14.46*	-2.70***	-40.92***	-0.52	-0.50	-13.89***	-168.20***
ly	-0.97	-3.79	-13.19***	-120.80***	-2.02	-3.88	-4.01***	-16.51***

continued

price change on economic activities, the lower bound ($\underline{c}$) of the grid search is set at the 50th percentile of all the points and, therefore, the search area ($\bar{c}$, $\underline{c}$) should be between the 85th percentile of the points (or the 75th percentile of the points) for the largest value, and the 50th percentile of the points for the smallest value.

Appendix Table 1. Results of Unit Root Tests (continued)

| | GB(T=400) | | | | US(T=424) | | | |
| | Level | | Difference | | Level | | Difference | |
	DF-GLS	MZ_α	DF-GLS	MZ_α	DF-GLS	MZ_α	DF-GLS	MZ_α
r	-1.45	-6.92	-19.56***	-144.36***	-2.80*	-12.78	-13.67***	-109.91***
lroilp	-1.63	-4.82	-5.97***	-125.44***	-1.40	-3.30	-16.78***	-205.77***
lrstkp	-1.42	-3.15	-13.53***	-151.33***	-1.14	-2.35	-3.38***	-86.59***
ly	-2.20*	-10.69	-3.15***	-10.97**	-2.60*	-11.88	-7.79***	-319.59***
	TW(T=292)							
	Level		Difference		Level		Difference	
	DF-GLS	MZ_α	DF-GLS	MZ_α	DF-GLS	MZ_α	DF-GLS	MZ_α
r	-1.40	-3.75	-13.96***	-72.15***				
lroilp	-1.57	-3.74	-5.70***	-61.83***				
lrstkp	-1.24	-4.26	-15.50***	-143.82***				
ly	-1.62	-3.92	-4.20***	-33.38***				

Note: r is the interest rate. lroilp, lrstkp and ly represent respectively the real oil price, real stock price, and industrial production index after logarithmic transformations; *, **, and *** represent respectively 10%, 5%, and 1% significance levels; T denotes sample size.

Appendix Table 2-1. Results of the Johansen Cointegration Test
(lroilp, r, lrstkp, ly model)

	H_0	Eigen value	Trace	p-value	Eigen value	Max-Eigen	p-value	I(1) Variable	Model
AT	r=0	0.05	47.31	0.06	0.05	22.60	0.19	r	VAR
	r≤1	0.03	24.70	0.17	0.03	14.39	0.33	lroilp	
	r≤2	0.02	10.31	0.26	0.02	10.29	0.19	lrstkp	
	r≤3	0.00	0.02	0.88	0.00	0.02	0.88	ly	
CA	r=0	0.04	21.47	0.33	0.04	15.09	0.28	lroilp	VAR
	r≤1	0.01	6.38	0.65	0.01	5.21	0.72	lrstkp	
	r≤2	0.00	1.17	0.28	0.00	1.17	0.28	ly	
DK	r=0	0.04	18.27	0.55	0.04	13.02	0.45	lroilp	VAR
	r≤1	0.01	5.25	0.78	0.01	3.86	0.87	lrstkp	
	r≤2	0.00	1.39	0.24	0.00	1.39	0.24	ly	
FI	r=0	0.07	46.53	0.07	0.07	24.01	0.13	r	VAR
	r≤1	0.06	22.52	0.27	0.06	18.40	0.12	lroilp	
	r≤2	0.01	4.11	0.89	0.01	4.11	0.85	lrstkp	
	r≤3	0.00	0.00	0.99	0.00	0.00	0.99	ly	
FR	r=0	0.05	44.62	0.10	0.05	19.79	0.36	r	VAR
	r≤1	0.05	24.83	0.17	0.05	18.39	0.12	lroilp	
	r≤2	0.01	6.44	0.64	0.01	5.81	0.64	lrstkp	
	r≤3	0.00	0.63	0.43	0.00	0.63	0.43	ly	
IN	r=0	0.03	21.03	0.36	0.03	11.00	0.65	lroilp	VAR
	r≤1	0.03	10.04	0.28	0.03	10.02	0.21	lrstkp	
	r≤2	0.00	0.02	0.89	0.00	0.02	0.89	ly	
IL	r=0	0.91	581.99	0.00	0.91	564.81	0.00	r	VECM
	r≤1	0.05	17.18	0.63	0.05	11.44	0.60	lroilp	
	r≤2	0.01	5.73	0.73	0.01	3.57	0.90	lrstkp	
	r≤3	0.01	2.17	0.14	0.01	2.17	0.14	ly	
IT	r=0	0.06	46.69	0.06	0.06	25.01	0.10	r	VAR
	r≤1	0.03	21.68	0.32	0.03	10.66	0.68	lroilp	
	r≤2	0.02	11.02	0.21	0.02	8.70	0.31	lrstkp	
	r≤3	0.01	2.33	0.13	0.01	2.33	0.13	ly	
JP	r=0	0.05	41.17	0.18	0.05	21.95	0.22	r	VAR
	r≤1	0.03	19.22	0.48	0.03	13.27	0.43	lroilp	
	r≤2	0.01	5.94	0.70	0.01	5.09	0.73	lrstkp	
	r≤3	0.00	0.85	0.36	0.00	0.85	0.36	ly	
KR	r=0	0.07	39.91	0.23	0.07	23.21	0.16	r	VAR
	r≤1	0.04	16.71	0.66	0.04	12.84	0.47	lroilp	
	r≤2	0.01	3.86	0.91	0.01	2.71	0.96	lrstkp	
	r≤3	0.00	1.15	0.28	0.00	1.15	0.28	ly	
MY	r=0	0.05	24.21	0.19	0.05	14.52	0.32	lroilp	VAR
	r≤1	0.03	9.69	0.31	0.03	8.54	0.33	lrstkp	
	r≤2	0.00	1.16	0.28	0.00	1.16	0.28	ly	

continued

Appendix Table 2-1. Results of the Johansen Cointegration Test (lroilp, r, lrstkp, ly model) (continued)

	H_0	Eigen value	Trace	p-value	Eigen value	Max-Eigen	p-value	I(1) Variable	Model
MX	r=0	0.12	57.82	0.00	0.12	31.65	0.01	r	VECM
	r≤1	0.06	26.17	0.12	0.06	15.63	0.25	lroilp	
	r≤2	0.04	10.54	0.24	0.04	9.07	0.28	lrstkp	
	r≤3	0.01	1.47	0.23	0.01	1.47	0.23	ly	
NL	r=0	0.04	26.35	0.12	0.04	16.02	0.22	lroilp	VAR
	r≤1	0.02	10.33	0.26	0.02	9.65	0.24	lrstkp	
	r≤2	0.00	0.68	0.41	0.00	0.68	0.41	ly	
NO	r=0	0.07	51.94	0.01	0.07	28.45	0.02	lroilp	VECM
	r≤1	0.04	23.49	0.10	0.04	15.51	0.17	lrstkp	
	r≤2	0.02	7.98	0.25	0.02	7.98	0.25	ly	
PH	r=0	0.07	38.52	0.13	0.07	19.09	0.30	lroilp	VAR
	r≤1	0.05	19.44	0.26	0.05	13.54	0.29	lrstkp	
	r≤2	0.02	5.90	0.47	0.02	5.90	0.47	ly	
PT	r=0	0.13	41.80	0.16	0.13	26.30	0.07	r	VAR
	r≤1	0.06	15.50	0.75	0.06	11.71	0.58	lroilp	
	r≤2	0.01	3.79	0.92	0.01	2.42	0.98	lrstkp	
	r≤3	0.01	1.38	0.24	0.01	1.38	0.24	ly	
ZA	r=0	0.04	37.17	0.34	0.04	15.76	0.69	r	VAR
	r≤1	0.04	21.41	0.33	0.04	15.40	0.26	lroilp	
	r≤2	0.01	6.01	0.69	0.01	5.30	0.70	lrstkp	
	r≤3	0.00	0.71	0.40	0.00	0.71	0.40	ly	
ES	r=0	0.06	27.54	0.09	0.06	20.65	0.06	lroilp	VAR
	r≤1	0.01	6.89	0.59	0.01	4.56	0.80	lrstkp	
	r≤2	0.01	2.32	0.13	0.01	2.32	0.13	ly	
GB	r=0	0.09	56.93	0.01	0.09	35.95	0.00	r	VECM
	r≤1	0.04	20.98	0.36	0.04	13.80	0.38	lroilp	
	r≤2	0.02	7.18	0.56	0.02	6.33	0.57	lrstkp	
	r≤3	0.00	0.85	0.36	0.00	0.85	0.36	ly	
US	r=0	0.05	49.34	0.04	0.05	21.95	0.22	r	VAR
	r≤1	0.04	27.39	0.09	0.04	18.41	0.12	lroilp	
	r≤2	0.02	8.98	0.37	0.02	8.93	0.29	lrstkp	
	r≤3	0.00	0.05	0.83	0.00	0.05	0.83	ly	
TW	r=0	0.11	68.45	0.00	0.11	33.49	0.01	r	VECM
	r≤1	0.10	34.96	0.01	0.10	30.12	0.00	lroilp	
	r≤2	0.02	4.84	0.83	0.02	4.83	0.76	lrstkp	
	r≤3	0.00	0.01	0.93	0.00	0.01	0.93	ly	

Note: Trace = Trace statistics; Max-Eigen= Maximum eigenvalue Statistics; VAR=Vector Autoregressive Model; VECM=Vector Error Correction Model. r=number of cointegating relation(s).

Appendix Table 2-2. Results of the Johansen Cointegration Test
(*v* , *r, lrstkp, ly* model)

	H_0	Eigen value	Trace	p-value	Eigen value	Max-Eigen	p-value	I(1) Variable	Model
	r=0	0.05	32.93	0.34	0.05	21.28	0.18	r	
AT	r≤1	0.02	11.65	0.84	0.02	8.75	0.75	lrstkp	VAR
	r≤2	0.01	2.90	0.89	0.01	2.90	0.89	ly	
	r=0	0.05	24.61	0.07	0.05	16.74	0.12	lrstkp	VAR
CA	r≤1	0.02	7.87	0.26	0.02	7.87	0.26	ly	
	r=0	0.03	19.27	0.27	0.03	11.64	0.45	lrstkp	VAR
DK	r≤1	0.02	7.62	0.28	0.02	7.62	0.28	ly	
	r=0	0.07	38.88	0.12	0.07	21.50	0.17	r	
FI	r≤1	0.04	17.37	0.39	0.04	13.46	0.29	lrstkp	VAR
	r≤2	0.01	3.92	0.75	0.01	3.92	0.75	ly	
	r=0	0.04	27.73	0.09	0.04	15.12	0.28	r	
FR	r≤1	0.03	12.61	0.13	0.03	12.36	0.10	lrstkp	VAR
	r≤2	0.00	0.25	0.62	0.00	0.25	0.62	ly	
	r=0	0.03	11.31	0.19	0.03	10.32	0.19	lrstkp	VAR
IN	r≤1	0.00	0.99	0.32	0.00	0.99	0.32	ly	
	r=0	0.90	556.49	0.00	0.90	548.53	0.00	r	
IL	r≤1	0.03	7.96	0.47	0.03	7.38	0.45	lrstkp	VECM
	r≤2	0.00	0.58	0.45	0.00	0.58	0.45	ly	
	r=0	0.03	21.74	0.31	0.03	11.60	0.59	r	
IT	r≤1	0.02	10.14	0.27	0.02	9.97	0.21	lrstkp	VAR
	r≤2	0.00	0.18	0.68	0.00	0.18	0.68	ly	
	r=0	0.03	17.22	0.62	0.03	11.48	0.60	r	
JP	r≤1	0.01	5.74	0.73	0.01	4.28	0.83	lrstkp	VAR
	r≤2	0.00	1.46	0.23	0.00	1.46	0.23	ly	
	r=0	0.07	29.23	0.06	0.07	24.06	0.02	r	
KR	r≤1	0.01	5.17	0.79	0.01	4.41	0.81	lrstkp	VECM
	r≤2	0.00	0.76	0.38	0.00	0.76	0.38	ly	
	r=0	0.03	8.75	0.39	0.03	7.99	0.38	lrstkp	VAR
MY	r≤1	0.00	0.76	0.38	0.00	0.76	0.38	ly	
	r=0	0.11	34.55	0.01	0.11	29.46	0.00	r	
MX	r≤1	0.02	5.10	0.80	0.02	4.45	0.81	lrstkp	VECM
	r≤2	0.00	0.65	0.42	0.00	0.65	0.42	ly	
	r=0	0.03	13.79	0.09	0.03	13.04	0.08	lrstkp	VAR
NL	r≤1	0.00	0.75	0.39	0.00	0.75	0.39	ly	
	r=0	0.03	19.89	0.23	0.03	13.15	0.32	lrstkp	VAR
NO	r≤1	0.02	6.73	0.37	0.02	6.73	0.37	ly	

continued

**Appendix Table 2-2. Results of the Johansen Cointegration Test
(v , r, *lrstkp*, *ly* model) (continued)**

	H_0	Eigen value	Trace	p-value	Eigen value	Max-Eigen	p-value	I(1) Variable	Model
PH	r=0	0.04	14.72	0.60	0.04	10.80	0.53	lrstkp	VAR
	r≤1	0.01	3.92	0.75	0.01	3.92	0.75	ly	
PT	r=0	0.10	26.43	0.12	0.10	21.54	0.04	r	
	r≤1	0.02	4.89	0.82	0.02	3.89	0.87	lrstkp	VAR
	r≤2	0.01	1.00	0.32	0.01	1.00	0.32	ly	
ZA	r=0	0.04	28.16	0.08	0.04	15.41	0.26	r	
	r≤1	0.03	12.75	0.12	0.03	10.84	0.16	lrstkp	VAR
	r≤2	0.00	1.90	0.17	0.00	1.90	0.17	ly	
ES	r=0	0.05	19.53	0.01	0.05	19.14	0.01	lrstkp	VECM
	r≤1	0.00	0.39	0.53	0.00	0.39	0.53	ly	
GB	r=0	0.07	38.31	0.00	0.07	29.85	0.00	r	
	r≤1	0.02	8.46	0.42	0.02	7.61	0.42	lrstkp	VECM
	r≤2	0.00	0.85	0.36	0.00	0.85	0.36	ly	
US	r=0	0.05	28.09	0.08	0.05	21.06	0.05	r	
	r≤1	0.02	7.03	0.57	0.02	6.95	0.49	lrstkp	VAR
	r≤2	0.00	0.07	0.79	0.00	0.07	0.79	ly	
TW	r=0	0.11	45.74	0.00	0.11	31.27	0.00	r	
	r≤1	0.05	14.47	0.07	0.05	13.98	0.06	lrstkp	VECM
	r≤2	0.00	0.49	0.48	0.00	0.49	0.48	ly	

Note: Trace=Trace statistics; Max-Eigen= Maximum eigenvalue Statistics; VAR=Vector Autoregressive Model; VECM=Vector Error Correction Model. r=number of cointegating relation(s).

Appendix Table 3-1. Results of the Tsay Nonlinear C(d) Test
($\Delta lroilp$, Δr, $\Delta lrstkp$, Δly model)

	d=1	d=2	d=3	d=4	d=5	d=6	d=7	d=8	d=9	d=10	d=11	d=12
AT	269.19	281.24	253.69	206.65	253.98	269.50	248.54	257.82	302.08	253.02	238.64	256.87
	(0.09)	(0.03)	(0.26)	(0.94)	(0.26)	(0.09)	(0.34)	(0.20)	(0.00)	(0.27)	(0.51)	(0.22)
CA	320.89	292.10	283.83	339.81	285.03	296.26	313.26	318.69	263.85	290.06	225.83	267.50
	(0.00)	(0.01)	(0.03)	(0.00)	(0.02)	(0.01)	(0.00)	(0.00)	(0.14)	(0.01)	(0.74)	(0.11)
DK	431.45	280.01	277.76	243.40	263.83	397.10	262.59	312.80	467.03	294.55	237.72	234.46
	(0.00)	(0.04)	(0.05)	(0.43)	(0.14)	(0.00)	(0.15)	(0.00)	(0.00)	(0.01)	(0.53)	(0.59)
FI	258.04	310.05	283.19	248.94	235.10	274.53	261.73	251.92	281.34	281.48	263.71	256.99
	(0.20)	(0.00)	(0.03)	(0.33)	(0.58)	(0.06)	(0.16)	(0.29)	(0.03)	(0.03)	(0.14)	(0.22)
FR	297.63	405.93	300.11	266.78	284.63	292.70	228.20	217.99	297.62	282.65	235.51	255.51
	(0.01)	(0.00)	(0.01)	(0.11)	(0.03)	(0.01)	(0.70)	(0.84)	(0.01)	(0.03)	(0.57)	(0.23)
IN	383.54	350.19	293.87	305.30	284.50	337.83	357.42	280.58	341.07	351.37	306.16	253.48
	(0.00)	(0.00)	(0.01)	(0.00)	(0.03)	(0.00)	(0.00)	(0.04)	(0.00)	(0.00)	(0.00)	(0.26)
IL	258.63	323.55	284.76	254.49	361.78	355.76	288.59	268.47	326.23	371.55	323.15	325.98
	(0.25)	(0.00)	(0.04)	(0.31)	(0.00)	(0.00)	(0.03)	(0.14)	(0.00)	(0.00)	(0.00)	(0.00)
IT	260.79	337.50	241.84	215.22	256.95	310.17	237.34	236.56	233.06	224.61	260.98	230.88
	(0.17)	(0.00)	(0.45)	(0.87)	(0.22)	(0.00)	(0.54)	(0.55)	(0.61)	(0.75)	(0.17)	(0.65)
JP	324.23	333.89	287.43	236.84	274.74	261.23	311.94	274.82	221.74	264.38	205.43	278.40
	(0.00)	(0.00)	(0.02)	(0.55)	(0.06)	(0.17)	(0.00)	(0.06)	(0.80)	(0.13)	(0.95)	(0.04)
KR	307.72	355.52	275.08	260.09	277.27	366.03	353.74	327.97	300.06	273.40	227.47	304.89
	(0.00)	(0.00)	(0.06)	(0.18)	(0.05)	(0.00)	(0.00)	(0.00)	(0.01)	(0.07)	(0.71)	(0.00)
MY	301.18	405.25	285.23	352.91	350.43	374.32	253.18	263.13	296.45	293.60	271.59	287.72
	(0.00)	(0.00)	(0.02)	(0.00)	(0.00)	(0.00)	(0.27)	(0.15)	(0.01)	(0.01)	(0.08)	(0.02)
MX	325.36	345.79	325.12	361.74	413.80	343.51	353.87	271.48	315.23	292.80	252.28	289.35
	(0.00)	(0.00)	(0.00)	(0.00)	(0.00)	(0.00)	(0.00)	(0.11)	(0.00)	(0.02)	(0.34)	(0.02)
NL	283.30	241.84	266.23	237.33	250.81	261.11	217.27	226.90	285.29	225.93	254.86	254.31
	(0.03)	(0.45)	(0.12)	(0.54)	(0.30)	(0.17)	(0.85)	(0.72)	(0.02)	(0.73)	(0.24)	(0.25)
NO	292.30	349.14	291.53	279.07	260.35	344.10	286.88	219.59	301.61	268.70	292.44	232.69
	(0.02)	(0.00)	(0.02)	(0.06)	(0.23)	(0.00)	(0.03)	(0.87)	(0.01)	(0.13)	(0.02)	(0.69)
PH	333.42	451.17	283.34	258.82	255.52	310.12	409.29	315.74	398.33	341.04	339.47	234.75
	(0.00)	(0.00)	(0.03)	(0.19)	(0.23)	(0.00)	(0.00)	(0.00)	(0.00)	(0.00)	(0.00)	(0.58)
PT	314.44	297.94	284.36	278.57	296.79	330.77	265.62	259.55	328.32	321.32	267.65	263.93
	(0.00)	(0.01)	(0.03)	(0.04)	(0.01)	(0.00)	(0.12)	(0.18)	(0.00)	(0.00)	(0.11)	(0.14)
ZA	302.90	274.57	242.27	230.22	338.26	253.77	304.78	224.58	269.13	250.99	261.84	265.80
	(0.00)	(0.06)	(0.45)	(0.66)	(0.00)	(0.26)	(0.00)	(0.75)	(0.10)	(0.30)	(0.16)	(0.12)
ES	257.33	316.11	284.65	272.64	275.78	310.16	247.01	267.13	215.40	235.39	215.73	265.57
	(0.21)	(0.00)	(0.03)	(0.07)	(0.06)	(0.00)	(0.36)	(0.11)	(0.87)	(0.57)	(0.87)	(0.12)
GB	269.14	354.96	263.88	296.23	282.22	268.06	253.97	291.87	281.27	283.64	255.33	216.97
	(0.13)	(0.00)	(0.18)	(0.01)	(0.05)	(0.14)	(0.32)	(0.02)	(0.05)	(0.04)	(0.30)	(0.89)
US	293.75	324.70	375.66	322.59	281.59	233.69	268.42	260.67	307.00	282.91	281.80	284.41
	(0.01)	(0.00)	(0.00)	(0.00)	(0.03)	(0.60)	(0.10)	(0.17)	(0.00)	(0.03)	(0.03)	(0.03)
TW	282.37	331.58	282.87	278.37	293.33	308.40	296.57	300.97	322.96	259.92	233.13	245.07
	(0.05)	(0.00)	(0.04)	(0.06)	(0.02)	(0.00)	(0.01)	(0.01)	(0.00)	(0.23)	(0.68)	(0.47)

Appendix Table 3-2. Results of the Tsay Nonlinear C(d) Test
($\Delta lroilp$, Δr, $\Delta lrstkp$, Δly model)

	d=1	d=2	d=3	d=4	d=5	d=6	d=7	d=8	d=9	d=10	d=11	d=12
AT	248.88	306.30	242.79	262.72	279.77	245.85	278.60	256.98	307.11	285.37	268.19	271.73
	(0.33)	(0.00)	(0.44)	(0.15)	(0.04)	(0.38)	(0.04)	(0.22)	(0.00)	(0.02)	(0.10)	(0.08)
CA	323.78	286.92	273.90	264.80	279.44	299.11	294.64	327.22	301.73	296.67	242.78	331.44
	(0.00)	(0.02)	(0.07)	(0.13)	(0.04)	(0.01)	(0.01)	(0.00)	(0.00)	(0.01)	(0.44)	(0.00)
DK	418.93	241.24	298.29	235.46	235.79	255.33	260.36	256.25	250.12	235.01	238.20	219.39
	(0.00)	(0.47)	(0.01)	(0.57)	(0.56)	(0.24)	(0.18)	(0.22)	(0.31)	(0.58)	(0.52)	(0.83)
FI	284.87	298.64	327.66	278.39	243.13	267.08	316.64	207.89	265.85	229.90	227.55	265.63
	(0.02)	(0.01)	(0.00)	(0.04)	(0.43)	(0.11)	(0.00)	(0.93)	(0.12)	(0.67)	(0.71)	(0.12)
FR	283.18	281.56	335.69	281.67	274.92	302.04	275.64	197.41	286.00	316.42	224.09	307.24
	(0.03)	(0.03)	(0.00)	(0.03)	(0.06)	(0.00)	(0.06)	(0.98)	(0.02)	(0.00)	(0.76)	(0.00)
IN	299.57	293.46	328.52	312.76	313.81	372.20	325.36	323.47	341.80	344.09	332.04	277.16
	(0.01)	(0.01)	(0.00)	(0.00)	(0.00)	(0.00)	(0.00)	(0.00)	(0.00)	(0.00)	(0.00)	(0.05)
IL	223.93	298.09	297.32	296.06	288.55	302.49	302.07	254.46	300.21	359.37	343.48	309.94
	(0.82)	(0.01)	(0.01)	(0.01)	(0.03)	(0.01)	(0.01)	(0.31)	(0.01)	(0.00)	(0.00)	(0.00)
IT	218.61	277.12	234.44	211.90	241.36	274.14	228.76	234.32	250.53	288.19	273.83	244.65
	(0.84)	(0.05)	(0.59)	(0.90)	(0.46)	(0.06)	(0.69)	(0.59)	(0.31)	(0.02)	(0.07)	(0.40)
JP	262.80	316.70	262.60	241.61	325.53	253.14	282.72	256.60	258.63	237.79	203.98	282.42
	(0.15)	(0.00)	(0.15)	(0.46)	(0.00)	(0.27)	(0.03)	(0.22)	(0.20)	(0.53)	(0.96)	(0.03)
KR	315.91	288.53	283.64	265.20	234.27	296.28	276.74	295.88	258.47	228.65	246.47	294.50
	(0.00)	(0.03)	(0.04)	(0.17)	(0.66)	(0.01)	(0.07)	(0.01)	(0.25)	(0.75)	(0.44)	(0.01)
MY	291.54	341.98	324.59	347.30	338.37	285.88	272.87	331.74	328.91	262.98	300.54	331.11
	(0.01)	(0.00)	(0.00)	(0.00)	(0.00)	(0.02)	(0.07)	(0.00)	(0.00)	(0.15)	(0.00)	(0.00)
MX	320.24	329.19	360.80	369.97	413.49	295.30	341.85	286.30	311.16	291.42	269.11	270.08
	(0.00)	(0.00)	(0.00)	(0.00)	(0.00)	(0.01)	(0.00)	(0.03)	(0.00)	(0.02)	(0.13)	(0.12)
NL	216.81	254.84	240.61	261.15	223.08	245.97	203.39	207.79	279.36	287.46	248.24	279.51
	(0.86)	(0.24)	(0.48)	(0.17)	(0.78)	(0.38)	(0.96)	(0.93)	(0.04)	(0.02)	(0.34)	(0.04)
NO	289.87	284.40	262.00	229.42	215.26	341.19	296.15	270.32	281.67	240.82	263.76	277.92
	(0.02)	(0.03)	(0.16)	(0.68)	(0.87)	(0.00)	(0.01)	(0.09)	(0.03)	(0.47)	(0.14)	(0.05)
PH	374.07	410.33	284.47	241.57	237.83	274.42	377.63	366.60	403.58	257.73	222.49	254.74
	(0.00)	(0.00)	(0.03)	(0.46)	(0.53)	(0.06)	(0.00)	(0.00)	(0.00)	(0.21)	(0.78)	(0.25)
PT	284.02	278.28	257.55	286.72	253.61	336.76	285.60	269.61	307.66	288.10	301.94	282.65
	(0.03)	(0.05)	(0.21)	(0.02)	(0.26)	(0.00)	(0.02)	(0.09)	(0.00)	(0.02)	(0.00)	(0.03)
ZA	252.19	274.06	281.36	316.33	236.90	237.69	263.92	197.39	253.41	242.74	268.40	297.59
	(0.28)	(0.06)	(0.03)	(0.00)	(0.54)	(0.53)	(0.14)	(0.98)	(0.26)	(0.44)	(0.10)	(0.01)
ES	275.13	287.12	295.83	254.28	265.68	251.15	236.26	313.45	245.20	221.76	238.08	280.87
	(0.08)	(0.03)	(0.01)	(0.31)	(0.16)	(0.36)	(0.63)	(0.00)	(0.47)	(0.84)	(0.59)	(0.05)
GB	310.06	338.53	197.31	269.51	262.69	248.07	254.31	283.01	262.06	265.76	267.61	302.59
	(0.00)	(0.00)	(0.99)	(0.13)	(0.20)	(0.42)	(0.31)	(0.04)	(0.20)	(0.16)	(0.14)	(0.01)
US	233.71	283.22	288.78	300.43	244.36	230.69	247.70	227.22	242.35	329.60	294.99	303.81
	(0.60)	(0.03)	(0.02)	(0.00)	(0.41)	(0.66)	(0.35)	(0.71)	(0.45)	(0.00)	(0.01)	(0.00)
TW	312.75	287.04	315.66	297.36	308.11	306.92	310.78	288.60	332.28	269.14	251.95	277.19
	(0.00)	(0.03)	(0.00)	(0.01)	(0.00)	(0.00)	(0.00)	(0.03)	(0.00)	(0.13)	(0.35)	(0.07)

Note: d = the delay periods of threshold variable; numbers inside ( ) are p-value.

Defence and Peace Economics, 2005,
Vol. 16(1), February, pp. 45–57

DEFENSE SPENDING AND ECONOMIC GROWTH ACROSS THE TAIWAN STRAITS: A THRESHOLD REGRESSION MODEL

CHUNG-NANG LAI[a], BWO-NUNG HUANG[b†] and CHIN-WEI YANG[c]

[a]*Armed Forces College, National Defense University, Taoyuan, Taiwan;* [b]*Department of Economics and Center for IDAF, National Chung-Cheng University, Chia-Yi, Taiwan 621;* [c]*Department of Economics, Clarion University of Pennsylvania, Clarion, PA 16214–1232, USA*

(Received 15 July 2003; in final form 5 October 2004)

This paper employs both linear and non-linear models to investigate the relationship between national defense spending and economic growth for Taiwan and China. Using data from 1953–2000 on defense spending, GDP, import, export and capital, we find that China's defense spending leads that of Taiwan. There exists the phenomenon of an arms race between both countries when official Chinese data are used. On the one hand, feedback relations prevail between economic growth and defense spending growth in Taiwan. On the other hand, China's national defense is found to lead economic growth.

Keywords: Arms race; Multivariate threshold models; Defense spending; Economic growth

INTRODUCTION

The long separation of Taiwan from China bred hostility, which led to a significant increase in defense spending to GDP ratios across the straits before the 1980s. In terms of game theory, an arms race causes 'positional externalities'. That is, competitive spending on national defense diverts resources away from private sectors to the military, and hence slows down economic growth. Empirical results of arms races can be found in Kollias and Paleologou (2002). However, they did not discuss the impact of national defense on GDP simultaneously in their model. Such an impact may well vary for developed, developing and underdeveloped countries. In a developed economy, defense spending is only a small part of the total economy. As such, its positive effect on the rest of the economy is relatively weak and technological spillover is rather limited. In contrast, it contributes more in a developing economy through the externality effect, such as technological spin-offs, enhancement in human capital and social infrastructure.[1]

---

† Corresponding author. E-mail: ecdbnh@ccunix.ccu.edu.tw
[1] The suggestion by a reviewer on this issue is greatly appreciated.

ISSN 1024-2694 print/ISSN 1476-8267 online © 2005 Taylor & Francis Ltd
DOI: 10.1080/1024269052000323542

The literature dates back to Benoit (1973) who first identified that military spending can positively impact economic growth for developing countries. Since then there has been a considerable literature on the topic (e.g. Biswas, 1993; Atesoglu and Mueller 1990; Ram, 1986; Smith, 1980 and Faini *et al.*, 1984). Early empirical studies rely heavily on simple unconditional correlation coefficients (Benoit, 1973, 1978). Later methodology focuses on using cross-sectional multiple regression models to investigate the impact of national defense spending on economic growth (Lim, 1983; Faini *et al.* 1984; Biswas and Ram, 1986; Deger, 1986; Grobar and Porter, 1989; Cohen *et al.* 1996; Antonakis, 1997 and Heo 1997). Recently, the Granger causality model has been employed. For instance, Joerding (1986), applying the model to 57 less-developed countries (LDCs), finds that economic growth leads defense spending growth. The message is that LDCs ought to focus on economic growth before increasing defense spending. Applying a similar model to 55 developing countries, Chowdhury (1991) finds (i) national defense growth leads economic growth in 15 countries, (ii) economic growth leads national defense growth in seven countries and (iii) there is a feedback relationship between the two variables in three countries. No relationship between economic growth and defense spending was found for the remaining 30 out of 55 countries. The result by Lacivita and Frederiksen (1991) indicates that most of the 62 developing countries exhibit feedback relationships. Similar results (feedback relationships) were reported by Khilji and Mahmood (1997). They apply a multivariate model to Pakistan using data from 1972–1995 and find negative relationships between the two variables. Most recently, Chang *et al.* (2000) find (i) feedback relationships in the case of Taiwan and (ii) that economic growth leads national defense growth for China. In a nutshell, there has been no agreement on whether national defense spending leads economic growth for developing countries.

Needless to say, defense spending cannot be the sole determinant of economic growth, especially for export-led countries such as Taiwan. Beyond that, one cannot ignore the role of capital formation in economic growth. It is also reasonable to take imports into consideration (a proxy for the outflow of domestic resources). In addition, the long-standing hostility between China and Taiwan breeds the externality of an arms race across the straits. What is the relationship between economic growth and defense spending in the presence of an arms race? The purpose of this paper is to explore such relationships via both linear and non-linear models.[2] Finally, this paper provides empirical findings in the rare case of the 55-year-old arch rivalry between China and Taiwan.

The organization of this paper is as follows. The next section presents data and the model, The final section includes a conclusion.

DATA, UNIT ROOT, COINTEGRATION AND CAUSALITY TEST

Prior studies often apply a bivariate Granger causality model to investigate the relationship between economic growth and defense spending (e.g. Chang *et al.*, 2000; Dakurah *et al.*, 2001). Generally speaking, three problems arise from using the bivariate model. First, as is often the case, other important variables such as capital, imports and exports are excluded from the model. Second, the traditional model may very well have neglected long-term cointegration. Finally, a rival's defense spending is needed to account for the presence of an arms race. These shortcomings relating to omitted variables can render estimated coefficients biased. To overcome this problem, our model takes into consideration a rival's spending and other important factors in the framework of an error correction model. Well-known in the

---

[2] The advantage of the non-linear model lies in its ability to detect shifting interactions between defense spending and economic growth when rival's defense spending growth exceeds a threshold value.

neo-classical production function, both labor L_t and capital stock K_t are pivotal factors. However, the investment-output ratio is a good proxy for capital stock, which is difficult to obtain. In addition, imports (IM_t), exports (EX_t) and defense spending (GD_t) per capita are included in the aggregate production function:

$$Y_t = f\left(\frac{I_t}{Y_t}, EX_t, IM_t, GD_t\right) \tag{1}$$

where Y_t is per capita output. Taking into consideration a rival's defense spending, GDR_t, and taking logarithmic transformation (lower case letters), we have the following model:

$$y_t = f\left(I_t/Y_t, ex_t, im_t, gd_t, gdr_t\right) \tag{2}$$

An appropriate lead–lag relation can be formulated either as a six-variable VAR model or VECM (in the case of a cointegration). Annual data from 1953 through 2000 are employed: from AREMOS data bank for Taiwan and China Statistics Year Book for China.[3] All data are deflated into real terms and converted into US dollars for comparison purpose.

As in many time series analyses, we need to examine if the variables are stationary or of $I(0)$ process. In the absence of $I(0)$ process for some variables, we must detect if a cointegration exists among $I(1)$ variables. Table I presents the unit root results from using the Phillips–Perron model for the six variables.

An examination of Table I indicates that we fail to reject the hypotheses of $I(1)$ for all the variables except the investment output ratio. However, we can reject the hypothesis of $I(1)$ for all the variables in first difference indicating the $I(1)$ process in imports, exports, output and defense spending. To detect the potential cointegration relation proposed by Engle and Granger (1987), we employ the trace statistic by Johansen (1988), and Johansen and Juselius (1990). Furthermore, rival defense spending (gdr_t) is included in the model to account for the long-standing China–Taiwan hostility.

TABLE I   Unit Root Results of Taiwan and China's Five-variable Model

Country	Variable Name	Logarithmic value in level	First-difference model
Taiwan	y	−3.09	−3.54*
	gd	−2.45	−3.99*
	ex	−2.61	−3.02*
	im	−3.00	−2.62*
	I/Y	−3.51**	−5.05*
China	y	−1.72	−5.24*
	gd	−1.40	−5.17*
	ex	−1.88	−3.99*
	im	−1.99	−4.56*
	I/Y	−4.26*	−6.54*

Notes: * = 1% significance level. Drift term and time trend are used (in logarithms) in the level model, but are not included in first-difference model. y = real GDP, gd = real defense spending, ex = real exports, im = real imports, I/Y = investment–output ratio.

---

[3] According to Waller (1997), China's defense spending may well be underestimated as a result of budget falsification on the PLA's (People Liberation Army) revenue and the purchasing power parity of yuen. The accuracy of the spending from the official publication and SIPRI Yearbook cannot be determined either. As such, we first test the model using the figures from China's yearbook followed by that from SIPRI Yearbooks.

It is to be pointed out that the potential existence of cointegration relation(s) among the five variables (y_t, ex_t, im_t, gd_t, gdr_t) is to be examined in the presence of an arms race. The literature does not lack models of an arms race. For example, Wolfson (1990) and Anderton (1993) apply Richardson or Lanchester types of arms race models to explore the relationship between defense spending and economic growth. Similarly, Kollias and Paleologou (2002) make use of a Granger causality model to detect the existence of an arms race between Greece and Turkey. Again, the bivariate causality models suffer from biased estimates due to omitted variables. The omitted variable problem cannot just be ignored, given the 55-year rivalry between China and Taiwan. Thus, we formulate a six-variable model in the presence of an arms race. Before estimating it, it is necessary to examine the existence of a potential cointegration relation among the five $I(1)$ variables (y_t, ex_t, im_t, gd_t, gdr_t), I/Y being of $I(0)$. The results shown in Table II are obtained from applying Johansen's cointegration test.

As is evident from Table II, there exists only one set of cointegration relations for China or Taiwan. The cointegration vector indicates that Taiwan's defense spending has no discernable impact on China's spending in the long run. In contrast, China's spending exerts a significant impact on Taiwan's spending (see the equations below Table II). In addition, China's economic growth positively impacts on its spending growth, whereas the relation is insignificant for Taiwan. The effects of imports and exports on defense spending are found to be significant but different. Given the results from the cointegration test, we can investigate the relationship between economic growth and defense spending within the framework of the six-variable vector error correction model (VECM) below:

$$
\begin{bmatrix} \Delta y_t \\ (I/Y)_t \\ \Delta im_t \\ \Delta ex_t \\ \Delta gd_t \\ \Delta gdr_t \end{bmatrix} = \begin{bmatrix} \alpha_{11}(L)\alpha_{12}(L)\alpha_{13}(L)\alpha_{14}(L)\alpha_{15}(L)\alpha_{16}(L) \\ \alpha_{21}(L)\alpha_{22}(L)\alpha_{23}(L)\alpha_{24}(L)\alpha_{25}(L)\alpha_{26}(L) \\ \alpha_{31}(L)\alpha_{32}(L)\alpha_{33}(L)\alpha_{34}(L)\alpha_{35}(L)\alpha_{36}(L) \\ \alpha_{41}(L)\alpha_{42}(L)\alpha_{43}(L)\alpha_{44}(L)\alpha_{45}(L)\alpha_{46}(L) \\ \alpha_{51}(L)\alpha_{52}(L)\alpha_{53}(L)\alpha_{54}(L)\alpha_{55}(L)\alpha_{56}(L) \\ \alpha_{61}(L)\alpha_{62}(L)\alpha_{63}(L)\alpha_{64}(L)\alpha_{65}(L)\alpha_{66}(L) \end{bmatrix} \begin{bmatrix} \Delta y_{t-1} \\ (I/Y)_{t-1} \\ \Delta im_{t-1} \\ \Delta ex_{t-1} \\ \Delta gd_{t-1} \\ \Delta gdr_{t-1} \end{bmatrix} + \begin{bmatrix} \gamma_1 \cdot ecm_{t-1} \\ \gamma_2 \cdot ecm_{t-1} \\ \gamma_3 \cdot ecm_{t-1} \\ \gamma_4 \cdot ecm_{t-1} \\ \gamma_5 \cdot ecm_{t-1} \\ \gamma_6 \cdot ecm_{t-1} \end{bmatrix} + \begin{bmatrix} \varepsilon_{1t} \\ \varepsilon_{2t} \\ \varepsilon_{3t} \\ \varepsilon_{4t} \\ \varepsilon_{5t} \\ \varepsilon_{6t} \end{bmatrix} \quad (3)
$$

where $\alpha(L)$ is a polynomial lag operator with optimal lags determined by AIC, and ecm_{t-1} denotes the cointegration vectors shown below Table II. To find out if economic growth Granger-causes defense spending growth, is equivalent to testing the null hypothesis H_0: α_{51}

TABLE II   Cointegration Results of the Five-variable Model for Taiwan and China

Null hypothesis	Trace statistic (China)	Trace statistic (Taiwan)
$H_0 : r = 0$	73.17**	80.99*
$H_0 : r \leq 1$	37.95	45.63
$H_0 : r \leq 2$	19.74	25.99
$H_0 : r \leq 3$	7.91	12.91
$H_0 : r \leq 4$	0.99	3.38

Notes: ** = 5% significance level. * = 1% significance level. The optimal lag length based on the AIC equals one in the five-variable VAR model for both Taiwan and China. The five-variable cointegration model is formulated as below. r = number of cointegration vectors.
China:
$ecm_{t-1} = gd_{t-1} + 6.44 - 2.20y_{t-1} - 2.09im_{t-1} + 3.09ex_{t-1} + 0.08gdr_{t-1}$
   (−9.53)   (−6.05)   (8.11)   (0.32)
Taiwan:
$ecm_{t-1} = gd_{t-1} - 4.64 - 1.39y_{t-1} + 16.03im_{t-1} - 13.92ex_{t-1} + 1.99gdr_{t-1}$
   (−0.37)   (3.71)   (−5.51)   (1.95)
t statistics are in parentheses. gdr_t = rival's defense spending for both Taiwan and China.

$(L) = 0$ and $\gamma_5 = 0$. Conversely, to find out if defense spending Granger-causes economic growth, we test the null hypothesis of H_0: $\alpha_{15}(L) = 0$ and $\gamma_1 = 0$. In a similar vein, to test whether a rival's spending growth Granger-causes home country's spending growth (H_0: $\Delta gdr_{t-1} \not\rightarrow \Delta gd_t$) translates into testing H_0: $\alpha_{56}(L) = 0$ and $\gamma_5 = 0$. By the same token, H_0: $\Delta gd_{t-1} \not\rightarrow \Delta gdr_t$ corresponds to H_0: $\alpha_{65}(L) = 0$ and $\gamma_6 = 0$. Note the optimal lag determined by AIC is one for both China and Taiwan. Table III reports the results of the six-variable VECM.[4]

As indicated in Table III, China's defense spending leads economic growth. In terms of the arms race, China's defense spending is found to lead Taiwan's spending but not vice versa. The above mentioned relation can be attributed to the existence of the significant error correction terms in the Δy_t and Δgdr_t equations. In the case of Taiwan, feedback relations are found to prevail between economic and defense growth as a result of the significant error correction terms in Δy_t and Δgd_t equations. As mentioned before, the unilateral causation – China's military spending Granger-causes Taiwan's spending – can be explained by the size of the PLA, which is far greater than Taiwan's military forces. An increase in China's defense spending typically gives rise to a corresponding increase in Taiwan's spending in order to achieve some kind of equilibrium. On the other hand, China's spending is also determined by the military spendings of neighboring nations – Russia, India, Vietnam, and so on, – and, as such, China's spending is less sensitive to that of Taiwan.

The results of China's growth on defense spending leads economic growth, and the feedback relations existing between Taiwan's defense and economic growth are in agreement with the outcome by Chang *et al.* (2000), who used a three-variable model (domestic economic growth, defense spending and rival's spending). No cointegration relation was found for either country and, as such, their causality is a short-run phenomenon instead of a long-term cointegration relation. Our model also takes into consideration an arms race, especially in the presence of recently escalating hostility.

The linear model thus far ignores a non-linear threshold effect: only when Taiwan's defense spending exceeds some threshold level, would China's spending increase as a result. The rival's threshold spending that needs to be addressed for such a potential non-linear relationship is not considered in the conventional linear model. Consequently, we employ the threshold regression developed by Tong (1978) and Tong and Lim (1980) in order to explore the relationship between defense spending and economic growth under different levels of an arms race.

Based on the previous six-variable VAR model, we employ the technique of a multivariate threshold regression (MVTAR) developed by Tsay (1998) to investigate the impacts as shown below:[5]

$$
\begin{bmatrix}
\Delta y_t \\
(I/Y)_t \\
\Delta im_t \\
\Delta ex_t \\
\Delta gd_t \\
\Delta gdr_t
\end{bmatrix}
= \phi_1(L)
\begin{bmatrix}
\Delta y_{t-1} \\
(I/Y)_{t-1} \\
\Delta im_{t-1} \\
\Delta ex_{t-1} \\
\Delta gd_{t-1} \\
\Delta gdr_{t-1}
\end{bmatrix}
1(z_{t-d} \leq c) + \phi_2(L)
\begin{bmatrix}
\Delta y_{t-1} \\
(I/Y)_{t-1} \\
\Delta im_{t-1} \\
\Delta ex_{t-1} \\
\Delta gd_{t-1} \\
\Delta gdr_{t-1}
\end{bmatrix}
1(z_{t-d} > c) +
\begin{bmatrix}
\varepsilon_{1t} \\
\varepsilon_{2t} \\
\varepsilon_{3t} \\
\varepsilon_{4t} \\
\varepsilon_{5t} \\
\varepsilon_{6t}
\end{bmatrix}
\tag{4}
$$

in which $\phi_1(L)$ and $\phi_2(L)$ represent polynomial lags. $1(.)$ is the index variable: it equals one if the relation in the parenthesis is satisfied, zero otherwise. z_{t-d} is the threshold variable with

---

[4] Only three equations (Δgd_t, Δy_t, Δgdr_r) are reported, in order to analyze the causality between them.
[5] Refer to Tsay (1998) for details of the multivariate threshold regression model.

TABLE III Test Results of the Six-variable Model (Chinese Statistics Yearbook)

	Dependent					
	China			Taiwan		
Independent	Δgd_t	Δy_t	Δgdr_t	Δgd_t	Δy_t	Δgdr_t
ecm_{t-1}	11.19	11.15*	−24.22*	−3.04*	−3.04*	1.87
	(1.40)	(2.23)	(−3.50)	(−2.90)	(−3.77)	(1.49)
c	9.47	9.29	9.88	10.47	6.63	−4.27
	(0.90)	(1.41)	(1.08)	(1.34)	(1.10)	(−0.45)
Δgd_{t-1}	0.32	0.18	0.26	−0.32***	0.02	0.20
	(1.60)	(1.40)	(1.52)	(−1.65)	(0.15)	(0.87)
Δy_{t-1}	−0.20	−0.20	0.08	0.54	−0.17	−0.48
	(−0.60)	(−0.95)	(0.78)	(1.31)	(−0.55)	(−0.97)
Δim_{t-1}	−0.04	−0.04	−0.30	0.27	0.43**	0.48
	(−0.22)	(−0.34)	(0.10)	(1.05)	(2.19)	(1.53)
Δex_{t-1}	−0.01	0.32**	0.18	−0.65*	−0.43**	0.13
	(−0.04)	(1.95)	(0.78)	(−2.55)	(−2.18)	(0.41)
$(I/Y)_{t-1}$	−0.34	−0.28	−0.26	−0.37	−0.05	0.07
	(−0.88)	(−1.17)	(0.45)	(−0.89)	(−0.14)	(0.13)
Δgdr_{t-1}	0.24	0.18***	−0.29***	0.19	0.11	0.23
	(1.37)	(1.67)	(−1.93)	(1.47)	(1.12)	(1.49)
$\bar{R}^2$	0.12	0.35	0.15	0.16	0.30	0.05
$Q(1)$	0.10	0.19	0.00	0.00	1.15	0.66
	[0.76]	[0.66]	[1.00]	[0.99]	[0.28]	[0.42]
$Q(4)$	2.21	4.20	1.16	0.53	5.70	5.80
	[0.70]	[0.38]	[0.89]	[0.97]	[0.22]	[0.22]
$Q(6)$	4.15	5.31	2.09	2.61	5.78	6.06
	[0.53]	[0.50]	[0.91]	[0.86]	[0.45]	[0.42]
$H_0 : \Delta y_{t-1} \nrightarrow \Delta gd_t$	2.40			4.45**		
	[0.31]			[0.02]		
$H_0 : \Delta gd_{t-1} \nrightarrow \Delta y_t$		5.29*			7.16***	
		[0.01]			[0.08]	
$H_0 : \Delta gdr_{t-1} \nrightarrow \Delta gd_t$	1.40			4.42**		
	[0.26]			[0.02]		
$H_0 : \Delta gd_{t-1} \nrightarrow \Delta gdr_{t-1}$			6.17*			1.45
			[0.00]			[0.25]

Notes: values in ( ) and [ ] are *t* statistics and *p* values respectively; $Q(1)$ is the Ljung–Box statistic with one lag; $\bar{R}_z^2$ = adjusted R^2; *, **, *** denotes 1%, 5% and 10% significance level respectively; *ecm* = cointegration vector; Δ = first difference; *gd, y, im, ex, I/Y, gdr* represent defense spending, output, imports, exports, investment–output ratio and rival's defense spending respectively; χ^2 statistic is used to test the null H_0 in which $\nrightarrow$ denote 'does not Granger-cause'.

delay *d* and a threshold value *c*. The selection of the threshold value hinges on the minimization of the variance–covariance matrix (logarithmic value of the determinant, log det $|\Sigma|$, where Σ is the variance–covariance matrix of equation (4)) of the above equation (Weise, 1999). Following the arranged regression model by Tsay (1989, 1998), the existence of a threshold effect can be detected via the $C(d)$ test. The $C(d)$ test obeys chi-square distribution with degrees of freedom $k(pk + 1)$ where *k* is the number of explanatory variables and *p* is the number of the lag length.

TABLE IV $C(d)$ Test of the Multivariate Threshold Model (Chinese Statistics Yearbook)

z_t	China (six-variable) d			Taiwan (six-variable) d		
	1	2	3	1	2	3
Change in Rival's GD/Y ratio	20.89 (0.05)	33.47 (0.00)	34.72 (0.00)	36.28 (0.00)	46.92 (0.00)	50.54 (0.00)

Notes: z_t = threshold variable; d = delay; values in parentheses are p values; GD = national defense spending; Y = GDP. One set of cointegration in VAR-ECM is used in the test for China and Taiwan.

In the presence of an arms race, a change in the rival's defense spending plays a key role in the model and we use it as the threshold variable.[6] In Taiwan's six variable VAR model, we use China's change in defense spending (Δgdr_{t-d}) as the threshold variable. Likewise, Taiwan's change in defense spending (Δgdr_{t-d}) is the threshold variable in China's six-variable VAR model. To ensure adequate degrees of freedom in annual data, we choose the value of d delay between 1 and 3 in the two-regime framework.

As shown from the $C(d)$ statistics in Table IV, the rival's spending expenditure (with one to three lags) can be used as a threshold variable to exhibit the significant non-linear relation in the six-variable multivariate threshold autoregressive (MVTAR) model.

A glance at Table V suggests the optimal threshold value for China is Taiwan's defense growth rate (lagged by three periods) at 5%. That is, 5% is used as the threshold value that divides the sample into two regimes. When Taiwan's spending growth is less than 5% (regime I), we witness that China's defense spending leads her own economic growth (the lower part of Table V) as is confirmed in the previous linear model. Likewise China's spending growth leads (by three periods) that of Taiwan, irrespective of Taiwan's spending growth. Much like that in the linear model, the converse does not hold: Taiwan's spending growth does not lead that of China. On the other hand, we can partition the data into two regimes by using China's spending growth at 3% as the threshold variable. The result is similar to that of the linear model: feedback relations prevail between Taiwan's economic and spending growth, except that the relations can now be attributed to the previous spending growth (lag by one) in regime II instead of the error correction terms as in the linear model. Furthermore, China's spending growth is found to lead Taiwan's in either regime in the presence of an arms race, but the causal relation is more significant in regime II. Notice that when China's spending growth exceeds 3%, it leads Taiwan's spending growth. In a similar vein, Taiwan's spending growth is found to lead that of China. That is, there exist feedback relations (an arms race) between Taiwan and China.

As Waller (1997) points out, Chinese military spending is indeed much (around four times) higher than is reported by official figures as a result of budgetary falsification, the revenue of PLA and the purchasing power parity of yuan. To circumvent this problem, we obtain data from the SIPRI Yearbook and compare them with those of the Chinese Statistical Yearbook (CSY). Clearly, figures after 1986 from the CSY are noticeably underestimated (Figure 1). Will such underestimations invalidate the previous results? Table VI reports the results of the linear VAR model for both countries.[7]

---

[6] We thank a reviewer who pointed out an important phenomenon: an arms race is occuring only in cases in which two defense expenditure rates are both on the rise. As such, using the rate of change in defense expenditure is a reasonable choice.

[7] The cointegration result remains unchanged, irrespective of using the SIPRI Yearbook or Chinese Statistics Yearbook: there exists only one set of cointegration.

TABLE V Causality Results between Defense Spending and Economic Growth of the Multivariate Threshold VAR Model (Chinese Statistics Yearbook)

	China						Taiwan					
	I	*II*	*I*	*II*	*I*	*II*	*I*	*II*	*I*	*II*	*I*	*II*
	Δgd_t	Δgd_t	Δy_t	Δy_t	Δgdr_t	Δgdr_t	Δgd_t	Δgd_t	Δy_t	Δy_t	Δgdr_t	Δgdr_t
c	15.17	−76.13	8.70	34.96	7.81	128.58*	20.73	105.08*	11.49	3.11	−19.58	19.05
	(1.46)	(−0.40)	(1.21)	(0.57)	(0.70)	(5.40)	(1.57)	(3.38)	(1.49)	(0.24)	(−1.30)	(0.52)
ecm_{t-1}	7.17	−97.47	6.08	−44.04	−19.02*	−67.81*	−11.23*	−51.14*	−13.10*	2.17	3.20	13.47
	(1.00)	(−0.97)	(1.22)	(−1.37)	(−2.44)	(−5.40)	(−2.21)	(−2.69)	(−4.42)	(0.27)	(0.55)	(0.59)
Δgd_{t-1}	0.53*	−1.14	0.30	−0.33	0.26	−0.58*	−0.39	−1.57*	0.07	0.54***	−0.04	0.13
	(2.39)	(−0.77)	(1.91)***	(−0.70)	(1.08)	(−3.14)	(−1.63)	(−2.29)	(0.50)	(1.88)	(−0.16)	(0.16)
Δy_{t-1}	−0.45	4.08	−0.34	2.78**	−0.37	3.14*	0.92	0.16	−0.45	−0.16	−0.35	0.23
	(−1.41)	(1.02)	(−1.52)	(2.16)	(−1.05)	(6.27)	(1.18)	(0.35)	(−0.99)	(−0.86)	(−0.40)	(0.42)
Δim_{t-1}	0.07	−4.00	−0.02	−2.68*	−0.20	−0.10	−0.69	4.07*	−0.34	−0.27	0.16	0.26
	(0.37)	(−1.09)	(−0.16)	(−2.27)	(−1.03)	(−0.22)	(−1.47)	(3.43)	(−1.22)	(−0.55)	(0.30)	(0.19)
Δex_{t-1}	−0.11	5.06	0.32***	2.87	0.48***	−1.60*	−0.11	−3.89*	0.37	0.22	0.34	−0.11
	(−0.43)	(0.91)	(1.85)	(1.61)	(1.75)	(−2.30)	(−0.25)	(−3.42)	(1.46)	(0.45)	(0.68)	(−0.08)
(I/Y_{t-1})	−0.45	1.72	−0.21	−1.68	−0.17	−5.25*	−0.87	−5.55*	−0.37	0.25	0.65	−0.48
	(−1.15)	(0.25)	(−0.77)	(−0.77)	(−0.40)	(−6.16)	(−1.35)	(−3.37)	(−0.97)	(0.36)	(0.88)	(−0.24)
Δgdr_{t-1}	−0.15	−0.96	−0.13	−0.24	−0.36	1.11*	0.16	1.17*	0.14	−0.05	0.20	0.38
	(−0.74)	(−0.02)	(−0.87)	(−0.32)	(−1.61)	(3.86)	(0.75)	(3.11)	(1.15)	(−0.33)	(0.82)	(0.85)
$\bar{R}^2$	0.29	−1.47	0.33	0.57	0.02	0.94	0.09	0.47	0.45	0.23	−0.11	0.20
$Q(1)$	0.45	0.42	0.62	0.42	1.76	0.42	0.41	0.66	0.07	0.15	1.97	0.19
	[0.51]	[0.52]	[0.43]	[0.52]	[0.18]	[0.52]	[0.52]	[0.42]	[0.79]	[0.70]	[0.16]	[0.66]
$Q(4)$	2.47	1.48	3.53	1.48	4.24	1.47	5.76	2.00	2.48	0.66	3.77	1.95
	[0.65]	[0.83]	[0.47]	[0.83]	[0.38]	[0.83]	[0.22]	[0.74]	[0.65]	[0.96]	[0.44]	[0.75]
$H_0: \Delta y_{t-1} \nrightarrow \Delta gd_t$	3.66	1.32					5.58***	7.99**				
	[0.16]	[0.52]					[0.06]	[0.02]				
$H_0: \Delta gd_{t-1} \nrightarrow \Delta y_t$			9.59*	1.91					19.62*	7.46*		
			[0.01]	[0.39]					[0.00]	[0.00]		
$H_0: \Delta gdr_{t-1} \nrightarrow \Delta gd_t$	2.34	0.94					5.09***	9.14*				
	[0.31]	[0.63]					[0.08]	[0.00]				
$H_0: \Delta gd_{t-1} \nrightarrow \Delta gdr_t$					7.98**	5.61***					1.01	4.82***
					[0.02]	[0.06]					[0.60]	[0.09]
z	Δgdr_{t-d}						Δgdr_{t-d}					
d	3						3					
c^*	5%						3%					

Notes: z = threshold variable; d = delay; c^* = optimal threshold value; I = regime I, which corresponds to the data set in which ($\Delta gdr_{t-d} \leq c^*$) holds; II = regime II, which corresponds to the data set in which ($\Delta gdr_{t-d} > c^*$) holds.

A perusal of Table VI reveals that the result of applying data from the SIPRI Yearbook remains qualitatively the same: China's defense spending growth leads her economic growth. In the case of Taiwan, the feedback relations still prevail, albeit becoming stronger. Additionally, China's spending growth is again found to lead that of Taiwan using the data from the SIPRI Yearbook. In sum, the results from the linear VAR model remains largely unchanged. In the case of the non-linearity test, we can reject the null of the linear model for both China (one lag) and Taiwan (one to three lags) as shown in Table VII. Results using the SIPRI Yearbook data for the two-regime VECM are reported in Table VIII.

billion US $

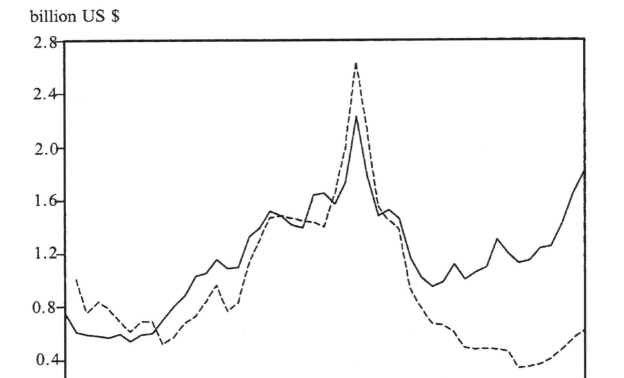

FIGURE 1   China's Defense Spending (Chinese Statistics Yearbook versus SIPRI Yearbook).

A comparison between Tables V and VIII indicates that (i) the optimal threshold level for China is Taiwan's spending growth at 5% regardless of which database is used, and (ii) the optimal threshold level for Taiwan is China's spending at 1.5% using the SIPRI Yearbook, compared with 3% using the CSY. Unlike the previous case, in which China's spending growth leads (i) her own economic growth and (ii) Taiwan's defense spending growth in regime I (Taiwan's spending growth falls short of 5%), only (ii) prevails using the SIPRI Yearbook data. In regime II, the result that China's defense growth leads Taiwan's remains qualitatively the same but with less significance, a probability value of 14%: a slight increase from 6%.

Taiwan's MVTAR model applying the SIPRI Yearbook data reveals that, as before, there exist feedback relations between Taiwan's defense and economic growth in regime I. Different from before, China's defense spending growth does not lead Taiwan's growth. In regime II, the feedback relations still persist between defense spending and economic growth. Unlike the previous model, in which feedback relations prevail between both countries' spending growth, China's spending growth unilaterally Granger-causes Taiwan's spending growth. Thus, data selection could make a slight difference in the conclusion. However, it can be concluded from using both datasets that China's defense spending leads that country's own economic growth. In addition, there exist feedback relations between Taiwan's spending and economic growth. Besides, China's spending growth leads Taiwan's spending growth when

TABLE VI   Test Results of the Six-variable Model (SIPRI Yearbook)

Independent	China			Taiwan		
	Δgd_t	Δy_t	Δgdr_t	Δgd_t	Δy_t	Δgdr_t
c	3.63	11.66***	1.59	9.97	6.26	0.88
	(0.45)	(1.67)	(0.15)	(1.26)	(1.01)	(0.13)
ecm_{t-1}	−4.79	−5.29***	7.39***	−8.62*	−8.28*	2.42
	(0.14)	(−1.92)	(4.83)	(−2.87)	(−3.55)	(0.93)
Δgd_{t-1}	0.05	0.17	0.23	−0.26	0.08	−0.06
	(0.26)	(1.09)	(0.99)	(−1.34)	(0.54)	(−0.34)
Δy_{t-1}	0.03	−0.10	0.21	0.33	−0.35	−0.23
	(0.14)	(−0.55)	(0.76)	(0.83)	(−1.13)	(−0.66)
Δim_{t-1}	0.02	−0.07	−0.10	0.29	0.46**	0.26
	(0.10)	(−0.56)	(−0.51)	(1.12)	(2.27)	(1.16)
Δex_{t-1}	−0.06	0.43*	−0.13	−0.50*	−0.31***	0.11
	(−0.35)	(2.84)	(−0.58)	(−2.22)	(−1.75)	(0.57)
$(I/Y)_{t-1}$	−0.04	−0.42***	0.03	−0.43	−0.10	−0.03
	(−0.35)	(−1.66)	(0.09)	(−1.00)	(−0.31)	(−0.08)
Δgdr_{t-1}	0.05	0.19	−0.21	0.32***	0.28***	0.09
	(0.41)	(1.63)	(−1.28)	(1.70)	(1.91)	(0.54)
$\bar{R}^2$	−0.07	0.30	−0.02	0.16	0.28	−0.06
$Q(1)$	0.00	0.22	0.14	0.02	0.77	0.04
	[1.00]	[0.64]	[0.70]	[0.90]	[0.38]	[0.84]
$Q(4)$	4.18	5.09	3.23	0.98	5.67	7.73
	[0.38]	[0.28]	[0.52]	[0.91]	[0.23]	[0.10]
$Q(6)$	9.23	6.15	5.18	2.55	5.82	13.09
	[0.16]	[0.41]	[0.52]	[0.86]	[0.44]	[0.04]
$H_0: \Delta y_{t-1} \nrightarrow \Delta gd_t$	2.46			8.70		
	[0.29]			[0.01]		
$H_0: \Delta gd_{t-1} \nrightarrow \Delta y_t$		6.23**			12.66*	
		[0.04]			[0.00]	
$H_0: \Delta gdr_{t-1} \nrightarrow \Delta gd_t$	2.29			8.97*		
	[0.31]			[0.01]		
$H_0: \Delta gd_{t-1} \nrightarrow \Delta gdr_{t-1}$			3.67			0.93
			[0.16]			[0.63]

Note: See Table III.

the former exceeds a threshold level. Of special importance is if the arms race (feedback relations) exists between Taiwan and China. The answer is yes when we use the CSY data set and when China's spending growth exceeds the threshold level. However, China's spending growth unilaterally Granger-causes Taiwan's spending growth when the former exceeds a threshold level (e.g. 1.5%) if the SIPRI data set is applied.

CONCLUSIONS

Long-standing hostility carves out the importance of defence spending on both sides of the Taiwan Straits. The paucity in the literature can be attributed to unreliable Chinese official

TABLE VII C(d) Test of the Multivariate Threshold Model (SIPRI Yearbook)

	China (six-variable) d			Taiwan (six-variable) d		
z_t	1	2	3	1	2	3
Rival's change in GDY	23.07 (0.03)	17.59 (0.13)	17.27 (0.14)	50.72 (0.00)	35.49 (0.00)	36.56 (0.00)

Note: z_t = threshold variable; d = delay; values in parentheses are p values; GD = national defense spending; Y = GDP. One set of cointegration in VAR-ECM is used in the test for China and Taiwan.

TABLE VIII Causality Results Between Defense Spending and Economic Growth of the MVTAR Model (SIPRI Yearbook)

	China						Taiwan					
	I	II	I	II	I	II	I	II	I	II	I	II
	Δgd_t	Δgd_t	Δy_t	Δy_t	Δgdr_t	Δgdr_t	Δgd_t	Δgd_t	Δy_t	Δy_t	Δgdr_t	Δgdr_t
c	2.83 (0.30)	−272.35 (−0.30)	13.54 (1.53)	−147.93 (−0.15)	14.11 (1.58)	327.31 (0.63)	28.46 (1.57)	−0.99 (−0.10)	14.21 (1.62)	−0.66 (−0.09)	−7.43 (−0.70)	11.69 (0.97)
ecm_{t-1}	−5.84*** (−1.66)	−2.60 (−0.14)	−3.58 (−1.09)	−2.87 (−0.14)	7.55** (2.27)	17.84*** (1.67)	−6.50 (−0.98)	12.82** (−2.37)	−8.51** (−2.65)	−13.01* (−3.30)	4.82 (1.24)	4.98 (0.77)
Δgd_{t-1}	−0.08 (−0.32)	0.90 (0.95)	0.13 (0.59)	0.20 (0.20)	0.01 (0.03)	−0.22 (−0.42)	−0.26 (−0.90)	−0.82** (−2.74)	0.26*** (1.90)	−0.05** (−2.31)	0.16 (0.94)	−0.44 (−1.21)
Δy_{t-1}	−0.02 (−0.05)	2.67 (0.30)	−0.04 (−0.14)	1.42 (0.15)	0.18 (0.62)	−2.12 (−0.43)	1.21 (1.37)	−0.00 (−0.00)	−0.34 (−0.80)	−0.14 (−0.23)	−0.60 (−1.16)	0.07 (0.07)
Δim_{t-1}	−0.05 (−0.28)	−2.02 (−0.29)	−0.17 (−0.94)	−1.01 (−0.14)	−0.06 (−0.33)	2.36 (0.60)	−0.57 (−1.06)	0.98** (2.35)	0.03 (0.11)	0.33 (1.08)	0.80** (2.58)	−0.00 (−0.01)
Δex_{t-1}	−0.03 (−0.13)	3.56 (−0.29)	0.52** (2.44)	1.18 (0.11)	−0.03 (−0.15)	−6.34 (−1.14)	−0.40 (−0.59)	−0.39 (−1.18)	−0.07 (−0.21)	0.18 (0.75)	−0.13 (−0.32)	0.09 (0.22)
$(I/Y)_{t-1}$	0.00 (0.01)	8.16 (0.29)	−0.49 (−1.52)	4.84 (0.16)	−0.38 (−1.17)	−9.84 (−0.57)	−1.26 (−1.35)	−0.59 (−1.16)	−0.33 (−0.73)	−0.34 (−0.90)	0.66 (1.20)	−0.75 (−1.19)
Δgdr_{t-1}	−0.01 (−0.07)	−0.60 (−0.26)	−0.09 (−0.53)	−0.11 (−0.04)	0.23 (1.42)	1.60 (1.24)	0.23 (0.36)	1.01** (2.51)	0.31 (0.98)	0.72** (2.47)	−0.05 (−0.12)	0.13 (0.27)
$\bar{R}^2$	−0.04	−1.68	0.28	−5.01	0.13	0.61	−0.03	0.34	0.26	0.51	0.26	0.004
$Q(1)$	0.16 [0.69]	0.02 [0.89]	0.12 [0.72]	0.02 [0.89]	4.38 [0.04]	0.02 [0.88]	0.00 [0.97]	0.60 [0.44]	0.70 [0.40]	0.05 [0.82]	0.00 [0.97]	0.02 [0.89]
$Q(4)$	3.70 [0.45]	2.17 [0.70]	2.63 [0.62]	2.17 [0.70]	5.08 [0.28]	2.17 [0.70]	2.95 [0.57]	3.34 [0.50]	4.15 [0.39]	0.57 [0.97]	0.11 [1.00]	0.99 [0.91]
$H_0 : \Delta y_{t-1} \nrightarrow \Delta gd_t$	2.84 [0.24]	0.09 [0.96]					5.05*** [0.08]	5.62*** [0.06]				
$H_0 : \Delta gd_{t-1} \nrightarrow \Delta y_t$			1.76 [0.41]	0.05 [0.97]					9.79* [0.01]	14.52* [0.00]		
$H_0 : \Delta gdr_{t-1} \nrightarrow \Delta gd_t$	3.33 [0.19]	0.07 [0.97]					1.14 [0.57]	8.22** [0.02]				
$H_0 : \Delta gd_{t-1} \nrightarrow \Delta gdr_t$					5.26*** [0.07]	2.81 [0.14]					2.68 [0.26]	2.34 [0.31]
z	Δgdr_{t-d}						Δgdr_{t-d}					
d	1						2					
c^*	5%						1.5%					

Note: See Table V.

data. Prior studies largely cater to a bivariate Granger causality model that did not take other factors, such as capital, imports and exports into consideration. Thus, failure to include a cointegration relation and a rival's defense spending (arms race) can also render estimates biased. To overcome these problems, this paper first applies a six-variable VAR model to explore the relationships between defense spending and economic growth in the presence of an arms race for both countries. The results of the linear model indicates (i) China's spending growth leads economic growth, (ii) Taiwan's spending and economic growth exhibit feedback relations, and (iii) China's spending growth is found to Granger-cause Taiwan's spending growth.

In addition, this paper uses rival's spending growth (with a lag of d) as the threshold variable to explore causal relations. Of particular importance is the finding that there exists feedback relations (an arms race) when China's spending growth exceed the 3% threshold value. A comparison of results from using both CSY and SIPRI Yearbook data indicates that the majority of the statistical relation remains qualitatively unchanged. In particular, China's spending growth is found to lead (one-way causality) Taiwan's spending growth irrespective of which data bank is used. In a similar vein, the results from using both data banks indicate (i) China's spending growth Granger-causes economic growth and (ii) there exist significant feedback relations between Taiwan's economic and defense growth.

ACKNOWLEDGEMENTS

An earlier draft of this paper was presented at the Third Annual Conference on Empirical Economics held in Pu-Li, Taiwan on April 21, 2002. We are grateful to Professor Lee Torng-Her for his helpful comments. Views expressed in this paper reflect those of the authors, not their respective institutions. We would like to thank three reviewers and editor for detailed comments and suggestions. All remaining errors are our own.

References

Anderton, C. H. (1993) Arms race modeling and economic growth. In *Defense Spending and Economic Growth,* edited by J. E. Payne and A. P. Sahu. Boulder, CO: Westview Press.

Antonakis, N. (1997) Military expenditure and economic growth in Greece, 1960–1990. *Journal of Peace Research* **34**(1) 89–100.

Atesoglu, H. S. and Mueller, M. J. (1990) Defense spending and economic growth. *Defense Economics* **2**(1) 19–27.

Benoit, E. (1973) *Defense and Economic Growth in Developing Countries.* Boston: D. C. Heath & Co.

Benoit, E. (1978) Growth and defense in developing countries, *Economic Development and Cultural Change* **26** 271–80.

Biswas, B. (1993) Defence spending and economic growth in developing countries. In *Defense Spending and Economic Growth,* edited by J. E. Payne, and A. P. Sahu, Boulder, CO: Westview Press.

Biswas, B. and Ram, R. (1986) Military expenditure and economic growth in less developed countries: an augmented model and further evidence. *Economic Development and Cultural Change* **34** 361–372.

Chang, T., Fang, W., Wen, L.F. and Liu, C. (2000) Defense spending, economic growth and temporal causality: evidence from Taiwan and mainland China, 1952–1995. *Applied Economics* **33**(10) 1289–1299.

Chowdhury, A. R. (1991) A causal analysis of defense spending and economic growth. *Journal of Conflict Resolution* **35** 80–97.

Cohen, J. S., Stevenson, R., Mintz, A. and Ward, M. D. (1996) Defense expenditures and economic growth in Israel: the indirect link. *Journal of Peace Research* **33** 341–352.

Dakurah A. H., Davies, S. P. and Sampath, R. K. (2001) Defense spending and economic growth in developing countries: a causality analysis. *Journal of Policy Modeling* **23** 651–658.

Deger, S. (1986) *Military Expenditure in Third World Countries,* London: RKP.

Engle, R. and Granger, C. W. J. (1987) Cointegration and error correction: representation, estimation and testing. *Econometrica* **35** 251–276.

Faini, R., Annez, P. and Taylor, L. (1984) Defense spending, economic structure, and growth: evidence among countries and over time. *Economic Development and Cultural Change* **32** 487–498.

Grobar, L. and Porter, R. (1989) Benoit revisited: defense spending and economic growth in less developed countries. *Journal of Conflict Resolution* **33** 318–345.

Heo, U. K. (1997) The political economy of defense spending in South Korea. *Journal of Peace Research* **34** 483–490.

Joerding, W. (1986) Economic growth and defense spending. *Journal of Development Economics* **21** 35–40.

Johansen, S. (1988) Statistical analysis of cointegration vectors. *Journal of Economic Dynamics and Control* **12** 231–254.

Johansen, S. and Juselius, K. (1990) Maximum likelihood estimation and inference on cointegration- with applications to the demand for money. *Oxford Bulletin of Economics and Statistics* **52** 169–210.

Khilji, N. M. and Mahmood, A. (1997) Military expenditures and economic growth in Pakistan. *The Pakistan Development Review* **36**(4II) 791–808.

Kollias C. and Paleologou, S. (2002) Is there a Greek–Turkish arms race? Some further empirical results from causality tests. *Defense and Peace Economics* **13**(4) 321–328.

Lacivita, C. J. and Frederiksen, P. C. (1991) Defense spending and economic growth, an alternative approach to the causality issues. *Journal of Development Economics* **35** 117–126.

Lim, D. (1983) Another look at growth and defense in less developed countries. *Economic Development and Cultural Change* **31** 377–384.

Ram, R. (1986) Government size and economic growth: a new framework and some evidence from cross section and time series data. *American Economic Review* **76**(1) 191–203.

Smith, R. (1980) Military expenditure and investment in OECD countries 1954–73. *Journal of Comparative Economics* **4** 19–32.

Tong, H. (1978) On a threshold model. In *Pattern Recognition and Signal Processing*, edited by C. H. Chen Amsterdam: Sijthoff & Noordhoff, 101–141.

Tong, H. and Lim, K. S. (1980) Threshold autoregressive, limit cycles, and data. *Journal of the Royal Statistical Society* B42 245–292 (with discussion).

Tsay, R. S. (1989) Testing and modeling threshold autoregressive process. *Journal of American Statistical Association* **84**(405) 231–240.

Tsay, R. S. (1998) Testing and modeling multivariate threshold models. *Journal of American Statistical Association* **93**(443) 1188–1202.

Waller, D. (1997) Estimating non-transparent military expenditures: the case of China (PRC). *Defense and Peace Economics* **8** 225–241.

Weise, C. L. (1999) The asymmetric effects of monetary policy: a nonlinear vector autoregression approach. *Journal of Money, Credit, and Banking* **31**(1) 85–108.

Wolfson, M. (1990) Perestroika and the quest for peace. *Defense Economics* **1**(3) 221–232.

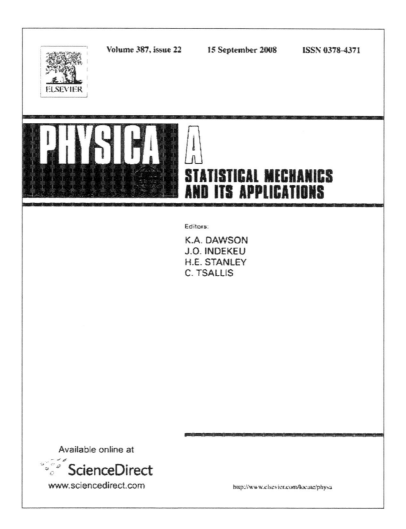

Volume 387, issue 22   15 September 2008   ISSN 0378-4371

ELSEVIER

PHYSICA A

STATISTICAL MECHANICS
AND ITS APPLICATIONS

Editors:
K.A. DAWSON
J.O. INDEKEU
H.E. STANLEY
C. TSALLIS

Available online at
ScienceDirect
www.sciencedirect.com

http://www.elsevier.com/locate/physa

265

Physica A 387 (2008) 5535–5542

Contents lists available at ScienceDirect

Physica A

journal homepage: www.elsevier.com/locate/physa

Tourism development and economic growth – a nonlinear approach

Wan-Chen Po [a], Bwo-Nung Huang [b],*

[a] Department of International Business, Kao Yuan University, Kaohsiung County 821, Taiwan
[b] Department of Economics & Center for IADF, National Chung Cheng University, Chia-Yi 621, Taiwan

ARTICLE INFO

Article history:
Received 11 February 2008
Received in revised form 15 April 2008
Available online 29 May 2008

Keywords:
Tourism development
Threshold regression
Degree of tourism specialization

ABSTRACT

We use cross sectional data (1995–2005 yearly averages) for 88 countries to investigate the nonlinear relationship between tourism development and economic growth when a threshold variable is used. The degree of tourism specialization (q_i, defined as receipts from international tourism as a percentage of GDP) is used as the threshold variable. The results of the tests for nonlinearity indicate that the 88 countries' data should be separated into three different groups or regimes to analyze the tourism-growth nexus. The results of the threshold regression show that when the q_i is below 4.0488% (regime 1, 57 countries) or above 4.7337% (regime 3, 23 countries), there exists a significantly positive relationship between tourism growth and economic growth. However, when the q_i is above 4.0488% and below 4.7337% (regime 2, 8 countries), we are unable to find evidence of such a significant relationship. Further in-depth analysis reveals that relatively low ratios of the value added of the service industry to GDP, and the forested area per country area are able to explain why we are unable to find a significant relationship between these two variables in regime 2's countries.

© 2008 Elsevier B.V. All rights reserved.

1. Introduction

The importance of tourism in the economic development of many countries is well documented and the contribution of tourism to an economy has long been a subject of great interest from a policy perspective. According to statistics compiled by the World Tourism Organization (WTO), international tourism receipts in 2003 accounted for approximately 6% of worldwide exports of goods and services (as expressed in US$). This figure is also expected to increase as more and more nations adopt tourism-oriented policies. In the meantime, the WTO's Web page also points out that when global economic growth exceeds 4%, the growth of tourism will tend to be higher. When GDP growth falls below 2%, the growth of tourism will tend to be even lower. However, Palmer and Riera [20] have pointed out that, while tourism may have many beneficial effects, we should not overlook the existence of a number of tourism-related economic, social and environmental costs that are normally not calculated and that should be taken into account when estimating the true social benefits of tourism. Usually the most important economic cost is the opportunity cost, due to the fact that the resources could be invested in other areas of production. Other costs include increased public expenditure which is caused by mass tourist arrivals due to the need to provide infrastructure and public services. There are also the cultural and environmental repercussions of tourism, and these costs may be regarded as externalities. Of course, whether or not the development of tourism will bolster the economy will depend on whether the benefits outweigh the negative externalities from tourism. In other words, the development of tourism does not always bring about economic growth. It is possible that the increase in tourism may or may not increase the economic growth.

---

* Corresponding author.

 E-mail addresses: dcbor@cc.kyu.edu.tw (W.-C. Po), ecdbnh@ccu.edu.tw (B.-N. Huang).

0378-4371/$ – see front matter © 2008 Elsevier B.V. All rights reserved.
doi:10.1016/j.physa.2008.05.037

The purpose of this paper is to use threshold regression to identify under which conditions the growth of tourism has had a positive impact on economic growth. By using the 1995–2005 yearly average cross sectional data for 88 countries and applying Barro's [5] growth equation and choosing the degree of tourism specialization (q_i, defined as international tourism receipts as a percentage of GDP) as the threshold variable, the results of our three-regime threshold regression reveal that when the q_i is below 4.0488% (regime 1, 55 countries) or above 4.7337% (regime 3, 23 countries), there exists a significantly positive relationship between tourism and economic growth. However, when the q_i is above 4.0488% and below 4.7337% (regime 2, 8 countries), we are unable to find a significant relationship between tourism and economic growth. The major contributions of this paper are as follows. Firstly, to the best of our knowledge, this is the first paper to apply the nonlinear model to a discussion of the tourism-growth nexus. Secondly, we are the first to point out that there exists a nonlinear relationship between tourism growth and economic growth when using the q_i as the threshold variable. Finally, we also find that tourism does not necessarily bring about economic growth under certain conditions.

This paper is organized as follows. Section 1 discusses the motivation behind the paper. Section 2 reviews the related literature. Section 3 introduces the data and econometric methods. Section 4 then analyzes and discusses the empirical results. The final section, Section 5, provides the concluding remarks.

2. Literature review

The Granger causality test has been widely used in the literature in analyzing the relationship between tourism and economic growth. Balaguer and Cantavella-Jorda [4] used a trivariate model of real GDP, real international tourism earnings, and the real effective exchange rate to discuss the relationship between tourism and economic growth. By taking the stationary and cointegration properties of the data into consideration, and using quarterly data for Spain for the period from 1975:Q1 to 1997:Q1, their results reveal that there is a cointegrating relationship between tourism and economic growth. This indicates that tourism positively affects Spanish economic growth. In addition, their results for the Granger causality test indicate that tourism affects Spain's economic growth unidirectionally. Therefore, their empirical results support the tourism-led growth hypothesis. Oh [19] adopted the Engle & Granger two-stage approach and a bivariate (real aggregate tourism receipts and real GDP) VAR model. Using Korean quarterly data from 1975:Q1 to 2001:Q1, the results of Oh's cointegration test indicate that there is no long-run equilibrium relationship between these two series. In addition, the results of Oh's Granger causality test imply the existence of a one-way causal relationship in terms of economic-driven tourism growth. The hypothesis of tourism-led economic growth, therefore, is not held in the Korean economy. Using annual data from 1963 to 2002, Gunduz and Hatemi-J [12] examined the role of the tourism-led economic growth hypothesis in Turkey. Instead of using traditional Granger causality, they used leveraged bootstrap causality for the reason that the data might exhibit non-normality and ARCH effects. They found that the tourism-led growth hypothesis is supported empirically in the case of Turkey. Dritsakis [9] examined the impact of tourism on the long-run economic growth of Greece using a similar method. One cointegrating vector was found among GDP, the real effective exchange rate and international tourism earnings from 1960:Q1~2000:Q4. Granger causality tests using the error correction model (ECM) indicated that there is a strong feedback relationship between tourism and economic growth. Unlike the previous literature, which only used bivariate or trivariate models, Durbarry [10] constructed a five-variable (including capital stock, human capital, real exports of sugar, real manufactures of exports and real tourism receipts) ECM to discuss the impact of tourism on Mauritius' economic development. Using 1952–1999 annual data, his results supported the view that tourism has promoted growth. Kim et al.. [16], on the other hand, continued to use a bivariate ECM to discuss the relationship between tourism and economic development in Taiwan. Using quarterly data from 1971–2003, their results for the Granger causality test suggest that there is a bi-directional causality between tourism and economic growth. Therefore, when examining Granger causality, the results of previous research appear to support the view that tourism leads economic growth.

However, the use of the Granger-causality approach in investigating the causal relationship between tourism and economic growth in the earlier literature has given rise to three possible problems: (a) whether or not the yearly data are sufficient to represent the long-term relationship between the two; (b) the inability of the yearly data to eliminate the problems of short-term fluctuations due to business cycles and structural change; and (c) the failure to delineate countries with special features in terms of different causal relationships.

Since the relationship between tourism and economic growth is inherently a long-term one, a biased estimate may be the result of an insufficiently large sample size in the time series, the existence of structural changes, or short-term economic fluctuations. To tackle the insufficient sample size problem, researchers have started to use panel data. Eugenio-Martin et al. [11] used panel data for 21 Latin American countries from 1985 until 1998. By applying the Arellano and Bond [3] GMM estimator for dynamic panels, their results reveal that growth in the number of tourists per capita is associated with low (3 countries) and medium (11 countries) levels of income per capita, but that there is a negative relationship between the two in high income (7 countries) Latin American countries. Sequeria and Campos [22] used the data derived from the Penn World Table and the World Development Indicators and considered four five-year periods between 1980 and 1999. They chose a broad sample of 509 observations and several smaller samples: Specialization in Tourism 1 (where receipts from tourism represent 10% or more of exports); Specialization in Tourism 2 (where receipts from tourism represent 20% or more of exports); Islands; Small countries; Rich countries; Poor countries; African countries; Asian countries; Latin American countries and European countries. By using tourism receipts as a percentage of exports and tourism receipts as a percentage of GDP as proxy variables for tourism, their results from pooled OLS, random effects and fixed effects estimators indicate

267

that growth in tourism is associated with economic growth only in African countries. In fact, a negative relationship is found between tourism and economic growth in Latin American countries, and in Specialization in Tourism 1 and 2 countries. Sequeria and Campos did not find a significant relationship between tourism and economic growth in the rest of the groups. Although the panel data approach fixed the problems of bias due to insufficient sample sizes, the problem associated with the failure to delineate countries with special features in terms of different relationships between tourism and growth remains unsolved. By grouping the panel data on the basis of per capita income, geographical groups or other specific characteristics, the problems associated with the failure to consider the heterogeneity among countries have only been partially solved. With sufficient time series data, a more appropriate approach to tackle the heterogeneity problem is to use the heterogeneous panel technique. By applying Pedroni's [21] fully-modified OLS for heterogenenous cointegrated panels, Lee and Chang[17] re-investigated the long-run comovements and causal relationships between tourism development and economic growth for OECD and non-OECD countries over the 1990–2002 period. Their results indicate that a cointegrated relationship between GDP and tourism development is substantiated. They also found that tourism development has a greater impact on GDP in non-OECD countries than in OECD countries, and that when the variable is tourism receipts, the greatest impact is in Sub-Saharan African countries. Their panel causality test reveals that there are unidirectional causality relationships from tourism development to economic growth in OECD countries, bidirectional relationships in non-OECD countries, and only weak relationships in Asian countries.

Although the use of either panel or heterogeneous panel data fixes the problems of insufficient sample sizes and takes into consideration the heterogeneity among countries or between groups, the problem of biased estimates remains unsolved. The bias may arise due to the inadequate representation of a long-run relationship, the inability to allow for structural change, or the presence of short-term fluctuations arising from business cycles when using annual data. In order to tackle these problems, many researchers employ average values over time for each cross-sectional unit.[1]

By using the cross-sectional data for 143 countries covering the period 1980–2003, Brau et al. [6] analyzed the empirical relationship between growth, country size and tourism specialization. By defining "small countries" as countries with an average population of less than one million during 1960–2003 and an average degree of tourism specialization (DTS, the average ratio of international tourism receipts to GDP) greater than 10% over the period 1980–2003, they used dummy regression analysis to compare the growth performance of small tourism specialization countries relative to the performance of a number of sub-sets of countries, namely, OECD, Oil, Small and LDCs. Their results revealed that small states are fast-growing only when they are highly specialized in tourism. Although Brau et al. [6] used cross-sectional data to fix the problems associated with the inability to allow for structural change, as well as the presence of short-term fluctuations from business cycles when using annual time series data, their selection of threshold variables (country size and DTS) was somewhat ad hoc.

To tackle the inability of the annual data to represent the long-run relationship between tourism and economic growth and to eliminate the short-term fluctuations due to business cycles and structural change, this paper intends to use cross-sectional data to study the nonlinear relationship between tourism and growth. We believe that there should be certain characteristics that separate the countries into different groups for each of which there should or should not exist different tourism-growth nexuses (e.g., small countries are used in Ref. [6]). Therefore we will use a threshold regression model originating from Refs. [24,23] to discuss the possible nonlinear relationship between tourism and growth in this paper.

3. Model specification, econometric method, and data

In common with the frequently-used growth models in the literature, for instance that of Ref. [5],[2] let the ith country's yearly average economic growth be Δy_i, the yearly average per capita international tourism real receipts growth be Δkte_i and other explanatory economic growth variables be referred to as x_i. In the literature, the x_i include inflation π_i, the proxy variable for the capital stock (gross fixed capital formation as a percentage of GDP, I/Y), and the exchange rate ($exch_i$). The exchange rate is the nominal exchange rate defined as local currency/$1 US dollar; therefore, an increase in the exchange rate denotes a depreciation in the local currency and vice versa. The output growth equation, which includes tourism growth, is shown below:

$$\Delta y_i = \alpha_0 ly0_i + \alpha_1 \Delta kte_i + \beta x_i + \varepsilon_i, \tag{1}$$

where Δ is the first difference operator, y_i is the logarithm of per capita real GDP for the ith country; $ly0_i$ represents the logarithm of initial income for the ith country, and kte_i denotes the log of per capita real tourism receipts (in US$). The purpose of this paper is to apply the threshold regression model to investigate the difference in the relationship between tourism and economic growth using certain threshold variables. The threshold regression model originated from the threshold autoregressive model in the time series model developed by Tong [24] and Tong and Lim [23]. By denoting

---

[1] For studies on the use of cross-sectional data using annual averages, the reader should refer to Ref. [15] on the relationship relationship between inflation and economic growth and to Ref. [18] on the relationship between banking and economic growth.

[2] The interested reader should refer to Ref. [5] for the growth-related models, to Ref. [18] for the relationship between monetary development and economic growth, to Ref. [1] for the relationship between defense spending and economic growth, and to Ref. [15] for the relationship between inflation and economic growth.

a threshold variable as q_i (in this paper, we will use the degree of tourism specialization as the threshold variable)[3] and an optimal threshold value as c_1^*, the two-regime threshold equation can be expressed as

$$\Delta y_i = (\alpha_{10}ly0_i + \alpha_{11}\Delta kte_i + \beta_{11}\pi_i + \beta_{12}(I/Y)_i + \beta_{13}exch_i) \cdot I(q_i \leq c_1^*)$$
$$+ (\alpha_{20}ly0_i + \alpha_{21}\Delta kte_i + \beta_{21}\pi_i + \beta_{22}(I/Y)_i + \beta_{23}exch_i) \cdot I(q_i > c_1^*) + \varepsilon_i, \tag{2}$$

where ε_i is assumed to be i.i.d. and to follow the white noise process. I (.) is an indicator function. If the relationship in (.) is present, then I (.) is 1, otherwise, I (.) is 0. Eq. (2) is a simple two-regime model delineated by the value of a threshold variable. It indicates that the relationship between the explanatory variables and economic growth is represented by $B_1 = (\alpha_{10}, \alpha_{11}, \beta_{11}, \beta_{12}, \beta_{13})$when q_i (the threshold variable) is less than or equal to a threshold value c_1^* (regime 1), but by $B_2 = (\alpha_{20}, \alpha_{21}, \beta_{21}, \beta_{22}, \beta_{23})$ when q_i is greater than the threshold value c_1^* (regime 2). Before estimating Eq. (2), it is necessary to examine the existence of the nonlinear relationship. Only after the null hypothesis ($H_0 : B_1 = B_2$)or linear relationship is rejected can we estimate Eq. (2). Given the fact that the threshold value (c) is typically unknown, the traditional F-test is not appropriate for testing the null hypothesis. Under the assumption that the error term is i.i.d., Hansen [14] suggests employing a test method derived from Refs. [7,8,2], namely, a test with near-optimum power against an alternative distance from the null hypothesis. It takes the form of a standard F statistic:

$$F = n\left(\frac{\tilde{\sigma}_n^2 - \hat{\sigma}_n^2}{\hat{\sigma}_n^2}\right), \tag{3}$$

in which $\tilde{\sigma}_n^2$ is the residual variance of Eq. (1) and $\hat{\sigma}_n^2$ is the residual variance of Eq. (2) under c_1^*. However, the asymptotic distribution of Eq. (3) does not follow an F distribution since c is unknown. Hansen [13] circumvents the problem by using a bootstrap method to produce an asymptotic distribution for testing.[4]

Once Eq. (3) is employed, we may be able to reject the null hypothesis. That is, there is a nonlinear relationship for the data where q_i is used as a threshold variable as shown in Eq. (2). First, the data need to be arranged in ascending order in q_i in order to estimate Eq. (2). To accommodate the degrees of freedom problem, the smallest and largest 15% of observations are removed. The remaining sample space bounded by threshold values $[\underline{c}, \bar{c}]$ is partitioned into n grids. For each c_i ($i = 1, \ldots, n$), we obtain a $\hat{\sigma}(c_i)$. Next, we select the smallest $\hat{\sigma}(c_i)$ to be the optimal threshold value c_1^* via Eq. (2):

$$c_1^* = \arg\min\hat{\sigma}(c_i)^2, \quad i = 1, \ldots, n$$
$$c_1^* \in [\underline{c}, \bar{c}] \tag{4}$$

where $\hat{\sigma}(c_i)^2$ denotes the residual variance from Eq. (2) under the threshold value c_1^*. If the data exhibit a second threshold level (c_2^*), the three-regime growth model is shown as

$$\Delta y_i = (\alpha_{10}ly0_i + \alpha_{11}\Delta kte_i + \beta_{11}\pi_i + \beta_{12}(I/Y)_i + \beta_{13}exch_i) \cdot I(q_i \leq c_1^*)$$
$$+ (\alpha_{20}ly0_i + \alpha_{21}\Delta kte_i + \beta_{21}\pi_i + \beta_{22}(I/Y)_i + \beta_{23}exch_i) \cdot I(c_1^* < q_i \leq c_2^*)$$
$$+ (\alpha_{30}ly0_i + \alpha_{31}\Delta kte_i + \beta_{31}\pi_i + \beta_{32}(I/Y)_i + \beta_{33}exch_i) \cdot I(q_i > c_1^*) + \varepsilon_i. \tag{5}$$

To test the existence of the three regimes versus the two regimes, the null hypothesis is set up as H_0 : two-regime versus three-regime. The test statistic is similar to the F-statistic in Eq. (3) with the exception that $\tilde{\sigma}_n^2$ is the residual variance of Eq. (1) and $\hat{\sigma}_n^2$ is the residual variance of Eq. (5). Since the threshold level is unknown and the asymptotic F-statistic cannot be derived from the χ^2 distribution, a bootstrap procedure is needed to test the hypothesis.

As to which variable should be chosen as the threshold variable, Brau et al. [6] used the degree of tourism specialization (q_i), defined as the average ratio of international tourism receipts to GDP, as the cut-off point to select the targeted countries to be studied. They considered countries with an average q_i greater than 10% over the period 1980–2003 in their study without providing any statistical justification. However, using the degree of tourism specialization as the threshold variable (q_i) seems appropriate. As a country concentrates more on tourism, its economy will depend more on the receipts from tourism. It is, therefore, expected that if the degree of tourism specialization is higher than a certain threshold level, there will be a significant relationship between tourism growth and economic growth.

The annual data for 88 countries covering the period from 1995 to 2005 have been collected from the WDI Database published by the World Bank. The data set includes inflation (π), a capital stock proxy variable (I/Y), per capita real (in 2000 USD) GDP (Y), per capita tourism real receipts (kte), and the exchange rate ($exch$).

To investigate whether there exists a nonlinear long-term relationship between tourism and economic growth, one set of cross-sectional data encompassing yearly average data (1995–2005 average) is used based on Eq. (1). We use Eq. (2) to decipher the relationship under the two-regime or Eq. (5) to decipher the relationship under the three-regime model depending on the nonlinearity test's result from Eq. (3).

---

[3] We will explain why the degree of tourism specialization is an appropriate threshold variable later.

[4] See Ref. [14, p. 6] for the bootstrap procedure.

Table 1
Test results of the nonlinear threshold model

	F	Threshold value
H_0 : 1 vs. 2	5.15 [0.01]	$c_1 = 4.0488$
H_0 : 2 vs. 3	2.68 [0.05]	$c_1 = 4.0488$ $c_2 = 4.7337$

Notes: H_0:1 vs. 2 or linear vs. 2-regime; H_0: 2 vs. 3 or 2-regime vs. 3-regime; c_1, c_2 represent first and second threshold levels, respectively. Numbers inside [.] denote p-values. The threshold variable is DTS (international tourism receipts as a percentage of GDP).

Table 2
Estimated results of the linear and nonlinear regressions

q_i = DTS	Linear (1)	Three regimes (2)		
	R1	R1	R2	R3
	Δy_i	Δy_i	Δy_i	Δy_i
$ly0_i$	-0.5387 (−0.55)	−0.3213* (−3.08)	1.3980* (4.74)	0.1192 (1.49)
Δkte_i	0.1245* (3.90)	0.1154* (3.07)	0.1261 (1.56)	0.1243* (4.32)
π_i	0.0229 (0.63)	0.0027 (0.09)	0.2376* (3.83)	0.0576* (2.07)
$(I/Y)_i$	0.1086* (2.95)	0.2330* (5.45)	−0.4175* (−2.15)	0.0305 (1.20)
$exch_i$	-0.0002 (-0.49)	−0.0010* (−4.04)	−0.0445* (−3.86)	0.0002 (1.64)
N	88	57	8	23
s.e.	1.8107	1.5677		
White-F test	42.19 [0.00]	45.18 [0.06]		

Notes: R1= regime 1; R2= regime 2; R3= regime 3; $ly0_i$= initial income; Δkte_i= per capita international tourism real receipts growth, (I/Y)= gross fixed capital formation / GDP; π = inflation; $exch$= exchange rate (local currency/ USD); Δy_i is per capita GDP growth; * represents 1% significance level; values inside (.) are t-values; values inside [ ] are p-values; s.e.= standard errors of estimate; and White-F= F statistic using the White heteroskedasticity test.

4. Empirical results and discussion

Before applying the threshold regression, it is necessary to test for the existence of the nonlinear relationship in terms of the threshold variable (i.e. the q_i) for the 88 countries. Table 1 displays the nonlinear test results.

Based on Table 1, when the q_i is used as a threshold variable, we reject the one-regime model at the 1% level and reject the two-regime model at the 5% level. This indicates that there is a three-regime relationship. That is, the 88 countries are grouped into three regimes for analysis based on the threshold levels of $q_i = 4.0488\%$ and 4.7337%.

Table 2 displays both the linear and nonlinear results estimated using the q_i as the threshold variable. From column (1) or the linear model, it can be seen that there is a positive significant relationship between tourism growth and economic growth. This result is in line with the previous result in which a unidirectional Granger causality running from tourism growth to economic growth was found. The estimated results for the other explanatory variables indicate that there is a positive relationship between capital and economic growth, but no significant relationship between other explanatory variables (i.e., π_i, $exch_i$ and $ly0_i$). It is thus clear that during the 1995–2005 period, the most important factors affecting economic growth were tourism growth and capital stock in our 88 countries. Finally, according to the White F test results, there is a heteroskedasticity problem with the linear model. We therefore use White's heteroskedasticity-consistent covariance to correct this problem.

With respect to the nonlinear estimation when q_i is used as the threshold variable, our test results indicate that we cannot reject the null hypothesis of a three-regime relationship. If the q_i variable (column (2)) is less than 4.0488% (57 countries, regime 1), a 1% increase in per capita real tourism receipt growth may contribute to a 0.1154% increase in economic growth. If the q_i variable is greater than 4.0488% and less than 4.7337% (regime 2, 8 countries), there is no significant relationship between tourism growth and economic growth. However, when the q_i variable exceeds 4.7337% (23 countries, regime 3), there is also a significant positive relationship between tourism growth and economic growth.[5] In regime 3, a 1% increase in per capita real tourism receipt growth may bring about a 0.1243% increase in economic growth.

---

[5] See Appendix Table 1 for the names of the countries for each regime.

Table 3
Mean and standard deviation of some tourism-related data characteristics in three regimes

Regime	ky	Area	Service %	Forest %
1	7726.30 (10769.37)	1287 430 (2527 734)	55.3957 (11.5782)	31.10 (19.83)
2	5372.33 (10265.55)	218 165 (240 612)	49.8621 (7.7580)	25.03 (19.07)
3	4764.67 (5620.93)	224 759 (282 235)	57.5271 (11.6667)	36.70 (23.13)

Note: ky = per capita real GDP; area = country area in km^2; service % = % of value added in services to GDP; forest % = forest area as percentage of country area; values inside (.) are standard deviations.

In both regime 1 and regime 3, although tourism was found to have a significantly positive relationship with economic growth, the effects of the growth of tourism on economic growth were not the same. From column (2) of Table 2, we saw that the positive effect in regime 3 was higher and more significant than in regime 1. A 1% increase in tourism was found to account for a 0.1154% increase in economic growth in regime 1, while the same 1% increase in tourism accounted for a 0.1243% increase in output in regime 3. In addition, we also observed that there existed no significant relationship between tourism and economic growth in regime 2. This indicates that for those 8 countries (Botswana, Honduras, Georgia, Moldova, Kenya, Iceland, Ghana, and Trinidad & Tobago) with a q_i of between 4.0488%–4.7337%, an expansion in the tourism industry would not bring about economic growth. Unlike most previous research that found a significant relationship between tourism and growth, by using a nonlinear model we were able to find a regime in which there was no significant relationship between the two. To further analyze why some countries could promote economic growth through tourism while other countries could not, we collected data on per capita real GDP, country area, the percentage of value added of the service industry to GDP, and forest area as a percentage of country area (a proxy for tourism resources). Table 3 displays the means and standard deviations of these variables for each of the three different regimes.

Based on the country area and per capita income data in Table 3, it seems that the regime 1 countries had the largest average area and highest average per capita real income among the three regimes. This implies that 58 countries in regime 1 were large and wealthy countries. On the other hand, countries in regime 2 and regime 3 were medium to small in size and were relatively low income countries. Although the average country size and per capita real income in regime 2 were about the same as for the countries in regime 3, the standard deviation of the average per capita income in regime 2 was higher than that for the countries in regime 3 ($10,266 vs. $5,621). In other words, the 8 countries in regime 2 where there was no significant relationship between tourism and economic growth were the countries that were medium to small in size and that had a dispersed per capita real income. In addition, if we further took the average data for services/GDP and forest area/country area into consideration, we found that these two figures (49.86% and 25.03%, respectively) in regime 2 were the lowest among the three regimes. This seems to suggest that in a country that was medium to small in size, with relatively low income, a low services/GDP ratio, and a low forest area to country area ratio, there tended to be no significant relationship between tourism and economic growth. On the contrary, in a country with a high services/GDP ratio and a high forest area to country area regardless of its size and income level, the promotion of tourism was able to increase economic growth (regime 1 and regime 3). However, there was a fundamental difference between regime 1 and regime 3, i.e., in large countries with high income (regime 1), capital stock also played an important role in promoting economic growth, while in the small and low income countries, tourism was the only source of economic growth according to Table 2.

5. Conclusion

It is believed that the tourism industry not only can increase foreign exchange income, but can also create employment opportunities. Therefore, tourism can be used to stimulate overall economic growth. Hence, the question of whether or not tourism can lead economic growth has become an important empirical issue. While much research has already been done on this issue, the use of the Granger causality approach to investigate the causal relationship between tourism and economic growth in the previous literature led to three possible problems: (a) whether or not the yearly data were sufficient to represent the long-term relationship between the two; (b) the inability of the yearly data to eliminate the problems of short-term fluctuations due to business cycles and structural change; and (c) the failure to delineate countries with special features in terms of different causal relationships. To tackle these problems, we used the cross sectional data (1995–2005 yearly average data) for 88 countries to investigate the nonlinear relationship between the development of tourism and economic growth when the threshold variable was used. We also chose the degree of tourism specialization (q_i, defined as international tourism receipts as a percentage of GDP) as the threshold variable. The results of the nonlinearity test indicated that we should separate the data for the 88 countries into three different groups to analyze the tourism-growth nexus. The results of the threshold regression showed that when the q_i was below 4.0488% (regime 1, 55 countries) or above 4.7337% (regime 3, 23 countries), there existed a significantly positive relationship between tourism growth and economic growth. However, when the q_i was above 4.0488% and below 4.7337% (regime 2, 8 countries), we were unable to find a significant relationship between tourism and economic growth. Further in-depth analysis revealed that a relatively low ratio of value

271

Table A.1
Names of the countries under each regime

Regime-1 (57)		Regime-2 (8)	Regime-3 (23)
Japan	Mexico	Botswana	Portugal
Nigeria	Azerbaijan	Honduras	Spain
Brazil	Kazakhstan	Georgia	Egypt, Arab Rep.
Iran, Islamic Rep.	Bolivia	Moldova	Panama
Venezuela, RB	Mali	Kenya	Slovenia
Pakistan	Sweden	Iceland	Tanzania
Estonia	Netherlands	Ghana	Greece
Togo	United Kingdom	Trinidad and Tobago	Austria
Kuwait	Ukraine		Malaysia
Belarus	Guatemala		Morocco
United States	Italy		Thailand
Romania	Latvia		Albania
Germany	Philippines		Costa Rica
Cameroon	Armenia		Cambodia
Argentina	Kyrgyz Republic		Bulgaria
China	Slovak Republic		Guyana
Colombia	Zambia		Tunisia
Macedonia, FYR	South Africa		Cape Verde
Korea, Rep.	Sri Lanka		Benin
Peru	Ethiopia		Jordan
Paraguay	Australia		Mauritius
Norway	Madagascar		Croatia
Canada	El Salvador		Seychelles
Chile	Switzerland		
Finland	Poland		
Ecuador	Lithuania		
Malawi	Swaziland		
Gabon	Senegal		
	Uruguay		

added of the service industry to GDP and forest area per country area (as a proxy for tourism resources) could explain why we were unable to find a significant relationship between these two variables in regime 2's countries.

What are the policy implications derived from this study? In light of our results, the main implication is that it is not necessary for all countries to benefit through the promotion of tourism growth. In countries with a degree of tourism specialization of between 4.7337% and 4.0488% and which have a relatively low ratio of value added of the service industry to GDP, an increase in tourism is unable to increase economic growth. Hence, our results suggest that a policy other than tourism in promoting economic growth seems more suitable for this group of countries. As to the other two groups of countries, the development of the tourism industry should be a part of their economic growth strategy.

Acknowledgments

The valuable comments from the editor and the referees are greatly appreciated.

Appendix

See Table A.1

References

[1] J. Aizman, R. Glick, Military expenditure, threats and growth, NBER Working paper no. 9618, 2003.
[2] D.W.K. Andrews, W. Ploberger, Optimal tests when a nuisance parameter is present only under the alternative, Econometrica 62 (1994) 1383–1414.
[3] M. Arellano, S.R. Bond, Some tests of specification for panel data: Monte Carlo evidence and an application to employment equations, Review of Economic Studies 58 (1991) 277–297.
[4] J. Balaguer, M. Cantavella-Jorda, Tourism as a long-run economic growth factor: The Spanish case, Applied Economics 34 (2002) 877–884.
[5] R.J. Barro, Government spending in a simple model of endogenous growth, Journal of Political Economy 98((5) (1990) 103–126.
[6] R. Brau, A. Lanza, F. Pigliaru, How fast are small tourism countries growing? The 1980-2003 evidence, Milan, Italy, Fondazione Eni Enrico Mattei Nota di Lavoro, No. 1, 2007.
[7] R.B. Davies, Hypothesis testing when a nuisance parameter is present only under the alternative, Biometrika 64 (1977) 247–254.
[8] R.B. Davies, Hypothesis testing when a nuisance parameter is present only under the alternative, Biometrika 74 (1987) 33–43.
[9] N. Dritsakis, Tourism as a long-run economic growth factor: An empirical investigation for Greece using causality analysis, Tourism Economics 10 (2004) 305–316.
[10] R. Durbarry, The economic contribution of tourism in Mauritius, Annals of Tourism Research 29 (2002) 862–865.
[11] J. Eugenio-Martin, N. Morales, R. Scarpa, Tourism and economic growth in Latin American Countries: A panel data approach, Milan, Italy, Fondazione Eni Enrico Mattei Nota di Lavoro, No. 26, 2004.
[12] L. Gunduz, A. Hatemi-J, Is the tourism-led growth hypothesis valid for Turkey, Applied Economics Letters 12 (2005) 499–504.
[13] B.E. Hansen, Inference when a nuisance parameter is not identified under the null hypothesis, Econometrica 64 (1996) 413–430.

[14] B.E. Hansen, Inference in TAR models, Studies in Nonlinear Dynamics and Econometrics 2 (1) (1997) 1–14.
[15] M.S. Khan, A.S. Senhadji, Threshold effects in the relationship between inflation and growth, IMF Staff Papers 48 (1) (2001) 1–21.
[16] H.J. Kim, M.H. Chen, S.C. Jang, Tourism expansion and economic development: The case of Taiwan, Tourism Management 27 (5) (2006) 925–933.
[17] C.C. Lee, C.P. Chang, Tourism development and economic growth: A closer look at panels, Tourism Management 29 (2008) 180–192.
[18] R. Levine, S. Zervos, Stock markets, banks, and economic growth, American Economic Review 88 (3) (1998) 537–558.
[19] Chi-Ok. Oh, The contribution of tourism development to economic growth in the Korean economy, Tourism Management 26 (2005) 39–44.
[20] T. Palmer, A. Riera, Tourism and environmental taxes: With special reference to the Balearic ecotax, Tourism Management 24 (2003) 665–674.
[21] P. Pedroni, Full modified OLS for heterogeneous cointegrated panels, in: Nonstationary Panels, Panel Cointegration and Dynamic Panels, in: Advances in Econometrics, vol. 15, JAI Press, 2000, pp. 93–130.
[22] T.N. Sequeira, C. Campos, International tourism and economic growth: A panel data approach, Milan, Italy, Fondazione Eni Enrico Mattei Nota di Lavoro, No. 141, 2005.
[23] H. Tong, K.S. Lim, Threshold autoregressions, limit cycles, and data, Journal of the Royal Statistical Society B 42 (1980) 245–292. with discussion.
[24] H. Tong, On a threshold model, in: C.H. Chen (Ed.), Pattern Recognition and Signal Processing, Sijthoff & Noordhoff, Amsterdam, 1978, pp. 101–141.

IV. Indirect Granger Causality, Random Walk, Long-term Memory, Volatility of Stock Market and Other Econometric Models

Like traveling at a large city, causality may take a sudden detour. The paper "Industrial Output and Stock Price Revisited: An Application of Multivariate Indirect Causality Model " employs a five-variable vector autoregressive model to investigate if industrial output predicts (causes) directly stock price for Canada, France, Japan, Taiwan and the US. The five variables are stock return, industrial production, money supply, inflation and interest rate.

Although no direct causal relationship was found for any country, two indirect causalities manifested themselves conspicuously. For the US, industrial output Granger-caused interest rate, which in turn Granger-caused stock price. For Taiwan, the indirect causality took the form of the following: Industrial output Granger-caused money supply. This can be expected: more export leads or Granger-causes increased reserve currencies at the central bank, which controls money supply. As is well known, money supply Granger-caused stock price.

The proverbial question lingers on: do stock returns represent a random walk or they have long term memory? The paper of " The Fractal Structure in Multinational Stock Returns " sheds some light on the issue for the US , the UK and the following Asian economies: Hong Kong, Indonesia, Japan, South Korea, Malaysia, the Philippines, Singapore, Thailand and Taiwan. We employed modified rescaled range (modified R/S) model by Lo (1991). While the traditional R/S model is robust to the normality assumption, it is very sensitive to short term autocorrelation.

The results of using the modified R/S model indicated that (1) the UK exhibited non-periodical cyclical patterns regardless of whether daily, weekly or monthly data were used. (2) There existed long term memory in Taiwan, the Philippines and Indonesia when traditional R/S was applied. But only the Philippines had the long term memory when the modified R/S was used. No other economies reveal the pattern of long term memory, which more often than not can manifest itself in a slow-decaying autocorrelation function.

The Taiwan Stock Market is quite different from the Western markets largely due to different regulations (e.g., inter-day price limit was set at plus and minus 7%). We applied the modified R/S model to both Taiwan and New York Stock Exchange to explore the market differences. In the paper of " A Comparative Statistical Analysis of the Taiwan Market and the New York Stock Exchange Using 5- Minute Data " , we first found that the means and standard deviations of Taiwan stock returns were not U-shaped as was found by McInish and Wood (1991) in the US market. The standard deviation spiked 5 minutes after the market open, peaked around 9:30, fluctuated randomly starting at 10:30 and rebounded at market close in the Taiwan Stock Market.

Mean return of the Taiwan Market in a typical bull market started with a trough, but gradually climbed up for 20 minutes until it reached pinnacle at 10:00 where it fluctuates randomly. At 11:00, mean stock return declined and never rebounded for the day. In the case of a bear market, stock returns were found to be consistent with the efficient market hypothesis except for the first 30 to 60 minutes after the market open (holding period of one week). The efficient market hypothesis in the US is supported in majority of the time with the exception of NYSE in 1987 and NASDAQ in 1992 using 15-minute data.

In an important paper, Professor Huang (1995) investigated do Asian stock markets follow random walk hypothesis? The variance ratio test with the homoscedasticity and heteroscedasticity-consistent error variance by Lo and MacKinlay (1988) was used to test the random walk hypothesis on Asian markets. The results suggested the random walk hypothesis was rejected for the markets of South Korea and Malaysia for all different holding periods. Similarly, by using the heteroscedasticity-consistent variance ratio test, the random walk hypothesis was rejected as well.

Another aspect of the stock markets is the volatility of its price which creates opportunity for profit due to some undervalued asset prices. Since the financial market liberalization, it is generally believed the volatility of the market has increased. The paper " The Impact of Financial Liberalization on Stock Price Volatility in Emerging Markets " applied the generalized error distribution model by Box and Tiao (1973), Harvey (1981) and GARCH model by Bollerslev (1986) to ten emerging markets. It is well-known that financial liberalization gives rise to stock price volatility due to the presence of major international investors. The results reveal that the markets of South Korea, Mexico and Turkey experienced increased volatility whereas the markets of Argentina, Chile, Malaysia and the Philippines experienced diminished volatility. There are no discernible patterns in the rest of the emerging markets.

The 1997 Asian financial debacle changed currency regime fundamentally: from fixed to flexible exchange rate system. In the paper "An Analysis of Exchange Rate Linkage Effect: An Application of Multivariate Correlation Analysis ", we investigated changes in correlation coefficients with standard error from a bootstrapping method. We first applied the Phillips and Perron unit root procedure (1988) to reject the null hypothesis that the variable after taking logarithm is of I (1).The dynamic conditional correlation is calculated based on forecast error of the vector autoregressive model by

Den Han (2000). For the 1-10 day window periods, the Philippines witnessed 7 positive currency interactions, each of Thailand, Taiwan, Indonesia and Singapore witnessed at least 5 times, clearly speaking to the increased currency linkage effect.

What are the characteristics of the Taiwan Stock Market in which 92% of investors are individuals with limited fund? In another paper " An Empirical Investigation of Trading Volume and Return Volatility of the Taiwan Stock Market ", we examine the Mixed Distribution Hypothesis by using 5-minutes interval stock returns of the Taiwan Stock Index. It was found the persistence of stock volatility remained dominant even after the stochastic mixing variable was included in the variance equation. The stochastic mixing variable in this case is the rate of information arrival (Lamoureux and Lastrapes 1990). The result pointed to the interesting phenomenon: the Mixing Distribution Hypothesis could not explain the ARCH effect. The persistence of the return volatility in the Taiwan market could be attributed to the fact that the great majority of the investors (92%) were individual investors who paid little attention to fundamental analysis coupled with the fact that there is indeed a price ceiling on inter-day stock prices.

The paper "The Impact of Settlement Time on the Volatility of Stock Market Revisited: An Application of the Iterated Cumulative Sum of Squares Detection Method for Changes of Variance " examined an important issue: can one detect a structural change based on the variance change? In this paper, we applied the iterated cumulative sums of squares method by Incan and Tiao to the Shanghai and Shenzhen Stock Markets. We found that by using the modified Levine statistic, before January 1 there were 3 large variance changes in the Shanghai market and 4 changes in the Shenzhen market. Hence one can expand his or her repertoire of econometrics by including this model as one of tool kits in detecting the coming of major events.

The economic miracle of Taiwan was analyzed in detail in the paper " The US and Taiwan Trade Balance Revisited : the Comparison of the Instrumental Variable and the VAR model "by Huang et al. (1999). The single equation estimation was based on the instrumental variable approach. Before proceeding to the estimation procedure, the Augmented Dickey Fuller (1979), Phillips and Perron (1988) and Perron tests with structural break (1989) were used to test the stationarity of all the variables. Among others, labor productivity, real exchange rate, domestic and foreign money supply and interest rate were of I (1) and hence first difference was taken before running the regression model.

Then a 5-variable VAR model came handy to simulate US-Taiwan bilateral trade balance: tariff rate, volatility of domestic money supply M2, private sector interest rate, lagged terms of the US-Taiwan trade balances. The simple 5-variable VAR model explained 80% of changes in the trade balances. It simply shows a properly formulated VAR model in the right form can provide excellent forecasts for policy makers.

Perhaps two of the best known models regarding random walk or simple efficiency market hypothesis are variance ratio (VR) and rescaled range (R/S) models. The purpose of the original R/S model by Hurst (1951) is to locate the existence of long term persistence in the sequence of water discharge data in river Nile. The R/S model is robust to non-normal return assumptions or infinite variance. However the power of the R/S is affected by the presence of short-term autocorrelation with especially high frequency data. To overcome the deficiencies, Lo (1991) modified the R/S model. The VR model developed by Lo and MacKinlay (1988, 1989) is found to be free from the problems caused by nuisance parameters.

To compare the VR and modified R/S models, 6 Asian emerging markets, 4 Central America countries, 3 Europeans markets and 1 Middle East market were taken in the study. 1174 daily prices from January 1 1988 to June 30, 1992 were used to test the Simple Efficiency Market Hypothesis. For the 14 Indexes, the results indicated the VR model tends to reject the efficiency hypothesis more frequently than does the modified R/S. Only in the Philippines, Mexico and Greece do the two models yield the consistent results. Via using the Monte Carlo simulations along with 4 data generating processes, the following results are obtained: both VR and modified R/S models correctly reject the null hypothesis if data comes from a fractionally integrated time series. If the data are generated from MA (1), VR rejects the null hypothesis while the modified R/S model does not. Since MA (1) does not necessarily imply market inefficiency, VR models have a tendency to erroneously reject the null hypothesis.

The paper "Are Mathematics, Economics and Accounting Courses Important Determinants in Financial Management : A Rank Order Approach " examined if Mathematics, Microeconomics and Accounting courses are important in determining letter grades in Financial Management ? Since letter grades are ordinal in nature: all we know is letter grade A is better than B, which in turn is better than C etc. Traditional regression analysis is not appropriate for such modeling and as a result, an ordered probit model is employed with data of 1542 students from Clarion University of Pennsylvania.

The result is rather surprising. First, grade point average which reflects student's effort is only marginally important. It is indicative of the possibility that it does not take much effort to get a good grade in the Financial Management course. Second, two required Mathematics courses are not as important as were expected. It reveals that the course wasn't taught with enough mathematic rigor. Third, both microeconomics and accounting courses are important because they are inseparable parts in financial management. The paper points to an important aspect of teaching economics or finance in state-supported public universities: quantitative skills are woefully lacking in middle and high schools. The inadequacy trickles down to Economics and Finance courses in state-supported universities as was abundantly shown in our large sample rank-ordered regression in the northwestern part of Pennsylvania.

The Manchester School Vol 72 No. 3   June 2004
1463–6786   347–362

INDUSTRIAL OUTPUT AND STOCK PRICE REVISITED: AN APPLICATION OF THE MULTIVARIATE INDIRECT CAUSALITY MODEL*

by

BWO-NUNG HUANG
Providence University and National Chung-Cheng University, Taiwan
and
CHIN-WEI YANG[†]
Clarion University of Pennsylvania

This paper presents an analysis of the empirical relationship between stock returns, industrial production, money supply, inflation and interest rates across five countries—Canada, France, Japan, Taiwan and the USA. Specifically, we estimate a five-variable vector autoregression model in order to answer the question: does industrial production predict stock returns directly or indirectly (i.e. does industrial production help predict a variable that itself predicts stock returns)? The key result is that there is no direct and significant statistical relationship in any of the five countries, but there is strong evidence of an indirect relation in Taiwan (via money supply) and another indirect relation in the USA (via interest rate). This indirect causality is verified by examining the relative predictability of stock returns both with and without the additional information. Predictability increases when the indirect relationship is exploited.

1   INTRODUCTION

As indicated by Professor Hamilton (1994), no time series should Granger-cause stock prices; Granger-causality tests should not be used to infer a direction of causation for any time series that reflects forward-looking behavior such as stock prices and interest rates. Is it true that Granger causality is lacking in any time series and as such cannot be used to infer a direction of causation for any forward-looking variables? The first assertion hinges on the assumption that asset prices follow a random walk. However, this may not hold true in an asset market with time-varying risk premiums. The second assertion ignores the role played by pricing kernels. Hansen and Singleton (1983) indicate that asset returns may be predictable through the prediction of a pricing kernel. In that sense, variables that help predict the pricing kernel could also help predict asset returns. Thus, not only may Granger-causality tests be used to infer a direction of causation, but indirect relationships from

*Manuscript received 8.3.02; final version received 21.8.03.
[†]Valuable suggestions from a reviewer are greatly appreciated. We also thank Clive W. J. Granger for helpful comments, but any error that remains is ours.

other variables can be of great theoretical importance as they imply that the causing variables are actually triggering the pricing kernel. Modern empirical asset pricing theorists claim that everything can be traced back to the basic pricing equation $p = E(mx)$ where m is the pricing kernel (or stochastic discount factor) and x is payoffs.[1] In this light, it is of primary interest to look for variables that directly cause the pricing kernel, which in turn causes the asset returns x/p.[2] The main purpose of this paper is to identify variables that predict the pricing kernel before predicting asset returns.

Prior studies focused primarily on relationships between macroeconomic variables and asset returns. Most results indicate that money supply and/or changes in interest rate lead asset returns. Lastrapes (1998) shows that, in all the G8 nations (except France), changes in money supply were found to impact (positively) changes in the real price of stocks (also see Mukherjee and Naka, 1995). Similar results have been verified by others (Cramer, 1986; Domian *et al.*, 1996; Huang, 1999). In some cases, stock price changes were found to lead macroeconomic variables: Lee's results (1992) indicate that stock price changes in the USA lead real activity. Furthermore, Fama and French (1996) and Geske and Roll (1983) pointed out a positive relationship between stock returns and real activity without resorting to formal statistical techniques. It seems to be out of the ordinary that money supply alone can predict asset price. More often than not, stock prices (or their change) largely adjust to the information disclosed by financial statements. That is, variables affecting a company's fundamental value may be used to help predict future stock returns.[3]

Based on the indirect causality model pioneered by Dufour and Renault (1998), this paper provides some evidence that industrial production, a proxy for fundamental values, directly Granger-causes change in interest rates (or money supply), which in turn Granger-causes stock price changes. This is in agreement with a recent report (*The Wall Street Journal*, 10 January 2003) that a lower interest rate (1.84 per cent on two-year Treasury notes) may well have contributed to higher stock prices. Since investors at large are not equipped with sophisticated forecasting skills, we propose, in place of the frequently used vector autoregression (VAR) model, the indirect causality methodology to forecast stock returns. In terms of out-of-sample forecast errors, we find that the indirect causality model yields more accurate results. The organization of the paper is as follows. In Section 2 we discuss the data and the empirical results. Section 3 formulates the indirect causality model

---

[1] See Cochrane (2001, Ch. 1) for details.

[2] We thank a reviewer for the useful suggestion which improved the quality of our paper greatly.

[3] On 28 February 2001, the US Department of Commerce announced a disappointing economic growth of 1.1 per cent for the fourth quarter of 2000. The announcement triggered a 157-point drop in the Dow Jones to 10479. Viewed in this context, past economic growth can certainly impact current stock prices. Likewise, a better-than-expected manufacturing production for December 2002 lifted the Dow Jones by 3.6 per cent on 2 January 2003.

along with its univariate forecast framework before comparing the results. Section 4 concludes.

2 DATA AND EMPIRICAL RESULTS

The essence of the indirect causality model by Dufour and Renault (1998) lies in the establishment of causality from Y to X via an auxiliary variable Z. Put simply, given an information set of past X and Z variables, the indirect causality can be ascertained when the auxiliary variable Z causes X in period h and Y causes Z in period 1.[4] In the USA, the prospects of the economy can certainly impact interest rates, be it through economic force or by Greenspan. Clearly, changing interest rates can in turn affect stock prices, as is the case most recently. As a result, the interest rate may be considered to be a candidate for the auxiliary variable. The main objective of this paper is to investigate whether changes in industrial output Granger-cause changes in stock price via an auxiliary variable. Since a direct causality is lacking from prior studies, we intend to search for an indirect causality through the interest rate or money supply. To ensure the robustness of our study, we employ the five-variable model with monthly macroeconomic data and stock prices of the USA, Canada, France, Japan and Taiwan. The sample data are obtained from (i) the International Financial Statistics (IFS) Databank in the case of Canada, France, the USA and Japan, and (ii) the *Taiwan Economic Journal* databank. This sample represents the most recently available data set from the databanks and as such includes the longest bull market in US history. In addition, the industrial production index is used as a proxy for output. Beyond that, M2 is used for money supply and the Treasury bill rate (r) represents the interest rate.[5] We chose these variables because of a well-known macroeconomic phenomenon: the growth rate of money supply and hence interest rates can certainly impact changes in stock price (e.g. Friedman, 1988; Mookerjee and Yu, 1997). Presumably, for stock prices to pick up momentum, there needs to be some dynamic force (which can be generated from increased capital or money supply). Thus, we anticipate a positive correlation between money supply and the stock index. From the perspective of intertemporal models, increased money supply normally leads to lower interest rates, which render investment loans more abundant. As such, we expect negative correlations between interest rates and changes in stock price. The industrial production index, reflecting output of an economy, is a good proxy for macroeconomic fundamentals. Growing output generally mirrors increased revenues of firms, which in turn signal higher potential dividends.

---

[4]Refer to the paper by Dufour and Renault (1998) for details regarding the indirect causality test.
[5]Private promissory note rate is used in the case of Japan, and 60-day commercial paper rate for Taiwan.

Greater potential dividend is then incorporated into the stock price today, which gives rise to higher future price of the stock. The relationship between inflation and stock price changes is not as explicit. Common wisdom suggests that an increase in inflation increases the nominal risk-free rate, and thus raises the discount rate in valuation models. However, Fama (1981) points out that the negative relationships between stock returns and inflation are proxies for positive relationships between stock returns and real variables. In other words, the negative return–inflation relationship can be decomposed into negative inflation–real activity relationships and positive return–real activity relationships via the capital expenditure process.

In order to understand the impacts of money supply, inflation, interest rate and industrial production on stock price changes (excess return or stock price changes minus riskless monthly interest rate), we take the logarithmic difference to measure rate of change (Δ or percentage change) for the five variables:[6]

$$\Delta y_t = \log y_t - \log y_{t-1} \qquad y_t = (\text{cpi}_t, \text{ip}_t, \text{M2}_t, \text{stk}_t) \tag{1}$$

$$\Delta r_t = \log(1+r_t) - \log(1+r_{t-1}) \tag{2}$$

where cpi_t is the consumer price index (IFS Databank, line 64.f), ip is industrial production (IFS Databank, line 66.f), M2 is the money supply (see the Appendix) and stk is the stock index (IFS Databank, line 62.f). All the IFS data obtained from Datastream and their definitions are reported in the Appendix.

To ensure the stationarity of rates of change of the five variables, we apply the Phillips–Perron (PP) unit root test and report the results in Table 1. An examination of Table 1 immediately indicates the stationary nature of these variables (in logarithmic difference form) via the PP approach. For a traditional direct Granger-causality model, a linear bivariate system of equations is used as the base for the test (F statistic):

$$\Delta y_{1t} = \alpha_0 + \sum_{i=1}^{k} \alpha_i \Delta y_{1t-i} + \sum_{j=1}^{k} \beta_j \Delta y_{2t-j} + e_{1t}$$

$$\Delta y_{2t} = \gamma_0 + \sum_{i=1}^{k} \gamma_i \Delta y_{1t-i} + \sum_{j=1}^{k} \rho_j \Delta y_{2t-j} + e_{2t} \tag{3}$$

When one fails to reject the hypothesis $H_0: \beta_1 = \beta_2 = \ldots = \beta_k = 0$, it implies that y_{2t} does not Granger-cause ($\nrightarrow$) y_{1t}. Similarly, if one fails to reject the hypothesis $H_0: \gamma_1 = \gamma_2 = \ldots = \gamma_k = 0$, it suggests that $y_{1t} \nrightarrow y_{2t}$. Since optimal

---

[6]Applying the five-variable model to G8 countries, Lastrapes (1998) fails to find a set of cointegration. Since our focus is on short-term direct and indirect causality between stock price and the macroeconomic variables, we do not consider their potential cointegration relations.

TABLE 1
UNIT ROOT AND BIVARIATE CAUSALITY RESULTS

	Name	Model	*y*	Δy	*Bivariate Granger-causality test result*
			Unit root test (PP)		
Canada	ip	τ_τ	−1.9875	−16.5525*	dcpi → dip*
	r	τ_τ	−2.6230	−14.9484*	dcpi → dm2*
	stk	τ_τ	−2.9337	−13.1957*	dip → dr*
	m2	τ_τ	−2.1676	−20.5215*	dr → dm2***
	cpi	τ_τ	−2.8930	−9.0151*	dr → dstk**
France	ip	τ_τ	−1.9881	−23.3841*	dcpi → dm2*
	r	τ_τ	−1.6650	−11.3356*	dr → dcpi*
	stk	τ_τ	−2.1552	−15.2123*	dm2 ↔ dip*
	m2	τ_τ	−2.9337	−19.5714*	dm2 → dstk**
	cpi	τ_τ	−7.8990*	−10.0814*	
Japan	ip	τ_τ	−1.5124	−22.1983*	dcpi ↔ dm2*
	r	τ_τ	−2.0282	−12.7810*	dcpi → dr*
	stk	τ_τ	−1.3915	−10.6327*	dm2 ↔ dip*
	m2	τ_τ	−0.4618	−29.0185*	dm2 → dr*
	cpi	τ_τ	−0.7635	−12.8906*	
Taiwan	ip	τ_τ	−2.0407	−34.8725*	dcpi → dr*
	r	τ_μ	−3.0121**	−11.7910*	dip → dcp**
	stk	τ_τ	−1.3165	−9.8894*	dcpi → dm2*
	m2	τ_τ	−0.8079	−9.6816*	dip → dr*
	cpi	τ_τ	−2.1320	−16.3702*	dr → dm2*
					dstk ↔ dr***
					dm2 ↔ dcp*
					dm2 → dstk**
USA	ip	τ_τ	−1.9629	−11.5050*	dcpi → dstk*
	r	τ_τ	−1.8667	−12.8430*	dr → dcpi*
	stk	τ_τ	−2.4784	−10.5547*	dip → dm2***
	m2	τ_τ	−3.1176	−9.9808*	dip → dstk***
	cpi	τ_τ	−0.9948	−5.5381*	dip → dr*
					dr → dm2*

Notes: ip, industrial production index; *r*, proxy for the interest rate (promissory note rate for Japan, Treasury bill rate for the USA, private commercial paper rate for Taiwan); stk, stock price index; m2, money supply (m1b for Taiwan); interest rate change is obtained from the first difference of $\log(1 + r)$; *y*, original variable level; Δy, difference of logarithmic variables; *, 1 per cent significance level; **, 5 per cent significance level; ***, 10 per cent significance level; ↛, does not Granger-cause; one-period lag is used in this study except for stock price and interest rate changes in Japan and Taiwan; the prefix d denotes logarithmic difference.

lag periods (*k*) are often selected based on the Schwartz Bayesian criterion (SBC) or the Akaiki information criterion, we employ the SBC to determine the optimal *k*. The Granger-causality test results between changes in stock price and those of other macroeconomic variables are also shown in Table 1.

An inspection of Table 1 suggests that (i) interest rate changes led stock price changes in Canada, (ii) changes in money supply led changes in stock price in France and Taiwan, (iii) changes in industrial production led stock price changes in the USA, and (iv) changes in stock prices led changes in commercial paper rates in Taiwan. These findings of bivariate causality (lead–lag relationships) are by and large in agreement with existing literature and as such do not help much in predicting stock prices. Consistent with previous findings, changes in industrial production are not found to lead (except in the USA at the 10 per cent significance level) stock prices for any of the four economies in the bivariate model. In reality, however, interaction of more than two variables cannot be ignored, especially in deciphering the inflation–stock price relationship. The results could be quite different once money supply or interest rate is added to the analysis. For this reason, we formulate a five-variable model (stock price change (dstk), industrial production change (dip), money supply change (dm2), inflation (dcpi) and interest rate change (dr)) in the framework of a VAR model to illustrate the potential caused relationship:

$$
W_t = \begin{bmatrix} \text{dcpi}_t \\ \text{dip}_t \\ \text{dm2}_t \\ \text{dstk}_t \\ \text{dr}_t \end{bmatrix} = \begin{bmatrix} \pi_{11_j} & \pi_{12_j} & \pi_{13_j} & \pi_{14_j} & \pi_{15_j} \\ \pi_{21_j} & \pi_{22_j} & \pi_{23_j} & \pi_{24_j} & \pi_{25_j} \\ \pi_{31_j} & \pi_{32_j} & \pi_{33_j} & \pi_{34_j} & \pi_{35_j} \\ \pi_{41_j} & \pi_{42_j} & \pi_{43_j} & \pi_{44_j} & \pi_{45_j} \\ \pi_{51_j} & \pi_{52_j} & \pi v_{53_j} & \pi_{54_j} & \pi_{55_j} \end{bmatrix} \begin{bmatrix} \text{dcpi}_{t-1} \\ \text{dip}_{t-1} \\ \text{dm2}_{t-1} \\ \text{dstk}_{t-1} \\ \text{dr}_{t-1} \end{bmatrix} + \begin{bmatrix} e_{1t} \\ e_{2t} \\ e_{3t} \\ e_{4t} \\ e_{5t} \end{bmatrix} \quad (4)
$$

where $\pi_{ik_j} = \Sigma_{j=0}^{p}\pi_{ik_j}L^j$, for $1 \le p < \infty$. In this VAR representation, the Granger (1969) causality runs from dip_t to dstk_t if $\pi_{42_j} \ne 0$. Again the optimal lag is found to be 1 via the SBC. Within the framework of the five-variable model, we then perform the multivariate causality test and report the outcome in Table 2.

From Table 2, the causal relation of dip → dstk is not found for the four economies. In the cases of Canada, France, Japan and the USA, the fact that $\pi_{42}(L) = 0$ and $\pi_{24}(L) = 0$ (failure to reject H_0) implies a lack of Granger causality between dstk and dip. Industrial production changes are found to lead stock price changes in the case of Taiwan. Much to the disappointment of practitioners, the anticipated causal relationship between fundamentals (industrial production) and stock price changes is not identified. Is it because such a relationship does not exist? Or is it masked by an indirect causality? We now employ the indirect causality model due to Dufour and Renault (1998) to test the existence of such an indirect causality. According to Dufour and Renault, for multivariate models where a vector of auxiliary variables Z is used in addition to the variables of interest X and Y, it is possible that Y does not cause X but can still help to predict X several periods ahead. For example, the values of Y up to time t may help to predict X_{t+2}, even though they are useless to predict X_{t+1}. This is due to the fact that Y may help to predict Z one period ahead, which in turn has an effect on X in a subsequent period.

TABLE 2
ESTIMATED RESULTS OF FIVE-VARIABLE VAR MODELS (OPTIMAL LAG 1)

	dcpi$_t$	dip$_t$	dm2$_t$	dstk$_t$	dr$_t$
Canada					
dcpi$_{t-1}$	0.36*	−0.56*	0.79*	0.38	1.13*
	7.32[a]	−3.47	4.89	0.60	2.52
dip$_{t-1}$	−0.01	−0.17*	−0.03	0.13	0.24***
	−0.71[a]	−3.22	−0.55	0.63	1.64
dm2$_{t-1}$	0.03**	0.01	−0.15*	−0.19	0.14
	1.97[a]	0.13	−2.96	−0.96	1.02
dstk$_{t-1}$	0.00	0.02	0.01	0.04	0.10*
	−0.09[a]	1.51	0.65	0.67	2.72
dr$_{t-1}$	0.01	0.02	0.02	−0.16*	0.47*
	1.42[a]	1.43	1.37	−2.45	9.97
France					
dcpi$_{t-1}$	0.72*	−0.46*	0.58*	0.23	0.89**
	17.17[a]	−2.37	4.27	0.21	1.98
dip$_{t-1}$	0.01	−0.33*	0.09*	0.42	0.09
	0.59[a]	−5.77	2.16	1.28	0.68
dm2$_{t-1}$	0.06*	0.35*	−0.08	1.07*	0.08
	2.97[a]	3.73	−1.18	2.01	0.36
dstk$_{t-1}$	0.00	0.01	−0.00	0.01	−0.04
	0.62[a]	0.67	−0.08	0.13	−1.56
dr$_{t-1}$	0.01*	0.04	−0.01	0.02	0.38*
	2.61[a]	1.42	−0.47	0.14	6.54
Japan					
dcpi$_{t-1}$	0.30*	−0.11	−0.09	−0.03	−0.09
	6.34[a]	−0.97	−0.79	−0.09	−0.36
dip$_{t-1}$	0.01	−0.27*	0.15*	−0.03	0.06
	0.30[a]	−5.27	2.77	−0.25	0.47
dm2$_{t-1}$	0.07*	0.07	−0.25*	0.28*	−0.22**
	3.34[a]	1.58	−5.06	2.24	−1.92
dstk$_{t-1}$	−0.01	0.02	−0.01	0.34*	−0.04
	−0.85[a]	0.94	−0.31	6.89	−0.81
dr$_{t-1}$	0.04*	−0.00	−0.07*	0.00	0.36*
	3.99[a]	−0.06	−3.14	0.00	7.27
Taiwan					
dcpi$_{t-1}$	−0.06	−0.54	−0.24	0.25	−0.07
	−0.91[a]	−0.92	−1.50	0.39	−1.50
dip$_{t-1}$	0.02*	−0.66*	0.12*	0.15***	0.03*
	2.24[a]	−9.38	6.20	1.94	4.61
dm2$_{t-1}$	0.00	−1.09*	0.58*	0.79*	0.07*
	0.07[a]	−4.28	8.26	2.85	3.52
dstk$_{t-1}$	−0.01	0.13*	−0.01	0.31*	0.00
	−0.82[a]	2.28	−0.60	4.87	0.86
dr$_{t-1}$	0.16***	−1.75*	−0.53*	0.92	0.24*
	1.74[a]	−2.21	−2.44	1.06	3.93
USA					
dcpi$_{t-1}$	0.61*	−0.24*	0.14	−1.71*	0.86
	14.54[a]	−2.08	1.43	−3.02	1.05
dip$_{t-1}$	−0.01	0.36*	0.01	−0.36	0.99*
	−0.83[a]	7.38	0.30	−1.51	2.83

TABLE 2 (*continued*)

	dcpi$_t$	dip$_t$	dm2$_t$	dstk$_t$	dr$_t$
dm2$_{t-1}$	0.01	0.05	0.07	−0.29	0.99*
	0.62^a	0.77	1.27	−0.96	2.23
dstk$_{t-1}$	0.00	0.01	0.01	0.20*	0.21*
	0.07^a	0.59	1.49	3.85	2.72
dr$_{t-1}$	0.00	0.01	−0.01*	−0.10*	0.30*
	1.25^a	1.37	−2.34	−2.87	5.73

Notes: The numerical values in rows with superscript a are *t* statistics; *, 1 per cent significance level; **, 5 per cent significance level; ***, 10 per cent significance level; the optimal lag period is 1 based on the SBC; dstk, dip, dm2 and dr denote rates of change in stock price, industrial production, money supply M2 and interest rates respectively.

In order to illustrate the indirect causality test, we need to establish the following multivariate VAR(p) model:[7]

$$W_t = \mu + \sum_{\ell=1}^{p} \pi_\ell W_{t-\ell} + a_t \qquad t = 1, \ldots, T \qquad (5)$$

in which $W_t = (\text{dcpi}_t, \text{dip}_t, \text{dm2}_t, \text{dstk}_t, \text{dr}_t)'$ is a 5×1 random vector, μ is a constant and

$$E\left[a(s)\, a(t)'\right] = \Omega \qquad \text{if } s = t$$
$$= 0 \qquad \text{if } s \neq t$$

$$\det(\Omega) \neq 0$$

Equation (5) represents a VAR(p) model at horizon 1. With a slight modification, it could be expanded to become the autoregressive process at horizon h, or

$$W_{t+h} = \mu^{(h)} + \sum_{\ell_1=1}^{p} \pi_{\ell_1}^{(h)} W_{t+1-\ell_1} + \sum_{\ell_2=0}^{h-1} \varphi_{\ell_2} a_{t+h-\ell_2} \qquad t = 0, \ldots, T-h \quad (5')$$

where $h < T$, $\varphi_0 = I$. Note that according to Dufour and Renault (1998) the following relations hold:

$$\pi_k^{(h+1)} = \pi_{k+h} + \sum_{\ell=1}^{h} \pi_{h-\ell+1} \pi_k^{(\ell)} = \pi_{k+1}^{(h)} + \pi_1^{(h)} \pi_k \qquad \pi_1^{(0)} = I, \pi_k^{(1)} = \pi_k$$

If we wish to test the hypothesis that dip$_t$ does not cause dstk$_t$ at horizon h (dip$_t \nrightarrow_h$ stk$_t$), it is essential that we test the following:

$$H^{(h)}: \pi_{42j}^{(h)} = 0 \qquad j = 1, 2, \ldots, p$$

[7]Refer to Dufour *et al.* (2003, pp. 2–9) for details.

TABLE 3
THE χ^2 TEST RESULT

h	Canada	France	Japan	Taiwan	USA
1	1.1476	0.0535	0.4923	8.0518	4.1040
	(0.2840)	(0.8172)	(0.4829)	(0.0045)	(0.0428)
2	0.5775	0.0050	0.7705	0.0341	0.0439
	(0.4473)	(0.9436)	(0.3801)	(0.8535)	(0.8340)
3	0.0135	0.1249	0.2460	0.0215	1.2205
	(0.9073)	(0.7238)	(0.6199)	(0.8834)	(0.2693)
4	0.0005	0.1806	0.3598	0.5305	0.7115
	(0.9828)	(0.6708)	(0.5486)	(0.4664)	(0.3989)

Note that numbers in parentheses are simulated p values. h, forecast horizon.

where $\pi_j^{(h)} = [\pi_{42j}^{(h)}]$, $j = 1, 2, \ldots, p$. One problem in applying the Wald test is the existence of a nonpositive definite variance–covariance matrix.[8] Within the framework of the multivariate VAR model with a moderate p and number of variables, one can use the heteroskedasticity-autocorrelation consistent estimator developed by Newey and West (1987) to obtain a positive semi-definite variance–covariance matrix. However, when p is large coupled with many variables, one needs to employ a Monte Carlo or bootstrap method to derive the consistent estimate. As the optimal lag is found to be 1 based on the SBC, h assumes a value of 4 according to Proposition 4.5 in Dufour and Renault (1998). Following Dufour *et al.* (2003), we present probability values on simulated test statistics (Wald test) in Table 3.

An inspection of Table 3 reveals that the indirect causality between changes in industrial production and stock price is lacking even with the money supply and interest rates as the auxiliary variables in the case of Canada, Japan and France up to horizon $h = 4$. However, such an indirect causality is found for Taiwan and the USA at horizon $h = 1$. The indirect causality from changes in industrial production to those in stock price exist via money supply changes with a probability value of 0.5 per cent for Taiwan. Likewise, when change in interest rates is used as an auxiliary variable, the probability value is about 4 per cent for the USA.

If the primary objective for central banks is to target (stabilize) the growth rate of money supply, then it comes as no surprise that the causality from industrial production to stock prices is found via money supply (e.g. Japan and Taiwan). On the other hand, interest rate targeting has been implemented by Greenspan to 'fine tune' the stock market. Thus, based on the empirical results, we can establish the causality from industrial production

---

[8]Noncausality restrictions at higher horizons (greater than or equal to 2) are intrinsically nonlinear. When applying standard test statistics such as Wald's test, such an estimation can easily lead to an asymptotically singular covariance matrix. Thus, the standard asymptotic theory could not be applied to such an estimator.

to stock price via the interest rate for the USA. This lends support to the hypothesis of indirect causality from industrial production change to stock price change. In this light, a less than satisfactory output growth could probably impact the future stock price, albeit indirectly.

How do interest rates affect expected return and risk? An interest rate change normally affects cost of capital (including bond rate), which leads to a re-ranking of various projects through recalculating expected returns and net present values of a firm. In mean–variance space, an investor's portfolio may well be altered in favor of a less risky asset (e.g. bonds) when interest rates go up.

Enough ink has been spread on the relationship between inflation and stock price changes. From our sample study, only in the USA do we find a negative correlation between them (the relationship could be explained by the proxy hypotheses (Fama, 1981)). Some studies suggest that changes in output or inflation affect interest rates, which in turn cause stock prices to change. In our analysis, a negative relationship is found between changes in interest rates and stock price for the USA and Canada. In the case of the USA, output changes are found to impact interest rates, which transmit changes to stock price at a 10 per cent significance level. On the other hand, inflation also affects interest rates, which transmit impacts to the stock market at a 6.5 per cent significance level for Canada.[9] Note that, in both cases, interest rate changes cause stock prices to change on the surface, but the hidden factors are rather different.

3 UNIVARIATE FORECASTING MODELS AND PERFORMANCE EVALUATION

Will such findings via the auxiliary variables improve the prediction of stock prices? To examine this issue, we first formulate an $ARMA(p, q)$ model (Model 1) to forecast dr_t or $dm2_t$:

$$dr_t(dm2_t) = \mu_t + v_{1t} \tag{6}$$

Next, we add a lagged variable of industrial production change (dip_{t-1}), which is found to lead dr_t or $dm2_t$ in the previous model. The forecasting model (Model 2) takes the form

$$dr_t(dm2_t) = \mu_t + \beta dip_{t-1} + v_{2t} \tag{7}$$

where μ_t represents an autoregressive moving average (ARMA) model such that (6) and (7) are free of serial correlation. If an indirect causality relationship exists, Model 2 (equation (7)) is expected to generate the interest rate change ($f2dr_t$) or the money supply change ($f2dm2_t$) that predicts stock price

---

[9] The null hypothesis here is H_0: $\pi_{45i}\pi_{51j} + \pi_{45i}\pi_{52j} = 0$, $j = 1, 2, \ldots, k$.

change better than does the corresponding $f1dr_t$ or $f1dm2_t$. In other words, we substitute the interest rate change and the money supply change into the following equations for performance evaluation.

$$dstk_t = \mu_t + \gamma_1 \cdot f1dr_t \text{ (or } f1dm2_t) + \varepsilon_{1t} \tag{8}$$

$$dstk_t = \mu_t + \gamma_2 \cdot f2dr_t \text{ (or } f2dm2_t) + \varepsilon_{2t} \tag{8'}$$

in which the μ_t represent an ARMA model such that ε_{1t} and ε_{2t} meet the requirements of independent and identical distribution. Let $fd1stk_t$ be the estimated value from (8) via (6) and $fd2stk_t$ from (8′) via (7). We then follow the generalized testing method in comparing the two forecast results (Diebold and Mariano, 1995). Let u_{1t} denote the forecast error $(dstk_t - fd1stk_t)$ of Model 1 or from equation (6) and u_{2t} the forecast error $(dstk_t - fd2stk_t)$ of Model 2 or from equation (7). The discrepancy in forecast can then be calculated as

$$f_t = |u_{1t}| - |u_{2t}| \tag{9}$$

Given $\mu \equiv Ef_t$, a failure to reject the null hypothesis of $\mu = 0$ suggests that the two models have similar predictive power. In the case of $\mu > 0$, Model 2 seems to be superior in forecasting stock price changes, and vice versa for $\mu < 0$. In particular, Diebold and Mariano (1995) prove the following property:

$$\sqrt{P}(f - \mu) \xrightarrow{a} N(0, S_{ff}) \tag{10}$$

in which $S_{ff} \equiv \Sigma_{-\infty}^{\infty} \Gamma_j$, $\Gamma_j \equiv E(f_t - \mu)(f_{t-j} - \mu)$. For actual estimation, one can use the Newey–West estimator to approximate $\hat{S}_{ff} \equiv \Sigma_{-\infty}^{\infty} \hat{\Gamma}_j$.

A recursive estimation can be made. First, estimate (8) and (8′) in the period from $t = 1$ to $t = R$. Second, perform the estimation for $t = R + 1$ based on $f1dr_t$ $(f1dm2_t)$ or $f2dr_t$ $(f2dm2_t)$ from both models. Third, from $fd1stk_{R+1}$ and $fd2stk_{R+1}$, it follows immediately that $u_{1,R+1} = dstk_{R+1} - fd1stk_{R+1}$ and $u_{2,R+1} = dstk_{R+1} - fd2stk_{R+1}$. In a similar vein, we can obtain P $(P = T - R + 1)$ forecast errors before substituting them into (9) and (10) for comparison purposes. We focus on Taiwan and the USA where a 5 per cent significance level is found in the indirect causality test. The in-sample estimation results are reported in Table 4 for both countries.

With the addition of changes in industrial production, the adjusted R^2 has increased in value from 0.1359 to 0.1566, a 14.38 per cent improvement in the case of the USA versus 26.10 per cent (from 0.30 to 0.3783) in the case of Taiwan.[10] Furthermore, serial correlation is purged from all the models based on the values of the Q statistic (Table 4). By substituting $f1dr_t$, $f2dr_t$, $f1dm2_t$ and $f2dm2_t$ into (8) and (8′), we can evaluate the forecasting performance from both models (Table 5).

---

[10]Harvey (1995) employs adjusted R^2 to investigate the predictability of stock for both emerging and developed markets.

TABLE 4
IN-SAMPLE REGRESSION RESULT WITH AND WITHOUT INDUSTRIAL OUTPUT

USA
Model 1
$$dr_t = 0.33dr_{t-1} - 0.15dr_{t-2} - 0.18dr_{t-6} + \varepsilon_t$$
$$(6.44) \quad (-2.99) \quad (-3.79)$$
$$\bar{R}^2 = 0.1359 \qquad Q(5) = 0.53[0.99] \qquad Q(10) = 7.43[0.68]$$

Model 2
$$dr_t = 0.30dr_{t-1} - 0.16dr_{t-2} - 0.18dr_{t-6} + 0.99dip_{t-1} + \varepsilon_t$$
$$(5.88) \quad (-3.25) \quad (-3.81) \quad (2.99)$$
$$\bar{R}^2 = 0.1566 \qquad Q(5) = 0.93[0.97] \qquad Q(10) = 7.61[0.67]$$

Taiwan
Model 1
$$dm2_t = 0.19dm2_{t-1} - 0.12dm_{t-5} - 0.46dm_{t-12} + \varepsilon_t$$
$$(3.29) \quad (2.04) \quad (7.56)$$
$$\bar{R}^2 = 0.3000 \qquad Q(5) = 1.55[0.91] \qquad Q(10) = 2.96[0.98]$$

Model 2
$$dm2_t = 0.41dm2_{t-1} - 0.13dm_{t-5} - 0.34dm_{t-12} + 0.10dip_{t-1} + \varepsilon_t$$
$$(5.93) \quad (2.41) \quad (5.75) \quad (5.24)$$
$$\bar{R}^2 = 0.3784 \qquad Q(5) = 2.39[0.79] \qquad Q(10) = 4.39[0.93]$$

Notes: t statistics are in parentheses; dr, rate of change in interest rate; dm2, rate of change in money supply M2; dip, rate of change in industrial production. $Q(k)$ is the Ljung–Box statistic with lag length k; numbers in brackets are p values.

A perusal of Table 5 suggests that the addition of the industrial production change (one lag) improves predictive power for interest rate change in the USA and money supply change for Taiwan. In terms of in-sample performance (to June of 1998), the adjusted R^2s have increased by 3.6 per cent (USA) and 38.2 per cent (Taiwan), illustrating the added explanatory power derived from the information of industrial production. In terms of out-of-sample performance (from July 1998 to December 2000), we can readily compare the forecasting results based on (8) and (8′).

To compare the out-of-sample forecasting accuracy with a prediction interval of 30 months from July 1998 to December 2000, we make use of (9) and (10) with the heteroskedasticity-consistent standard error by Newey and West. Potential discrepancies in predictive power between the two models are shown in Table 6.

An examination of Table 6 suggests immediately that we can reject H_0: $\mu = 0$ in the case of Taiwan as the expected value of $f_t = |u_{1t}| - |u_{2t}|$ is 0.77, a value much greater than 0 at a 9.96 per cent significance level. For the USA, the significance level is approximately 15.45 per cent, a value indicating that the addition of industrial output as the explanatory variable has noticeably improved the forecast ability in stock price changes. This result is also very much in line with the criterion set forth by Ashley *et al.* (1980): to ascertain a true causality, both in-sample and out-of-sample predictive powers of a causality model are expected to be superior to that of a noncausality model.

TABLE 5
IN-SAMPLE RESULTS OF STOCK RETURN ESTIMATION MODELS

USA
Model 1
$dstk_t = 0.59 + 0.26dstk_{t-1} - 0.34f1dr_t + \varepsilon_{1t}$
    (3.19) (5.27)   (−3.87)
$\bar{R}^2 = 0.1109$ $Q(5) = 2.92[0.71]$ $Q(10) = 5.96[0.82]$

Model 2
$dstk_t = 0.67 + 0.26dstk_{t-2} - 0.34f2dr_t + \varepsilon_{2t}$
    (3.60) (5.33)   (−4.11)
$\bar{R}^2 = 0.1153$ $Q(5) = 3.45[0.63]$ $Q(10) = 6.21[0.80]$

Taiwan
Model 1
$dstk_t = 0.40dstk_{t-1} - 0.13dstk_{t-2} + 0.89f1dm2_t + \varepsilon_{1t}$
    (5.79)   (−1.93)   (2.10)
$\bar{R}^2 = 0.1588$ $Q(5) = 2.55[0.77]$ $Q(10) = 7.77[0.65]$

Model 2
$dstk_t = 0.39dstk_{t-1} - 0.14dstk_{t-2} + 1.04f2dm2_t + \varepsilon_{2t}$
    (5.63)   (−2.06)   (2.73)
$\bar{R}^2 = 0.1707$ $Q(5) = 2.67[0.75]$ $Q(10) = 8.46[0.58]$

Notes: dstk denotes logarithmic stock returns. Model 1 uses estimated $f1dr_t$ and $f1dm2_t$ obtained from (6). Model 2 uses estimated $f2dr_t$ and $f2dm2_t$ obtained from (7).

TABLE 6
COMPARISON OF OUT-OF-SAMPLE FORECAST RESULTS

	Mean (t statistics)	*Significance level for rejecting* H_0: $\mu = 0$
USA	0.0380	15.45%
	(1.4634)	
Taiwan	0.7699	9.96%
	(1.7012)	

Notes: The out-of-sample prediction interval is from July 1998 to December 2000. $f_t = |u_{1t}| - |u_{2t}|$. u_{1t} is the discrepancy between $dstk_t$ and $fd1stk_t$, the latter being derived from substituting $f1dr_t$ or $f1dm2_t$ into (8) via (6). u_{2t} is similarly derived via (7). $\mu > 0$ implies $|u_{1t}|$ exceeds $|u_{2t}|$. Standard errors are heteroskedasticity consistent, as developed by Newey and West (1987).

4 CONCLUSION

According to the paradigm of efficient markets, no past information can be used to predict stock prices today. In other words, Granger causality does not exist between economic variables and stock returns. Prior analyses primarily emphasized direct causal relations between financial variables—money supply or interest rate—and stock returns. A significant relationship between real variables such as industrial production and stock returns has not been found in the literature. It is very likely that stock price changes depend at least partly on past profit levels. More specifically, modern asset pricing models (Cochrane, 2001) focus on the key role played by the pricing kernel in influencing stock prices. Pricing kernels are in turn impacted by some real variables (e.g. output). As such, the causal relationship may well be hidden and indirect. In an attempt to verify the indirect causality, we employ monthly data on stock returns, industrial production, interest rates and the money supply for the USA, Canada, France, Japan and Taiwan in our model. Applying the indirect causality model due to Dufour and Renault (1998), we detect such an indirect causality for the USA, Japan and Taiwan: industrial production changes indirectly cause stock return. In the case of the USA, industrial production changes Granger-cause interest rate changes, which in turn Granger-cause stock return. In the case of Japan and Taiwan, it is the money supply change rather than the interest rate change that serves as the intermediary in the indirect causality model.

According to Ashley *et al.* (1980), both in-sample and out-of-sample predictive accuracy in the Granger-causality model is expected to be superior to that without using the causal relation. For an average investor, we propose a simple ARMA formulation as the benchmark model. We then add an indirect causality relationship via money supply or interest rate change to the ARMA formulation. Based on the testing procedure developed by Diebold and Mariano (1995), it is apparent that the ARMA model with the indirect or hidden variables outperforms the simple ARMA model. In other words, despite the fact that industrial production does not Granger-cause stock return directly, it indirectly impacts stock returns via changes in interest rates or money supply.

APPENDIX: VARIABLE DEFINITIONS AND THEIR DATASTREAM CODES

Canada, money M1 (*Banking Survey*), CURN	CNI34...A
Canada, quasi-money, CURN	CNI35...A
Canada, Treasury bill rate	CNI60C..
Canada, share price index, NADJ	CNI62...F
Canada, CPI NADJ	CNI64...F
Canada, industrial production, SADJ	CNI66..CE
Canada, industrial production, VOLA	CNI66..IG
US money M2, CURN	USI59MB.A
US money M2, CURA	USI59MBCB
US Treasury bill rate	USI60C..
US share price index, NADJ	USI62...F
US CPI NADJ	USI64...F
US industrial production, SADJ	USI66..CE
US industrial production, VOLA	USI66..IG
France, money M2 (national definition), CURN	FRL39MB.A
France, money M2 (national definition), CURA	FRL39MBCB
France, Treasury bill rate	FRI60C..
France, share price index, NADJ	FRI62...F
France, CPI NADJ	FRI64...F

France, industrial production, SADJ	FRI66..CE
France, industrial production, VOLA	FRI66..IG
Japan, money M1 (*Banking Survey*), CURN	JPI34...A
Japan, quasi-money, CURN	JPI35...A
Japan, money market rate (Federal Funds)	JPI60B..
Japan, three-month LIBOR: offer, London	JPI60EA.
Japan, share price index, NADJ	JPI62...F
Japan, CPI NADJ	JPI64...F
Japan, industrial production, SADJ	JPI66..CE
Japan, industrial production, VOLA	JPI66..IG

Note that all of the data are obtained from the IFS data bank of Datastream.

REFERENCES

Ashley, R., Granger, C. W. J. and Schmalensee, R. (1980). 'Advertising and Aggregate Consumption: An Analysis of Causality', *Econometrica*, Vol. 48, No. 5, pp. 1149–1168.

Cochrane, J. H. (2001). *Asset Pricing*, Princeton, NJ, Princeton University Press.

Cramer, J. S. (1986). 'The Volume of Transactions and the Circulation of Money in the United States, 1950–1979', *Journal of Business and Economic Statistics*, Vol. 4, No. 2, pp. 225–232.

Diebold, F. X. and Mariano, R. S. (1995). 'Comparing Predictive Accuracy', *Journal of Business and Economic Statistics*, Vol. 13, No. 3, pp. 253–263.

Domian, D. L., Gilster, J. E. and Louton, D. A. (1996). 'Expected Inflation, Interest Rates, and Stock Returns', *Financial Review*, Vol. 31, No. 4, pp. 809–830.

Dufour, J.-M. and Renault, E. (1998). 'Short Run and Long Run Causality in Time Series: Theory', *Econometrica*, Vol. 66, No. 5, pp. 1099–1125.

Dufour, J.-M., Pelletier, D. and Renault, E. (2003). 'Short Run and Long Run Causality in Time Series: Inference', *Working Paper*, Université de Montréal.

Fama, E. F. (1981). 'Stock Returns, Real Activity, Inflation and Money', *American Economic Review*, Vol. 71, No. 4, pp. 545–565.

Fama, E. F. and French, K. R. (1996). 'Multifactor Explanations of Asset Pricing Anomalies', *Journal of Finance*, Vol. 51, No. 1, pp. 55–84.

Friedman, M. (1988). 'Money and the Stock Market', *Journal of Political Economy*, Vol. 91, No. 2, pp. 221–245.

Geske, R. and Roll, R. (1983). 'The Fiscal and Monetary Linkage Between Stock Returns and Inflation', *Journal of Finance*, Vol. 38, No. 1, pp. 1–33.

Granger, C. W. J. (1969). 'Investigating Causal Relations by Econometric Models and Cross-spectral Methods', *Econometrica*, Vol. 37, No. 3, pp. 424–438.

Hamilton, J. (1994). *Time Series Analysis*, Princeton, NJ, Princeton University Press.

Hansen, L. P. and Singleton, K. J. (1983). 'Stochastic Consumption, Risk Aversion and the Temporal Behavior of Asset Returns', *Journal of Political Economy*, Vol. 91, No. 2, pp. 249–265.

Harvey, C. R. (1995). 'Predictable Risk and Returns in Emerging Markets', *Review of Financial Studies*, Vol. 8, No. 3, pp. 773–816.

Huang, B.-N. (1999). 'Taiwan Stock Price and Macroeconomics Variables', *Review of Securities and Futures Markets*', Vol. 10, No. 4, pp. 89–109.

Lastrapes, W. D. (1998). 'International Evidence on Equity Prices, Interest Rates and Money', *Journal of International Money and Finance*, Vol. 17, No. 3, pp. 377–406.

Lee, B.-S. (1992). 'Causal Relations Among Stock Returns, Interest Rates, Real Activity and Inflation', *Journal of Finance*, Vol. 47, No. 4, pp. 591–603.

Mookerjee, T. and Yu, Q. (1997). 'Macroeconomic Variables and Stock Prices in a Small Open Economy: the Case of Singapore', *Pacific-Basin Finance Journal*, Vol. 5, No. 3, pp. 377–388.

Mukherjee, T. K. and Naka, A. (1995). 'Dynamic Relations between Macroeconomic Variables and the Japanese Stock Markets: an Application of a Vector Error Correction Model', *Journal of Financial Research*, Vol. 18, No. 2, pp. 223–237.

Newey, W. K. and West, K. D. (1987). 'A Simple Positive Semi-definite, Heteroskedasticity and Autocorrelation Consistent Covariance Matrix', *Econometrica*, Vol. 55, No. 3, pp. 703–708.

Applied Economics Letters, 1995, **2**, 67–71

The fractal structure in multinational stock returns

BWO-NUNG HUANG and CHIN W. YANG[‡]

*Institute of International Economics, National Chung Cheng University, Chia-Yi,
621 Taiwan*
‡Department of Economics, Clarion University of Pennsylvania, Clarion, 16214, USA

Received 26 September 1994

The essence of fractal analysis is seeking for a pattern that is independent of scale. This paper examines the existence of long-term memory in nine Asian stock markets together with US and UK indices using the modified rescaled-ranged (R/S) statistic. The modified R/S statistic is robust not only with respect to the normality assumption, but also to short-term autocorrelation. The data in the sample range from 1 January 1988 to 30 June 1992 and are arranged in daily, weekly and monthly returns. In most cases, the phenomenon of long-term memory is not found; hence the random walk hypothesis cannot be rejected. The UK market, however, exhibits some long-term memory for various data frequencies and lags. The result of this paper provides directions for future research.

I. INTRODUCTION

One of the dominant themes in the literature of capital-market equilibrium since the 1960s has been the concept of an efficient capital market; a scenario in which security prices fully reflect all available information. The predictability of security prices has become the focal point of investment decisions involving multinational stock markets. The results of earlier studies are in general supportive of the random-walk hypothesis in security prices. That is, the returns of securities are often found to exhibit the property of white noise. Consequently, investors cannot predict stock prices simply on the basis of previous values.[1] However, several recent studies have shown that some component of stock returns can be effectively explained by the following models. First, chaos (Hsieh, 1991) or neural-network models began to shed new light on the role of deterministic returns. Second, the models using variance ratio (Lo, 1988; 1989) and the rescaled-range (Hurst, 1951) emphasize the importance of long-term dependence in stock returns. Third, the weekend effect (Keim and Stambough, 1984; Huang *et al.*, 1994) and the January effect, especially for small firms (Roll, 1982; Reinganum, 1983; and Keim, 1983), have been cited in the literature in different stock markets. It is worthwhile pointing out that if any of the above three models is supported by appropriate statistical tests, at least some portion of stock returns can be explained and predicted. In that case, the efficient-market hypothesis may need to be re-examined more rigorously. Care must be exercised, however; even if a part of

stock returns is inconsistent with the random-walk hypothesis, this does not necessarily imply a violation of the efficient-market hypothesis. Moreover, the empirical results, more often than not, hinge on either the power of the statistical test(s) or the choice of the national stock index. Previous studies have normally concentrated on the market indices of industrialized economies, i.e. the USA, UK and Japan. Only scant attention has been paid to the emerging markets (e.g. Pacific Rim stock markets). A study of these emerging stock markets is therefore warranted since even though they are not as well developed and established, these economies are becoming noticeably more important. There are signs that the centre of gravity in the financial market is shifting to Asia (Van Horne, 1990). In addition, a more refined statistical test is needed to overcome some estimation problems in the previous studies. We start by discussing some recent statistical tests of the efficient-market hypothesis.

From the late 1980s the variance ratio (VR) test began to receive attention for its capacity to detect long-term dependence in stock returns (e.g. Fama and French, 1987; Lo and MacKinlay, 1988, 1989; Cochrane, 1988). Lo and MacKinlay (1988) show approaches normality under some regularity assumptions. The variance ratio test is also free from the problems caused by nuisance parameters.[2] Although the VR test has desirable properties, it remains an unsettled issue, especially with a non-normal sampling distribution or in the presence of a short-term autocorrelation. As an alternative to the variance ratio test, Hurst (1951) developed a rescaled-range or (R/S) approach to test the

---

[1]See the seminal work by Fama (1970, 1991) on the efficient-market hypothesis.

1350–5851 © *1995 Chapman & Hall*

existence of long-term persistence in the long sequence of water-discharge data from the Nile. Unlike the VR test, the R/S approach is free from the normality assumption. None the less, the power of the R/S test is affected by the presence of short-term autocorrelation. To circumvent this problem, Lo (1991) proposed a modified R/S statistic to test for the potential existence of long-term memory.

If capital assets exhibited a pattern of long-term dependence, the choice of optimum length of time period (horizon) in consumption, investment and saving would be a critical issue. In addition, the existence of a long-term non-linear dependence in security returns would certainly invalidate the results of well established models in finance, i.e. capital-asset-pricing and arbitrage-pricing models. The purpose of this paper is, by using both the standard and modified R/S approach, to test the existence of long-term memory in the stock markets of the USA, UK, Japan and the emerging Asian countries. The next section examines the modified R/S test; Section III discusses data and empirical results; and the last section gives summaries and conclusions of the paper.

II. MODIFIED RESCALED-RANGE TEST

Hurst first developed the R/S statistic to compute the reservoir storage required to yield an average flow by computing the cumulative sums of the departures of the annual totals from the mean annual total discharge. The range of the maximum and the minimum of these cumulative totals is taken as the required storage (Hurst, 1951, p. 770). Similarly, the range of partial cumulative sums of deviations of a sequence of security returns from their mean of various lengths, rescaled by the standard deviation, constitutes the traditional R/S statistic:

$$\tilde{Q}_n \equiv \frac{1}{S_n} \left[\underset{1 \leq k \leq n}{Max} \sum_{j=1}^{k} \left(r_j - \bar{r}_n\right) - \underset{1 \leq k \leq n}{Min} \sum_{j=1}^{k} \left(r_j - \bar{r}_n\right) \right] \quad (1)$$

where r1, r2 ... rn represents security returns over n periods; $\bar{r}_n = \frac{1}{n} \sum_j r_j$ is the sample mean return; and

$$S_n \equiv \left[\frac{1}{n} \sum_j \left(r_j - \bar{r}_n\right)^2 \right]^{1/2}$$

is an estimator of the standard deviation.

For each subperiod of length $n > 3$, an average $\tilde{Q}_n$ over different starting-points is computed for the standard R/S analysis. From one perspective, the R/S statistic is superior to other traditional time-series approaches such as autocorrelation, variance ratio and spectral decomposition models in detecting long-term dependence. For instance, via Monte Carlo simulations, Mandelbrot and Wallis (1969) indicated that the R/S statistic is robust in capturing long-range dependence even in the presence of greater magnitudes or skewness and kurtosis in non-Gaussian distributions. Furthermore,

---

²For details, see footnote 6 in Lo and MacKinlay (1988).

the R/S statistic possesses the almost certain convergence property in the stochastic process with infinite variance: a property that is not found in the tests of the variance ratio or autocorrelation (Mandelbrot, 1972, 1975). The second advantage of the R/S statistic, according to Mandelbrot (1972), is its capability of testing the existence of non-periodic cycles. Despite the strength of the R/S statistic, Lo (1991) shows that the test result is extremely sensitive with time-series data manifesting short-range dependence.

To account for the potential effect of the short-range dependence in the R/S test, Lo (1991) proposes the modified R/S statistic as follows:

$$Q_n \equiv \frac{1}{\hat{\sigma}_n(q)} \left[\begin{array}{c} \underset{1 \leq k \leq n}{Max} \sum_{j=1}^{k} \left(r_j - \bar{r}_n\right) - \\ \underset{1 \leq k \leq n}{Min} \sum_{j=1}^{k} \left(r_j - \bar{r}_n\right) \end{array} \right] \quad (2)$$

$$\hat{\sigma}_n(q) \equiv \frac{1}{n} \sum_{j=1}^{n} \left(r_j - \bar{r}_n\right)^2 + \frac{2}{n} \sum_{j=1}^{q} \omega(q)$$

$$\left\{ \sum_{i=j+1}^{n} \left(r_j - \bar{r}_n\right)\left(r_{i-j} - \bar{r}_n\right) \right\} \quad (3)$$

where q is number of periods lagged; and the weight is defined as $\omega_j(q) \equiv 1 - \frac{j}{q+1}$ for $q < n$, so that the modified standard deviation estimator in Equation 3 is positive (see Newey and West, 1987). Note that the modified standard deviation estimator also contains weighted autocovariance up to lag q. Andrews (1991) suggests a data-dependent rule to determine the optimum value for q based on some asymptotic property. However, the choice becomes rather arbitrary in the case of a finite sample.

The asymptotic property of the modified R/S or Q_n statistic can be alternatively given by

$$V_n(q) \equiv \frac{1}{\sqrt{n}} Q_n \overset{a}{\sim} V \quad (4)$$

in which v has the following probability distribution (see Kennedy, 1976; Siddiqui, 1976):

$$F_{\bar{v}}(v) = 1 + 2 \sum_{k=1}^{\infty} \left(1 - 4k^2 v^2\right) \exp^{\left(-2(kv)^2\right)} \quad (5)$$

The critical values can therefore be computed from Equation 5; and some frequently used values were reported by Lo (1991).

IV. DISCUSSION OF DATA AND THE EMPIRICAL RESULTS

Economic applications of the R/S test started in 1977 on the stock market (Greene and Fielitz, 1977). Several more empirical studies can be found on gold price (Booth and Karen, 1979), futures contracts (Helms *et al.*, 1984) and stock and bond prices (Peters, 1989). Most recently, Ambrose *et al.* (1992) applied the R/S statistics to study the long-term memory phenomenon in US real estate investment trust and security returns. In general, only the standard R/S test or Equation 1 was employed in their studies. Hence the test statistic may be biased due to the presence of short-term autocorrelation. (Lo (1991) and Ambrose *et al.* (1992) applied the modified R/S test, but failed to find any long-term persistence in the US security-returns data.

However, such a study on the stock markets of emerging Pacific Rim economies has, thus, far, to the best of our knowledge, eluded the literature. The purpose of our paper is to fill this void. Nine stock indices of the emerging Pacific economies including Hong Kong (HK), Indonesia (IND), Japan (JPN), Korea (KOA), Malaysia (MAL), Philippines (PHI), Singapore (SIG), Thailand (THA) and Taiwan (TWN) are obtained from the Morgan–Stanley data tape. To compare the results with those of well developed and established markets, we include in the study the market indices (no dividend) of the USA, Japan and UK. There are 1174 daily, 234 weekly and 54 monthly returns from January 1988 to June 1992 for each country. First difference of logarithmic stock returns is used in calculating Q_n and $\tilde{Q}_n$ statistics.

Via the software package (RATS),[3] Q_n and $\tilde{Q}_n$ or alternatively

$$\tilde{V}_n = \frac{1}{\sqrt{n}} \tilde{Q}_n \text{ and } V_n = \frac{1}{\sqrt{n}} Q_n \text{ are calculated with an appropriate}$$

lag period q. The values of q are dependent on the sample size of the market data, i.e. q equals 35, 7 and 3 for daily, weekly and monthly returns respectively. The estimation results are reported in Tables 1 to 3. An examination of the tables suggests the following implications.

First, the UK market index, regardless of whether daily, weekly or monthly returns are used, exhibits a non-periodic cyclical pattern. Such a phenomenon is also found for different values of q. That is, the null hypothesis that a short-term memory exists is rejected by the modified R/S test.

Second, the traditional R/S test suggests that there exists some long-term persistence in the market indices of the Philippines, Indonesia and Taiwan. However, the persistence in the market indices disappears except in the Philippines once the modified R/S test is applied. In the case of Philippine market, the null hypothesis can be rejected for a handful of q values at about the 5% significant level. Our empirical finding clearly indicates that using the traditional R/S approach may well lead to an erroneous conclusion about long-term persistence in the data.

Last, except for the Philippines stock market, we have found the results are fairly consistent for the set of q values. The modified R/S tests reveals that the V_n statistic from the tables rejects or accepts the null hypothesis simultaneously for various values of q.

---

[3]RATS stands for Regression Analysis of Time Series by Estima.

IV CONCLUDING REMARKS

Do stock markets have a non-periodic long-term memory? The modified R/S test provides a more accurate answer since it takes deterministic non-linearity into consideration by reducing the size of disturbance term. One implication of the presence of a long-term memory is that the autocorrelation functions decay at a slower rate than if the data contain short-term memory. Such a phenomenon is consistent with the stochastic process in which slow decay of the autocorrelation function is witnessed via the fractional difference (Granger, 1980; Granger and Joyeux, 1980). Traditionally, the R/S approach can be used to detect whether time-series data are consistent with a fractionally differenced stochastic process. However, it is extremely sensitive to the presence of the short-term dependence, which the modified R/S approach can purge and thus better detect the existence of long-term persistence in the data set.

The empirical findings of our paper suggest that long-term memory is not found by the modified R/S test in most of the emerging Asian markets, except Philippines. However, the long-term memory phenomenon appears to prevail in the UK stock market. More research needs to be done on the causes and timing of the critical events giving rise to the presence of such long-term memory in the UK market.

REFERENCES

Ambrose, B. W., Ancel, E. and Griffiths, M. D. (1992) The fractal structure of real estate investment trust returns: the search for evidence of market segmentation and nonlinear dependency, *Journal of the American Real Estate and Urban Economics Association,* **20**(1), 25–54.

Andrews, D. (1991) Heteroskedasticity and autocorrelation consistent covariance matrix estimation, *Econometrica,* **59**, 817–58.

Booth, G. and Kaen, F. (1979) Gold and silver spot prices and market information efficiency, *Financial Review,* **14**, 21–6.

Cochrane, J. H. (1988) How big is the random walk in GNP?, *Journal of Political Economy,* **96**(5), 893–920.

Fama, E. F. (1970) Efficient capital markets: a review of theory and empirical work, *Journal of Finance,* **25**, 383–417.

Fama, E. F. (1991) Efficient capital markets: II, *Journal of Finance,* **56**(5), 1575–1617.

Fama, E. F. and French, K. R. (1988) Permanent and temporary components of stock prices, *Journal of Political Economy,* **96**(2), 246–73.

Granger, C. (1980) Long memory relationships and the aggregation of dynamic models, *Journal of Econometrics,* **14**, 227–38.

Granger, C. and Joyeux, R (1980) An introduction to long-memory time series-models and fractional differencing, *Journal of Time Series Analysis,* **1**, 15–29.

Greene, M. and Fielitz, B. (1977) Long-term dependence in common stock returns, *Journal of Financial Economics,* **4**, 339–49.

Helms, B., Kaen, F. and Rosenman, R. (1984) Memory in commodity futures contracts, *Journal of Futures Markets,* **4**, 559–67.

Hsieh, D. A. (1991) Chaos and nonlinear dynamics: application to financial markets, *Journal of Finance,* **46**(5), 1839–77.

Huang, B. N., Liu, Y. J. and Yang, C. W. (1994) A comparative study of the weekend effects in the United States, British and the Pacific Rim stock markets: an application of the Arch model, *Advances* in Pacific Basin Business, Economics and Finance, forthcoming.

Hurst, H. (1951) Long term storage capacity of reservoirs, *Transactions of the American Society of Civil Engineers*, **116**, 770–99.

Keim, D. B. and Stambaugh, R. F. (1984) A further investigation of the weekend effect in stock returns, *The Journal of Finance*, **39**(3), 819–40.

Keim, D. B.(1983) Size related anomalies and stock return seasonality: further empirical evidence, *Journal of Financial Economics*, **12**.

Kennedy, D. (1976) The distribution of the maximum Brownian excursion, *Journal of Applied Probability*, **13**, 371–6.

Lo, A. W. and MacKinlay, A. C. (1988) Stock market prices do not follow random walks: evidence from a simple specification test, *The Review of Financial Studies*, **1**(1), 41–66.

Lo, A. W. and MacKinlay, A. C. (1989) The size and power of the variance ratio test in finite samples: a Monte Carlo investigation, *Journal of Econometrics*, **40**, 203–38.

Lo, A. W. (1991) Long-term memory in stock market prices, *Econometrica*, **59**(5), 1279–1313.

Mandelbrot, B. (1972) Statistical methodology for non-periodic cycles: from the covariance to R/S analysis, *Annals of Economic and Social Measurement*, **1**, 259–90.

Mandelbrot, B. (1975) Limit theorems on the self-normalized range for weakly and strongly dependent processes, *Z. Wahrscheinlichkeitstheorie verw. Gebiete*, **31**, 271–85.

Mandelbrot, B. and Wallis, J. (1969) Computer experiments with fractional Gaussian noises: Parts 1, 2, 3, *Water Resources Research*, **5**, 228–67.

Newey, W. and West, K. (1987) A simple positive definite, heteroscedasticity and autocorrelation consistent covariance matrix, *Econometrica*, **55**, 703–5.

Peters, E. E. (1989) Fractal structure in the capital markets, *Financial Analysts Journal*, 32–7.

Reinganum, M. R. (1983) The anomalous stock market behavior of small firms in January: empirical tests for tax-loss selling effect, *Journal of Financial Economics*, **12**(1).

Roll, R. (1982) The turn of the year effect and the return premium of small firms, *Journal of Portfolio Management*, **x**, 0.

Siddiqui, M. (1976) The asymptotic distribution of the range and other functions of partial sums of stationary processes, *Water Resources Research*, **12**, 1271–6.

Van Horne, J. C. (1990) Changing world and Asian financial markets, in *Pacific-Basin Capital Markets Research* (ed.) S. G. Rhee and R. P. Chang, Amsterdam, North-Holland, pp. 65–80.

Table 1. *Daily stock returns' rescaled-range statistics*

	$\tilde{V}_n$	$V_n(q)$						
		5	10	15	20	25	30	35
HK	1.091	1.215	0.975	1.019	1.043	1.129	1.204	1.250
IND	2.127**	1.460	1.544	1.423	1.399	1.367	1.371	1.406
JPN	1.197	1.163	1.099	1.163	1.039	1.098	1.125	1.086
KOA	1.467	1.468	1.439	1.378	1.336	1.387	1.356	1.359
MAL	1.313	1.199	1.144	1.135	1.262	1.253	1.304	1.182
PHI	2.392**	2.031**	1.940**	1.755***	1.626	1.627	1.535	1.476
SIG	1.148	1.076	1.057	1.012	1.155	1.091	1.122	0.969
THA	1.409	1.269	1.286	1.267	1.198	1.126	1.071	0.992
TWN	1.811***	1.553	1.484	1.323	1.368	1.414	1.280	1.244
UK	0.699*	0.621*	0.616*	0.622*	0.661*	0.692*	0.726**	0.768
USA	0.977	1.046	1.141	1.046	1.116	1.237	1.363	1.148

NOTE: $\tilde{V}_n$ and $V_n(Q)$ denote the traditional rescaled-range statistics and modified rescaled range, respectively. 1% significant level are in the interval [0.721,2.098], 5% significant level are in the interval [0.809,1.862], and 10% significant level are in the interval [0.861,1.747] (Lo(1990)). *, **, *** indicate significant at 1%, 5%, and 10%.

Table 2. *Weekly stock returns' rescaled-range statistics*

	$\tilde{V}_n$	$V_n(q)$						
		1	2	3	4	5	6	7
HK	1.065	0.990	0.921	0.915	0.977	1.039	1.112	1.143
IND	1.519	1.550	1.453	1.418	1.398	1.379	1.405	1.388
JPN	1.166	1.118	1.051	0.999	0.996	1.025	1.093	1.100
KOA	1.398	1.455	1.326	1.258	1.304	1.374	1.344	1.323
MAL	1.129	1.117	1.065	1.137	1.194	1.258	1.191	1.058
PHI	1.978**	1.891**	1.771***	1.607	1.526	1.542	1.460	1.423
SIG	1.004	1.012	0.968	1.040	1.087	1.084	1.037	0.889
THA	1.292	1.223	1.225	1.156	1.114	1.087	1.036	0.952
TWN	1.599	1.466	1.348	1.345	1.364	1.316	1.292	1.188
UK	0.636*	0.596*	0.598*	0.608*	0.648*	0.676*	0.709*	0.754**
USA	0.895	0.952	0.962	0.879	0.998	1.185	1.128	1.071

NOTE: See Table 1.

Table 3. *Monthly stock returns' rescaled-range statistics*

	$\tilde{V}_n$	$V_n(q)$		
		1	2	3
HK	0.936	1.156	1.290	1.111
IND	1.612	1.312	1.338	1.302
JPN	1.059	0.988	0.990	1.075
KOA	1.407	1.252	1.283	1.221
MAL	1.151	1.092	1.192	1.180
PHI	1.761***	1.497	1.403	1.318
SIG	1.022	0.964	0.976	1.061
THA	1.080	0.926	0.927	0.955
TWN	1.394	1.303	1.192	1.032
UK	0.525*	0.640*	0.754**	
USA	0.896	0.953	1.148	

NOTE: See Table 1.

[Asian Economic Journal 1999, Vol. 13 No. 3]

A Comparative Statistical Analysis of the Taiwan Market and the New York Stock Exchange Using Five-Minute Interval Data*

Bwo-Nung Huang

National Chung Cheng University

Chin-Wei Yang

Clarion University of Pennsylvania

This paper compares some important statistics of the stock returns between the Taiwan market and the NYSE. The result of our analysis indicates that (i) mean and standard deviation of the stock returns in the Taiwan market are not U-shaped; a shape which was first observed by McInish and Wood (1991), in the US market, and (ii) we fail to reject the simple efficiency hypothesis for the Taiwan market whereas we reject the simple efficiency hypothesis during noon and at market close for the NYSE. The sharp contrast suggests that further analysis of trading hours and composition of investors are needed to provide a better explanation of the different market behavior of Taiwan.

I. Foreword

One of the most travelled grounds in finance literature is, without doubt, statistical tests on random-walk hypothesis. Since the seminal work by Fama (1970), there have been a plethora of empirical testings on the hypothesis (Fama, 1991). As is implied in the random-walk hypothesis, the great majority of the empirical studies support the efficient-market hypothesis, i.e., no abnormal profit can be realized in the long run. Since most prior studies have used the price or volume data at the time when a market closed, they have failed to mirror the statistical properties during a given trading day. Consequently, such data may well misrepresent the underlying statistical characteristics for that trading day. Naturally, a support for the simple efficiency hypothesis using market-close data does not necessarily imply that shrewd investors cannot make some abnormal profit during the day.

* We are grateful to an anonymous referee for valuable suggestions. However, we wish to take full responsibility for any remaining errors. An earlier version of this paper was read at the 2nd Annual Global Finance Conference, San Diego, 8–10 May, 1995.

Abnormal profit may be realized by taking advantage of the so called weekend effect (Keim and Stambaugh, 1984) although its effect is insignificant. As far as investors are concerned, proper timing to make a profit during a trading day is of the utmost importance in their investment decision-making process. Seen in this perspective, a test on the weak-form market efficiency needs to be re-examined by using intraday data with a more advanced statistical technique.

Prior studies using intraday data include works by Wood et al. (1985), Harris (1986), Jain and Joh (1988), McInish and Wood (1990) and Lockwood and Linn (1990). In particular, Stoll and Whaley (1990) find much greater variability in both stock returns and trading volume at market open or market close. Hence the return-generating process for either market open or market close may very well differ drastically from other hours in a trading day. Furthermore, McInish and Wood (1991), using the intraday stock index of the NYSE, find that first-order autocorrelation coefficients exhibit a U-shaped relation as do the mean stock returns and their standard deviations. However, most recently, Huang et al. (1998),[1] employing 15-minute return data of the NYSE and NASDAQ, detect several violations of the random-walk hypothesis during a trading day. Such a finding implies that the long established weak-form hypothesis in stock markets needs to be re-evaluated in a new light.

To date, most of the intraday market analyses are focused on the US (e.g., Wood et al., 1985) and Canadian markets (McInish and Wood, 1990). For the Asian markets, the intraday studies have just begun. For example, Ho and Cheung (1991) analysed the Hang Seng Index of Hong Kong; Chang et al. (1993a) discuss some statistics of the Tokyo Stock Exchange; and Chang et al. (1993a, 1993b) analyse the intraday returns of the Kuala Lumpur Stock Exchange and the Korea Stock Exchange respectively. However, an intraday analysis of the Taiwan market has been lacking. Apart from being an emerging market, the Taipei Stock Exchange (TSE) has its unique characteristics. One of these pertains to the length of trading hours: the market is not open after 12:00 noon[2] as compared with other markets whose trading hours pass beyond noon. It is speculated that the stock behaviour at noon is markedly different from that of other time segments. Therefore it is interesting to analyse the TSE in terms of some important descriptive statistics. It should be pointed out that prior studies using intraday data did not discuss the possibility of random walk: an important topic in the efficient-market hypothesis. The purpose of this paper is to (i) investigate the distribution of the descriptive statistics, and (ii) test the random-walk hypothesis of the TSE using 5-minute interval data. In order to facilitate the comparison with a mature financial market, we perform the same analysis on the 1992 NYSE data as well. The organization of the paper is as follows. The next section provides background information on the TSE. Section III describes the data and

---

1. Using the minute-by-minute transactions from the 1987 NASDAQ and 1992 NYSE data, Huang and Yang (1994) also find a similar U-shaped pattern.
2. The market of the TSE closes at 11:00 on Saturday.

methodology used in this study. Section IV discusses the empirical results. Finally, Section V contains the conclusions.

II. Background Information on the Taipei Stock Exchange[3]

Apart from its unusual trading hours, the TSE also contains some idiosyncratic microstructures. Basically, the TSE is a market of order-driven, electronic broker systems without market makers. The major participants are individual traders who account for approximately 95% of total transaction volumes. Among the numerous individual traders are a handful of investors who are capable to some extent of influencing the market. Surprisingly, institutional investors, according to the annual report of the Grand Cathy Corp. (1992), account for only 5% of transaction volumes. On the other hand, individual traders comprise approximately 30% of the total transaction volume in the US market. Seen in this perspective, the TSE is extremely sensitive to virtually all news announcements. It comes as no surprise that the turnover rate of the TSE is the highest in the world.[4]

In contrast, the NYSE is characterized by an order-driven floor system in which specialists act in some passive capacity to stabilize the market in the presence of order imbalance. More specifically, it possesses a continuous auction trading mechanism for trades initiated after market opening. The resulting prices are determined by the information of the call markets gathered by the specialists prior to market opening. In addition, there is no price limit set during the transaction, As with the NYSE, the price of orders in the TSE is determined through the call mark. Beyond that, the market of the TSE functions as a combination of call and continuous trading. Orders are then matched, on average, every two minutes during which computers sequentially take the order, and the match is made through the call market.[5] As markets close, each stock is matched for the last time and other late orders will be removed from the book.

Individual traders account for the lion's share of transactions owing to the fact that there are no market makers. The exchange of the market functions as a stabilizing force by setting an interday price limit at ±7% and an intraday price which mandates stock price to move within two ticks of the bid, ask or transaction price determined from the previous match. A trader can obtain quasi-market orders via limiting orders which adhere to the interday price limit despite the fact that a formal market order is technically disallowed. Given the different (short) trading hours, microstructure regarding matching practice, and the existence of the price ceiling, it is interesting to explore the statistical properties using the intraday data of the TSE in comparison with that of the NYSE.

---

3.   This section draws largely on Chen et al. (1994), whom we gratefully thank for letting us share the information.

4.   According to a recent issue of the *Economist* (19–25 March 1994), the turnover rate of the TSE is 35% higher than that of the Korea Stock Exchange.

5.   The length of time for a match is entirely dependent on the volume and there is no hard and fast rule to determine it.

Figure 1   Time Series of the Taiwan Stock Index

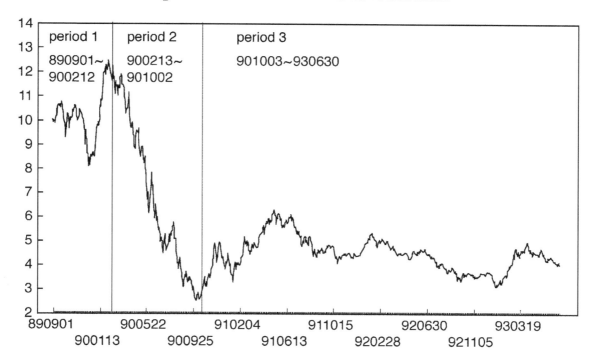

III.   Source of Data and Methodology

The stock returns of the NYSE are obtained from the 1992 CRSP (Center for Research in Security Prices) tape with the trading hours from 9:30 am till 4:00 pm or 390 minutes (6.5 hours). There were 254 trading days in 1992 with 78 five-minute intervals each day. Thus, we have 19812 five-minute-interval stock returns in the sample. In the case of the TSE, the sample period covers from 1 September 1989 to 30 June 1993. The trading hour started from 9:00 to 12:00 or three hours a day Monday to Friday; it lasts from 9:00 am to 11:00 am on Saturday. To make trading hours consistent throughout a week, we delete all the Saturday data from the sample in the case of the TSE.[6] Hence, there are 36 five-minute interval stock returns each day Monday to Friday. To take bull and bear markets into consideration, we segment the sample period into three subperiods (see Figure 1). Period 1 runs from 1 September 1989 to 12 February 1990 or bull market. Period 2 runs from 13 February 1990 to 2 October 1990 or bear market. Finally, period 3 runs from 3 October 1990 to 30 June 1993 or bull-bear market.

First, some descriptive statistics of the TSE for the subperiods as well as the whole period are reported in Figure 2 to Figure 9 while those of the NYSE are

---

6.   As is well-known from the prior studies, the stock returns at market close differ drastically from that during a trading day. Hence the stock returns at 11:00 am on Saturday will differ markedly from those of the other five weekdays. To compare the returns at 11:00 am on Saturday with that of 12:00 noon on weekdays will render some time segments unused.

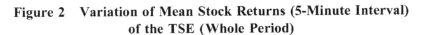

**Figure 2   Variation of Mean Stock Returns (5-Minute Interval)
of the TSE (Whole Period)**

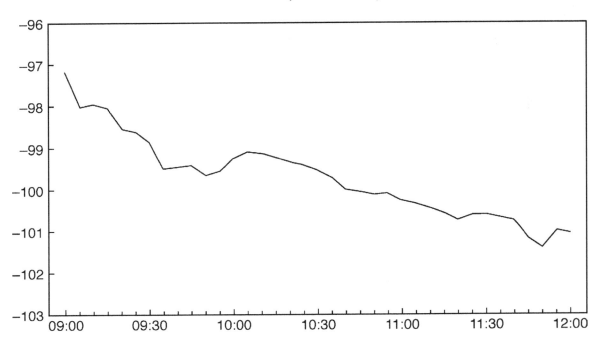

**Figure 3   Variation of the Standard Deviation of Stock Returns (5-Minute Interval)
of the TSE (Whole Period)**

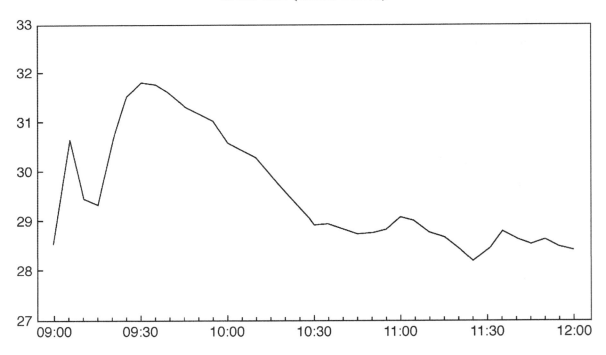

**Figure 4   Variation of Mean Stock Returns (5-Minute Interval)
of the TSE (Period 1)**

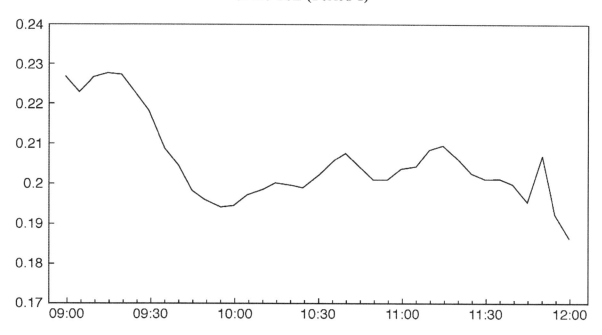

**Figure 5   Variation of the Standard Deviation of Stock Returns (5-Minute Interval)
of the TSE (Period 1)**

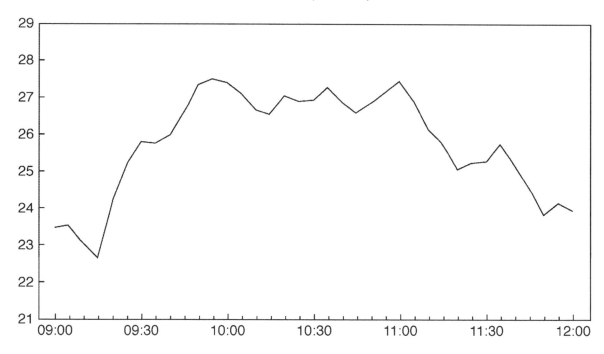

**Figure 6   Variation of Mean Stock Returns (5-Minute Interval)
of the TSE (Period 2)**

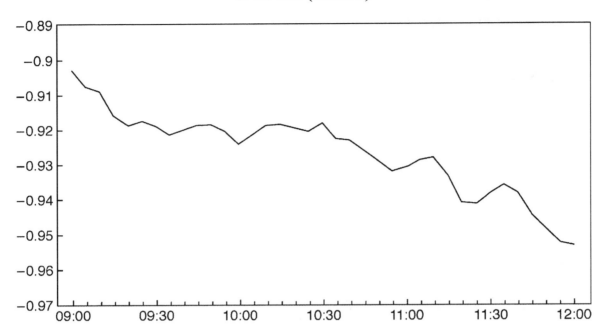

**Figure 7   Variation of the Standard Deviation of Stock Returns (5-Minute Interval)
of the TSE (Period 2)**

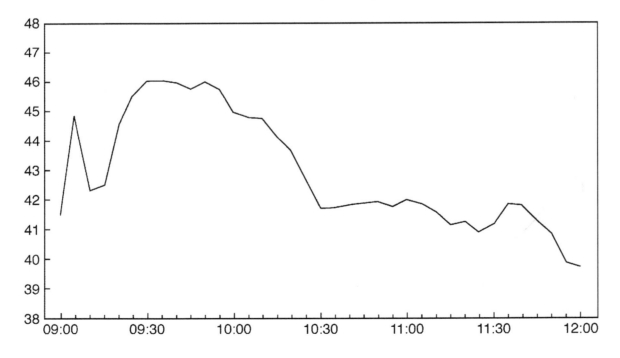

**Figure 8   Variation of Mean Stock Returns (5-Minute Interval)
of the TSE (Period 3)**

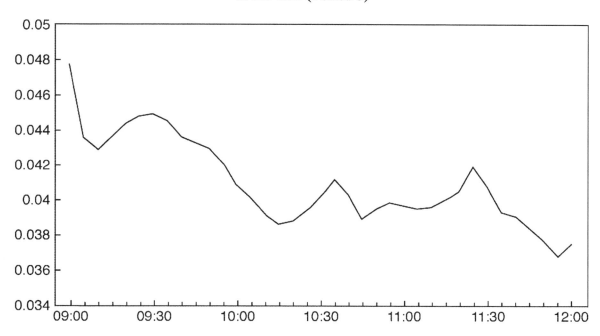

**Figure 9   Variation of the Standard Deviation of Stock Returns (5-Minute Interval)
of the TSE (Period 3)**

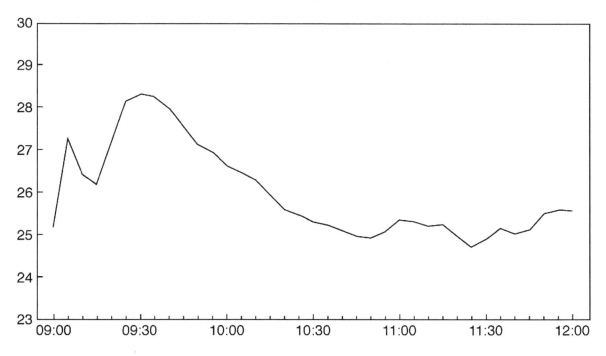

shown in Figures 10 and 11.[7] Next, we examine a critical issue: do the stock returns have long-term memory? Using the technique developed by Lo (1991), we test this hypothesis using the modified rescaled range (R/S) approach.

In his classical paper, Hurst (1951) first developed the R/S statistic to compute reservoir storage required to yield the average flow by computing the cumulative sums of the departures of the annual totals from the mean annual total discharge. Thus the range of the maximum and the minimum of these cumulative totals is taken as the required storage (Hurst, 1951, p. 770). Using the same analogy, the range of partial cumulative sums of deviations of a sequence of security returns from its mean of various lengths, rescaled by its standard deviation constitutes the traditional R/S statistic:

$$\tilde{Q}_n \equiv \frac{1}{S_n}\left[\underset{1\le k\le n}{Max}\sum_{j=1}^{k}(r_j - \bar{r}_n) - \underset{1\le k\le n}{Min}\sum_{j=1}^{k}(r_j - \bar{r}_n)\right] \qquad (1)$$

where $r_1, r_2, \ldots, r_n$ represents security returns over n periods; $\bar{r}_n = \frac{1}{n}\sum_j r_j$ is the sample mean return; and $S_n \equiv \left[\frac{1}{n}\sum_j(r_j - \bar{r}_n)^2\right]^{1/2}$ is an estimator of the standard deviation.

Note that for each subperiod of length $n > 3$, an average $\tilde{Q}_n$ over different starting point is computed for the standard R/S analysis. From some perspectives, the R/S statistic is superior to other traditional time series approaches such as autocorrelation, variance ratio, and spectral decomposition models in detecting long-term dependence. For instance, through Monte Carlo simulations, Mandelbrot and Wallis (1969) show that the R/S statistic is robust in capturing long-range dependence even in the presence of greater magnitudes of skewness and kurtosis in non-Gaussian distributions. Furthermore, the R/S statistic possesses the almost-sure convergence property in the stochastic process with infinite variance: a property that is not found in the tests of the variance ratio or autocorrelation (Mandelbrot, 1972, 1975). The second advantage of the R/S statistic, according to Mandelbrot (1972), is its ability to test for the existence of nonperiodic cycles. Despite the strength of the R/S statistic, Lo (1991) shows that the test result is extremely sensitive with respect to the time series data manifesting short-range dependence.

---

7. These descriptive statistics include mean, standard deviation, skewness, kurtosis quantile, and the J-B Statistic. Being too numerous, we present only the first two statistics graphically. Other statistics are available upon request. Despite the nonnormality problem, most prior studies focused on discussion of mean and variance (Bekaert and Harvey, 1997). We take the same path in this paper. In addition, Wood et al. (1985) dissects the time period into different segments in a trading day to study the trading behaviour through the examination of the changing means and variances.

**Figure 10 Variation of Mean Stock Returns (5-Minute Interval):
the NYSE (1992)**

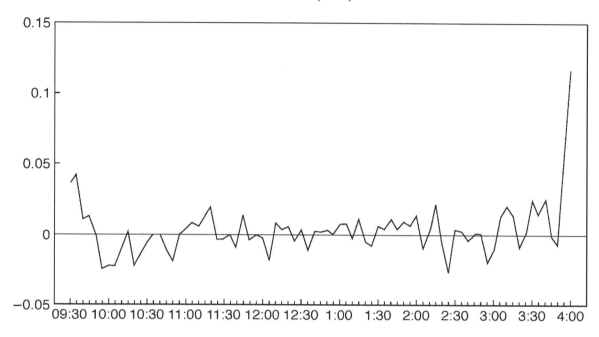

**Figure 11 Variation of the Standard Deviation of Stock Returns (5-Minute Interval):
the NYSE (1992)**

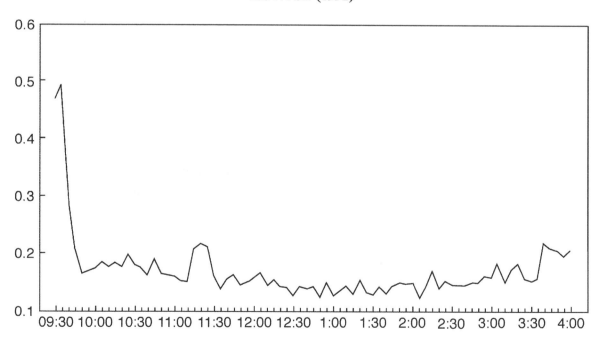

To account for the potential effect of the short-range dependence in the R/S test, Lo (1991) proposes the modified R/S statistic as follows:

$$\tilde{Q}_n \equiv \frac{1}{\hat{\sigma}_n(q)} \left[\underset{1 \leq k \leq n}{Max} \sum_{j=1}^{k} (r_j - \bar{r}_n) - \underset{1 \leq k \leq n}{Min} \sum_{j=1}^{k} (r_j - \bar{r}_n) \right] \tag{2}$$

$$\hat{\sigma}_n \equiv \frac{1}{n} \sum_{j=1}^{n} (r_j - \bar{r}_n)^2 + \frac{2}{n} \sum_{j=1}^{q} w_j(q) \left\{ \sum_{i=j+1}^{n} (r_i - \bar{r}_n)(r_{i-j} - \bar{r}_n) \right\} \tag{3}$$

where q is number of periods lagged; and the weight is defined as $w_j \equiv 1 - j\big/(q+1)$ for $q < n$, so that the modified standard deviation estimator in Equation (3) is positive (see Newey and West, 1987). It should be pointed out that the modified standard deviation estimator also contains weighted autocovariance up to lag q. In particular, Andrews (1991) suggests a data-dependent rule to determine the optimum value for q based on some asymptotic property. However, the choice becomes rather arbitrary in the case of a finite sample.

The asymptotic property of the modified R/S or Q_n statistic can be alternatively given by

$$V_n(q) \equiv \frac{1}{\sqrt{n}} Q_n \overset{a}{\longrightarrow} V \tag{4}$$

in which v has the following probability distribution (see Kennedy, 1976; Siddiqui, 1976):

$$F_{\bar{v}}(v) = 1 + 2 \sum_{k=1}^{\infty} (1 - 4k^2 v^2) e^{-2(kv)^2} \tag{5}$$

The critical values can therefore be computed from Equation (5), and some often-used values are reported by Lo (1991). In our study, the holding period equals one week ($q = 5$) and three weeks ($q = 15$) respectively. In the case of the TSE, the modified R/S statistics are reported from Figure 12 to 19 while those of the NYSE are produced in Figure 20 and 21.

IV. Discussion of The Empirical Results

Wood et al. (1985) find a U-shaped relation in terms of the mean intraday stock returns and their standard deviations for the NYSE. Likewise, McInish and Wood (1990) detect a similar pattern in the Toronto Stock Exchange. It comes as no surprise that the U-shaped pattern is observed in both markets because they share a similar microstructure. However, the U-shaped relation is also found by Cheung (1993) in the Hong Kong Market whose structures differ from those of the

Figure 12 Variation of the Modified R/S Statistics with 5-Minute Interval ($q = 5$) of the TSE (Whole Period)

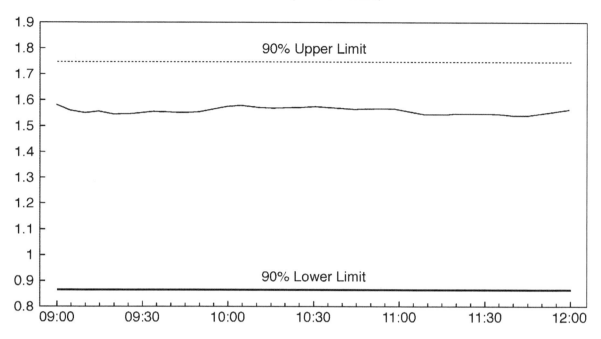

Figure 13 Variation of the Modified R/S Statistics with 5-Minute Interval ($q = 15$) of the TSE (Whole Period)

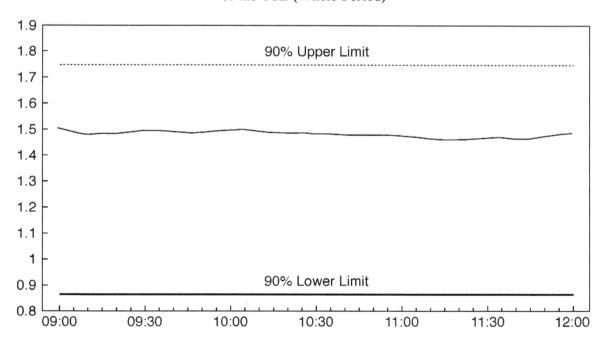

**Figure 14   Variation of the Modified R/S Statistics with 5-Minute Interval ($q = 5$)
of the TSE (Period 1)**

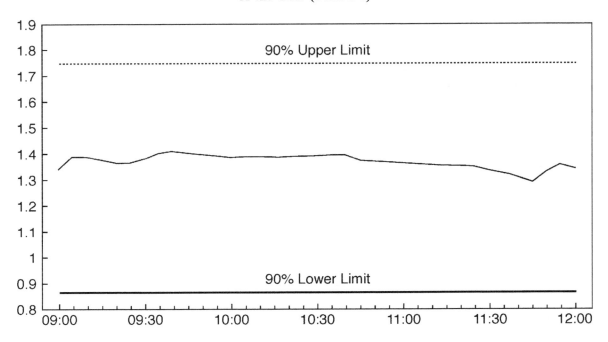

**Figure 15   Variation of the Modified R/S Statistics with 5-Minute Interval ($q = 15$)
of the TSE (Period 1)**

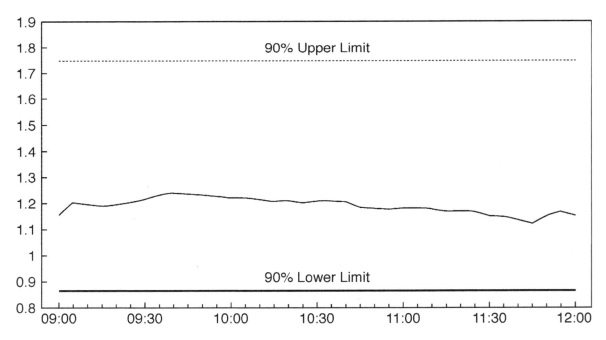

**Figure 16   Variation of the Modified R/S Statistics with 5-Minute Interval ($q = 5$)
of the TSE (Period 2)**

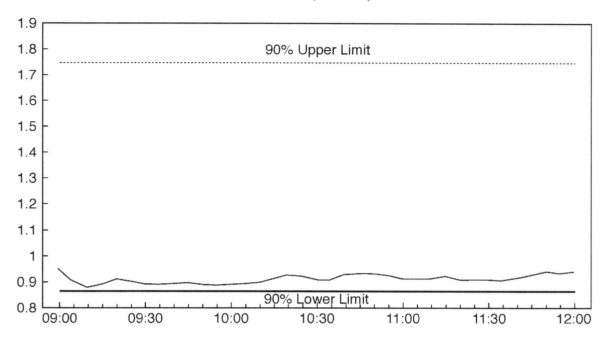

**Figure 17   Variation of the Modified R/S Statistics with 5-Minute Interval ($q = 15$)
of the TSE (Period 2)**

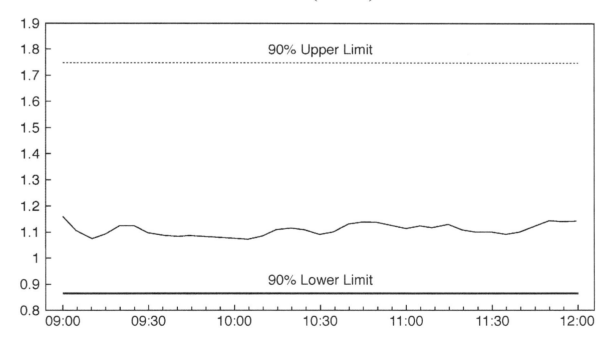

Figure 18   Variation of the Modified R/S Statistics with 5-Minute Interval ($q = 5$)
of the TSE (Period 3)

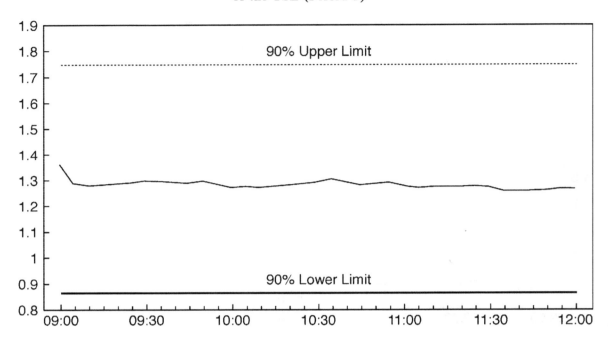

Figure 19   Variation of the Modified R/S Statistics with 5-Minute Interval ($q = 15$)
of the TSE (Period 3)

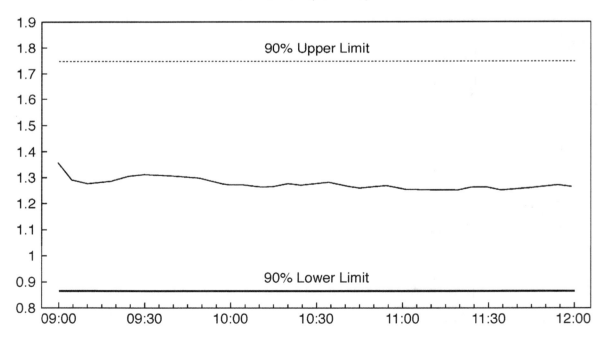

Figure 20   Variation of the Modified R/S Statistics with 5-Minute Interval ($q = 5$)
the NYSE (1992)

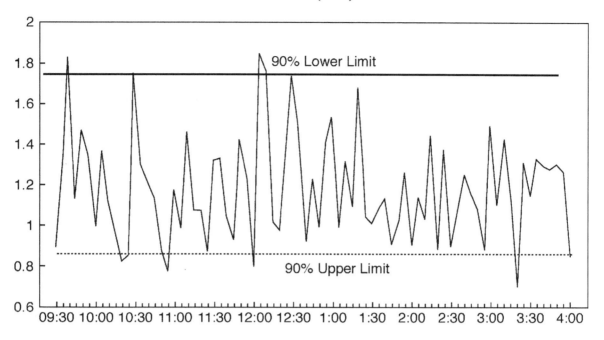

Figure 21   Variation of the Modified R/S Statistics with 5-Minute Interval ($q = 15$)
the NYSE (1992)

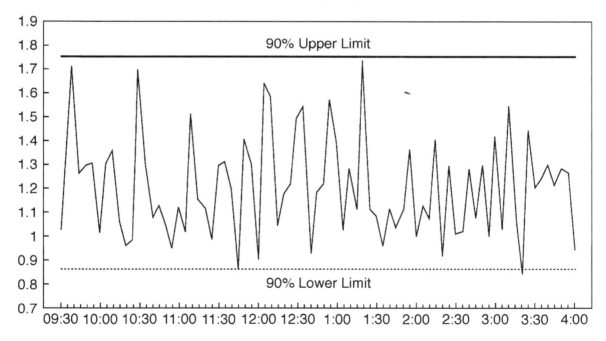

NYSE.[8] Likewise Chang et al. (1993b) confirm the same pattern in the Malaysian market. Although, U-shaped relations have been found in these stock markets which differ markedly either in structure or in institutional factors, these markets do have some common denominators. First their trading hours extend beyond noon. Second, they do not seem to have price ceilings. Conversely, the TSE has price ceilings and extends its transactions no later than noon. Would such unique characteristics coupled with the fact that the market is dominated by individual traders cause the behaviour of the TSE to differ from what has been reported in the literature?

Surprisingly, a U-shaped relation does not exist for mean returns; instead, it is J curved (or an incomplete U relation). This relation is closer to the bid and ask spread result by McInish and Wood (1992). However, the J curve phenomenon of the TSE is not identical to that of McInish and Wood. In particular, the mean return at market close is no higher than the mean returns of other time segments. In contrast, the mean return at market open is the highest in a typical trading day. After that, the mean returns are in general decline until five minutes before market close. There were some minor spikes during a trading day. In the bull market and bull-bear market, the mean returns at various time segments show greater differences. More specifically, 25 minutes after market open, the mean return remains approximately at peak in the bull market. It bottoms at about 10:00 am. After that, a minor spike takes place at 10:40 am. It then slides down until around 11:15 and 11:50 when another two small upswings occur. The mean return at market close is generally the smallest during the trading hours in the TSE. In the bull market, the overall mean return remains positive.

Consistent with the result by Wood et al. (1985), the five-minute interval mean return of 1992 NYSE declines to negative values after market open. It fluctuates around zero mean return during a typical trading day but it rises appreciably at market close. Evidently, the behaviour of the mean returns of the TSE differs distinctively from that of the NYSE and other Asian markets.

As with the McInish and Wood model (1992), the behaviour of standard deviation of stock returns of 1992 NYSE exhibits a pattern of incomplete U or J curve. That is, the standard deviation reaches its peak at market open owing to the greater interday risk. The intraday risk rises at market close reflecting slightly higher standard deviations. On the other hand, the TSE exhibits rather different behaviour in terms of standard deviations in all periods except the bull market. That is, the standard deviation at market open does not reach its maximum; it spikes five minutes after market open. It reaches the maximum point around 9:30 am (see Figure 3). The standard deviation of the stock returns fluctuates randomly around a lower level starting at 10:30 am. Although the standard deviation slightly increases its value at market close in period 3, the same phenomenon is not found in other two periods.

---

8. Miller (1989) argues that specialists normally set the prices higher at market close so as to better defend their position next day. However, in many Asian markets, there is no such role for specialist to play.

The behaviour of the standard deviations of stock returns is rather different in the case of the bull market of the TSE. In general, the standard deviation reached its trough at market open, but it began to increase twenty minutes after. Around 10:00 am, it reaches a pinnacle for a given trading day. After that, the stock returns fluctuate randomly around at a high level until 11:00 am. It then declines and never rebounds back appreciably at market close.

Evidently, the behaviour of mean and standard deviation of the TSE differs markedly from that of the NYSE especially in terms of standard deviations of the stock returns. It is to be noted that the result of Cheung (1993) indicates that the patterns of means and standard deviations of the Hong Kong market are very similar to those of the NYSE although they have different institutional factors. There is also the existence of a lunch break (the trading hours are 10:00 am~12:30 pm and 2:30 pm~3:00 pm) in the Hong Kong market. Note that both markets extend beyond noon. On the other hand, institutional investors dominate both market in contrast to the 95% of investors who are individual traders in the TSE where trading hours do not extend beyond noon. Viewed in this light, the different patterns of means and standard deviations of the TSE may be attributed to these two factors: the overabundance of individual traders and the fact that the market closes at noon. However, a microstructure model is needed to reach useful conclusions before applying more rigorous econometric techniques. While this provides an interesting and challenging idea for future research, it falls beyond the scope of the present paper.

Beyond the fact that means and standard deviations of the TSE have distinctive patterns, it is of great interest to test the existence of the simple-efficiency hypothesis using five-minute interval data from the TSE market. Recently, the emphasis of the efficient-market hypothesis has shifted from the unit-root tests and variance-ratio approach to the modified R/S statistic (Lo, 1991).[9] More strikingly, Huang et al. (1994), using fifteen-minute return data from 1987 NYSE and 1992 NASDAQ, detect that the stock returns within some time segments are not consistent with the efficient-market hypothesis. Likewise, we have found that the stock returns of 1992 NYSE are not in conformity with the efficient-market hypothesis at time interval 11:00 am and 12:00 am. This is shown in Figures 20 and 21 with the holding period of one week, i.e., the modified R/S statistic exceeds its 90% critical value. If we extend the holding period to three weeks ($q = 15$), the stock returns at 11:45 am, 12:00, 1:15 pm and 3:20 pm are found to be inconsistent with the simple-efficiency hypothesis. This result is very similar to that of Huang et al. (1998). Naturally, the same test on the efficient-market hypothesis is also of great interest for the TSE. Rather surprisingly, the only R/S statistic to approach the 90% critical value occurred at 30–60 minutes after market opening in the period of the bear market with a holding period of one week. The R/S statistic at all other time intervals in all other types

9. After Lo's paper (1991), applications of the modified R/S model can be found in Huang and Yang (1995), Huang et al. (1997), Batten and Ellis (1996), Chow et al. (1996) and Coggin (1998).

of markets is well within the critical values, hence the stock returns of the TSE do not have long-term memory in the sense that we fail to reject the null hypothesis.

V. Conclusion

Since the work by Wood et al. (1985), a number of studies have focused on the patterns of some basic statistics using intraday data. Virtually unanimously, they have found either a U-shaped relation or J-curve pattern. More surprisingly, such relations are also found in the Hong Kong market whose institutions and trading formats are quite different from others. However, this study has shown that the behaviour of stock returns in the TSE is strikingly different from the other markets in terms of means, standard deviations, and the simple efficiency test. Given the background information about the TSE, we suspect that the unique behavior of the Taiwan market may be because trading closes at noon and because 95% of investors are individual traders whose extreme sensitivity to news announcements may reduce the rationality of their actions in the stock market. Whether the anomalous behaviour of the Taiwan stock market can be accounted for by the two explanations provides a wide field for future research.

References

Andrews, D., 1991, Heteroskedasticity and autocorrelation consistent covariance matrix estimation. *Econometrica*, **59**, pp. 817–858.

Batten, Jonathan and Craig Ellis, 1996, Fractal structures and naive trading system: evidence from the spot US dollar/Japanese yen. *Japan & the World Economy*, **8**(4), pp. 411–421.

Bekaert, Geert and Campbell R. Harvey, 1997, Emerging equity market volatility. *Journal of Financial Economics*, **43**, pp. 29–77.

Chang, Rosita P., Toru Fukuda, S. Ghon Rhee and Makoto Takano, 1993a, Intraday and behavior of TOPIX. *Pacific Basin Finance Journal*, **1**, pp. 67–95.

Chang, Rosita P., Jun-Koo Kang and S. Ghon Rhee, 1993b, *The Behavior of Malaysian Stock Prices*. Working Paper, Department of Finance and Insurance, University of Rhode Island.

Chen, Meng-Hsiu, Edward H. Chow, Victor W. Liu and Yu-Jane Liu, 1994, *Intraday Stock Returns of Taiwan: An Examination of Transaction Data*. Paper read at the First NTU International Conference on Finance, 27–28 June 1994, Taipei, Taiwan.

Cheung, Yan-Leung, 1993, *Intraday Returns and the Day-end Effect: Evidence from the Hong Kong Equity Market*. Working Paper, Department of Finance and Insurance, City University of Hong Kong.

Chow, K., Victor, Mong-Shium Pan, and Sakauo Ryoichi, 1996, On the long-term or short-term dependence in stock prices: evidence from international stock markets. *Review of Quantitative Finance and Accounting*, **6**(2), pp. 181–194.

Coggin, T. D., 1998, Long-term memory in equity style indexes. *Journal of Portfolio Management*, **24**(2), pp. 37–46.

Economist, 1994, 12–25 March.

Fama, Eugene F., 1970, Efficient capital markets: a review of theory and empirical work. *Journal of Finance*, **25**, pp. 383–417.

Fama, Eugene F., 1991, Efficient capital markets: II. *Journal of Finance*, **56**(5), pp. 1575–1617.

Grand Cathy Corp., 1992, Rest of reference to follow.

Harris, L., 1986, How to profit from intraday stock returns. *Journal of Portfolio Management*, **13**, pp. 61–64.

Ho, Yan-Ki and Yan-Leung Cheung, 1991, Behaviour of intra-daily stock return on an Asian emerging market – Hong Kong. *Applied-Economics*, **23**(5), pp. 957–966.

Huang, Bwo-Nung and Chin W. Yang, 1994, *An Examination of the Random Walk Hypothesis of the New York Stock Exchange Using One-minute Interval*. Working Paper, Department of Economics, National Chung-Cheng University.

Huang, Bwo-Nung and Chin W. Yang, 1995, The fractual structure of the multinational stock markets. *Applied Economics Letter*, **2**(3), pp. 67–71.

Huang, Bwo-Nung, Chin W. Yang and Walter C. Labys, 1997, The random walk hypothesis of the emerging stock markets revisited: a comparison of test power of the variance ratio test and re-scaled range models. *Journal of Financial Studies*, **5**(2), pp. 59–86.

Huang, Bwo-Nung, D. B. Means, Chin Wei Yang and Robert Van Ness, 1998, The test of simultaneous efficient and inefficient markets: an application of the modified R/S model with intraday stock returns. *Global Business and Financial Review*, **3**(2), pp. 49–54.

Hurst, H., 1951, Long-term storage capacity of reservoirs. *Transactions of the American Society of Civil Engineers*, **116**, pp. 770–799.

Jain, P. and G. Joh, 1988, The dependence between hourly prices and trading volume. *Journal of Financial and Quantitative Analysis*, **23**, pp. 269–284.

Keim, D. B. and R. F. Stambaugh, 1984, A further investigation of the weekend effect in stock returns. *Journal of Finance*, **39**(3), pp. 819–840.

Kennedy, D., 1976, The distribution of the maximum brownian excursion. *Journal of Applied Probability*, **13**, pp. 371–376.

Lo, Andrew W., 1991, Long-term memory in stock market prices. *Econometrica*, **59**(5), pp. 1279–1313.

Lockwood, L. J. and S. C. Linn, 1990, An examination of stock market return volatility during overnight and intraday periods 1964–1989. *Journal of Finance*, **45**, pp. 591–601.

Mandelbrot, B., 1972, Statistical methodology for non-periodic cycles: from the covariance to R/S analysis. *Annals of Economic and Social Measurement*, **1**, pp. 259–290.

Mandelbrot, B., 1975, Limit theorems on the self-normalized range for weakly and strongly dependent processes. *Z. Wahrscheinlichkeitstheorie verw. Gebiete*, **31**, pp. 271–285.

Mandelbrot, B. and J. Wallis, 1969, Computer experiments with fractional gaussian noises. Parts 1, 2, 3, *Water Resources Research*, **5**, pp. 228–267.

McInish, Thomas H. and Robert A. Wood, 1990, A transaction data analysis of the variability of common stock returns during 1980–1984. *Journal of Banking and Finance*, **14**, pp. 99–112.

McInish, Thomas H. and Robert A. Wood, 1991, Autocorrelation of daily index returns: intraday-to-intraday versus close-to-close intervals. *Journal of Banking and Finance*, **15**, pp. 193–206.

McInish, Thomas H. and Robert A. Wood, 1992, An analysis of intraday pattern in bid/ask spreads for NYSE stocks. *Journal of Finance*, **87**, pp. 753–764.

Miller, E. M., 1989, Explaining intra-day and overnight price behavior. *Journal of Portfolio Management*, **15**, pp. 10–16.

Newey, W. and K. West, 1987, A simple positive definite, heteroscedasticity and autocorrelation consistent covariance matrix. *Econometrica*, **55**, pp. 703–705.

Siddiqui, M., 1976, The asymptotic distribution of the range and other functions of partial sums of stationary processes. *Water Resources Research*, **12**, pp. 1271–1276.

Stoll, Hans R. and Robert E. Whaley, 1990, Stock market structure and volatility. *Review of Financial Studies*, **3**, pp. 37–71.

Wood, R. A., T. H. McInish and J. K. Ord, 1985, An investigation of transactions data for NYSE stock. *Journal of Finance*, **40**, pp. 723–739.

Applied Financial Economics, 1995, **5**, 251–256

Do Asian stock market prices follow random walks? Evidence from the variance ratio test

BWO-NUNG HUANG

Institute of International Economics, National Chung Cheng University, Chia-Yi 621, Taiwan

Using a basis of the variance ratio statistics with both homoscedastic and hetero-scedatic error variances (Lo and Mackinlay, 1988) the random walk hypothesis of the Asian stock markets is tested. Of the developed and emerging markets, it is found that the random walk hypothesis for the markets of Korea and Malaysia is rejected for all different holding periods. In addition, the random walk hypothesis is also rejected for the Hong Kong, Singapore, and Thailand markets using the heteroscedasticity-consistent variance ratio estimator.

I. INTRODUCTION

Whether the stock prices follow the random walk hypothesis has great implications in financial theories and statistical modelling. If stock prices follow random walk, as is implied by many empirical studies, the market is said to be efficient in the sense that is discounts all available public information and manifests itself in subsequent cash flows.[1] That is, not being able to reject the random walk hypothesis will render the stock returns unpredictable. Recently, however, several studies have uncovered empirical evidence which suggests that stock returns contain predictable components (e.g. Fama and French, 1988; Lo and Mackinlay, 1988; Keim and Stampbaugh, 1986).

Despite these startling findings, their resutls of failing to reject the random walk hypothesis applies only to the stock markets of industrialized nations. Consequently, it is of great interest to explore if the similar patterns can be identified in the Asian stock markets. Besides, newer and more powerful testing procedures are needed to better understand the Asian stock markets including some well-developed markets (e.g. the Hong Kong and Japan markets). Altogether the random walk hypothesis will be tested in the nine

Asian stock markets. The organization of the paper is as follows. The next section describes the data and the corresponding statistical properties. Section III presents the variance ratio (VR) and Dickey–Fuller (DF) unit root test. Section IV discusses the empirical results. The conclusion is contained in the last section.

II. DATA AND THE STATISTICAL PROPERTIES

The weekly stock returns are from Morgan Stanley Stock Index Database including the markets of Hong Kong (HK), Indonesia (IND), Japan (JPN), Korea (KOA), Malaysia (MAL), Philippines (PHI), Singapore (SIG), Thailand (THA) and Taiwan TWN). The nine stock price indices are weighted by market values in US dollars. To avoid the weekend effect, we take the closing price on Wednesday as the representative price for the week.[2] Sample period spans from 1 January 1988 to 30 June 1992 with the index of 1 January 1988 as the base, i.e., it equals 100. In total, there are 234 weekly returns for each market and their characteristics are reported in Table 1. It should be pointed out that

---

[1] For example, see Fama (1965) for detail.
[2] If a Wednesday price is not available, we replace it with the Tuesday price. The Thursday price will be used if the Tuesday price is not available.

0960–3107 © *1995 Chapman & Hall*

Table 1. *Characteristic of the Asian stock markets*

	HK	IND	MAL	KOA	JPN	SIG	PHI	TWN	THA
T.R.	4732	40.6	94.3	172.2	33.4	89.8	25.1	235.5	91.5
Ranking	(18)	(25)	(5)	(2)	(30)	(7)	(37)	(1)	(6)
Capitalization[a]	385 247	32 953	220 328	139 420	2 999 756	132 742	40 327	195 198	130 510
as % of world	(2.73)	(0.23)	(1.56)	(0.99)	(21.27)	(0.94)	(0.29)	(1.38)	(0.93)
Ranking	(6)	(31)	(9)	(15)	(2)	(17)	(29)	(13)	(18)
Value traded[b]	131 550	9158	163 661	211 710	954 341	81 623	6785	346 487	86 934
as % of world	(1.79)	(0.12)	(2.09)	(2.88)	(13)	(1.11)	(0.09)	(4.72)	(1.18)
Ranking	(13)	(32)	(15)	(7)	(3)	(31)	(30)	(18)	(17)
Listed firms	450	174	410	693	2155	178	180	285	347
Ranking	(13)	(32)	(15)	(7)	(3)	(31)	(30)	(18)	(17)

Source: *Emerging Stock Markets Fact book*, 1994, Published by International Financial Corporation.
Notes: HK = Hong Kong, IND = Indonesia, MAL = Malaysia, JPN = Japan, SIG = Singapore, PHI = Philippines, TWN = Taiwan, THA = Thailand.
T.R. = turnover rate, as % of world = as percentage of world total.
[a]total world capitalization is 14 100 763 million US dollars for 1993.
[b]total world traded value is 7 342 265 billion for 1993.

we employ weekly data to circumvent the bias caused by non-trading the bid-ask spread, and nonsynchronous prices.[3]

First, we calculate the rate of stock returns by taking the first difference of stock returns measured in logarithmic prices of

$$r_t = \log P_t - \log P_{t-1} \qquad (1)$$

where P_t is the current stock price index and r_t is the rate of stock return. An examination of Table 1 readily reveals the important role the Asian markets plays in the world. In 1993, the nine stock markets account for approximately 30.33% of world total capitalization value and 27% of world total traded value respectively. Needless to say, while the market of Japan dominates the Asian markets, other emerging economies (Hong Kong, Malaysia and South Korea) begin to witness bubbling trading activities in their financial markets. The Asian markets manifest themselves in the top 35 ranks (in most cases) in the categories reported in Table 1.

Several fundamental statistics, listed in Table 2, shed some light on the Asian markets. First, with the exception of the Indonesia markets, all markets have positive mean stock returns. In particular, the markets of Hong Kong and Korea have the two highest mean stock returns. In the case of variance of stock returns, the markets of Hong Kong, Malaysia, Korea and Philippines manifest themselves in greater fluctuations. As indicated by the Jarque–Bera statistic, barring the Thailand market, none of the markets has a normally distributed return. More specifically, they are characterized by leptokurtoses of varying degrees. Aside

from these properties, it is of great concern to investigate if the stock returns, measured by the first difference of the logarithmic prices, exhibit the phenomenon of autoregressive conditional heteroscedasticity (ARCH).[4] The χ^2 test statistic with one lag period indicates that all the markets excluding that of Korea, Japan and Taiwan unravel the existence of the ARCH(1). In this case, we shall employ the heteroscedasticity-consistent VR statistic to circumvent the bias caused from using the conventional VR test statistic.

III. THE VARIANCE RATIO AND THE DICKEY–FULLER TESTS

As shown by Lo and Mackinlay (1988), the VR statistic is derived from the assumption of linear relations in observation interval regarding the variance of increments.[5] If a time series follows a random walk process, the variance of a kth-difference variable is k time as large as that of the first-difference variable. Hence, for equally spaced intervals, we partition the stock price into $nk + 1$ segments denoting them by $y_0, y_1 \cdots y_{nk}$. For a time series characterized by random walks, one kth of the variance of $P_t - P_{t-k}$ is expected to be the same as the variance of $P_t - P_{t-1}$ or

$$VR(k) = \frac{\sigma_k^2}{\sigma_1^2} \qquad (2)$$

where σ_k^2 is the unbiased estimator of one kth of the variance of $\ln P_t - \ln P_{t-k}$ and σ_1^2 is the unbiased estimator of the variance of $\ln P_t - \ln P_{t-1}$. These estimators can be conve-

---

[3]See Lo and Mackinlay (1988) for detailed discussions.
[4]Such a phenomenon was discussed by Hamao *et al.* (1990), Theodossiou and Lee (1993) and Koutmos *et al.* (1994).
[5]See Cochrane (1988).

Table 2. *Fundamental statistics of the Asian stock markets*

	HK	IND	MAL	KOA	JPN	SIG	PHI	TWN	THA
Mean	0.6282	− 0.0092	0.3042	0.4768	0.2580	0.3954	0.3615	0.2605	− 0.1728
S.D.	7.2616	3.6891	6.7996	4.1956	2.9650	3.1466	4.1841	2.6968	3.2261
Skewness	5.3820	0.6069	− 0.5656	− 0.1410	− 0.5810	7.9937	− 0.5639	− 0.7788	− 0.1319
Kurtosis	66.0132	6.8039	3.8350	5.1788	5.2508	14.9981	5.7464	6.3242	3.2611
Median	0.3289	0.0417	1.3998	0.3158	0.2152	0.5485	0.2820	0.1860	− 0.1030
J-B	9673.32*	154.77*	19.19*	46.86*	62.29*	1579.00*	85.57*	130.83*	1.34
AH(1)	13.45*	25.25*	7.07*	0.18	0.15	13.30*	5.63*	0.18	12.74*

Notes: S.D. = Standard deviation, J-B = Jarque-Bera χ^2 statistic for testing normality with df = 2. AH(1) = 1st order ARCH χ^2 statistic with df = 1. Rate of stock return is defined as the difference of two logarithmic prices. *significant at 5%.

niently calculated as following

$$\sigma_k^2 = \frac{1}{k(T-k+1)(1-\frac{k}{T})} \sum_{t=k}^{T} (Y_t - Y_{t-k} - k\hat{\mu})^2 \quad (3)$$

$$\sigma_1^2 = \frac{1}{T-1} \sum_{t=1}^{T} (Y_t - Y_{t-1} - \hat{\mu})^2 \quad (4)$$

in which T is the sample size and $\hat{\mu} = \frac{1}{T}(y_T - y_0)$. With the assumption of homoscedasticity, the asymptotic variance of the VR statistic is shown to be

$$\Phi(k) = \frac{2(2k-1)(k-1)}{3kT} \quad (5)$$

The VR statistic (Lo and Mackinlay, 1988) asymptotically approaches normality or

$$Z(k) = \frac{VR(k)-1}{[\Phi(k)]^{1/2}} \xrightarrow{a} N(0,1) \quad (6)$$

where $\xrightarrow{a}$ denotes that the distributional equivalence is asymptotic.

As in well-documented in the literature, variances of most stock returns are conditionally heteroscedastic with respect to time (Hamao *et al.*, 1990; Theodossiou and Lee, 1993; Koutmos *et al.* 1993, 1994). As a result; there may not exist a linear relation over the observation intervals. To overcome this difficulty, Lo and Mackinlay (1988) derive the heteroscedasticity-consistent variance estimator $\Phi^*(k)$

$$\Phi^*(k) = \sum_{j=1}^{k-1} \left[\frac{2(k-j)}{k} \right] \hat{\delta}(j) \quad (7)$$

in which

$$\delta(j) = \frac{\sum_{t=j+1}^{T} (S_t - S_{t-1} - \hat{\mu})^2 (S_{t-j} - S_{t-j-1} - \hat{\mu})^2}{\left[\sum_{t=1}^{T} (S_t - S_{t-1} - \hat{\mu})^2 \right]^2} \quad (8)$$

Thus, the variance ratio test statistic can be standardized asymptotically to a standard normal variable or

$$Z^*(k) = \frac{VR(k)-1}{[\Phi^*(k)]^{1/2}} \xrightarrow{a} N(0,1) \quad (9)$$

Another important alternative approach to examining the random walk hypothesis is the Dickey–Fuller (DF) unit root test. More specifically, the augmented Dickey–Fuller (ADF) test is often used to model the time series data that are not generated by the pure AR(1) process and the data which are fraught with non-white noise error terms. Typically, the ADF test is based on the following formulation:

$$\Delta Y_t = \alpha + \rho Y_{t-1} + \sum_{i=1}^{k} \beta_i \Delta Y_{t-i} + \varepsilon_t \quad (10)$$

where $\Delta y_t = y_t - y_{t-1}$, α is a drift term with the null hypothesis $H_0: \rho = 0$ and its alternative hypothesis $H_0: \rho < 0$. Note that failing to reject H_0 implies the time series has the property of random walk.

IV. EMPIRICAL RESULTS

Based on the ADF test (Equation 10) on the nine stock returns, we cannot reject the null hypothesis (Table 3). That is, after taking first differences on the stock price indices, it appears that there exists some evidence of random walks in nine stock price indices. However, the existence of random walk components in stock price does not necessarily imply that stock returns are unpredictable. If stock returns are characterized by a white noise process, the corresponding price indices are said to follow the random walk. In that case, the stock returns are considered to be unpredictable. On the other hand, if stock returns do not follow white noise, or they are integrated of order one or I(1), there exists some predictable components. The purpose of the VR approach is to detect if the short-term fluctuations dominate the stochastic trend components, while the ADF approach is formulated to examine only the existence of stochastic trend components. Viewed in this perspective, the result from applying the ADF test on the nine Asian stock returns

Table 3. *Results of the Dickey–Fuller unit root test*

	HK	IND	JPN	KOA	MAL	PHI	SIG	THA	TWN
X	− 2.1970	− 0.9342	− 2.1871	− 1.2948	− 2.0868	− 2.4349	− 1.6521	− 0.5092	− 2.1785
dX	− 6.2321*	− 7.1869*	− 7.6068*	− 5.7772*	− 6.2189*	− 5.8948*	− 7.4222*	− 6.4268*	− 7.1570*

Notes: X = stock price indices, dX = first difference of X. *significant at 5% based on Equation 10 with $k = 4$.

Table 4. *Result of the VR tests of the nine Asian stock markets*

	$k = 4$	$k = 6$	$k = 8$	$k = 10$	$k = 20$	$k = 30$	$k = 40$
HK	1.08	1.15	1.19	1.24	1.43	1.60	2.03
	(0.62)	(0.90)	(0.98)	(1.07)	(1.34)	(1.49)	(2.21*)
	(0.95)	(1.56)	(1.77)	(2.14*)	(3.27*)	(3.77*)	(5.96*)
IND	1.07	1.07	1.07	1.10	1.17	1.36	1.47
	(0.56)	(0.44)	(0.35)	(0.46)	(0.53)	(0.90)	(1.00)
	(0.95)	(0.74)	(0.58)	(0.78)	(0.89)	(1.45)	(1.66)
JPN	1.06	1.00	1.02	1.06	1.03	0.84	0.72
	(0.47)	(− 0.03)	(0.11)	(0.27)	(0.08)	(− 0.41)	(− 0.59)
KOA	1.23	1.41°	1.55°	1.65°	2.25°	2.64°	2.81°
	(1.92)	(2.52*)	(2.84*)	(2.97*)	(3.85*)	(4.07*)	(3.85*)
MAL	1.28	1.35°	1.45°	1.55°	1.99°	2.15°	1.98
	(2.27*)	(2.15*)	(2.33*)	(2.51*)	(3.05*)	(2.84*)	(2.09*)
	(3.54*)	(3.38*)	(3.47*)	(3.65*)	(4.32*)	(4.03*)	(2.98*)
PHI	1.14	1.25	1.40	1.47	1.52	1.18	1.05
	(1.14)	(1.53)	(2.05*)	(2.15*)	(1.61)	(0.44)	(0.10)
	(1.31)	(1.80)	(2.41*)	(2.59*)	(1.96*)	(0.57)	(0.13)
SIG	1.24	1.21	1.15	1.10	1.01	0.73	0.60
	(1.94)	(1.31)	(0.78)	(0.45)	(0.02)	(− 0.67)	(− 0.86)
	(2.03*)	(1.56)	(0.95)	(0.57)	(0.03)	(− 1.05)	(− 1.37)
THA	1.18	1.25	1.23	1.24	1.02	0.85	0.79
	(1.46)	(1.53)	(1.20)	(1.07)	(0.07)	(− 0.36)	(− 0.45)
	(2.12*)	(2.19*)	(1.73)	(1.54)	(0.09)	(− 0.54)	(− 0.67)
TWN	1.02	0.98	1.03	1.08	1.01	0.83	0.73
	(0.12)	(− 0.15)	(0.17)	(0.37)	(0.02)	(− 0.41)	(− 0.58)

Notes: the first entry for each country is the VR statistic; the second and third entry are $Z(k)$ and $Z^*(k)$ respectively. *significant at 5%. k = holding period. Since Japan (JPN); Korea (KOA) and Taiwan (TWN) are free of ARCH(1) phenomenon (see Table 2), they do not have $Z^*(k)$ statistics. °significant at 5% based on Table 5.

indices does not tell if the short-term fluctuations dominate the stochastic trend components. To take it into consideration, we apply both VR statistics with homoscedastic and heteroscedastic error terms, denoted by $Z(k)$ and $Z^*(k)$ respectively, to the nine Asian stock price indices. A perusal of Table 4 indicates that, using $Z(k)$, we can reject the null hypothesis of random walk for all holding periods for the Korea and Malaysia markets. In addition, we can also reject the null hypothesis for (a) the Philippines market with holding periods of 8 and 10 weeks and (b) the Hong Kong market with a holding period of 40 weeks. Since the validity of $Z(k)$ depends, to some extent, on its asymptotic distribution towards normality, one would prefer a large sample.[6]

Given that Lo and Mackinlay (1988) have 1216 observations in their sample, the 233 observations in our study is far from being adequate. To overcome the difficulty, we generate a set of critical values via Monte Carlo simulations (Table 5). Table 5 reveals immediately that we cannot reject the null hypothesis for (a) the Philippines market with the holding periods of 8 and 10 weeks, (b) the Hong Kong market with a holding period of 40 weeks, and (c) the Korea and the Malaysia markets with a holding period of 4 weeks. Clearly, the UV statistic does not approach normality with the sample size of 233 observations.

As discussed in the previous section, only the Japan, Korea and Taiwan markets are free of the ARCH

[6]Large samples are preferred as pointed out by Lo and Mackinlay (1988, p. 50).

Table 5. *The empirical critical percentile of the variance ratio test*

Percentile	$k = 2$	$k = 4$	$k = 6$	$k = 8$	$k = 10$	$k = 20$	$k = 30$	$k = 40$
99.5	1.17	1.34	1.48	1.61	1.70	2.19	2.63	2.96
97.5	1.13	1.25	1.35	1.43	1.50	1.80	2.06	2.28
95.0	1.11	1.21	1.28	1.35	1.40	1.65	1.81	1.98
90.0	1.08	1.16	1.22	1.26	1.30	1.45	1.59	1.70
50.0	1.00	1.00	0.99	0.99	0.98	0.96	0.94	0.90
10.0	0.92	0.85	0.80	0.76	0.73	0.61	0.53	0.46
5.0	0.90	0.81	0.75	0.71	0.67	0.54	0.44	0.38
2.5	0.87	0.77	0.71	0.66	0.62	0.47	0.38	0.33
1.0	0.83	0.71	0.64	0.58	0.52	0.37	0.29	0.24
Mean	1.0009	1.0014	1.0020	1.0028	1.0037	1.0068	1.0088	1.0110
Std.	0.0645	0.1221	0.1633	0.1968	0.2261	0.3416	0.4357	0.5160
Skewness	0.0335	0.2361	0.3715	0.4619	0.5373	0.8511	1.0849	1.2994
Kurtosis	3.0454	3.1011	3.2080	3.3239	3.4459	4.1340	4.8349	5.7021

Notes: the data generating process is based on the Gaussian Random Walk model with the replication of 10 000. Std = standard deviation.

phenomenon.[7] Hence, the $Z(k)$ with homoscedastic error terms may not provide an accurate test in the presence of heteroscedasticity. To generalize the result, Lo and Mackinlay (1988) developed a Heteroscedasticity-consistent VR statistic $Z^*(k)$ or Equation 9. The results from Table 4 using $Z^*(k)$ suggest that (a) $Z^*(k)$ of the Malaysia market becomes more significant, i.e., the null hypothesis of random walk is to be rejected, (b) the null hypothesis can be rejected for the Hong Kong market with the holding periods of 10, 20, 30 and 40 weeks, and (c) the null hypothesis can be rejected for the Philippines market with the holding periods of 8, 10 and 20 weeks. Similarly, we are able to reject the null hypothesis for the Singapore market (4 weeks), and the Thailand market (4 and 6 weeks).

There is little doubt that more evidence has turned up recently in rejecting the random walk hypothesis (Shiller and Perron, 1985; Summers, 1986; Poterba and Summers, 1988). In particular, Lo and Mackinlay (1988) demonstrated that the NYSE stock price index was not consistent with the random walk hypothesis. In an attempt to minimize the bias introduced by different microstructures of the nine Asian markets, we utilize weekly data of the stock returns. However, as pointed out by Cohen (1983), infrequent and non-synchronous tradings are considered the culprit of spurious correlation found in the stock returns. In general, such artificial correlations can be attributed to the infrequent trading of small-capitalization stocks. As new information is impounded into the stocks of large capitalization, it trickles down to stocks of small capitalization with lags. Thus a positive serial correlation is generated through the diffusion process.[8] Such an induced positive serial correlation, according to Lo and Mackinlay (1988), causes more impacts on the equal-weighted stock index than value-weighted index. Even in the case of using equal-weighted index, it is difficult to ascertain that the positive serial correlation is a direct consequence of infrequent tradings.

The VR statistic, as is found in this study, exceeds one in the markets of Korea, Malaysia, Hong Kong, Thailand and Philippines, indicating the existence of positive serial correlation in the stock returns. The infrequent tradings of the small capitalization stocks do not seem to be the major cause for positive serial correlation.[9] It should be pointed out that rejections of the random walk hypothesis of the Asian markets do not necessarily imply that the markets are not efficient. In particular, since the VR statistic leading to the rejection of the null hypothesis often exceeds one, it is indicative of the existence of a positive serial correlation which manifests itself in short term fluctuations. Unlike the case of a negative serial correlation, it does not gives rise to mean reversion. More detailed statistical analyses are needed to address this problem in the future.

V. CONCLUSION

This paper applying the VR approaches on nine Asian stock markets examines the random walk hypothesis. To circumvent the problems caused by infrequent and nonsynchronous tradings, we employ the weekly stock price data in the study. While the Dickey–Fuller test is both convenient

---

[7] We cannot reject the null hypothesis of the ARCH(2) for the Korea market at $\alpha = 5\%$. In addition the markets of Japan and Taiwan are free of the ARCH(2).

[8] It implies that the VR statistic exceeds one.

[9] Our return data are value-weighted weekly index. In addition, the infrequent trading does not lead to rejection of the null hypothesis (Lo and Mackinlay, 1988).

and effective in detecting the existence of random walk components in a time series, it cannot distinguish the serial correlation components from the short-term fluctuations. Furthermore, Lo and Mackinlay (1989) also demonstrate via Monte Carlo simulations the superiority of the VR over DF in terms of statistical power. Altogether six of the nine stock markets exhibit various degrees of the ARCH phenomenon. As a result, we employ both VR statistics with homoscedastic and heteroscedastic error terms to test the random walk hypothesis. It is interesting to know that the markets of Malaysia and Korea for all holding periods manifest in themselves various degrees of positive serial correlations. Similarly, positive serial correlations in stock returns are also found of the markets of Hong Kong, Singapore and Thailand for some holding periods. While the existence of such positive serial correlations is not an issue of microstructure of the Asian stock markets, it provides potential avenues for future studies.

REFERENCES

Cochrane, J. H. (1988) How big is the random walk in GNP?, *Journal of Political Economy*, **96**(5), 893–920.

Cohen, K. (1983) Friction in the trading process and the estimation of systematic risk, *Journal of Financial Economics*, **12**, 263–78.

Dickey, D. A. and Fuller, W. A. (1981) Likelihood ratio statistics for autoregressive time series with a unit root, *Econometrica*, **49**, 1057–72.

Fama, E. F. (1965) Random walks in stock market prices, *Financial Analyst Journal*, **21**(5), 55–59.

Fama, E. F. and French, K. R. (1988) Permanent and temporary components of stock prices, *Journal of Political Economy*, **96**(2), 246–73.

Hamao, Y., Masulis, R. W. and Ng, V. (1990) Correlation in price changes across international stock markets, *Review of Financial Studies*, **3**, 281–307.

Keim, D. and Stambaugh, R. (1986) Predicting returns in stock and bond markets, *Journal of Financial Economics*, **17**, 357–90.

Koutmos, G., Negakis, C. and Theodossiou, P. (1993) Stochastic behaviour of the Athens stock exchange, *Applied Financial Economics*, **3**, 119–126.

Koutmos, G., Lee, U. Theodossiou, P. (1994) Time-varying betas and volatility persistence in international stock markets, *Journal of Economics and Business*, **46**, 101–12.

Lo, A. W. and MacKinlay, A. C. (1988) Stock market prices do not follow random walks: evidence from a simple specification test, *Review of Financial Studies*, **1**(1), 41–66.

Lo, A. W. and MacKinlay, A. C. (1989) The size and power of the variance ratio test in finite samples: a Monte Carlo investigation, *Journal of Econometrics*, **40**, 203–38.

Poterba, J. and Summers, L. (1988) Mean reversion in stock returns: evidence and implications, *Journal of Financial Economics*, **22**, 27–59.

Shiller, R. J. and Perron, P. (1985) Testing the random walk hypothesis: power versus frequency of observation, *Economics Letters*, **18**, 381–86.

Summers, L. H. (1986) Does the stock market rationally reflect fundamental values? *Journal of Finance*, **41**, 591–600.

Theodossiou, P. and Lee, U. (1993) Mean and volatility spillovers across major national stock markets: further empirical evidence, *Journal of Financial research*, **16**, 337–50.

Journal of Comparative Economics **28**, 321–339 (2000)

doi:10.1006/jcec.2000.1651, available online at http://www.idealibrary.com on **IDE**L®

The Impact of Financial Liberalization on Stock Price Volatility in Emerging Markets[1]

Bwo-Nung Huang[2]

National Chung-Cheng University, Chia-Yi, Taiwan 621

E-mail: ecdbnh@ccunix.ccu.edu.tw

and

Chin-Wei Yang

Clarion University of Pennsylvania, Clarion, Pennsylvania 16214-1232

Received February 5, 1999; revised December 13, 1999

Huang, Bwo-Nung, and Yang, Chin-Wei—The Impact of Financial Liberalization on Stock Price Volatility in Emerging Markets

Is financial market liberalization, i.e., increased accessibility for international investors, in emerging economies a conduit to greater stock price volatility? If so, capital controls may be a useful policy option. Using daily returns for 10 emerging markets coupled with a world index, we find that South Korea, Mexico, and Turkey suffered from greater volatility, Argentina, Chile, Malaysia, and the Philippines experienced diminished volatility, and no definitive pattern can be discerned for the other countries after market liberalization. Our result is important because our sample period includes the Asian financial crises. *J. Comp. Econ.*, June 2000, **28**(2), pp. 321–339. National Chung-Cheng University, Chia-Yi, Taiwan 621; Clarion University of Pennsylvania, Clarion, Pennsylvania 16214-1232. © 2000 Academic Press

Journal of Economic Literature Classification Numbers: G150, F30.

1. INTRODUCTION

Beginning in the early 1990's, a massive amount of capital flowed into Southeast Asia as restrictions were lifted gradually. International capital invest-

[1] We thank Clive W. J. Granger for helpful comments. Generous financial assistance from the National Science Council (Taiwan) and Fulbright Scholarship Program are gratefully acknowledged. This paper was written while B.-N. Huang was a visiting scholar at UCSD. We are grateful to the Editor and two anonymous referees for valuable suggestions. However, we wish to absorb all the culpability.

[2] To whom all correspondence should be addressed.

0147-5967/00 $35.00

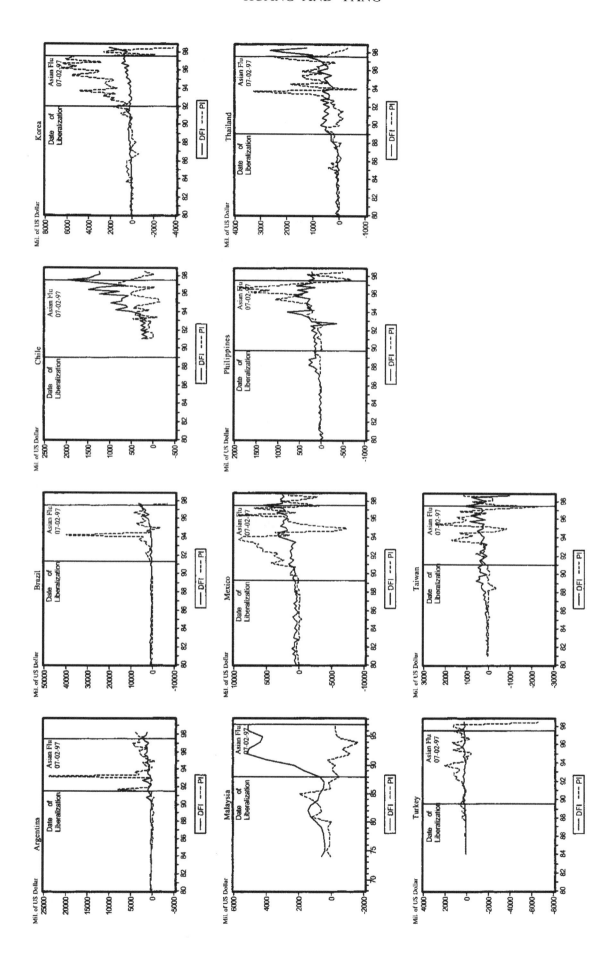

ment accounts for sizable proportions of national GDPs in these countries, e.g., 11% for Malaysia, 10% for Thailand, 5% for Indonesia, and 4% for South Korea. Generally speaking, these capital inflows can be classified into two categories, long-term investments or direct foreign investments (DFI) and short-term hot money or portfolio investments (PI). While long-term investment dominated in Malaysia and Chile, hot money accounted for the lion's share of investment in Thailand, South Korea, Mexico, Argentina, Brazil, the Philippines, Turkey, and Taiwan (*The Economist,* Jan. 14, 1998, p. 70).

Examination of Fig. 1 indicates that PI exceeded DFI after liberalization in all markets except Chile and Malaysia. Furthermore, PI decreased around the period of the Asian financial crisis. The pummeling of the Thai baht on July 2, 1997, started a domino effect in both stock and exchange markets in these countries, and the financial avalanche continued until March 8, 1998. General sentiment in Asia links capital flight to the Asian economic crisis. The velocity and the frequency with which short-term capital enters and leaves can be quite disruptive. As pointed out by Granger et al. (1999), short-term capital outflow caused the market crisis, to a large extent. As such, a legitimate question is whether liberalization of stock markets leads to greater volatility. Opponents of liberalization hold the view that foreign investments toward emerging markets are extremely volatile and depend on changing economic conditions. A consequence of volatile investment flows is high volatility in stock prices. The empirical implication is that market volatility should increase after liberalization. In contrast, Domowitz et al. (1998) show that liberalization may induce greater participation by foreign investors, whose entry can reduce price volatility. In other words, new investors broaden the market, which in turn dampens the shocks on prices from order transactions. Quite interestingly, foreign participators may also make prices more efficient by increasing the precision of public information regarding fundamental values (De Santis and Imrohoroglu, 1997, p. 575).

Needless to say, predicting volatility in emerging capital markets is important for determining the cost of capital and for evaluating direct investment and asset allocation decisions. Because higher volatility implies higher capital costs and, as such, may also increase the option value of waiting, it delays investments. As Kyle (1985) points out, more information is revealed in the volatility of stock prices than in the prices themselves. In particular, the study by Koutmos (1998) speaks to several important phenomena. First, stock returns are found to be positively autocorrelated. Second, the unconditional distributions appear to be excessively leptokurtic with short-term stock returns invariably exhibiting volatility clustering. Third, the stylized fact that changes in stock prices tend to be negatively related to changes in volatility is observed. Based on monthly stock

---

FIG. 1. Time series of direct foreign investment and portfolio investment of the 10 markets. Source: IFS database. DFI, direct foreign investment; PI, portfolio investment. All data are quarterly observations except for Malaysia, where annual data are used.

prices, the analysis by Bekaert and Campbell (1997) suggests at least four distinguishing features of emerging market returns, higher sample average return, low correlations with stock returns of developed markets, more predictable returns, and higher volatility. Their empirical findings indicate that the unconditional variance has decreased noticeably in the markets of Brazil, Mexico, Taiwan, and Portugal two years after market liberalization. Pakistan is an exception to this relationship. Similarly, De Santis and Imrohoroglu (1997) find decreasing unconditional variances for the markets of India, Taiwan, and Argentina. However, Kim and Rogers (1995) find that volatility remains relatively unchanged in the Korean stock market. The study by Schwert (1998) reveals diminished volatility, both in monthly and daily returns, for the industrialized nations except for Japan after the 1987 market crash. Despite these findings of decreased variance, the results must be taken with a grain of salt for three reasons. First, important information may be averaged out due to aggregation. Hence, daily data provide better results than do monthly data. Second, the sample size of only two years after liberalization may coincide with the lowest volatility period; thus it suffers from a selection bias. A longer time span is needed to include the period of greater volatility, such as the one that recently occurred in Asia. Third, it is commonly recognized that liberalization has intensified the degree of market integration across nations. Such integration will inevitably enhance market volatility, which needs to be removed from the analysis.

To advance the analysis further, we employ daily returns for 10 emerging markets, and apply the generalized error distribution (GED) with the Generalized Autoregressive Conditional Heteroscedasticity (GARCH) model.[3] The results indicate that, while unconditional variances in South Korea, Mexico, and Turkey have increased after liberalization, they have decreased significantly in Argentina, Chile, Malaysia, and the Philippines. The structural change in the volatility of stock prices in the remaining three markets cannot be determined using the implied unconditional variance (iuv). However, in terms of conditional variance (cv), the markets of Brazil, Korea, Thailand, and Turkey exhibit increases in cv while the remaining six markets exhibit decreases after liberalization. In the next section, we discuss the data and the basic statistics. We present the statistical model in Section 3. Estimation results are analyzed in Section 4, and Section 5 concludes with some policy implications.

2. DATA AND BASIC STATISTICS

Daily returns are employed for the 10 emerging stock markets, South Korea (KOA), Malaysia (MAL), the Philippines (PHI), Thailand (THA), Taiwan (TWN), Turkey (TUK), Argentina (ARG), Brazil (BRA), Chile (CHL), and Mexico (MEX). The sample period ranges from January 5, 1988, to April 2,

---

[3] A problem of convergence for the India market prevents us from including India although it was in the original 11-country study.

TABLE 1

Liberalization of Financial Markets in Emerging Economies

Country	Opening date	Degree of openness
Korea	January 1992	10% of capital of listed companies, 25% after July 1992
Malaysia	December 1988	30% for banks and institutions, 100% for remaining stocks
Philippines	October 1989	investable up to 40%
Taiwan	January 1991	investable up to 10%
Thailand	December 1988	investable up to 49%
Turkey	August 1989	fully open
Argentina	October 1991	fully open
Brazil	May 1991	100% of non-voting preferred stock, 49% of voting common stock
Chile	December 1988	25% of shares of listed companies
Mexico	May 1989	30% of banks, 100% for other stocks

Source. Santis and Imrohoroglu (1997, p. 577).

1998, for Turkey, Argentina, Brazil, Chile, and Mexico. The rest of the countries have earlier starting dates from January 5, 1986, forward. These daily data, i.e., five days per week, yield totals of 2674 and 3194 observations for these two sets of countries.[4] As is customary in financial analysis, the data are transformed into a logarithmic scale. The trends and dates of liberalization are indicated by the position of vertical lines shown in Fig. 2 and in Table 1.

As is evident from Table 1, the date of liberalization and the degree to which liberalization occurred are quite different across the countries. Thailand, Malaysia, and Chile were among the first to open their markets, starting at the end of 1988, compared with South Korea, which did not open until 1992. The degree of openness varies drastically from complete openness to foreign investment in Argentina and Turkey to an allowance of only 10% in Taiwan, with the rest of the countries in between these amounts.

An examination of Fig. 2 indicates a general upward trend in each market. However, the Asian markets suffered from noticeable collapses to varying degrees after July 1997. Greater volatility in stock returns in the emerging markets are manifested as well in Fig. 2. We report in Table 2 basic statistics for the entire sample period, before and after market liberalization, based on conventional logarithmic difference,

$$\Delta y_{i,t} = (\log y_{i,t} - \log y_{i,t-1}) \times 100, \tag{1}$$

[4] We are extremely grateful to the economics department of UCSD for providing the data from Datastream.

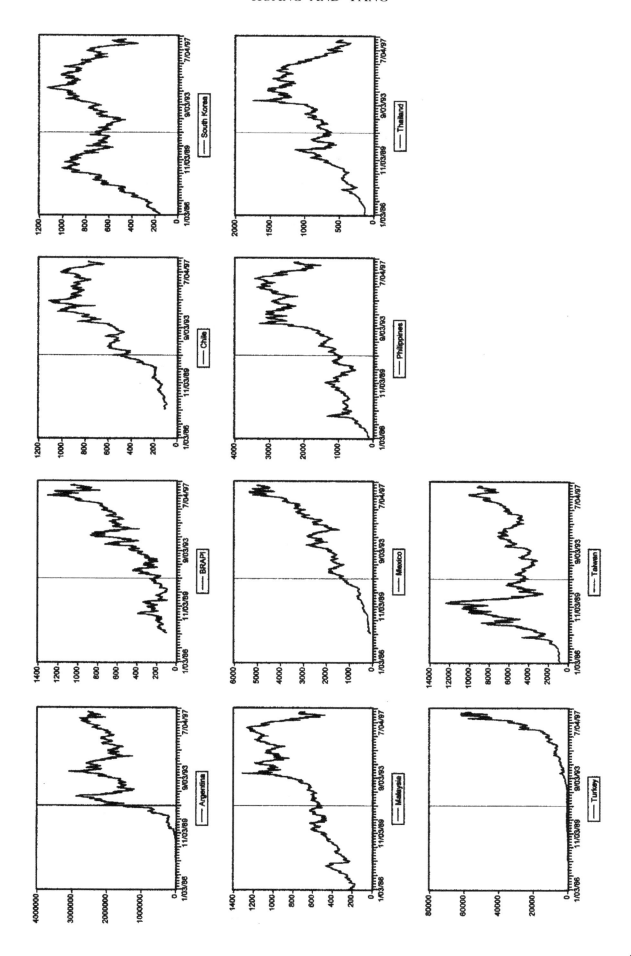

where the natural logarithm (log) is applied to $y_{i,t}$, the stock price at t of country i, and $\Delta y_{i,t}$ denotes the rate of change of $y_{i,t}$. This represents the continuously compounded return or log return of an asset (Campbell et al., 1997, p. 11).

A perusal of Table 2 indicates that the mean rate of log return ranges between 0.033 and 0.38% for these countries. In comparison with the mean rate for the Dow Jones Index in the United States of 0.055%, all the emerging markets except South Korea, Malaysia, and Thailand register higher rates of log return for the entire sample period. We note that the average rates of log return reported by Bekaert and Harvey (1997) are larger because their sample period did not include the recent stock market debacles in Asia. As we segment the entire sample period into before and after liberalization, it becomes immediately clear that all the emerging markets, except Turkey, had mean rates of log return greater than that of the United States before the liberalization of their capital markets. In contrast, the five Asian markets experienced lower rates after market liberalization because of the financial crises. In short, the picture in emerging markets is not as rosy as that in the United States after 1992. Using the t statistic, one could examine if liberalization caused significant changes in the log return. A perusal of the second to the last row in Table 2 indicates that four indexes, i.e., South Korea, Thailand, Philippines, and Argentina, experienced statistically significant decreases while Turkey exhibited an increase. The differences between log returns before and after liberalization in the remaining five markets were not statistically significant.

In terms of standard deviation for the entire sample, the markets in Argentina and Brazil have the largest values while Chile, South Korea, and Malaysia experience the lowest values although they are greater than that in the United States. Before liberalization, the greatest standard deviation is found in the Argentinian market, whereas the smallest occurs in Thailand and South Korea. After liberalization, the Chilean market exhibits the smallest standard deviation, followed by the markets in Malaysia and Mexico. The Brazilian market assumes the greatest value after liberalization. In general, the standard deviations are found to be greater than that in the U.S. market. This finding is consistent with that of Bekaert and Harvey (1997). In addition, the changes in the standard deviation are mixed; three markets, South Korea, Thailand, and Turkey, experience an increase while the rest exhibit diminished volatility of the log return after liberalization.[5] One of the stylized facts about emerging markets is that the

---

[5] The comparison is made based on the size of the statistics instead of on formal statistical analyses, which will be performed in the next section. Assuming normally distributed stock returns before and after liberalization, F statistics are convenient for testing a significant change in the volatility of the markets. As indicated in Table 2, we see significant increases in volatility in South Korea, Thailand, and Turkey and an insignificant change in Brazil. The remaining six markets experienced appreciable decreases to varying degrees.

FIG. 2. Trends of stock prices and dates of market liberalization.

TABLE 2

Basic Statistics of Daily Stock Price Changes (%) in Emerging and U.S. Markets

	KOA	MAL	PHI	THA	TWN	TUK	ARG	BRA	CHL	MEX	USA
Mean	0.0330	0.0350	0.0878	0.0384	0.0745	0.2378	0.3793	0.0877	0.0767	0.1455	0.0550
S.D.	1.5477	1.5542	1.9050	1.6502	2.0953	2.7147	3.8424	3.4401	1.2928	1.6896	1.0488
Skewness	0.0904	−0.1295	0.0923	−0.1273	−0.1069	−0.1004	−0.2975	−1.9278	−0.7161	−0.0753	−4.8831
Kurtosis	8.0797	28.6809	12.9428	9.8653	5.7745	4.9673	44.7171	22.2327	17.2417	11.1701	121.8387
J-B	3438.4	87778.7	13161.0	6281.1	1030.6	435.7	193940.1	42869.0	22826.7	7436.8	1892181
LB(10)	24.40*	85.98*	112.42*	135.90*	70.95*	105.49*	121.82*	59.17*	137.37*	161.53*	32.78*
$LB^2(10)$	1905.50*	726.11*	883.80*	1214.60*	2201.90*	469.04*	141.18*	116.25*	72.21*	613.73*	151.44*
N	3194	3194	3194	3194	3194	2674	2674	2674	2674	2673	3194
					Before liberalization						
Mean ($\bar{X}_1$)	0.0862	0.0568	0.2228	0.1419	0.1294	0.0306	0.9937	0.0529	0.0826	0.2743	0.0458
S.D. (s_1)	1.4193	1.8245	2.3858	1.3638	2.5545	2.0553	5.7345	3.5587	1.6593	2.3159	1.2869
Skewness	0.0745	−2.0982	0.0706	−1.3773	−0.1296	−0.4995	−0.5115	−1.2196	−3.7350	0.2134	−5.2541
Kurtosis	5.2920	23.7241	13.2346	14.3651	4.6263	9.2168	24.6991	16.7358	41.3710	10.2858	108.4497
J-B	343.1	14102.2	4251.8	4313.4	146.9	677.3	19170.8	7030.7	15154.0	763.4645	73.0892
LB(10)	14.54	52.28*	26.18*	107.72*	66.77*	31.99*	37.03*	87.16*	13.16	101.54*	24.43*
$LB^2(10)$	291.47*	597.23*	280.83*	476.42*	954.41*	17.44**	30.07*	194.57*	2.10	135.81*	70.81*
$N(n_1)$	1561	757	974	757	1300	410	975	867	238	344	1562
					After liberalization						
Mean ($\bar{X}_2$)	−0.0163	0.0278	0.0289	0.0058	0.0364	0.2778	0.0222	0.1052	0.0768	0.1266	0.0639
S.D. (s_2)	1.6597	1.4608	1.6479	1.7292	1.7108	2.8154	2.0016	3.3833	1.2515	1.5763	0.7538
Skewness	0.1235	1.0578	−0.0076	0.0865	−0.1160	−0.0896	−0.2554	−2.3190	−0.0209	−0.2482	−0.6572
Kurtosis	9.2550	30.3731	7.6930	9.0592	6.1581	4.5875	6.7307	25.3796	9.1293	9.8095	10.9652
J-B	2663.0	76475.4	2036.3	3728.0	790.5	240.6	1002.6	39307.3	3811.8	4521.8	4431.63
LB(10)	28.05*	69.81*	134.34*	81.72*	19.34**	82.40*	43.03*	37.71*	140.24*	64.55*	25.80*
$LB^2(10)$	1231.9*	156.95*	304.67*	840.00*	484.64*	408.18*	481.08*	29.34*	255.58*	349.08*	180.35*
$N(n_2)$	1631	2435	2219	2435	1892	2262	1697	1806	2435	2328	1632
t^a	−1.87***	−0.45	−2.65*	−1.98**	−1.23	1.70***	−6.43*	0.37	−0.07	−1.51	0.49
F^b	1.37*	0.64	0.48	1.61*	0.45	1.88*	0.12	0.90	0.57	0.46	0.34

Note. (i) Rate of change in stock prices is calculated according to $\Delta y_{i,t} = (\log y_{i,t} - \log y_{i,t-1}) \times 100$, where natural logarithm (log) is applied to $y_{i,t}$ (stock price at t of country i) and $\Delta y_{i,t}$ denotes rate of change of $y_{i,t}$. (ii) S.D., Standard Deviation; J-B, Jarque–Bera normality test. (iii) KOA, South Korea; MAL, Malaysia; PHI, Philippines; THA, Thailand; TWN, Taiwan; TUK, Turkey; ARG, Argentina; BRA, Brazil; CHL, Chile; MEX, Mexico. (iv) Sample period extends from January 6, 1986 (or January 6, 1988) to April 2, 1998. (v) The dates for the market liberalization are reported in Table 1. (vi) The U.S. data are segmented into two subsamples with December 31, 1991, as the breaking point. (vii) LB(10) denotes the Ljung–Box statistic with a lag of 10 days; $LB^2(10)$ represents LB statistics of squared stock returns with a lag of 10 days.

[a] The t-statistics are based on $t = (\bar{X}_2 - \bar{X}_1)/s_p(1/n_1 + 1/n_2)^{1/2}$, where $S_p = ((n_1 - 1)s_1^2 + (n_2 - 1)s_2^2)/(n_1 + n_2 - 2)$.

[b] The F statistics are based on s_2^2/s_1^2.

* 1% significance level.

** 5% significance level.

*** 10% significance level.

predictable portion of log return is large (Bekaert and Harvey, 1997). A typical way to detect the predictable portion of log return is to use the Ljung–Box statistic. We present the LB(10) statistics in Table 2 using lag of ten trading days or two weeks.

An inspection of Table 2 reveals that, before liberalization and with the exception of South Korea and Chile, all remaining eight markets have LB(10) statistics significantly greater than that of the United States. All the LB(10)s are significantly greater than that of the U.S. market except for Taiwan, and its LB(10) is significant but less than that of the United States. These LB(10) statistics suggest a strong linear dependence as well as higher-order relationships as manifested in volatility clustering by the Autoregressive Conditional Heteroscedasticity (ARCH) model (Engle, 1982). In particular, we employ $LB^2(10)$ from the squared estimated log returns to detect the existence of second-moment dependence. As shown in Table 2, most of the $LB^2(10)$ statistics are significant and greater than the corresponding LB(10). This is indicative of the prevalence of second-moment dependence in log returns. Our result is compatible with that in Bekaert and Harvey (1997) and in Koutmos (1998).

3. STATISTICAL MODELS

Since the log returns are known to be leptokurtic and exhibit the volatility clustering phenomenon as shown in Table 2 (Koutmos, 1998), an alternative model may be preferred. More often than not, turbulent periods with large and frequent changes in stock prices, tend to be followed by tranquil periods with small and infrequent stock price changes. The ARCH model, introduced by Engle (1982), takes into account several such characteristics of financial data. First, it allows for disturbances that are serially uncorrelated but show volatility and clustering. Second, it accounts for distributions with heavy tails as exhibited by the unconditional distribution. As such, the ARCH or ARCH-related models may offer better estimates of market volatility before and after the liberalization. Based on the GARCH model (Bollerslev, 1986), we apply the Generalized Error Distribution (GED) model developed by Box and Tiao (1973) and Harvey (1981) to the leptokurtic log returns of the emerging markets as shown below:

$$\Delta y_t = b_0 + \sum_{j=1}^{p} b_j \Delta y_{t-j} + \sum_{k=1}^{q} c_k u_{t-k} + d_w \Delta y_{w,t} + u_t \quad u_t | I_{t-1} \sim \text{GED}(0, h_t, v)$$

$$h_t = \gamma + \alpha u_{t-1}^2 + \beta h_{t-1} + \kappa u_{t-1}^2 d_{t-1}$$

$$f(u_t) = \frac{v \exp[-(1/2)|u_t h_t^{1/2}/\lambda|^v]}{\lambda 2^{[(v+1)/v]}\Gamma(1/v)} h_t^{-1/2}, \quad \text{with } \lambda = \left[\frac{2^{(-2/v)}\Gamma(1/v)}{\Gamma(3/v)}\right]^{1/2}. \quad (2)$$

Here I_{t-1} is the information set available at the beginning of time t, and $\Delta y_{w,t}$ represents the world index in logarithmic differences. Such an index is used to account for the portion of the stock price variations due to changes in the world market. Variable d_{t-1} is defined to be unity if u_{t-1} is negative or is zero otherwise. Thus, if the parameter $\kappa > 0$, a stock price decrease would have a greater effect on subsequent volatility than would a stock price increase of the same magnitude. This parameterization of the leverage effect was first proposed by Glosten et al. (1993); henceforth it is referred to as the GJR model.[6]

As shown in Table 3, the relatively greater first-order autocorrelation in the emerging markets requires the use of the Autoregressive and Moving Average [ARMA(p, q)]model.[7] The choice of a GED density is dictated by the inability of Gaussian GARCH processes to accommodate the leptokurtosis of many log return series, an issue that is likely to be even more relevant when using emerging market data.[8] Note that $\Gamma(\cdot)$ in (2) is the gamma function, and ν is a parameter measuring the thickness of tails, which is equal to two for the normal density. For $\nu > 2$, the distribution of u_t has thinner tails than that of the normal. For example, when $\nu = \infty$, u_t is uniformly distributed on the interval $[-3^{1/2}, 3^{1/2}]$ and, for $\nu < 2$, the distribution of u_t has thicker tails than that of the normal; e.g., if $\nu = 1$, u_t has a double exponential distribution (Nelson, 1991).

The empirical estimations are based on Eq. (2). First, we estimate the coefficients for the entire sample period. Then, we segment the sample into two periods, before and after liberalization. Sample coefficients and iuv are estimated for each market in order to evaluate the changes that occurred before and after financial liberalization.

4. DISCUSSION OF EMPIRICAL RESULTS

Based on Eq. (2), we perform the estimation for the 10 emerging stock markets in three different sample periods in Table 3.[9] In addition to these estimates, we employ the Ljung–Box (LB) statistics on standardized residuals $(uh^{-1/2})$ and squared standardized residuals $(u^2 h^{-1})$ to test the existence of autocorrelation up to order 10. As indicated in Table 3, only for Thailand after liberalization do we find the estimate moderately significant at slightly greater than 10%. Lack of autocorrelation is confirmed in terms of $Q_{12}(uh^{-1/2})$ and $Q_{12}(u^2 h^{-1})$ for the rest of the markets.

The sizes of the tail-thickness parameter ν, with a value ranging between 0.7 and 1.49, suggest that emerging markets have log returns characterized by thicker tail distributions. Regarding the leverage effect parameter (κ), a positive κ value

[6] The leverage effect refers to an explanation of the tendency of stock markets to be more volatile during periods of decline. Black (1976) and Christie (1982) suggest that, during market declines, the debt/equity ratio rises and, hence, the riskiness (volatility) of the market also increases.

[7] The complexity of the autocorrelation in emerging markets leaves many open issues for future time series analysis.

[8] The smallest kurtosis is 5.0 for Turkey while the greatest is found to be 44.7 for Argentina. The GED-GARCH is especially appropriate for such fat-tailed distributions.

[9] All the estimations are made using RATS.

indicated that a stock price decrease would have a greater effect on subsequent volatility than would a stock price increase of the same magnitude. Examination of the table indicates that three of the ten markets have insignificant κ values (Brazil, Chile, and Turkey) while the rest have significantly positive values for the whole period. The κ values are insignificant for Malaysia, Philippines, Thailand, Argentina, Brazil, and Chile before liberalization, but the leverage effect becomes significant after liberalization except for Brazil, Chile, and Turkey. The impacts of the world log return index on the regional markets are reflected by d_w. Excepting the insignificant coefficient for Chile and the smaller but significant coefficients for Taiwan and South Korea, the coefficients for the other seven markets become both numerically greater and statistically significant after liberalization.

As is well known in the literature, the inclusion of a first-order Moving Average (MA(1)) model in the log return equation can overcome first-order autocorrelation problems caused by the non-synchronousness of the data (Scholes and Williams, 1977). However, only in the cases of Mexico and Malaysia does the MA(1) model describe log returns adequately. Higher order ARMA models are needed in the mean equation. That is, the MA(1) model fails to remove the autocorrelation of the standardized residuals in most emerging markets. Interestingly, all the estimated coefficients (α and β) of the GARCH equations are significant with β greater than α in the conditional variance equation for the whole period. This result implies that large market surprises induce only smaller revisions in future volatility. As for the persistence parameter, i.e., $\alpha + \beta + \kappa/2$, the values of the persistent parameter lie between 0.93 and 0.99 except for Argentina, Brazil, and Thailand, whose persistent parameters are larger than unity. In the cases of Argentina, Brazil, and Thailand, the implied unconditional volatility (iuv) cannot be estimated.[10] The iuv of the remaining markets are estimated and reported in Table 3, with the greatest value of 9.2300 for Turkey and the smallest value of 2.1563 for Chile.

Market liberalization is modeled by dividing the sample into two subperiods, i.e., before and after liberalization. Using the estimated values for γ, β, α, and κ, we can calculate the sizes of the iuv before and after liberalization. Table 3 indicates that the values of iuv become smaller in Argentina, Chile, and Malaysia, while these values become greater in Korea, Turkey, and Mexico after liberalization. However, the values of iuv are found to be negative and, hence, are not comparable for the Philippines, Taiwan, Thailand, and Brazil, due to the fact that their persistent parameters are greater than one. In the Philippines, where the leverage effect was insignificant before liberalization, we may calculate the persistent parameter based on $\alpha + \beta$. By doing so, we find that its iuv decreases

---

[10] The implied unconditional volatility is defined as $\gamma/(1 - \alpha - \beta - \kappa/2)$; see De Santis and Imrohoroglu (1997) and Cuddington and Liang (1998) for details. For $\alpha + \beta + \kappa/2 > 1$, the denominator of $\gamma/(1 - \alpha - \beta - \kappa/2)$ is negative. Thus iuv is negative, which is an unreasonable value.

TABLE 3

Estimates of the GED-GARCH-ARMA Model for the 10 Emerging Markets

| | Argentina | | | | | | Turkey | | | | | | Brazil | | | | | |
| | All period | | Before | | After | | All period | | Before | | After | | All period | | Before | | After | |
	Estimate	t-value	Estimate	t-value	Estimate	t-value	Estimate	t-value	Estimate	t-value	Estimate	t-value	Estimate	t-value	Estimate	t-value	Estimate	t-value
b_0	0.0531	1.6519	0.1840	1.5846	0.0156	0.4385	0.1294	2.5768	−0.4713	−2.1813	0.1766	3.3905	0.1133	3.0369	0.1040	1.1198	0.1011	2.5355
c_1	0.0917	4.7626	0.0738	2.5123	0.1038	4.0436	0.1768	8.9632	0.2669	2.5939	0.1685	8.3453	0.1040	5.9815	0.1597	5.5531	0.0735	3.4098
c_2	−0.0468	−2.4681	−0.0573	−1.9475	−0.0557	−2.2758												
c_3	0.0450	2.3742	0.0556	1.9030	0.0351	1.4193	0.0310	1.5948	0.0402	0.4605	0.0303	1.5226						
c_9													0.0299	1.8057	0.0499	1.9150	0.0189	0.9157
d_w	0.7222	13.3979	0.2835	1.9900	0.7998	13.5407	0.1323	2.1229	0.0791	0.3086	0.1432	2.2086	0.2061	3.4193	−0.0059	−0.0538	0.3190	4.3563
γ	0.0278	2.5227	0.4460	2.3200	0.0833	4.1848	0.3632	4.3183	0.7304	1.5701	0.3608	4.0962	0.0385	2.1165	0.9430	2.2330	0.0266	1.7703
β	0.8842	86.7090	0.8917	42.0347	0.8433	45.2681	0.7944	32.4220	0.4463	2.1231	0.8043	32.5364	0.8962	85.0025	0.8283	14.6272	0.9009	74.9006
α	0.1034	7.2066	0.1140	4.2350	0.0569	3.2853	0.1797	6.4368	0.2482	1.2244	0.1749	6.2499	0.1202	6.3004	0.0788	2.2000	0.1095	4.5815
ν	1.2629	37.1336	1.1059	22.2297	1.4864	22.3259	1.3377	32.0674	1.7732	4.6580	1.3333	31.0126	1.0977	37.6526	1.0301	22.4592	1.1318	30.0261
κ	0.0488	2.4748	−0.0407	−1.0185	0.1534	5.5626	−0.0269	−0.8716	0.3575	1.1801	−0.0353	−1.1536	−0.0114	−0.5614	0.0052	0.1356	−0.0041	−0.1566
$\alpha + \beta + \kappa'2$	1.0120		0.9854		0.9769		0.9607		0.8733		0.9616		1.0107		0.9097		1.0084	
iuv	−2.3167		30.4437		3.6061		9.2300		5.7625		9.3836		−3.5981		10.4430		−3.1856	
$Q_{12}(uh^{-1/2})$	12.911	(0.391)	14.164	(0.224)	8.859	(0.546)	17.402	(0.134)	8.820	(0.549)	14.850	(0.138)	14.718	(0.257)	18.617	(0.045)	11.732	(0.303)
$Q_{12}(u^2h^{-1})$	10.432	(0.578)	2.935	(0.983)	14.038	(0.171)	11.356	(0.499)	8.705	(0.560)	9.994	(0.441)	7.111	(0.850)	6.088	(0.808)	8.863	(0.545)
$\log L$	−6024.9		−2807.8		−3159.6		−6133.7		−285.6		−5857.2		−6297.1		−2111.4		−4161.6	

| | Chile | | | | | | Malaysia | | | | | | South Korea | | | | | |
| | All period | | Before | | After | | All period | | Before | | After | | All period | | Before | | After | |
	Estimate	t-value	Estimate	t-value	Estimate	t-value	Estimate	t-value	Estimate	t-value	Estimate	t-value	Estimate	t-value	Estimate	t-value	Estimate	t-value
c_1	0.2832	14.8963	0.3995	5.8313	0.2793	14.0696	0.1231	7.8936	0.0915	2.7128	0.1292	7.3915	0.0495	2.7718	0.0628	2.4630	0.0477	1.8780
c_2	0.0351	1.8997	0.1079	1.7355	0.0323	1.6548												
b_2													0.0368	2.1356	0.0641	2.6820	0.0180	0.7283
d_w	0.0039	0.1505	−0.0427	−0.4723	−0.0004	−0.0138	0.2607	12.5674	0.0024	0.0592	0.3158	13.3190	0.1579	5.2454	0.1644	4.6703	0.1142	2.0504
γ	0.1380	5.4519	0.2301	1.4313	0.1147	4.8354	0.0627	7.2002	0.2503	2.7509	0.0691	7.1876	0.1165	5.4054	0.3625	3.7336	0.0545	3.2801
β	0.6933	20.5785	0.5846	3.0474	0.7267	22.8900	0.8156	51.8108	0.7219	10.4808	0.7997	41.7053	0.8146	38.7813	0.6340	8.8502	0.8775	41.2219
α	0.2243	6.9254	0.3941	1.6780	0.1908	6.2784	0.1174	5.5010	0.1443	2.3218	0.1002	4.0607	0.0857	4.8910	0.1114	2.9522	0.0549	3.0932
ν	1.1396	41.6916	1.0409	9.3562	1.1618	39.9690	1.0221	48.5374	1.0431	20.1008	1.0001	43.1250	1.3140	28.7261	1.2424	20.6540	1.3994	19.6879
κ	0.0368	0.8573	−0.0723	−0.2528	0.0534	1.3500	0.0942	3.0335	0.0895	1.1697	0.1430	3.4168	0.1014	3.6130	0.1693	2.4567	0.0887	3.0436
$\alpha+\beta+\kappa/2$	0.9360		0.9426		0.9442		0.9801		0.9110		0.9714		0.9510		0.8301		0.9768	
iuv	2.1563		4.0052		2.0556		3.1508		2.8108		2.4161		2.3776		2.1330		2.3441	
$Q_{12}(uh^{-1/2})$	9.807	(0.633)	5.383	(0.864)	7.989	(0.630)	10.300	(0.590)	14.364	(0.167)	13.502	(0.197)	13.604	(0.327)	11.775	(0.300)	7.835	(0.645)
$Q_{12}(u^2h^{-1})$	3.394	(0.992)	6.089	(0.808)	2.205	(0.995)	1.821	(0.999)	1.879	(0.997)	1.063	(1.000)	8.570	(0.739)	15.258	(0.123)	15.391	(0.119)
log L	−3943.3		−351.2		−3596.2		−4767.0		−1297.9		−3446.6		−5392.2		−2595.9		−2762.3	

TABLE 3—*Continued*

| | Taiwan | | | | | | Thailand | | | | | | Philippines | | | | | |
| | All period | | Before | | After | | All period | | Before | | After | | All period | | Before | | After | |
	Estimate	t-value	Estimate	t-value	Estimate	t-value	Estimate	t-value	Estimate	t-value	Estimate	t-value	Estimate	t-value	Estimate	t-value	Estimate	t-value
b_0	0.0341	1.4566	0.0928	1.8788	0.0034	0.1501	0.0269	1.6463	0.0378	1.8712	0.0102	0.4518						
b_1	0.0345	2.1355	0.1352	4.7646	-0.0283	-1.6536	0.1295	7.7921	0.2167	6.2846	0.1157	5.9905	0.1635	11.1033	0.0684	3.4992	0.2027	10.8053
b_2	0.0460	2.9014	0.0674	2.3675	0.0146	0.8604												
b_3	0.0390	2.4586	0.0951	3.4641	0.0038	0.2239												
b_8	0.0431	2.8265	0.0472	1.8350	0.0168	1.0470							0.0269	1.9685	0.0247	1.4692	0.0190	1.0686
b_{10}	0.0445	2.8922	0.0638	2.3108	0.0190	1.2160												
b_{12}													0.0314	2.3913	0.0014	0.0925	0.0468	2.6841
b_{13}							0.0406	2.8575	0.0249	0.9027	0.0447	2.7231						
d_w	0.1648	4.4771	0.1989	3.3871	0.0749	1.7926	0.0461	1.9976	-0.0175	-0.6748	0.1803	5.0404	0.1314	4.3817	-0.0002	-0.0101	0.2192	5.7030
γ	0.0377	3.3653	0.0350	2.3143	0.0922	3.0448	0.0373	4.0988	0.0090	1.5200	0.1015	4.5078	0.0717	4.5365	0.0989	2.3110	0.0470	3.4316
β	0.9072	83.9003	0.9009	52.4725	0.8829	40.7801	0.8340	60.1367	0.8016	34.6853	0.8163	40.0040	0.8612	59.5475	0.8480	25.9951	0.8769	59.2137
α	0.0663	4.8715	0.0603	2.8499	0.0639	3.1409	0.1319	6.8547	0.2543	4.6284	0.0943	5.0595	0.0928	6.0362	0.1495	3.2651	0.0688	4.4568
ν	1.1682	28.6804	1.4985	15.8864	0.9455	22.7536	1.1397	31.6831	1.0900	15.2729	1.2019	27.3033	1.0299	39.4864	0.7634	19.7969	1.1815	29.7889
κ	0.0442	2.5224	0.0808	2.6086	0.0515	1.9262	0.0791	2.9758	-0.0275	-0.3969	0.1140	3.7088	0.0695	2.8839	0.0394	0.5573	0.0831	3.6448
$\alpha + \beta + \kappa/2$	0.9956		1.0016		0.9726		1.0055		1.0422		0.9676		0.9888		1.0172		0.9873	
iuv	8.5682		—		3.3588		-6.8440		-0.2135		3.1327		6.3733		-5.7500		3.6863	
iuv^a			21.8750												39.5600		0.8655	
$Q_{12}(uh^{-1/2})$	11.311	(0.501)	4.934	(0.896)	15.658	(0.110)	9.350	(0.675)	14.965	(0.115)	15.963	(0.103)	17.108	(0.144)	11.867	(0.293)	10.933	(0.363)
$Q_{12}(u^2h^{-1})$	9.241	(0.682)	15.382	(0.119)	10.432	(0.403)	8.256	(0.765)	4.161	(0.940)	11.791	(0.299)	1.676	(1.000)	0.790	(1.000)	6.622	(0.761)
log L	-6197.5		-2745.5		-3373.0		-5252.0		-978.5		-4230.5		-5699.38		-1858.4		-3790.7	

Mexico

	All period		Before		After	
	Estimate	t-value	Estimate	t-value	Estimate	t-value
b_0	0.0772	2.9004	0.2022	2.3308	0.0737	2.6311
c_1	0.2388	12.8061	0.4428	10.7591	0.2058	10.1668
d_w	0.4851	14.1648	0.0103	0.0913	0.5443	15.1625
γ	0.1662	6.1134	0.0334	2.3168	0.1921	5.7452
β	0.7751	32.0543	0.9518	45.6651	0.7593	26.5442
α	0.0841	4.1946	0.0779	2.7817	0.0645	3.2219
ν	1.2400	34.0186	1.3113	8.5008	1.2861	32.3236
κ	0.1477	4.4938	−0.1021	−3.0775	0.1851	5.1124
$\alpha + \beta + \kappa/2$	0.9331		0.9787		0.9164	
iuv	2.4824		1.5644		2.2965	
$Q_{12}(uh^{-1/2})$	12.727	(0.389)	11.241	(0.339)	7.463	(0.681)
$Q_{12}(u^2h^{-1})$	4.135	(0.981)	3.343	(0.972)	3.875	(0.953)
log L	−4525.5		−593.92		−3901.80	

Note. (i) $Q_{12}(uh^{-1/2})$ and $Q_{12}(u^2h^{-1})$ are the standardized residuals and their squared terms of the Ljung and Box statistics (order 10). (ii) Log L denotes the maximum likelihood function value. (iii) Figures in parentheses are p values. (iv) Here, "iuv" denotes implied unconditional variance. [a]The iuv values are calculated without taking κ into consideration.

TABLE 4

Descriptive Statistics of Conditional Variance for Each Market

	ARG	BRA	CHL	KOA	MAL	MEX	PHI	THA	TUK	TWN
					Before liberalization					
Mean	34.4702	11.3614	2.8736	2.0035	3.2020	3.3471	6.9204	2.1365	4.3214	6.6450
Median	20.3263	8.7054	1.1809	1.5511	1.7526	1.4092	3.3052	0.8635	3.0740	3.8815
Maximum	385.4345	106.0463	87.5125	23.3126	89.0703	17.8970	88.6172	34.2839	21.4562	37.9133
Minimum	5.4410	6.1837	0.5950	1.0220	0.9568	0.7422	0.7421	0.0588	1.3985	0.5932
Std. dev.	42.2280	10.1491	9.0479	1.4075	7.7819	3.9759	10.6561	3.9740	3.8261	6.7737
					After liberalization					
Mean	3.7327	12.6026	1.5942	2.6084	1.9835	2.1939	2.5915	2.9037	7.6767	2.9462
Median	2.0813	6.6998	1.0814	1.5691	0.9202	1.5270	1.7864	1.7051	6.0849	2.2417
Maximum	26.8401	162.9150	33.2047	30.0215	45.2898	38.8653	15.8086	36.2324	63.1767	17.6712
Minimum	0.6822	0.5182	0.4355	0.6440	0.3757	0.8409	0.4909	0.6659	1.9036	0.9559
Std. dev.	3.9990	17.9855	1.8002	3.5745	3.8338	2.3620	2.2933	3.3826	5.3446	1.9846
Direction of change	−	+	−	+	−	−	−	+	+	−

Note. The conditional variance (cv) is calculated based on the variance equation (2) as $h_t = \gamma + \alpha u_{t-1}^2 + \beta h_{t-1} + \kappa u_{t-1}^2 d_{t-1}$, where h_t is the conditional variance; $+$ $(-)$ represents the increase (decrease) of the cv; Std. dev. indicates standard deviation.

in value after market liberalization. Unfortunately, for the other three markets, it is difficult to determine whether return volatility has changed based on the results in Table 3. In Taiwan and Thailand, the estimated γ increases in value along with a reduced value of $\alpha + \beta + \kappa/2$ after liberalization. The opposite is true for Brazil, where reduced γ is accompanied by increased $\alpha + \beta + \kappa/2$.

In order to evaluate further the return volatility in each market during the two periods, we calculate the descriptive statistics, mean, median, standard deviation, and maximum and minimum values of the conditional variance (cv), and report these in Table 4.

The markets in Argentina, Chile, Malaysia, Mexico, the Philippines, and Taiwan have experienced smaller cv since liberalization. This result is compatible with that from the iuv approach. Notice that the iuv increases its value while the value of cv diminishes in Mexico after liberalization. The cv value of the Taiwan market becomes smaller, but its iuv is not useful for comparison after liberalization. Broadly consistent with the iuv approach, the markets in Brazil, Korea, Thailand, and Turkey have greater cv after liberalization. Based on the iuv approach, we could not compare the volatility in Thailand for the two periods owing to its persistence parameter being greater than 1. Nonetheless, under the conditional variance approach, the cv values appear to have increased after liberalization. A similar phenomenon is observed in Brazil as the cv value increases after liberalization. Table 5 provides a brief summary of these two approaches.

From Table 5, four of the ten markets experienced reduced return volatility

TABLE 5

Summary of the Two Approaches to Measuring the Change in Volatility after Liberalization

Nation	iuv	cv
Argentina	−	−
Brazil	?	+
Chile	−	−
South Korea	+	+
Malaysia	−	−
Mexico	+	−
Philippines	−	−
Taiwan	?	−
Thailand	?	+
Turkey	+	+

Note. (i) The results are based on the estimates shown in Tables 3 and 4. (ii) Here, iuv denotes implied unconditional volatility, where iuv $= \gamma/(1 - \alpha - \beta - \kappa/2)$. (iii) Conditional variance is calculated from the variance equation (2). (iv) A question mark denotes a case where iuv cannot be calculated.

after liberalization while three markets exhibited greater volatility under the iuv approach. For the rest of markets, the comparison is indeterminate using this approach. Using the cv approach, six of the ten markets exhibited reduced volatility while four markets had higher volatility in stock returns. The major discrepancy occurs in Mexico, where the two approaches give different results. However, the iuv measures long-run effects whereas the cv reflects short-term phenomena. Our findings are different from those in previous studies. For example, Mexico, Argentina, Taiwan, and Brazil were found to exhibit significantly diminished market volatility after liberalization in the study by Bekaret and Harvey (1997), whereas only the Argentinian market exhibited such a phenomenon in our analysis. Using the iuv approach, De Santis and Imrohoroglu (1997) find lessened volatility after liberalization in Taiwan and Argentina and a non-comparability in Brazil because $\alpha + \beta > 1$ and $\gamma < 0$. In contrast, our study detects a reduced iuv value for Argentina and indeterminacy for Taiwan and Brazil. Furthermore, the unconditional volatility of South Korea, found by Kim and Rogers (1995) to be relatively unchanged, is increased in our estimate.[11] The discrepancy between our results and those in the literature may be due to the frequency of the data. We use daily returns, whereas De Santis and Imrohoroglu (1997) use weekly returns and Bekaret and Harvey (1997) use monthly returns. Despite the 18-market analysis by De Santis and Imrohoroglu (1997), changes in volatility due to liberalization are studied in only five markets due to inadequacy of the data. The main feature of our study is the use of daily returns, which

---

[11] Kim and Rogers (1995) did not take into consideration that the measure of volatility persistence could also vary in the process.

provide adequate sample size and capture the volatility that would have been averaged out in weekly or monthly data. Furthermore, our sample period covers both the Peso crisis and the Asian financial turmoil.

Capital flows in each country (Fig. 1) may explain some of our results. The fact that the inflow of DFI exceeded that of PI in Chile and Malaysia after liberalization contributes to the reduced volatility measured by either approach. On the other hand, the remaining markets experienced greater inflows of PI than DFI. Similarly, the volatility in exchange rates contributes to return volatility especially in the period of the Asian financial crisis (Granger et al., 1999). Presumably, stock returns in U.S. dollars are a more appropriate measure of volatility.[12] Owing to the paucity of data, we calculate stock returns in local currencies as do Koutmos (1998) and De Santis and Imrohoroglu (1997). However, studying volatility of returns in U.S. dollars is a useful research project.

5. CONCLUSION

Policymakers in emerging economies began to consider capital controls due to the recent string of market collapses in Asia. Does market liberalization provide a conduit to greater stock price volatility? By including the world stock price index in our analysis and applying the iuv approach, we find that four of the ten markets exhibited diminished volatility, while three markets witnessed increased volatility after liberalization. Six of the ten markets experienced smaller cv values based on the short-term conditional variance. This result differs noticeably from previous findings. We believe that the discrepancy is attributable to the use of daily data, the failure of previous work to include the period of the Asian financial crises, and the exclusion of the world index as an explanatory variable in previous studies. Since return volatility can increase in one market but decrease in another after liberalization, the argument for capital controls is not clear. However, the magnitude and velocity of capital flows should be consistent with the fundamentals of the local economy in order to minimize adverse effects. For emerging economies, a stable inflow of DFI is more beneficial than speculative PI, as demonstrated by Chile and Malaysia, where the return volatility did not change noticeably after market liberalization.

REFERENCES

Bekaert, Geert, and Campbell, R. Harvey, "Emerging Equity Market Volatility." *J. Finan. Econ.* **43,** 1:29–77, Jan. 1997.

Black, Fischer, "Studies of Stock Price Volatility Changes." In *1976 Proceedings of the Business and Economic Statistics Section,* pp. 177–181. Washington, DC: American Statistical Association, 1976.

Bollerslev, Tim, "Generalized Autoregressive Conditional Heteroscedasticity." *J. Econometrics* **31,** 3:307–327, Apr. 1986.

---

[12] We are grateful to one of the reviewers for this suggestion.

Box, George E. P., and Tiao, George C., *Bayesian Inference in Statistical Analysis.* Reading, MA: Addison–Wesley, 1973.

Campbell, John Y., Lo, Andrew W., and MacKinlay, Archie Craig, *The Econometrics of Financial Markets.* Princeton, NJ: Princeton Univ. Press, 1997.

Christie, Andrew A., "The Stochastic Behavior of Common Stock Variances: Value, Leverage and Interest Rate Effects." *J. Finan. Econ.* **10,** 4:407–432, Dec. 1982.

De Santis, Giorgio, and Imrohoroglu, Selahattin, "Stock Returns and Volatility in Emerging Financial Markets." *J. Int. Money Finance* **16,** 4:561–579, Aug. 1997.

Domowitz, Ian, Glen, Jack, and Madhavan, Ananth, "International Cross-Listing and Order Flow Migration: Evidence from an Emerging Market." *J. Finance* **53,** 6:2001–2027, Dec. 1998.

Engle, Robert, "Autoregressive Conditional Heteroscedasticity with Estimates of the Variance of United Kingdom Inflation." *Econometrica* **50,** 4:987–1006, July 1982.

Glosten, Lawrence R., Jagannathan, Ravi, and Runkle, David E., "On the Relation between the Expected Value and the Volatility of the Nominal Excess Return on Stocks." *J. Finance* **48,** 5:1779–1801, Dec. 1993.

Granger, Clive W. J., Huang, Bwo-Nung, and Yang Chin Wei, "A Bivariate Causality between Stock Prices and Exchange Rates: Evidence from Recent Asian Flu." *Quart. J. Econ. Finance,* forthcoming.

Harvey, Andrew C., *The Econometric Analysis of Time Series.* Oxford: Philip Allan, 1981.

Kim, Sang W., and Rogers, John H., "International Stock Price Spillovers and Market Liberalization: Evidence from Korea, Japan, and the United States." *J. Empirical Finance* **2,** 2:117–133, 1995.

Koutmos, Gregory, "Asymmetries in the Conditional Mean and the Conditional Variance: Evidence from Nine Stock Markets." *J. Econ. Bus.* **50,** 3:277–290, May–June 1998.

Kyle, Albert S., "Continuous Auctions and Insider Trading." *Econometrica* **53,** 6:1315–1335, Nov. 1985.

Nelson, Daniel B., "Conditional Heteroscedasticity in Asset Returns: A New Approach." *Econometrica* **59,** 2:347–370, March 1991.

Scholes, Myron, and Williams, Joseph T., "Estimating Betas from Nonsynchronous Data." *J. Finan. Econ.* **5,** 3:309–327, Dec. 1977.

Schwert, G. William, "Stock Market Volatility: Ten Years after the Crash." NBER Working Paper Series, No. 6381. Cambridge, MA: National Bureau of Economic Research, 1998.

Available online at www.sciencedirect.com

SCIENCE @ DIRECT°

JOURNAL
OF
ASIAN ECONOMICS

NORTH-HOLLAND

Journal of Asian Economics 14 (2003) 337–351

Short paper

An analysis of exchange rate linkage effect: an application of the multivariate correlation analysis

Bwo-Nung Huang[a,*], Chin-Wei Yang[b]

[a]*Department of Economics, Center for Industry Analysis, Development and Forecasting, National Chung-Cheng University, Chia-Yi 621, Taiwan*
[b]*Department of Economics, Clarion University of Pennsylvania, Clarion, PA 16214-1232, USA*

Received 18 October 2001; received in revised form 12 November 2002; accepted 20 January 2003

Abstract

This paper employs 10-nation daily exchange rates to analyze changes in correlation coefficients before (from 4 January 1994 to 30 June 1997) and after (from 1 July 1998 to 22 June 2001) the Asian financial crisis. Via the forecast errors (30-days forward) of the VAR model proposed by Den Haan [J. Monet. Econom. 46 (2000) 3], we calculate pairwise dynamic correlation coefficients with the standard error derived from a bootstrapping method. The result indicates the linkage between exchange rates of the Philippines, Indonesia, Thailand, Taiwan, and their adjacent economies has intensified after the financial crisis mostly due to the switch from fixed exchange rate to floating exchange rate system.
© 2003 Elsevier Science Inc. All rights reserved.

JEL classification: F310

Keywords: Multivariate correlation analysis; Exchange rates; Financial crisis; Currency devaluations; Spillover effects

1. Introduction

The Asian financial crisis ushered itself in the second half of 1997. Not only did the crisis inflict adverse effects on world economy, it also created a financial tsunami with plummeting stock and exchange rate markets. Quite a few studies have analyzed potential

---

* Corresponding author.
E-mail address: ecdbnh@ccunix.ccu.edu.tw (B.-N. Huang).

1049-0078/03/$ – see front matter © 2003 Elsevier Science Inc. All rights reserved.
doi:10.1016/S1049-0078(03)00021-6

correlations among regional stock or exchange rate markets. Others emphasize spread-out and spillover effects among national stock markets. For example, Baig and Goldfajn (1998) did not find more intensified interaction among the five Asian stock markets after the financial crisis. In a broader view, Granger, Huang, and Yang (2000) indicate the existence of significant feedback phenomena between exchange rate and stock markets during the crisis. Actually, the culprit of the Asian financial crisis is the spillover effect that emanated from competitive currency devaluations. Most Asian economies rely heavily on export to the US and thus competitive devaluation cannot possibly lead to noticeable increases in export for all countries. To study spillover effect, Baig and Goldfajn (1998) compute unconditional correlation coefficients for the five Southeast Asian currencies (Thailand, Malaysia, Indonesia, Korea, and the Philippines). Their results suggest (i) rising correlation coefficients in exchange rate markets but (ii) inconclusive outcome for stock markets. Using a Markov switching model, Cerra and Saxena (2002) identify the spillover effect on Indonesia rupiah from the devaluation of South Korean won and Thailand baht during the crisis. There are speculations that the devaluation of RMB of China as the culprit of the crisis. However, Fernald, Edison, and Loungaui (1999) in evaluating the linkage effect between China and emerging countries in the region, fail to find such a causality. As expected, competitive devaluations of the currencies can intensify export competition to the US and certainly slowdown China's economic growth. In contrast, Liu, Norland, Robinson, and Wang (1998) via a computable general equilibrium model find that the devaluation of RMB contributed, to some extent, to the crisis, but was not a major factor.

Although some prior studies investigated the linkage or interaction mechanism of the stock and/or exchange rate markets, the simultaneous analysis on the 10 Asian exchange rate markets in Southeast Asia during the period of a major crisis is scanty. Well-known in the literature, a significant historical event such as the Asian financial crisis is capable of rendering a significant blow to any regional financial framework. Thus, it is of utmost importance to find out if the exchange rate mechanism among the 10 markets has undergone substantial changes after the crisis. If it has, an accurate forecast is not plausible without considering the change especially in the presence of competitive devaluation. Applying the VAR forecast-error model developed by Den Haan (2000) and Den Haan and Summer (2001) to the 10 Southeast Asian economies, we find dynamic correlation coefficients in terms of exchange rate comovements have increased much more frequently after the crisis: six times for the Philippines peso, five times for Indonesia rupiah, four times for Taiwan dollar and three times for Thailand baht.

The organization of the paper is as follows. Section 2 discusses data and their time series properties. Section 3 presents methodology and empirical results. Section 4 includes a conclusion.

2. Data sources and unit root test

Currencies of the 10 nations examined in this study include China (CHN), Hong Kong (HKN), Indonesia (IND), Japan (JPN), Korea (KOA), Malaysia (MAL), the Philippines (PHI), Singapore (SIG), Thailand (THA), and Taiwan (TWN). Exchange rates are average values of ask and offer prices based on WM/Reuters closing spot rates (no. of local

Table 1
Results of the unit root test

	Before crisis		After crisis	
	y	Δy	y	Δy
CHN	-17.51^*	-30.15^*	-4.85^*	-35.15^*
HKN	-5.30^*	-29.11^*	-1.67	-35.41^*
IND	-3.62^{**}	-34.23^*	-2.92	-24.90^*
JPN	-2.28	-29.28^*	-1.50	-26.40^*
KOA	-0.55	-28.68^*	-1.51	-25.28^*
MAL	-1.97	-26.94^*	-3.98^*	-43.72^*
PHI	-2.08	-34.52^*	-1.64	-28.73^*
SIG	-1.74	-30.98^*	-2.23	-31.62^*
THA	-4.43^*	-30.88^*	-2.82	-27.74^*
TWN	-2.06	-29.63^*	-0.03	-26.66^*

CHN: China, HKN: Hong Kong, IND: Indonesia, JPN: Japan, KOA: Korea, MAL: Malaysia, PHI: the Philippines, SIG: Singapore, THA: Thailand, TWN: Taiwan. y is the logarithmic exchange rate (level), Δy is the first difference of logarithmic exchange rate.

[*] 1% significance level.
[**] 5% significance level.

currencies per US dollar). Sample period spans from 4 January 1994 to 22 June 2001: a total of 1949 daily data from Datastream Database.[1] In order to investigate the linkage effect before and after the Asian financial crisis, we take July 1 1997 as the demarcation date. This is to say, we have 912 daily prices before the crisis (4 January 1994–30 June 1997) and 778 daily prices after the crisis (1 July 1998–22 June 2001).[2] Their time series plots are exhibited in Fig. 1.

Evident from Fig. 1, except for Japan, Taiwan, and Singapore in which floating exchange rates system prevailed, the exchange rates of other seven economies were pegged to US dollar before the crisis. In addition, only China and Hong Kong retained fixed exchange rates during the crisis period, the rest switched to floating exchange rates regime. Note that Malaysia reverted back to pegged system after 1 September 1998. Even among the economies that maintained fixed exchange rate, the rates were occasionally adjusted depending on economic conditions. China's RMB is about the only currency whose value remained stable during the crisis. Note that after the crisis, ringgit of Malaysia and RMB of China are the only two currencies that hold onto fixed exchange rate regime. For stationarity test, we employ the method developed by Phillips and Perron (1988). The result of unit root test is reported in Table 1.

The null hypothesis of the Phillips and Perron (1988) test is that variables are of $I(1)$ against $I(0)$ in the alternative hypothesis. A time series data is said to exhibit stationary process when the null hypothesis is rejected. Table 1 illustrates that we can reject the null hypothesis of $I(1)$ for the currencies of China, Hong Kong, Indonesia, and Thailand before

---

[1] The data base is published by Datastream Inc.

[2] Well-known in econometrics theory, exceedingly volatile data can contaminate the result. To this end we exclude the crisis period (July 1997–June 1998). As such, the second period started from 1 July 1998 to 22 June 2001.

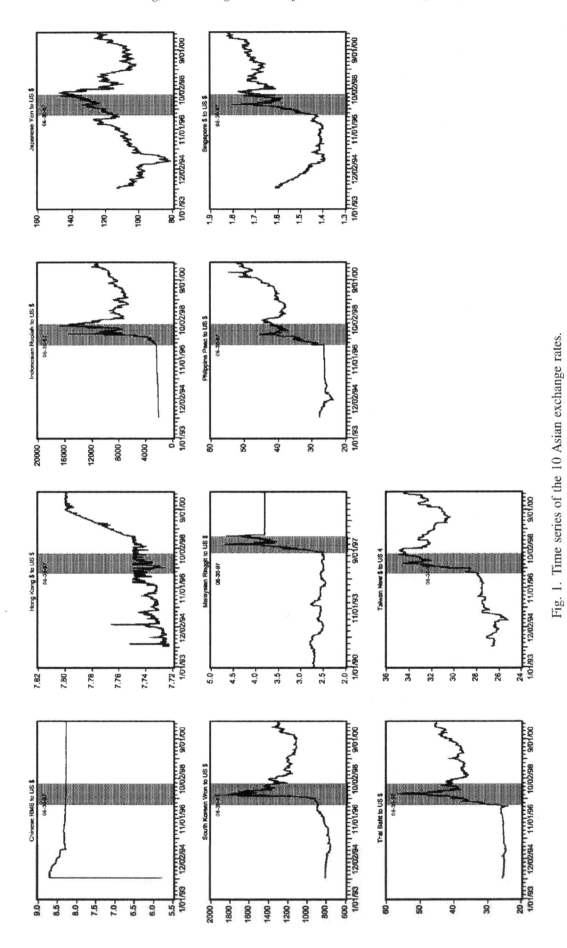

Fig. 1. Time series of the 10 Asian exchange rates.

the crisis. This was done by using logarithmic exchange rate at level. After the crisis, only in the case of China and Malaysia, do we reject the null hypothesis of $I(1)$. However, if we take first difference of the logarithmic variables, the null of $I(1)$ is rejected for all currencies before and after the crisis. It implies the logarithmic first difference is sufficient to ensure stationarity property in our data. For currencies adopting fixed exchange rate system— China, Hong Kong, Indonesia, and Thailand before the crisis as well as China and Malaysia after the crisis—the exchange rate (level) is clearly of $I(0)$. On the other hand, currencies under flexible exchange rate regime may be of $I(1)$. An appropriate level of data to use depends on the market one analyzes. Many prior studies indicated that change in exchange rate (in terms of logarithmic difference) was found to effectively mirror exchange rate exposure; hence, it may have better economic interpretation. Besides, using returns (logarithmic difference) rather than levels of exchange rates can accommodate more complex situations than is implied by simple unit root properties. For example, profitable trade generally depends on volatility in returns.[3] Since the purpose of this study is to investigate short-term exchange rate dynamic, we take logarithmic difference to reflect changes in exchange rate as shown below:

$$\Delta y_{i,t} = \log e_{i,t} - \log e_{i,t-1} \tag{1}$$

where $e_{i,t}$ denotes the nominal exchange rate for country i in terms of US dollar ($i =$ CHN, HKN,...,TWN)

3. Empirical estimates and discussion of results

To investigate the spillover effect, we calculate the dynamic conditional correlation based on forecast errors of the VAR model developed by Den Haan (2000) whose analysis focuses on the comovements between output and inflation in both long and short run. The purpose of this paper is to apply the dynamic conditional correlation method to exchange rate markets before and after the Asian financial crisis. We briefly present the dynamic conditional correlation approach as shown below.[4]

Consider a vector of random variables X_t of size N in which components can be of $I(0)$, $I(1)$, or $I(d)$. In the case of 10 Asian exchange rate markets, a VAR model can be formulated:

$$X_t = a + bt + ct^2 + \sum_{t=1}^{L} A_t X_{t-1} + \varepsilon_t \tag{2}$$

where A is the coefficient matrix of $N \times N$; a, b, and c are vectors of dimension N; ε is a vector of innovation of dimension N. Optimum lag periods are determined by the conventional SBC or AIC criterion. Let X_{1t+k} and $X_{1t+k,t}^{ue}$ denotes k-period forecast value and forecast error of X_1. Likewise, X_{2t+k} and $X_{2t+k,t}^{ue}$ denote the same for X_2. In addition, the covariance and correlation coefficients between variables (k-periods forward) are

---

[3] We are indebted to one of the referees for pointing out this possibility.
[4] See Den Haan (2000) and Den Haan and Summer (2001) for detail.

presented by COV(k) and COR(k). In other words, we can calculate the k-period-forward dynamic conditional correlation bases on its forecast errors of X_1 and X_2. If all of the variables in the vector of X_t are $I(0)$, the correlation is expected to converge to its unconditional value as k increases. If some variables are nonstationary, Den Haan (2000) proves that COV(k) and COR(k) can still be estimated consistently for a fixed K value.

However, an important assumption for deriving the consistency results is that Eq. (2) must be correctly specified. When X_t exhibits the integrated process, one might choose to estimate a VAR in first difference or an error-correction system. When the restrictions that lead to these systems are correctly specified, then imposing the restriction may give rise to more efficient forecasts in a finite sample. It is to be noted that while COR(k) is a pairwise correlation coefficient, it takes other influences (other exchange rates) into consideration as Eq. (2) is formulated in terms of multivariate VAR model. In order to calculate 45 sets of pairwise dynamic conditional correlations, we formulate a 10-country VAR model using Δy_t. The optimum lag is chosen until the serial correlation is purged of ε_t (Den Haan and Summer, 2001). We find the optimum lag of $L = 5$ (both before and after the crisis) in the 10-variable VAR model. Moreover, to fathom the comovements further, we adopt 30 days as forecast period (one and a half month). In order to explore significant changes before and after the crisis, we calculate standard deviations of the correlation coefficients via the bootstrap method before applying the conventional t test. That is, standard deviations of these pairwise conditional correlations (30-periods forward) are calculated via the bootstrap method (500 times) prior to testing the difference of correlation coefficients before and after the crisis. Fig. 2 illustrates the results for the 45 pairs of correlation coefficients from the 10 countries.

Evident from Fig. 2, the only currency experienced increasing correlation with China's RMB after the crisis is Hong Kong dollar. In the case of Hong Kong, the pairwise correlations with China, Indonesia, Malaysia, and the Philippines are found to be stronger. For Indonesia, stronger correlations are found with respect to Hong Kong, Japan, Korea, Malaysia, the Philippines, Singapore, Thailand, and Taiwan (or cohorts). For Japan, the cohorts are Indonesia and the Philippines; for Korea, the cohorts are Indonesia, the Philippines, and Taiwan; for Malaysia, the cohorts are Hong Kong, Indonesia, and Thailand. For the Philippines, the cohorts are Hong Kong, Indonesia, Japan, Korea, Singapore, Thailand, and Taiwan; for Singapore, the cohorts are Indonesia, the Philippines, and Taiwan; and for Taiwan, the cohorts are Indonesia, Korea, and the Philippines.[5]

Fig. 2, however, does not provide any statistical test on significant difference before and after the crisis. The correlation coefficients (30-periods forward) after the crisis, in some cases, exceed that before the crisis. In other cases, the opposite holds due to perhaps the large k value (forecast period ahead). To circumvent the problem, we divide the forecast period into short run (1–10 business days or 2 weeks), medium run (11–20 days or 4 weeks), and long run (21–30 days) and apply the t statistic to the 45 pairwise correlation coefficients as shown in Fig. 3.

---

[5] In some cases, the dynamic correlation coefficient after the crisis is greater than that before the crisis for some years. For other years, the reverse is true due to the relatively large forecast period (30-period). Under such circumstances, we do not regard it as an overall increase in correlation coefficient.

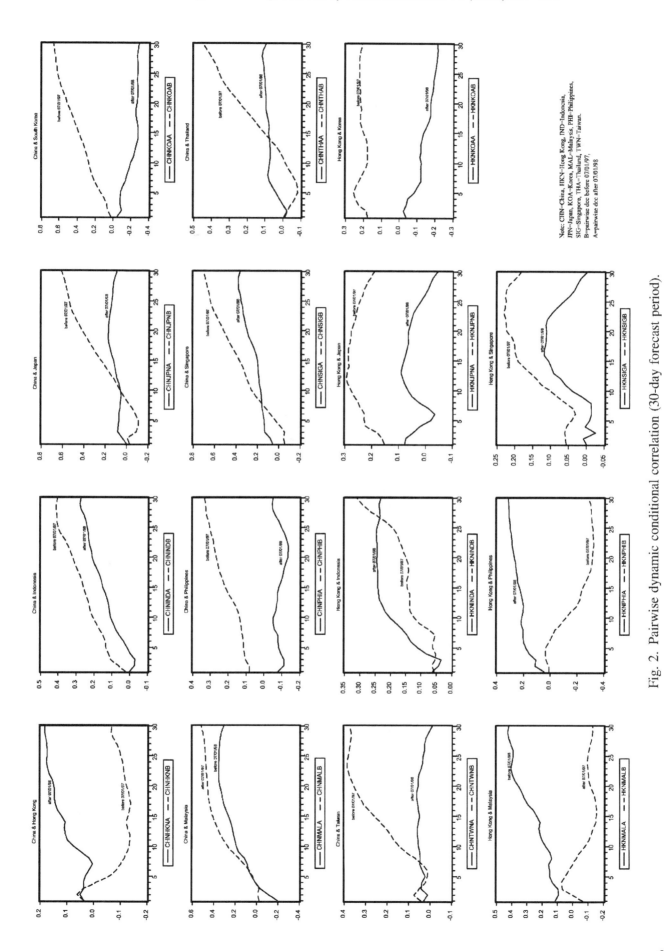

Fig. 2. Pairwise dynamic conditional correlation (30-day forecast period).

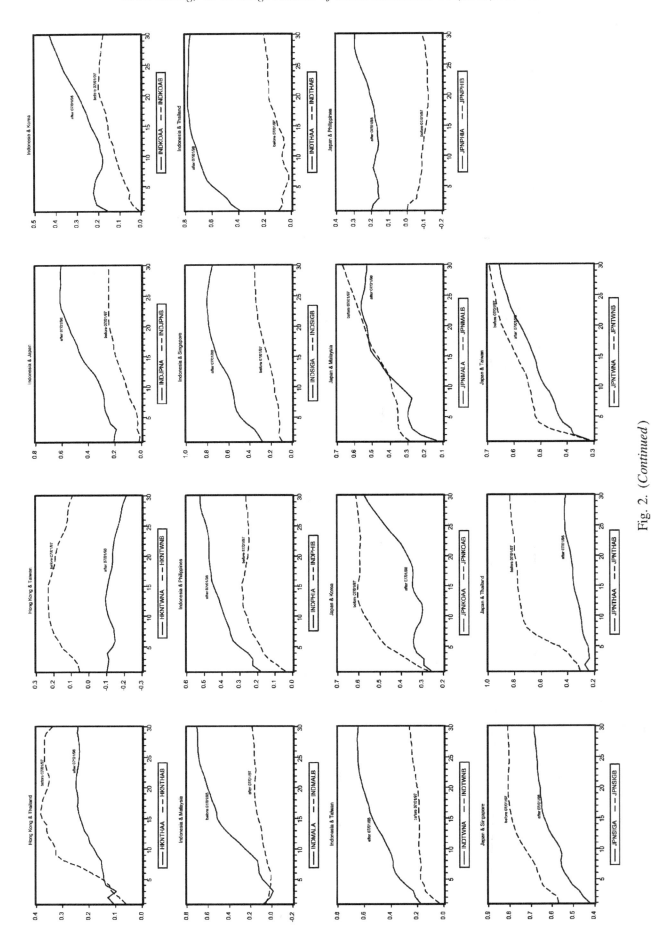

Fig. 2. (Continued)

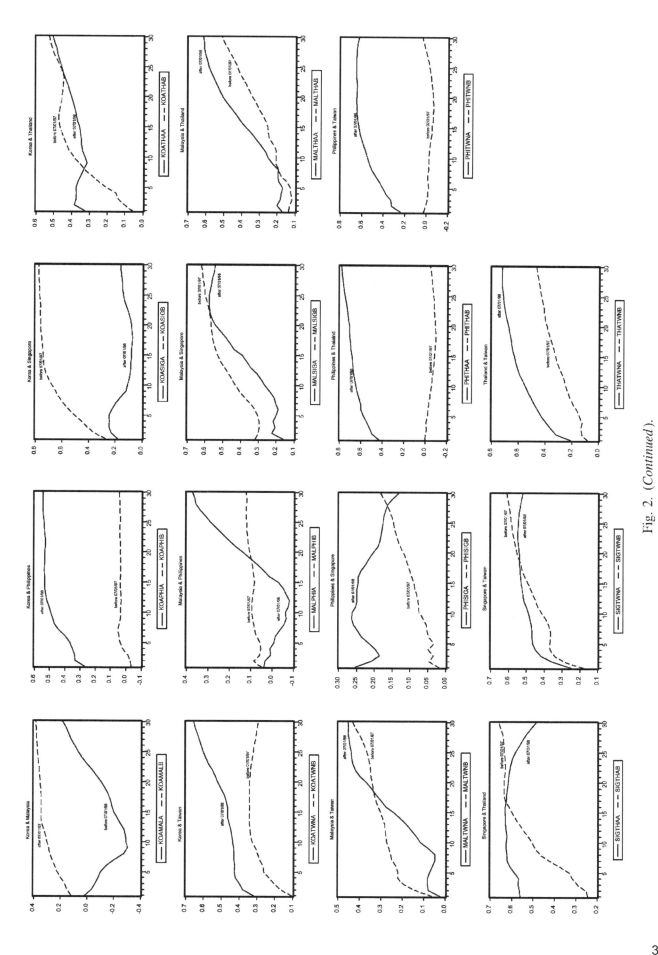

Fig. 2. (Continued).

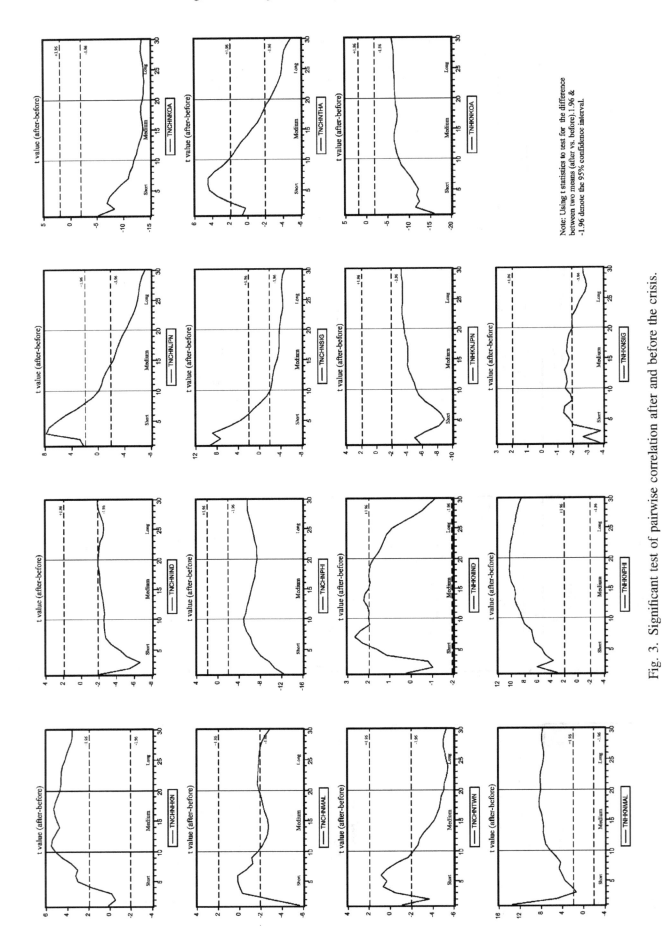

Note: Using t statistics to test for the difference between two means (after vs. before).1.96 & -1.96 denote the 95% confidence interval.

Fig. 3. Significant test of pairwise correlation after and before the crisis.

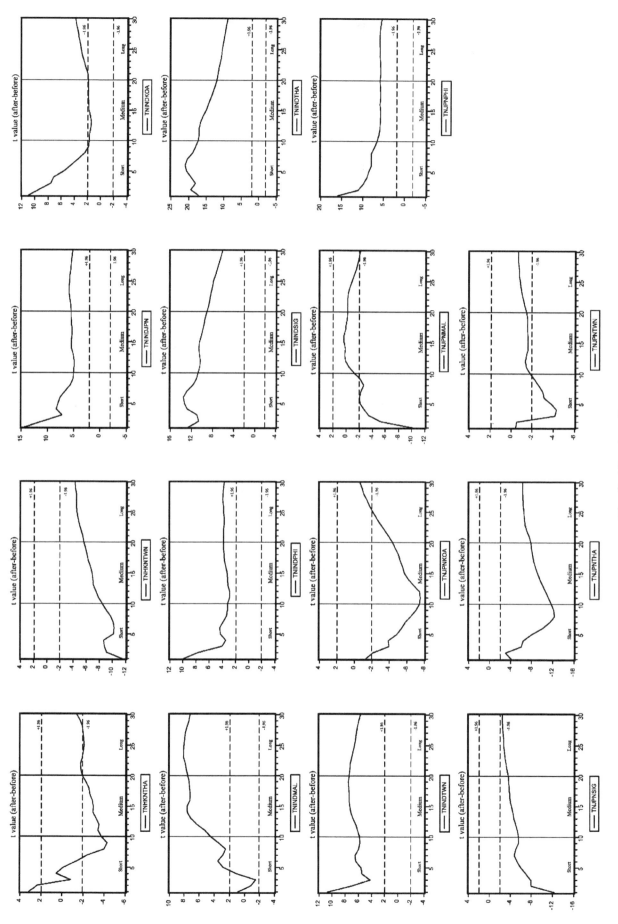

Fig. 3. (*Continued*)

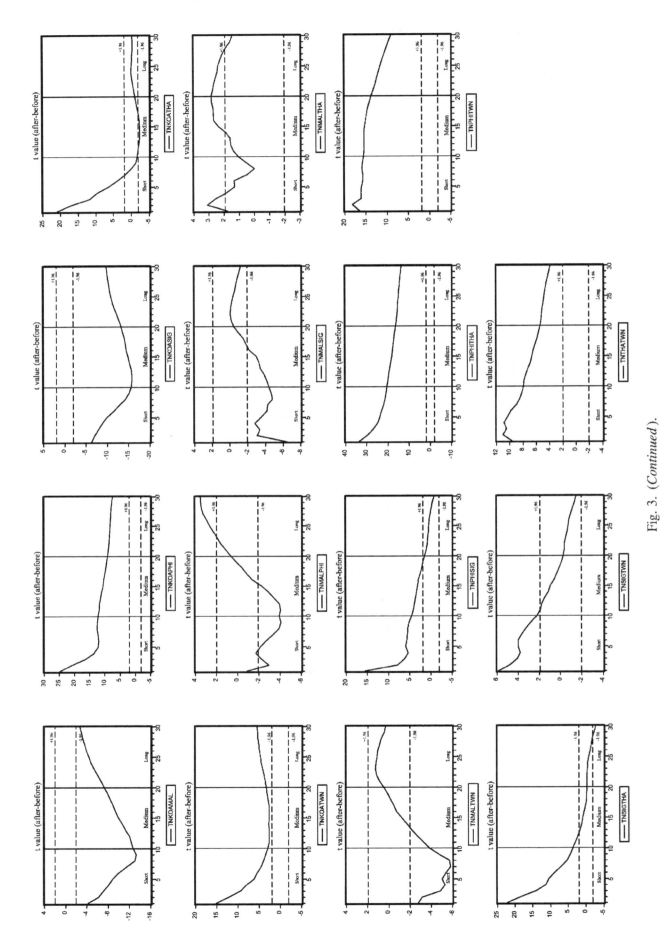

Fig. 3. (Continued).

Table 2
Summary of pairwise correlation coefficients before and after the crisis

	CHN			HKN			IND			JPN			KOA			MAL			PHI			SIG			THA			TWN			All windows (+)	All windows (−)
	S	M	L	S	M	L	S	M	L	S	M	L	S	M	L	S	M	L	S	M	L	S	M	L	S	M	L	S	M	L		
CHN				0	+	+	−	−	−	+	0	−	−	−	−	0	−	0	−	−	−	+	−	−	+	0	−	0	−	−	0	3
HKN	0	+	+				0	+	0	−	−	−	−	−	−	+	+	+	+	+	+	0	0	−	0	−	0	−	−	−	2	3
IND	−	−	−	0	+	0				+	+	+	+	0	+	0	+	+	+	+	+	+	+	+	+	+	+	+	+	+	5	1
JPN	+	0	−	−	−	−	+	+	+				−	−	0	−	0	0	+	+	+	−	−	−	−	−	−	−	0	0	2	3
KOA	−	−	−	−	−	−	+	0	+	−	−	0				−	−	−	+	+	+	−	−	−	+	−	0	+	+	+	2	4
MAL	0	−	0	+	+	+	0	+	+	−	0	0	−	−	−				−	−	+	−	−	0	0	+	+	−	0	0	1	1
PHI	−	−	−	+	+	+	+	+	+	+	+	+	+	+	+	−	−	+				+	+	0	+	+	+	+	+	+	6	1
SIG	+	−	−	0	0	−	+	+	+	−	−	−	−	−	−	−	−	0	+	+	0				+	0	0	+	0	0	1	2
THA	+	0	−	0	−	0	+	+	+	−	−	−	+	−	0	0	+	+	+	+	+	+	0	0				+	+	+	3	1
TWN	0	−	−	−	−	−	+	+	+	−	0	0	+	+	+	−	0	0	+	+	+	+	0	0	+	+	+				4	1
Each window (+)	3	1	1	2	4	3	6	7	7	3	2	2	4	2	3	1	3	4	7	7	7	5	2	1	6	4	4	5	4	4		
Each window (−)	3	6	7	3	4	4	1	1	1	6	4	4	5	6	4	5	4	1	2	2	1	3	4	4	1	3	2	3	2	2		

CHN: China, HKN: Hong Kong, IND: Indonesia, JPN: Japan, KOA: Korea, MAL: Malaysia, PHI: the Philippines, SIG: Singapore, THA: Thailand, TWN: Taiwan, CS: the window period of 1–10 days. M: the window period of 11–20 days. L: the window period of 21–30 days. 0 implies ($k < 5$) where k is the number of days in the window in which positive (+) or negative correlations prevailed. $+ (−)$ implies $k > 5$ where k is the number of days in which positive (negative) correlation prevailed for more than 5 days. All windows: frequency of the increased correlation coefficients for all three windows. The results are summarized from the t tests in Fig. 3.

The two dotted parallel lines in Fig. 3 represent upper and lower bounds for the 95% confidence interval. Note that if the t statistics fall within ± 1.96, one cannot reject the null hypothesis of no significant difference between the correlation coefficients before and after the crisis. If, however, the t statistics exceeds 1.96 (above the upper bound), the correlation coefficient after the crisis is significantly greater. The reverse is true if the sample t statistic falls short of -1.96. The three vertical lines demarcate three different forecast horizons: short, medium, and long runs. To facilitate comparison, we summarize 45 pairwise correlations into Table 2.

The 30-day-forward ($k = 30$) dynamic correlation coefficients shown in Fig. 1 could cloud the relationship in the sense that the correlations in some cases are not monotonic. To capture the time-varying dynamic correlations, we segment k values into three windows: $1 \leq k \leq 10$, $11 \leq k \leq 20$ and $21 \leq k \leq 30$ as before. If the t statistic is shown to be significantly different from zero for k value greater than 5 days, we consider the dynamic correlation coefficients drastically different before and after the crisis for the window. Based on this analysis from Table 2, we find the Philippines experienced the most intense exchange rate interactions (correlations) with its neighboring countries (6) after the crisis, followed by Indonesia (5), and Taiwan (4). China, on the other hand, had no significant interaction with any country (0), followed by Malaysia and Singapore (1). The result is not suprising as China and Malaysia adopt fixed exchange rate system after the crisis. While Singapore is shown to be closely related to only Indonesia, its dynamic correlations after the crisis exhibit more significant increases with neighboring countries especially in the short-run window. Also shown in Table 2 are the countries that experience decreasing correlations with neighboring countries after the crisis: Korea (4), China (3), Hong Kong (3), and Japan (3).

For an exchange rate encountering increased correlations with at least four other currencies, we define that currency as having experienced increased interaction. Based on this definition, we detect currencies of Indonesia, Korea, the Philippines, Singapore, Thailand, and Taiwan witnessed increased interaction with others in the short-run window. Similarly, currencies of Hong Kong, Indonesia, the Philippines, Thailand, and Taiwan experienced increased correlation in the medium-run window. Finally, currencies of Indonesia, Malaysia, the Philippines, Thailand, and Taiwan underwent stronger interaction in the long-run window. In sum, regardless of the window used or for the entire window, there existed undeniable evidence of greater currency interdependence after the crisis, partly due to the regime shift from fixed exchange rate to the flexible one (e.g. Indonesia and the Philippines). Our result confirms: during the crisis, pairwise correlations of the five countries (Thailand, Malaysia, Indonesia, Korea, and the Philippines) become significantly greater. Our result also points out an interesting phenomenon: The currencies of Malaysia and Korea witnessed increasingly more negative correlation with other currencies in the region. This anomalous result could be explained by (i) Malaysia returned to the fixed-rate exchange rate system on 1 September 1998 and (ii) Korea won appreciated its value drastically after the Asian financial crisis.

4. Conclusion

Using daily exchange rate data for the 10 markets in East Asia before and after the crisis and applying the dynamic correlation model by Den Haan (2000) and Den Haan and Summer (2001), we are able to calculate pairwise correlation coefficients with bootstrap-generated standard deviations. The results indicate increased interactions with other currencies for the Philippines (six times), Indonesia (five times), Taiwan, and Thailand (four times) in all three window periods. In the shortest window period of 1–10 days, increased frequencies of positive currency interactions were found in the Philippines (seven times), Thailand, Taiwan, Indonesia, and Singapore (at least five times). The increased magnitude of the currency spillover effects can be attributed to shift in exchange rate regime: from the fixed-rate to floating-rate system during or after the crisis.

It has been nearly 4 years since the last Asian financial crisis. However, starting from July 2001, currencies of East Asia began to experience competitive devaluation again. It remains to be seen if it is the beginning of another currency crisis. As US economy sputters recently, the strategy of competitive devaluation has rather limited effect due to increased currency interactions in East Asia. Such a regime shift no doubt has intensified currency spillover phenomenon. As a result, competitive devaluation will have greater adverse impacts in real term on Indonesia, the Philippines, Thailand, Taiwan, and Singapore at least in the short run. Our paper does not consider the weekday or weekend effect of exchange rates. It may lead to biased results. This remains, however, a future research topic.[6]

Acknowledgements

We are grateful to two anonymous referees for their valuable suggestions. The remaining errors, however, are authors' only.

References

Baig, T., & Goldfajn, I. (1998). *Financial market contagion in the Asian crisis* (pp. 98–155). International Monetary Fund Working Paper.

Cerra, V., & Saxena, S. C. (2002). Contagion, monsoons, and domestic turmoil in Indonesia. *Review of International Economics, 10*(1), 36–44.

Den Haan, W. J. (2000). The comovement between output and prices. *Journal of Monetary Economics, 46*, 3–30.

Den Haan, W. J., & Summer, S. (2001). *The covement between real activity and prices in the G7*. UCSD Discussion Paper 2001–2005.

Fernald, J., Edison, H., & Loungaui, P. (1999). Was China the first domino? assessing links between China and the rest of emerging Asia. *Journal of International Money and Finance, 18*(4), 515–535.

Granger, C. W. J., Huang, B. - N., & Yang, C. - W. (2000). A bivariate causality between stock prices and exchange rates: evidence from recent Asia flu. *Quarterly Journal of Economic and Finance, 40*, 337–354.

Liu, L., Norland, M., Robinson, S., & Wang, Z. (1998). *Asian competitive devaluations*. Working Paper 98-2, Institute for International Economics.

Phillips, P. C. B., & Perron, P. (1988). Testing for a unit root in time series regression. *Biometrika, 75*, 335–346.

---

[6] We thank an anonymous reviewer for pointing this out.

NORTH-HOLLAND

Global Finance Journal 12 (2001) 55–77

Global Finance
Journal

An empirical investigation of trading volume and return volatility of the Taiwan Stock Market

Bwo-Nung Huang[a],*, Chin-Wei Yang[b,1]

[a]*Institute of International Economics, National Chung Cheng University, Chia-Yi 621, Taiwan*
[b]*Department of Economics, Clarion University of Pennsylvania, Clarion, PA 16214, USA*

Received 15 May 2000; received in revised form 2 August 2000; accepted 21 September 2000

Abstract

This study examines the Mixed Distribution Hypothesis (MDH) using 5-min interval stock returns of the Taiwan Stock Index (TSI). Startlingly enough, the persistence of stock volatility remains dominant after the stochastic mixing variable was included in the variance equation. It implies that the MDH cannot explain away the ARCH phenomenon. We have found that the composition of participants (approximately 92% of participants are individual investors) in TSI is a major contributing factor to the persistent volatility. In addition, the existence of limits on price changes, to some extent, accounts for the persistence phenomenon. Similar results are also found for individual stocks in the sample. Interestingly enough, the explanatory power of trading volume exhibits a U-shaped pattern in explaining return volatility in Taiwan Stock Market. © 2001 Elsevier Science Inc. All rights reserved.

JEL classification: G15; F30

Keywords: Trading volume; Return volatility; Taiwan Stock Market

1. Introduction

Taiwan, the orphan of Asia, has had more than her share of political vicissitude throughout her 400-year-old history. Prior to the war of 1895 in which Taiwan was seceded to Japan, the staple of export had been deer hides and camphor products under the rule of Dutch and Manchu

---

* Corresponding author. Fax: +886-5-2720816.
E-mail address: ecdbnh@ccunix.ccu.edu.tw (B.-N. Huang).
[1] Fax: +1-814-2261910.

1044-0283/01/$ – see front matter © 2001 Elsevier Science Inc. All rights reserved.
PII: S1044-0283(01)00023-0

regimes. During Japanese occupation, the lion's share of export hinges on sugar industry. It was not until early 1960s that Taiwan develops her sound export-oriented trade policy under the nationalist government. Since then, the bubbling economy of Taiwan is considered an example of successful economic development especially for Asian emerging economies. The fact that the Taiwan Stock Index (TSI) is included in Morgan–Stanley Emerging Market Free Index, World Free Index, and Fareast (except Japan) Free Index is testimonial to an ever increasing financial role Taiwan plays in the Asian market. Prior to the announcement (April 3, 1996), the Taipei market had gained 95 points with the closing index of 5127.49 valued at US$30.86 billion. On April 6, TSI rocketed to 5377.19 (valued at US$76.97 billion) with a sizable gain of 201.44 points in just 2 h. Because Morgan–Stanley index is one of the major guiding principles for portfolio managers, the inclusion of TSI makes the Taiwan market an attractive choice in international financial market. However, in an attempt to prevent the first popular presidential election (March 23, 1996) in Taiwan, mainland China launched a series of missile exercises around the island. As a consequence, TSI took a plunge to about 4700 points, but bounced back to approximate 5500 points as more positive information arrived. Since then, other economic upheavals and political jitters have caused TSI to be very sensitive to newly arrived information. Recent stock turmoils in Hong Kong, Taiwan, and other Asian emerging economies inevitably transmitted its volatility to the record 554-point plunge and 337-point, next-day rebound of the Dow Jones industrial average in October 1997. Hence, a careful examination of the variance equation of TSI can be of critical importance in understanding the nature of volatility in an emerging market. As pointed out by Huang (1995), only 5% of participants in TSI were institutional investors, with a great majority of 95%, individual investors.[2] Such an unusual composition creates an environment, which is conducive to the formation of "irrational exuberance" in which individual investors focus primarily on short-term profit. As such, the great majority of TSI investors appears very jittery to the arrival of new information, while the fundamental aspects of firms play only a secondary role.

No sooner had TSI (77 individual stocks) been included in the Morgan–Stanley Emerging Market Free Index and the Dow Jones World Index than the bubbling Taiwan market became a financial safari for foreign investment companies. A conservative estimate of new international investment on TSI is US$30 billion in the first 2 years. The flip side is that an emerging market in general does not have a well-structured financial system and lacks maturity in terms of market scales and regulations. Needless to say, the bottom line for many investors is the profitability (abnormal profit) of the Taiwan market, which inevitably begs to the question of the weak form efficient market hypothesis. Majority of studies of the efficient market hypothesis on TSI (e.g., Huang 1995; Huang & Yang 1995) cannot find a serious deviation from the random walk hypothesis.[3] Their conclusion, based on the variance ratio (VR) test and rescaled range (R/S) test, does not necessarily suggest an efficient market for TSI. The discrepancy between statistical results and the widely-held view about the Taiwan market may

---

[2] According to 1995 data released by Taiwan Stock Exchange, the ratio is 8% and 92%, respectively.

[3] Huang and Yang (1995), using the modified R/S technique (Lo, 1991), fail to reject the null hypothesis with daily, and monthly stock returns. Similarly, Huang (1995) fails to reject the null hypothesis, using the heteroscedasticity-consistent VR statistic.

be attributed to (1) power of the statistics employed and/or (2) the limited information (only price variables are used). However, the R/S technique is known to be robust (with respect to Gaussian assumption), and the V/R technique is heteroscedasticity-consistent. They do not appear to have major flaw(s). Hence, it is likely that using only price data leaves out an important variable: quantity or trading volume, which may well lead to inadequate descriptions of the market. As Beaver (1968) put it, "An important distinction between the price and volume tests is that the former reflects change in the expectations of the market as a whole while the latter reflects changes in the expectations of individual investors." Viewed in this perspective, it is important to examine joint distribution of both price and volume variables in order to provide more accurate statistical inferences. Well known in the literature, empirical investigations on speculative prices have revealed kurtotic properties as compared to the normal distribution. The leptokurtic distribution of rates of return is a sampling consequence, when data are pooled from a mixture of distributions (MD) with varying conditional variances. This is to say that statistical tests employing both price and volume variables tend to support the MD hypothesis. In this light, price data can be viewed as a conditional stochastic process with a changing variance parameter that can be proxied by volume (Karpoff, 1987). Simultaneous consideration of both price and volume variables could shed new light on the understanding of the financial market.

Lamoureux and Lastrapes (1990) successfully used daily trading volume as the proxy variable for information arrivals.[4] Their model or (LL model) shows that autoregressive conditional heteroscedasticity (ARCH) phenomena tend to vanish if volume is considered in explaining the return volatility. Because the mixed stochastic variables are often the cause of the leptokurtic distribution of stock returns, an inclusion of trading volumes as the proxy can explain away much of the nonnormality. In particular, trading volumes are found to be a good proxy for information arrivals in the US market (Lamoureux & Lastrapes, 1990). In the case of emerging markets, previous studies in general fail to include volumes into the analysis (e.g., the GARCH model by Huang, Liu, & Yang, 1995). This being the case, the objective of this paper is to apply the LL model to the Taiwan Stock Market in which individual investors constitute the lion's share of the equity market. In particular, we will test the MD hypothesis in this unique market. Such tests are instrumental in supporting or refuting the MD hypothesis and have the potential to provide intuitively clear interpretations to many ARCH-related empirical findings. Beyond that, we employ intraday data of different time segments. This approach is important because the returns and volumes within time intervals in a given trading day are known to be quite different from that at market open or market close. The segmentation of daily transactions into various subintervals can provide precious information to practitioners and academicians as well. The organization of the paper is as follows. Section 2 provides a description of the LL model. Section 3 explains the data and some basic statistics. Section 4 discusses empirical results of the Taiwan Stock Market for various time segments. Section 5 investigates the same phenomenon for each individual stock and explains why the result of the Taiwan Stock Market is different. The conclusion is given in the Section 6.

---

[4] According to Clark's (1973) mixture of distribution models, informational flow is a latent common factor that affects both daily stock returns and trading volumes. Furthermore, trading volume is used as the proxy variable for daily information arrivals in the study by Baek and Brock (1992).

2. Mixed Distribution Hypothesis (MDH) and the LL model

Since the seminal work (ARCH) by Engle (1982), various hypotheses have been attempted to explain the phenomenon in asset returns. One plausible explanation is that daily returns seem to be generated by a MD. In particular, the rate of daily information arrivals can be viewed as a generating process by the stochastic mixing variable. As pointed out by Diebold (1986), a proper ARCH model can capture the time series properties of such mixing variables. We briefly present the LL model below to illustrate that the daily returns can be represented as a subordinated stochastic process:

$$r_t = u_{t-1} + \varepsilon_t \tag{1}$$

in which

$$\varepsilon_t \mid (\varepsilon_{t-1}, \varepsilon_{t-2}, \cdots) \sim N(0, h_t) \tag{2}$$

$$h_t = \alpha_0 + \alpha_1(L)\varepsilon_{t-1}^2 + \alpha_2(L)h_{t-1} \tag{3}$$

where r_t is the daily return, u_{t-1} denotes the mean of r_t conditional on past information, L represents lag operator, and $\alpha_0 > 0$. For positive parameters of $\alpha_1(L)$ and $\alpha_2(L)$, the shocks administered to return volatility persist over time. The magnitude of persistence is dependent on the size of these parameters. In the case of $\alpha_1(L) + \alpha_2(L) = 1$, the IGARCH model prevails (Engle & Bollerslev, 1993). In this case, shocks to the conditional variance are persistent, in the sense that they remain important for forecasts of all horizons (Bollerslev & Engle, 1986)

Given Eqs. (1)–(3), intraday information flows play an important role via the following definition [Eq. (4)]:

$$\varepsilon_t = \sum_{t=-1}^{n_t} \delta_{it} \tag{4}$$

in which δ_{it} represents the equilibrium price increment in the ith time segment of day t. The random variable n_t is the mixing variable, i.e., the rate of information flow into a market during a given day is considered to be stochastic. ε_t is drawn from a MD, where the variance of each distribution depends on information arrival time. This is referred to as ε_t, which is subordinated to δ_i, with n_t being the directing process (Lamoureux & Lastrapes, 1990). For a sufficiently large n_t and $E(\delta_{it}) = 0$ and $V(\delta_{it}) = \sigma^2$, the Central Limit Theorem implies $\varepsilon_t \mid n_t \sim N(0, \sigma^2 n_t)$. However, variation in n_t over the sample period could lead to a rejection of normality in the unconditional distribution even if the theorem applies (Lamoureux & Lastrapes 1990; Osborne, 1959).

Following Lamoureux and Lastrapes (1990), we assume that the number of information arrivals is serially correlated:

$$n_t = a + b(L)n_{t-1} + e_t \tag{5}$$

where a is a constant, $b(L)$ denotes a lag polynomial of order q, and e_t obeys white noise process. For $\Omega = E(\varepsilon_t | n_t)$, it follows that $\Omega = \sigma^2 n_t$, if the mixture model is valid. Substituting this relation into Eq. (5) readily yields:

$$\Omega_t = \sigma^2 a + b(L)\Omega_{t-1} + \sigma^2 e_t \tag{6}$$

Eq. (6) represents the persistence in terms of conditional variance that can be estimated by a GARCH model. Since the relation between daily return variance and the unobservable mixing variable n_t cannot be easily estimated, a proxy (daily trading volume) is required.

The estimation procedure of the LL model is first to estimate Eqs. (1)–(3). Including volume variable (V) to Eqs. (2) and (3) gives the following generalized variance specification:

$$\varepsilon_t \mid (V_t, \varepsilon_{t-1}, \varepsilon_{t-2}, \cdots) \sim N(0, h_t) \tag{2'}$$

$$h_t = \alpha_0 + \alpha_1 \varepsilon_{t-1} + \alpha_2 h_{t-1} + \alpha_3 V_t \tag{3'}$$

As is pointed out in the LL paper, a GARCH(1,1) specification is a parsimonious representation of conditional variance, while it fits comfortably with many economic time series (e.g., Bollerslev, 1987).

If V_t represents a reasonable proxy for information arrival and is serially correlated, estimation based on Eqs. (1), (2'), and (3') would yield $\alpha_3 > 0$, and values of α_1 and α_2 are significantly smaller than that when V_t is not included (Lamoureux & Lastrapes, 1990).[5] In other words, the mixing variable is statistically significant in explaining the volatility of stock returns. According to the MDH, the inclusion of the mixing variable is expected to rid the ARCH effect in stock returns. In terms of empirical estimates, it is manifested in the size of $\alpha_1 + \alpha_2$, a measure of persistence of shocks administered to the volatility. That is, $\alpha_1 + \alpha_2$ is expected to fall far below unity, and both tend to be statistically insignificant with the presence of V_t.

The empirical results of the LL paper indicates that on average of 20 individual stocks, $\alpha_1 + \alpha_2$ has decreased significantly from 0.728 to 0.073, while α_3 is significantly positive. It suggests that a satisfactory proxy has been included in the variance equation. Of particular interest to us is that: Can the inclusion of serially correlated proxy diminish the values of α_1 and α_2 significantly in the case of an emerging market? If not, the information content of V_t between a mature and an emerging market may be rather different. To this end, we test the LL model for TSI in the Section 3.

3. Data description and some basic statistics

Weighted price indices and trading volumes (in 5-min intervals) are obtained from *Commerical Times*. Sample period starts from September 1, 1989 to June 30, 1993.[6] Taiwan

---

[5] As pointed out by a referee, the inclusion of volume as a separate regressor in an ARCH framework can decrease the explanatory power of the lagged volatility variables even if the MDH is not appropriate. The reason is that volatility, lagged volatility, and volume are all correlated.

[6] The sample period ends on June 30, 1993 for this study.

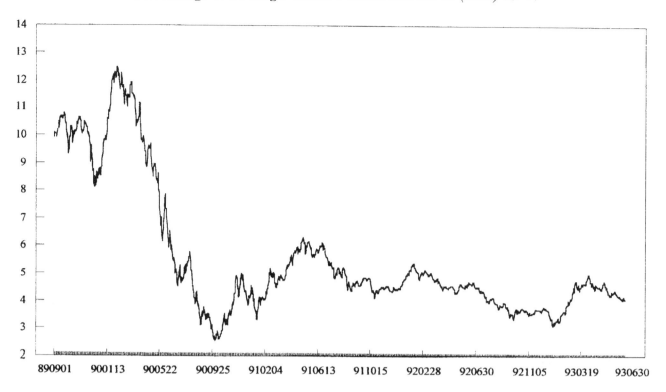

Fig. 1. Trend of the TSI. These indices are calculated based on the data at market close.

Stock Market opens from 9:00 a.m. to 12:00 noon, Monday–Friday, and from 9:00 to 11:00 a.m. on Saturday. In order to have consistent time intervals throughout the sample period, we delete all Saturday transaction data from the study. If we include the transaction data on Saturday, the 11:00 a.m. data represent transactions at market close on Saturday, but not for the rest of trading days in a week. In total, there are 971 daily transaction data (closing price index of the TSI) as shown in Fig. 1. Note that there are 36 volumes and 37 weighted price indexes of transactions each day.[7] Prior studies on stock markets typically take the transaction data at market close as proxy for the day. However, there has been growing amount of uneasiness about using the daily data at market close (e.g., Chang, Fukuda, Ghon Rhee, Takano, 1993; Chang, Kang, Ghon Rhee, 1993; Harris, 1986; Ho & Cheng 1991; Jain & Joh 1988; Lockwood & Lin 1990; McInish & Wood, 1990; Wood, McInish, & Ord, 1985).

One of the special features of the TSI is that there exist a maximum of ±7% price changes (a total of 14% range for price changes) during a trading day, but only a maximum of 7% of price changes is allowed at market close. Such a practice casts serious doubts on the suitability of taking the data at market close as daily proxy in the Taiwan market. In order to explore the return volatility and the role trading volume plays based on 5-minute intervals, we calculate the stock returns based on the following formula:

$$r_{t,s} = [\log i_{t,s} - \log i_{t-1,s}] \times 100 \qquad t = 1, \cdots, 917, \ s = 2, 3, \cdots, 37 \tag{7}$$

---

[7] There are thirty-six 5-minute intervals in the 3-hour period in each trading day. For instance, stock returns at 9:05 in each day is calculated by taking the logarithmic difference of two consecutive trading days (e.g., day t-1 and day t) as shown in Eq. (7).

where $i_{t,s}$ denotes the weighted price index in time segment (5-minute) s of the tth day in the sample. Hence $r_{t,s}$ is the daily rate of return during the sth time interval of day t. In addition, trading volume (in 10^7 s) at time interval s in day t is expressed as [Eq. (8)]:

$$V_{t,s} = \text{Vol}_{t,s}/10,000,000 \qquad (8)$$

$\text{Vol}_{t,s}$ is obtained by adding up the trading volumes from time interval $s+1$ in day $t-1$ to time interval s of day t. For instance, $\text{Vol}_{t,s}$ of interval 9:05 a.m. in day t is calculated by cumulating the trading volumes from 9:10 a.m. in day $t-1$ till 9:05 a.m. in day t.

Means and S.D. of $r_{t,s}$ and $V_{t,s}$ shown in Figs. 2 and 3 reveal that the absolute values of the means and the S.D. at market close were in general smaller than that of other time intervals in a given trading day. It clearly indicates the inappropriateness of using the data at market close as daily proxy. Furthermore, the mean daily return was positive only for the time interval starting at 9:10 a.m. in the entire sample period. On average, S.D. assumes the greatest value within 1 hour after market open. It is worth mentioning that the volume variable of TSI exhibits some interesting patterns. Mean volumes reached their trough both at approximately 10 minutes after market open (9:10 a.m.) and at market close. They remain fairly stable during the rest of a trading day. The S.D. reached its maximum value shortly after market open, but diminished gradually as time went by. After reaching the bottom at 11:10 a.m. it climbed up and eventually peaked at market close (Fig. 3). An examination of Fig. 3 suggests that, if volumes are good proxies for information arrivals, the volatility of information arrivals manifests clearly itself in a V-shaped pattern.[8]

Reported in Table 1 are autocorrelation coefficients (up to four lags) and $Q(12)$ statistics of intraday returns for each time segment. These statistics reveal an interesting pattern: The autocorrelation coefficient, while insignificant at market open, began to become significantly positive at 9:50 a.m. (with a lag of three periods) and remained significant as indicated by $Q(12)$ until 10:25 a.m. Starting at 11:10 a.m., the autocorrelation coefficients (one lag period) became significantly positive and peaked at 11:30 a.m., with a numerical value .1013. However, its value decreased somewhat but remained significant until 11:55 a.m. After that, the autocorrelation phenomenon vanished. The positive autocorrelation in various time segments of a trading day of TSI has one important implication. That is, using a conditional mean instead of a constant in Eq. (1) is preferred for the time intervals with significant autocorrelation.[9] Similarly, trading volumes exhibit a good deal of autocorrelation phenomenon (Table 2), which is consistent with the finding of the LL model or Eq. (5).

---

[8] The way $\text{Vol}_{t,s}$ is calculated suggests that the difference, for instance, of Vol between 9:05 and 9:10 a.m. very much hinges on the trading volumes at these two times. This could have contributed to the V-shaped pattern.

[9] Quite a few studies indicate that GARCH(1,1)–MA(1) model fits return data comfortably (e.g., Hamao et al., 1990; Wei et al., 1995). However, simulations based on the likelihood function suggest that traditional GARCH(1,1) is a better choice.

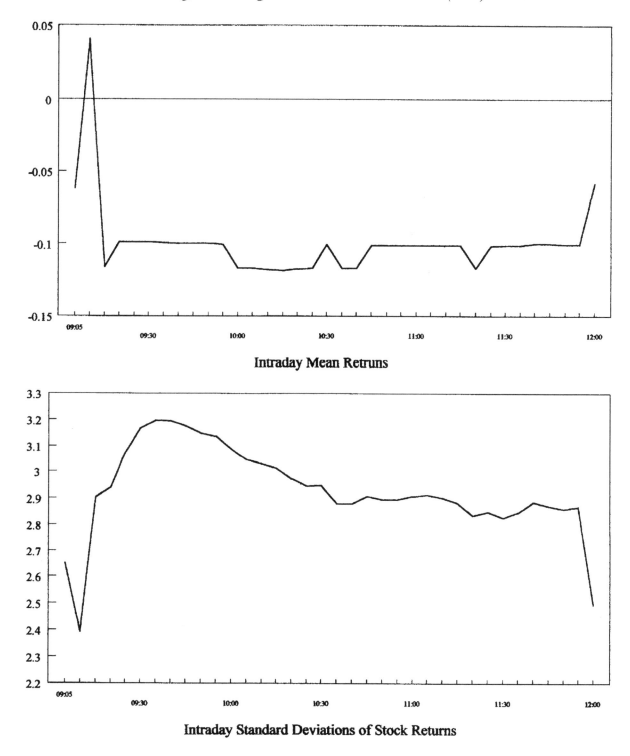

Fig. 2. Intraday mean and S.D. of stock returns of the Taiwan Stock Market.

4. Discussion of empirical results

As shown by the basic statistics, the data-generating process differs appreciably over various time segments in a given trading day. As such, it is important to investigate the price volatility for each time segment. Notice that when serial correlation exists during some time

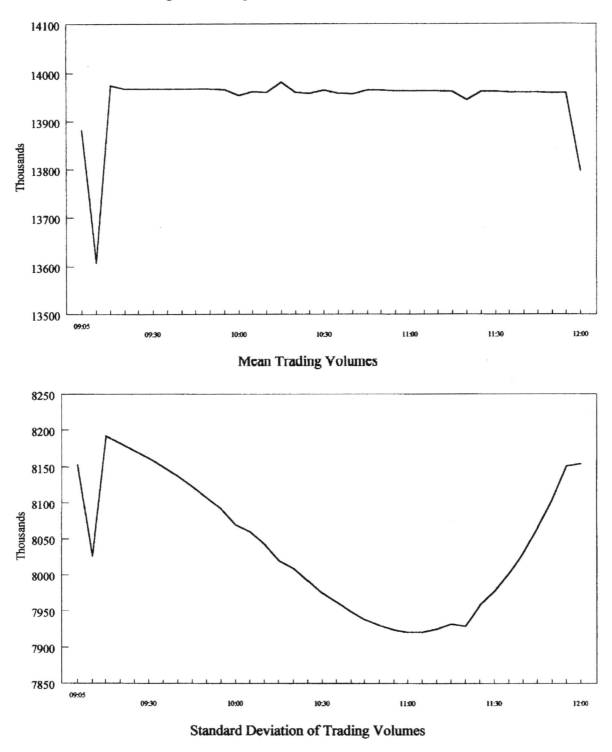

Fig. 3. Intraday mean and S.D. of trading volumes.

segment(s), the conditional mean model is preferred to that with a constant term. This implies that an ARMA model is to be used in estimating Eq. (1). We present the estimates of Eqs. (1)–(3) or Model I in the first row of Table 3 for each time segment while that of Eqs. (1), (2′), and (3′) or Model II are shown in the second row. Surprisingly, the ARCH phenomenon does not vanish over 36 intraday time segments with the inclusion of the proxy variable

Table 1
Autocorrelation coefficients (up to four lags) and $Q(12)$ statistics of intraday stock returns

	9:05	9:10	9:15	9:20	9:25	9:30
1	.0472 (.0333)	.0114 (.0338)	.0296 (.0331)	.0483 (.0330)	.0129 (.0330)	−.0126 (.0330)
2	−.0056 (.0333)	.0361 (.0338)	.0149 (.0331)	.0176 (.0331)	−.0083 (.0330)	−.0342 (.0330)
3	.0569 (.0333)	.0439 (.0339)	.0610 (.0331)	.0361 (.0331)	−.0083 (.0330)	.0709* (.0331)
4	−.0017 (.0334)	.0378 (.0339)	−.0369 (.0332)	−.0500 (.0332)	−.0436 (.0331)	−.0350 (.0333)
$Q(12)$	16.9128 (.15290)	19.4887* (.0774)	27.6681* (.0062)	21.5942* (.0423)	20.1880* (.0636)	23.7928* (.0217)

	9:35	9:40	9:45	9:50	9:55	10:00
1	−.0265 (.0330)	−.0234 (.0330)	−.0131 (.0330)	−.0014 (.0330)	.0050 (.0330)	.0079 (.0331)
2	−.0388 (.0331)	−.0447 (.0331)	−.0407 (.0330)	−.0428 (.0330)	−.0562 (.0330)	−.0537 (.0331)
3	.0840* (.0331)	.0916* (.0331)	.0804* (.03310)	.0738* (.0331)	.0826* (.0331)	.0809* (.03320)
4	−.0453 (.0333)	−.0587 (.0334)	−.0622 (.0333)	−.0602 (.0333)	−.0537 (.0334)	−.0459 (.0334)
$Q(12)$	29.1793* (.0037)	31.4097* (.0017)	25.8210* (.0114)	19.4326* (.0786)	22.3527* (.0338)	29.5184* (.0033)

	10:05	10:10	10:15	10:20	10:25	10:30
1	.0155 (.0331)	.0119 (.0331)	.0078 (.0331)	.0208 (.0331)	.0251 (.0331)	.0396 (.0330)
2	−.0462 (.0331)	−.0347 (.0331)	−.0175 (.0331)	−.0199 (.0331)	−.0119 (.0331)	−.0313 (.0331)
3	.0902* (.0331)	.0943* (.0331)	.0872* (.0331)	.0824* (.0331)	.0667* (.0331)	.0645 (.0331)
4	−.0528 (.0334)	−.0616 (.0334)	−.0501 (.0334)	−.0318 (.0333)	−.0190 (.0332)	−.0251 (.0333)
$Q(12)$	23.9108* (.0209)	24.9623* (.0150)	18.2660 (.1078)	17.4075 (.1349)	15.7689 (.2021)	16.9944 (.1498)

	10:35	10:40	10:45	10:50	10:55	11:00
1	.0445 (.0331)	.0386 (.0331)	.0409 (.0330)	.0496 (.0330)	.0517 (.0330)	.0486 (.0330)
2	-.0047 (.0331)	.0058 (.0331)	-.0055 (.0331)	-.0108 (.0331)	-.0090 (.0331)	-.0089 (.0331)
3	.0521 (.0331)	.0515 (.0331)	.0582 (.0331)	.0529 (.0331)	.0367 (.0331)	.0216 (.0331)
4	-.0171 (.0332)	-.0198 (.0332)	-.0261 (.0332)	-.0272 (.0332)	-.0193 (.0332)	-.0038 (.0331)
Q(12)	9.8511 (.6290)	8.4483 (.7492)	10.6562 (.5586)	11.0424 (.5253)	9.8071 (.6329)	8.3582 (.7565)

	11:05	11:10	11:15	11:20	11:25	11:30
1	.0548 (.0330)	.0678* (.0330)	.0826* (.0331)	.0808* (.0331)	.0888* (.0330)	.1013* (.0330)
2	-.0188 (.0331)	-.0204 (.0332)	-.0089 (.0333)	.0058 (.0333)	.0036 (.0333)	-.0063 (.0334)
3	.0194 (.0332)	.0230 (.0332)	.0108 (.0333)	.0091 (.0333)	-.0077 (.0333)	.0053 (.0334)
4	.0151 (.0332)	.0066 (.0332)	-.0008 (.0333)	.0013 (.0333)	.0012 (.0333)	-.0089 (.0334)
Q(12)	9.6332 (.6481)	13.4397 (.3379)	15.9805 (.1921)	13.7068 (.3198)	14.0560 (.2971)	16.1909 (.1826)

	11:35	11:40	11:45	11:50	11:55	12:00
1	.0993* (.0330)	.0914* (.0330)	.0904* (.0330)	.0841* (.0330)	.0798* (.0330)	.0279 (.0335)
2	-.0157 (.0334)	-.0220 (.0333)	-.0066 (.0333)	.0150 (.0333)	.0075 (.0333)	-.0011 (.0335)
3	.0074 (.0334)	.0008 (.0333)	-.0109 (.0333)	-.0143 (.0333)	-.0016 (.0335)	.0355 (.0335)
4	-.0200 (.0334)	-.0176 (.0333)	-.0118 (.0333)	-.0131 (.0333)	-.0140 (.0336)	-.0436 (.0336)
Q(12)	17.4654 (.1329)	16.9937 (.1498)	15.7837 (.2013)	13.4514 (.3371)	14.7230 (.2569)	10.6348 (.5604)

The numbers in parentheses are standard errors of autocorrelation coefficients; $Q(12)$ = Box–Pierce Statistics with lag period of 12; the numbers in parentheses corresponding to $Q(12)$ statistics are probability values.

* Denotes significance at 5% level.

Table 2
Autocorrelation coefficient (up to four lags) and $Q(12)$ statistics of intraday trading volumes

	9:05	9:10	9:15	9:20	9:25	9:30
1	.8532* (.0333)	.8466* (.0338)	.8576* (.0331)	.8603* (.0330)	.8631* (.0330)	.8664* (.0330)
2	.7989* (.0521)	.7944* (.0528)	.8032* (.0520)	.8051* (.0520)	.8071* (.0521)	.8092* (.0523)
3	.7553* (.0643)	.7479* (.0650)	.7545* (.0641)	.7564* (.0642)	.7583* (.0643)	.7607* (.0645)
4	.7259* (.0734)	.7215* (.0742)	.7267* (.0732)	.7280* (.0733)	.7295* (.0735)	.7312* (.0736)
$Q(12)$	5382.6593 (.0000)	5045.7198 (.0000)	5429.6276 (.0000)	5451.5771 (.0000)	5477.8857 (.0000)	5508.2371 (.0000)

	9:35	9:40	9:45	9:50	9:55	10:00
1	.8698* (.0330)	.8736* (.0330)	.8778* (.0330)	.8823* (.0330)	.8872* (.0330)	.8916* (.0331)
2	.8116* (.0524)	.8143* (.0525)	.8172* (.0527)	.8204* (.0528)	.8238* (.0530)	.8270* (.0532)
3	.7633* (.0647)	.7662* (.0649)	.7693* (.0651)	.7726* (.0653)	.7760* (.0655)	.7792* (.0658)
4	.7331* (.0738)	.7352* (.0741)	.7376* (.0743)	.7402* (.0746)	.7429* (.0749)	.7450* (.0752)
$Q(12)$	5542.0040 (.0000)	5579.7477 (.0000)	5621.6958 (.0000)	5666.4759 (.0000)	5714.3403 (.0000)	5749.6932 (.0000)

	10:05	10:10	10:15	10:20	10:25	10:30
1	.8972* (.0331)	.9024* (.0331)	.9073* (.0331)	.9128* (.0331)	.9177* (.0331)	.9227* (.0330)
2	.8312* (.0534)	.8349* (.0536)	.8381* (.0538)	.8423* (.0540)	.8458* (.0542)	.8490* (.0543)
3	.7854* (.0660)	.7889* (.0663)	.7918* (.0666)	.7958* (.0668)	.7992* (.0671)	.8009* (.0673)
4	.7495* (.0756)	.7526* (.0759)	.7550* (.0762)	.7589* (.0765)	.7621* (.0768)	.7641* (.0770)
$Q(12)$	5819.3778 (.0000)	5871.8299 (.0000)	5899.2182 (.0000)	5976.7466 (.0000)	6027.5823 (.0000)	6076.1647 (.0000)

	10:35	10:40	10:45	10:50	10:55	11:00
1	.9269* (.0331)	.9307* (.0331)	.9343* (.0330)	.9369* (.0330)	.9386* (.0330)	.9343* (.0330)
2	.8521* (.0545)	.8548* (.0546)	.8571* (.0548)	.8590* (.0548)	.8604* (.0549)	.8611* (.0549)
3	.8054* (.0675)	.8080* (.0677)	.8091* (.0678)	.8109* (.0680)	.8122* (.0681)	.8129* (.0681)
4	.7679* (.0773)	.7704* (.0775)	.7714* (.0777)	.7731* (.0778)	.7744* (.0779)	.7751* (.0780)
Q(12)	6121.3030 (.0000)	6161.3525 (.0000)	6198.0139 (.0000)	6225.8005 (.0000)	6245.4251 (.0000)	6255.5117 (.0000)

	11:05	11:10	11:15	11:20	11:25	11:30
1	.9393* (.0330)	.9380* (.0330)	.9356* (.0330)	.9318* (.0331)	.9272* (.0330)	.9211* (.0330)
2	.8611* (.0549)	.8604* (.0549)	.8589* (.0548)	.8564* (.0547)	.8532* (.0545)	.8490* (.0543)
3	.8127* (.0681)	.8118* (.0680)	.8102* (.0679)	.8064* (.0678)	.8044* (.0675)	.8002* (.0672)
4	.7752* (.0780)	.7747* (.0779)	.7735* (.0778)	.7697* (.0776)	.7689* (.0773)	.7654* (.0769)
Q(12)	6255.0351 (.0000)	6242.9984 (.0000)	6219.5759 (.0000)	6164.2143 (.0000)	6135.2378 (.0000)	6073.5981 (.0000)

	11:35	11:40	11:45	11:50	11:55	12:00
1	.9137* (.0330)	.9048* (.0330)	.8945* (.0330)	.8829* (.0330)	.8698* (.0330)	.8487* (.0335)
2	.8437* (.0540)	.8373* (.0537)	.8299* (.0533)	.8215* (.0529)	.8121* (.0524)	.7962* (.0524)
3	.7949* (.0668)	.7886* (.0664)	.7813* (.0659)	.7728* (.0653)	.7632* (.0647)	.7466* (.0646)
4	.7611* (.0765)	.7559* (.0759)	.7499* (.0753)	.7428* (.0746)	.7348* (.0739)	.7122* (.0736)
Q(12)	5997.2638 (.0000)	5906.6530 (.0000)	5802.3998 (.0000)	5683.9724 (.0000)	5551.2485 (.0000)	5154.3136 (.0000)

The numbers in parentheses are standard errors of autocorrelation coefficients; $Q(12)=$ Box–Pierce Statistics with lag period of 12; the numbers in parentheses corresponding to $Q(12)$ statistics are probability values.

* Denotes significance at 5% level.

Table 3
The estimated results of Models I and II

	α_1	α_2	α_3	β_1	$\alpha_1 + \alpha_2$
9:05					
I	.1376	.8156			.9532
II	.1068	.8282	.2717		.9350
9:10					
I	.0914	.8762			.9676
II	.0680	.8861	.1581		.9541
9:15					
I	.0949	.8860			.9809
II	.0738	.9020	.1169		.9758
9:20					
I	.0890	.8929			.9819
II	.0685	.9089	.1148		.9774
9:25					
I	.0969	.8841			.9810
II	.0707	.9042	.1483		.9749
9:30					
I	.1034	.8796			.9830
II	.0787	.8969	.1519		.9756
9:35					
I	.1130	.8704			.9834
II	.0842	.8892	.1776		.9734
9:40					
I	.1062	.8777			.9839
II	.0796	.8966	.1557		.9762
9:45					
I	.1055	.8791			.9846
II	.0807	.8964	.1534		.9771
9:50					
I	.1018	.8840			.9858
II	.0799	.8984	.1483		.9783
9:55					
I	.1067	.8802			.8969
II	.0835	.8946	.1525		.9781

(continued on next page)

Table 3 (*continued*)

	α_1	α_2	α_3	β_1	$\alpha_1 + \alpha_2$
10:00					
I	.1083	.8771			.9854
II	.0848	.8923	.1583		.9771
10:05					
I	.1145	.8668			.9813
II	.0882	.8845	.1748		.9727
10:10					
I	.1130	.8705			.9835
II	.0891	.8862	.1446		.9753
10:15					
I	.1084	.8752			.9836
II	.0857	.8904			.9761
10:20					
I	.1044	.8800			.9844
II	.0813	.8951	.1349		.9764
10:25					
I	.1093	.8743			.9836
II	.0854	.8896	.1437		.9750
10:30					
I	.1067	.8777			.9844
II	.0845	.8911	.1450		.9756
10:35					
I	.1069	.8721			.9790
II	.0872	.8849	.1543		.9721
10:40					
I	.1069	.8761			.9830
II	.0845	.8884	.1543		.9729
10:45					
I	.1011	.8835			.9846
II	.0801	.8954	.1433		.9755
10:50					
I	.0966	.8862			.9828
II	.0760	.8992	.1412		.9752

(continued on next page)

Table 3 (*continued*)

	α_1	α_2	α_3	β_1	$\alpha_1 + \alpha_2$
10:55					
I	.0965	.8851			.9816
II	.0755	.8984	.1502		.9739
11:00					
I	.0963	.8845			.9808
II	.0771	.8972	.1459		.9743
11:05					
I	.0971	.8825			.9796
II	.0780	.8950	.1492		.9730
11:10					
I	.0991	.8824			.9815
II	.0807	.8929	.1425		.9736
11:15					
I	.1010	.8796			.9806
II	.0825	.8903	.1379		.9728
11:20					
I	.0966	.8829			.9795
II	.0785	.8937	.1355		.9722
11:25					
I	.0970	.8819			.9789
II	.0788	.8926	.1344		.9714
11:30					
I	.0937	.8844		.0894	.9781
II	.0763	.8948	.1309		.9710
11:35					
I	.0977	.8809		.0844	.9786
II	.0800	.8907	.1337		.9707
11:40					
I	.1046	.8733			.9779
II	.0854	.8807	.1649		.9661
11:45					
I	.1006	.8768			.9774
II	.0814	.8836	.1709		.9650

(continued on next page)

Table 3 (*continued*)

	α_1	α_2	α_3	β_1	$\alpha_1 + \alpha_2$
11:50					
I	.0957	.8807			.9764
II	.0769	.8875	.1712		.9644
11:55					
I	.1027	.8702			.9729
II	.0807	.8797	.1936		.9604
12:00					
I	.1213	.8242			.9455
II	.0839	.8517	.2186		.9356

We first estimate $r_t = \mu + \beta_1 \varepsilon_{t-1} + \beta_2 \varepsilon_{t-2} + \varepsilon_t$. For Model I, we perform the estimation based on $h_t = \alpha_0 + \alpha_1 \varepsilon^2_{t-1} + \alpha_2 h_{t-1} + e_t$. For Model II, $h_t = \alpha_0 + \alpha_1 \varepsilon^2_{t-1} + \alpha_2 h_{t-1} + \alpha_3 V + e_t$, where V denotes trading volume, r_t is stock return, and e_t is white noise. All the statistics is significant at the 5% level.

(trading volume). The sums of α_1 and α_2 are fairly close to unity, and do not undergo noticeable change when compared to the model without the proxy variable. Furthermore, significant MA(1) estimates are found only at time period of 11:30 and 11:35 a.m. despite the fact that serial correlation prevails in intraday returns.[10]

Much like the result of the LL model, Fraser and Power (1996) reach the same conclusion: After including the proxy for daily information arrivals (Model II), the ARCH effect vanishes.[11] It suggests that at least part of the persistence of stock volatility can be explained away by information arrivals. The result that trading volumes over various time segments do not significantly explain the persistence of volatility of TSI is puzzling, to say the least. Moreover, their explanatory powers within intraday time segments are roughly similar, even though the basic statistics in these periods differ conspicuously. We would like to point out several factors, which may have contributed to the strikingly different result of TSI.

First, the content of daily information arrivals is unique in TSI. As was mentioned in Section 1, the lion's share of the participants in TSI is made of individual investors who focus primarily on the superficiality of such arrivals. Lacking the fundamental analysis, they are many times overreactive to news announcements.[12] As pointed out by Karpoff (1987), different level and intensity of information flow can give rise to different price–volume relationship. Short-term speculative behaviors combined with overreactions to news announcements establish a unique information flow pattern in TSI.

---

[10] Via MA(4) or ARMA(1,1) model, the estimates are found to be insignificant in most cases. In addition, both autocorrelation and heteroscedasticity phenomena are nonexistent using the hypothesis testing procedures.

[11] Fraser and Power (1996) employ unexpected innovations (in market value) in place of trading volume on the emerging markets.

[12] TSI, overflowed with speculative individual investors, is notorious for its oversensitivity with respect to news announcements.

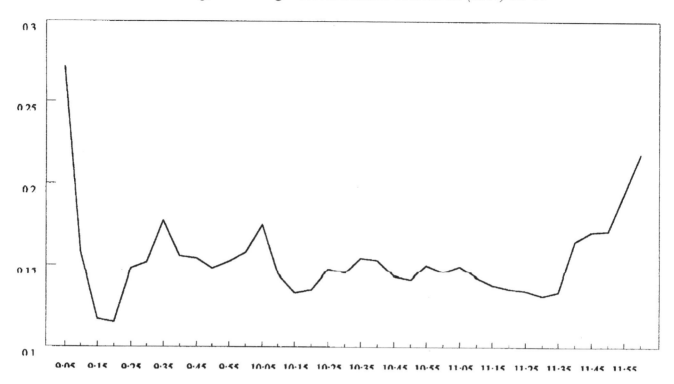

Fig. 4. Proportion of return volatility explained by volumes in a trading day (estimated by the authors based on Table 3).

Second, the maximum price changes allowed in a trading day will certainly inhibit, to some extent, the transmission mechanism, i.e., from news announcements to realized price changes. Since only part of information content is reflected in price changes, it is not unanticipated that our empirical results differ from prior studies. The maximum ceilings imposed on price changes coupled with frequent information arrivals certainly hinder the transmission mechanism.

Third, our use of weighted price index for the Taiwan market could have made the difference, because only individual stocks were analyzed in prior studies. To do the same, we shall employ data of individual stocks and reexamine the result in the Section 5.

Yet another interesting phenomenon arises despite that the MD hypothesis is not supported in our study. As shown in Fig. 4, the coefficient of trading volume (α_3) has a U-shaped pattern in explaining the return volatility. In particular, the proportions of return volatility that can be explained by trading volumes are considerably higher at both market open and market close. This result bears some degree of similarity to the basic statistics of intraday returns. In addition, after market open, the sum of $\alpha_1 + \alpha_2$ begins to decrease more obviously than that in other subperiods.

5. Discussion of empirical results using individual stocks

As mentioned in Section 4, the use of weighted price index of the Taiwan market could have contributed to the rather puzzling result. As such, we randomly select 17

Table 4
Estimates of Models I and II and other statistics for 17 individual stocks

Code Number	α_1	α_2	α_3	$\alpha_1 + \alpha_2$	corr(r,v), corr(r^2,v)	$H(n)$
1101						
I	.1155	.8579		.9734	.2067 .3543	2.56% (898)
II	.2038		.0019	.2038		
1107						
I	.1230	.8587		.9808	.1527 .4242	3.70% (890)
II	.2018	.1362	.0026	.3380		
1216						
I	.1773	.8004		.9777	.2125 .3390	1.66% (902)
II	.1433		.0028	.1433		
1303						
I	.1832	.8230		1.0062	.1503 .4373	2.65% (903)
II	.1380		.0015	.0434		
1310						
I	.1511	.7996		.9507	.1091 .3240	4.62% (887)
II	.1526	.0884	.0024	.2410		
1407						
I	.1438	.8186		.9624	.1623 .2829	4.83% (889)
II	.1380	.7410	.00004	.8790		
1432						
I	.1824	.7221		.9045	.1640 .3079	6.55% (884)
II	.1465		.0024	.1465		
1434						
I	.2219	.7765		.9984	.2102 .3458	2.22% (898)
II	.2222	.6698	.0004	.8920		
1502						
I	.1402	.8324		.9726	.1698 .2600	3.90% (896)
II	.1388	.1269	.0018	.2657		
1604						
I	.1194	.8438		.9632	.2552 .2089	4.26% (892)
II	.1056	.8561	.0001	.9617		
1707						
I	.1367	.8057		.9424	.2052 .3324	7.01% (883)
II	.1722	.3903	.0038	.5625		

(continued on next page)

Table 4 (*continued*)

Code Number	α_1	α_2	α_3	$\alpha_1 + \alpha_2$	corr(r,v), corr(r^2,v)	H (n)
1802						
I	.2176	.7437		.9613	.1543 .3476	1.88% (902)
II	.2109	.7099	.0004	.9208		
1902						
I	.1157	.8453		.9610	.2055 .2730	5.40% (889)
II	.1061	.8327	.0001	.9388		
2006						
I	.1230	.8422		.9652	.2071 .2848	3.68% (897)
II	.1038	.8496	.0001	.9534		
2102						
I	.1584	.7963		.9547	.0892 .3344	7.42% (875)
II	.2677	.2854	.0013	.5531		
2201						
I	.1455	.8239		.9694	.2152 .3209	3.72% (886)
II	.1265		.0012	.1265		
2303						
I	.1401	.8152		.9694	.2258 .2765	4.49% (889)
II	.2816	.2494	.0011	.5310		

The estimating procedure is similar to that of Table 3; corr(r,v) = correlation coefficient between stock returns and volumes; corr(r^2,v) = correlation coefficient of stock returns squared and volumes; H = percentage of times (based on sample sizes) when stock returns hit ±7% price limits; 1101, 1107,... are code numbers of 17 individual stocks; and n = sample size.

individual stocks from TSI and perform the same analysis (Models I and II). A perusal of Table 4 indicates that 9 of 17 stocks exhibit a high degree of persistence ($\alpha_1 + \alpha_2 > 0.5$) after volumes are included as the proxy. Most of the remaining eight individual stocks have somewhat significant statistics. Only one stock has the measure of persistence ($\alpha_1 + \alpha_2$) less than 0.1. This result clearly indicates that the persistence of return volatility does not vanish either for individual stocks even after volume variable is included in the variance equation

Naturally, one would conjecture that the ceilings on price changes and persistent volatility in TSI are in some way related. To answer this question, we investigate the size of persistence ($\alpha_1 + \alpha_2$) and percentage of times (in terms of sample size) when the price changes hit the ceilings (floors). The result reported in Table 4 in general suggests that the sum of α_1 and α_2 tends to be larger if the percentage of stock prices that hit the limits increases. However, this positive relationship is not very strong, since there are cases in which $\alpha_1 + \alpha_2$ are low while percentages of stock price reaching the limits are high. The relationship is, however, quite noticeable as shown in Fig. 5, and one may surmise that the price limits contribute, to some extent, to the persistent volatility of TSI.

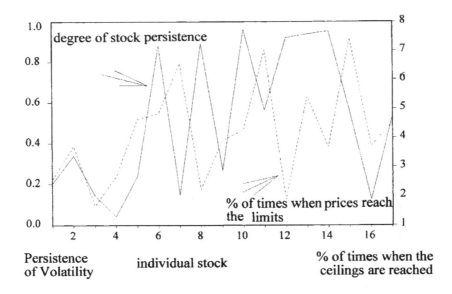

Fig. 5. Relationship between the degree of persistence of stock volatility and percentage of times when prices reach the limits. Note that degree of persistence in stock volatility is measured by $\alpha_1 + \alpha_2$ in Model II; numbers of times when prices reach the limits are graphed based on the raw data.

6. Concluding remarks

The Taiwan Stock Market is known for its unusually high turnover ratio. Other than that, the majority of prior studies on TSI have found that the Taiwan market is very much in line with other developed stock markets. Their results are, in most cases, based on price variables only. That is, half of the information needed for an equilibrium market is missing. In the price–volume model, it is known that the ARCH effect is an artifact generated by a mixture of returns, with the rate of information arrivals being the stochastic mixing variable. After taking information arrival into account, the stock returns are expected to be normally distributed. Lamoureux and Lastrapes (1990), among others, support the MDH for the US market. Similarly, Fraser and Power (1996) verify the MDH for the emerging markets. However, their choice of proxy (unexpected innovation of market value) is rather different from trading volumes often used as the proxy by most researchers. In either case, the persistent volatility of stock returns declines significantly once the proxy is included. Surprisingly, such is not the case for the Taiwan market. The empirical evidence from our result does not support the MDH, and the persistence of stock return volatility still remains strong.

In a nutshell, the persistent volatility of TSI perhaps has its root in the way it assimilates and disseminates information contents. Inasmuch as it prevails, the persistent volatility remains even after trading volume is included in the estimation. The failure to support the MDH in the Taiwan market can be attributed to two rather unique factors. First, the great majority of participants are short-term individual investors who fervently engage in speculative activities. Lacking fundamental analyses, their behavior can be characterized by overreaction to news announcements. Second, the price limits imposed by Taiwan Security Exchange Committee do contribute, to some extent, to the persistence of return volatility. Such price limits hinder the information transmission mechanisms; the price can only go down once the ceiling is reached. Hence it intensifies the return volatility. Combined with these two factors, history of Taiwan is replete with political and economic instabilities. The unfortunate past of Taiwan's history, to a great extent, accounts for overreactive and speculative behaviors in the market. Viewed in this perspective, the efficient market hypothesis for TSI needs to be reevaluated and reconsidered cautiously in the future.

Acknowledgments

Financial support by the National Science Council of Taiwan (Grant No. NSC 86-2415-H194-006) is gratefully acknowledged. We are grateful to the Editor for valuable suggestion. Any remaining errors are the author's own responsibility.

References

Baek, E., & Brock, W. (1992). A general test for nonlinear granger causality: bivariate model. Working Paper, Iowa State University & University of Wisconsin, Madison.

Beaver, W. H. (1968). Information content of annual earnings announcements. *Journal of Accounting Research*, *6*, 67–92.

Bollerslev, T. (1987). A conditionally heteroskedastic time series model for speculative prices and rates of return. *Review of Economics and Statistics, 69*, 542–547.

Bollerslev, T., & Engle, B. F. (1993). Common persistence in conditional variances. *Econometrica, 61* (1), 167–186.

Chang, R. P., Fukuda, T., Ghon Rhee, S., & Takano, M. (1993a). Intraday and behavior of TOPIX. *Pacific Basin Finance Journal, 1*, 67–95.

Chang, R. P., Kang, J.-K., & Ghon Rhee, S. (1993b). The behavior of Malaysian stock prices. Working Paper, Rhode Island University.

Clark, P. (1973). A subordinated stochastic process model with finite variance for speculative prices. *Econometrica, 41*, 135–155.

Diebold, F. X. (1986). Modeling the persistence of conditional variances: A comment. *Econometric Reviews, 5*, 51–56.

Engle, R. F. (1982). Autoregressive conditional heteroskedasticity with estimates of the variance of United Kingdom information. *Econometrica, 50*, 987–1007.

Engle, R. F., & Bollerslev, T. (1993). Modelling the persistence of conditional variances. *Econometric Reviews, 5*, 81–87.

Fraser, P., & Power, D. M. (1996). Conditional heteroscedasticity in the equity returns from emerging markets. In: T. Bos, & T. A. Fetherston (Eds.), *Advances in pacific basin financial markets, vol.* 2B331–347.

Hamao, Y., Masulis, R. W., & Ng, V. (1990). Correlations in price changes and volatility across international stock markets. *Review of Financial Studies, 3*, 281–307.

Harris, L. (1986). Cross-security test of the mixture of distributions hypothesis. *Journal of Financial and Quantitative Analysis, 21*, 39–46.

Ho, Y.-K., & Cheng, Y.-L. (1991). Behavior of intra-daily stock return on an Asian emerging market — Hong Kong. *Applied Economics Letters, 23*, 957–966.

Huang, B.-N. (1995). Do Asian stock market prices follow random walk? Evidence from the variance ratio test. *Applied Financial Economics, 5*, 251–256.

Huang, B.-N., Liu, Y. J., & Yang, C. W. (1995). A comparative study of the weekend effects in the United States, British and the Pacific Rim stock markets: an application of the ARCH model. *Advances in Pacific Basin Business Economics and Finance, 1*, 279–292.

Huang, B.-N., & Yang, C. W. (1995). The fractal structure in multinational stock returns. *Applied Economics Letters, 2*, 67–71.

Jain, P. C., & Joh, G. H. (1988). The dependence between hourly price and trading volume. *Journal of Financial and Quantitative Analysis, 23*, 269–284.

Karpoff, J. M. (1987). The relation between price changes and trading volume: a survey. *Journal of Financial and Quantitative Analysis, 22*, 109–126.

Lamoureux, C. G., & Lastrapes, W. D. (1990). Heteroskedasticity in stock return data: volume versus GARCH effects. *Journal of Finance, 45*, 221–230.

Lo, A. W. (1991). Long-term memory in stock market price. *Econometrica, 59*, 1279–1313.

Lockwood, L. J., & Lin, S. C. (1990). An examination of stock market return volatility during overnight and intraday periods 1964–4989. *Journal of Finance, 45*, 591–601.

McInish, T.H., & Wood, R. A. (1990). A transaction data analysis of the variability of common stock returns during 1980–1984. *Journal of Banking and Finance, 14*, 99–112.

Osborne, M. (1959). Brownian motion in the stock prices. *Operations Research, 7*, 145–173.

Wei, K. C. J., Liu, Y. -J., Yang, C. C., & Chuang, K. H. (1995). Volatility and price changes spillover effects across the developed and emerging markets. *Pacific-Basin Finance Journal, 3*, 113–136.

Wood, R. A., McInish, H., & Ord, J. K. (1985). An investigation of transaction data for NYSE stock. *Journal of Finance, 40*, 723–739.

Applied Economics Letters, 2001, **8**, 665–668

The impact of settlement time on the volatility of stock market revisited: an application of the iterated cumulative sums of squares detection method for changes of variance

BWO-NUNG HUANG* and CHIN-WEI YANG

Department of Economics, National Chung-Cheng University Chia-Yi, Taiwan 621 and *Department of Economics Clarion University of Pennsylvania, Clarion, PA 16214-1232, US*

Volatility changes before and after a major event cannot be effectively modelled without considering the impact of other events during the sample period. This paper reexamines the impact of settlement time changes on the volatility change in the Shanghai and Shenzhen Stock Exchange by Li *et al.* (1997) via the iterated cumulative sums squares (ICSS) method developed by Inclán and Tiao. This study detected three other events during the sample period (two before and one after the structural break). After removing these factors, it is found that change in settlement time does not impact the volatility of the stock returns in a noticeable way.

I. INTRODUCTION

Volatility of financial variables over time plays a pivotal role as it mirrors risk: a key factor for investors. More often than not, volatility change follows a major structural break (e.g., financial market liberalization of institutional adjustments). A typical way is to determine the break point priori before conducting statistical analysis. This method has at least two problems. First, prior determination of the breaking point without statistical inference may not reflect the actual time due to expectation by investors; hence, data itself can be self-adjustable. Second, impacts of other significant events need to be taken into consideration. For instance, failing to consider the impact of other significant event(s) may render erroneous conclusions. The objective of this paper is to reexamine the volatility changes in stock returns due to major structural break in People's Republic of China (PRC): settlement time is extended from that in the same day to that the following day starting 1 May 1995.

Li *et al.* (1997) detect decreased volatility of the stock returns after the settlement time change without considering other events in their sample period. To allow for such possibilities, the iterated cumulative sums of squares (ICSS) method, recently developed by Inclán and Tiao (1994), is employed. This study finds that no significant volatility change took place after the structural break. The organization of the paper is as follows. Section II presents data, methodology and empirical results. Section III discuss the results and offers a conclusion.

II. DATA AND METHODOLOGY

In order to compare these results with that by Li *et al.* (1997), the Shanghai Stock Exchange (A share) Index and the Shezhen Stock Exchange (A share) Index are employed. Figure 1 presents the time series plots of both indices.

* Corresponding author. E-mail: ecdbntt@ccunix.ccu.edu.tw

Applied Economics Letters ISSN 1350-4851 print ISSN 1466-4291 online © 2001 Taylor & Francis Ltd
http://www.tandf.co.uk/journals
DOI: 10.1080/13504850110036346

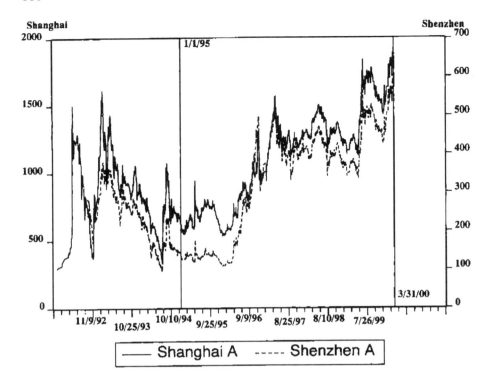

Fig. 1. *Time series plots of Shanghai A and Shenzhen A*

Both indices exhibited noticeable downward swings between 1992 and 1995 before bouncing upward in 1996. Such an upward movement continues despite a slight setback during the 1997–1998 Asian financial debacle. The sample period by Li *et al.* (1997) spans from 3 January 1994 to 31 August 1995 with the break point of 1 January 1995 when changes of settlement time occurred (see fig. 2). As indicated in fig. 2, a bear market seems to have prevailed before the break and on the other hand, a bull market manifests itself afterwards. Well-known in the literature, volatility of stock returns in a bear is generally than that in a bull market due to then leverage effect.

It is little wonder that they found diminished volatility after the settlement time change in both markets. It can also be explained by the arbitrary sample period: about one year before the break but only eight months afterwards. Such a selection ignores impacts of other events and as such may lead to biased result. As Li *et al.* (1997: Stated 690), '...during the earlier periods both the Shanghai and Shezhen Stock Exchange witnessed a rapid development and a frequent change of institutional setting that would prevent us from separating the settlement time impact from other impacts'. However, these institutional changes need to be removed for more accurate statistical inference.

The essence of the ICSS approach (Inclán and Tiao, 1994) lies in detecting significant change of variance in a

sample period and then determine the optimal sample period performing the analysis. First define the logarithmic stock return as

$$r_t = \log p_t - \log p_{t-1} \qquad (1)$$

where p_t denotes stock indices from either market (share A). A simple regression is needed to remove the constant term.

$$r_t = \mu + a_t \qquad (2)$$

in which a_t is an uncorrelated random variable with zero mean and variance σ_t^2, $t = 1, 2, T$. Define $C_k = \sum_{t=1}^{k} a_t^2$ as the cumulative sum of squares for a_t and consider the centered or normalized cumulative sum of squares:

$$D_k = \frac{C_k}{C_T} - \frac{k}{T}, \quad \text{for} \quad k = 1, 2, \dots T \text{ and } D_0 = D_T = 0 \quad (3)$$

Under the condition of variance homogeneity, $\sqrt{(T/2)}|D_K|$ behaves like a Brownian bridge asymptotically. Table 1 of Inclán and Tiao lists the critical values of $\sqrt{(T/2}|D_K|$ for different sample sizes. If $|D_k|$ is greater than the critical value, this study concludes the datum at that point (k^*) represents a structural break. This method can also be used to examine multiple break points as a major structure break can mask the impacts of other break points. For this reason, Inclán and Tiao suggest the following ICSS algorithm. Let $t_1 = 1$ as step zero before proceeding to step 1.

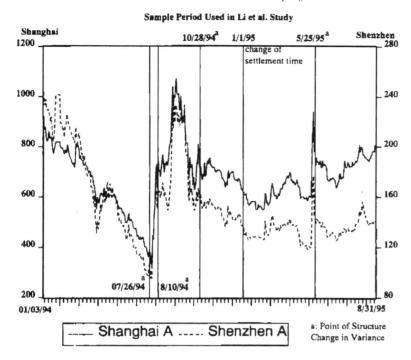

Fig. 2. *Time series plots of Shanghai and Shenzhen stock indices*

Step 1

First calculated was $D_k(a[t_1 : T])$, and let $k^*(a[t_1 : T])$ be the point at which max $imal|D_k(a[t_1 : T])|$ for k is obtained, and define

$$M(t_1 : T) = \max_{t_1 \le k \le T} \sqrt{\frac{(T - y_1 + 1)}{2}} |D_k(a(t_1 : T))|.$$

If $M(t_1 : T) > D^*$, it may be considered that there is a break point at $k^*(a[t_1 : T])$ and proceed to step 2a. The value of D' is D_{1-p} from Inclán and Tiao's Table 1 for a desired value of p.

However, if $M(t_1 : T) < D^*$ there is no evidence of significant variance change in the series. The algorithm then stops here.

Step 2a

Define $t_2 = k^*(a[t_1 : T])$ and evaluate $D_k(a[t_1 : t_2])$. For $M(t_1 : t_2) > D^*$, then there is a new point of variance change and should repeat step 2a until $M(t_1 : t_2) < D^*$. When this occurs, it can be said that there is no evidence of variance change in $t = t_1, t_2$, and therefore, the first point of the change is $k_{first} = t_2$.

Step 2b

Now perform a similar search starting from the first change point found in step 1, all the way to the end of the series.

Define a new value for $t_1 : t_1 = k^*(a[t_1 : T]) + 1$ and evaluate $D_k(a[t_1 : T])$. Repeat step 2b until $M(t_1 : T) < D^*$ let $K_{last} = t_1 - 1$.

Step 2c

If $k_{first} = k_{last}$, there appears to be only one change point. The algorithm stops here. If $k_{first} < k_{last}$, keep both values as potential change points, and repeat step 1 and step 2 form the middle part of the series: that is, $t_1 = k_{first} + 1$ and $T = K_{last}$.

Each time step 2a and 2b are repeated, the result can be one or two more such points. Refer to $\hat{N}_T$ as the number of change points found as far.

Step 3

If there are two or more potential change points, it is important that they are in ascending order. Let cp be the vector of all such possible change points found so far, we then define the two extreme values $cp_0 = 0$ and $cp_{N_{t+1}} = T$. Check each possible change point by calculating $D_k(a[cp_{j-1} + 1 : cp_{j+1}])$, $j = 1, \ldots \hat{N}_j$. If $M(cp_{j-1} + 1 : cp_{j+1}) > D^*$, then keep the point; otherwise, eliminate it. Repeat step 3 until the number of change points does not change and the points found in each new pass are sufficiently 'close' to those on the previous pass.

Based on the ICSS algorithm and sample data by Li *et al.* (1997), this study obtains three change points in the

Table 1. *Points of structural changes in variance.*

	Shanghai Stock Exchange (Share A)	Shenzhen Stock Exchange (Share A)
Date when the change occurred	07/26/94 10/28/94 05/25/95	07/22/94 08/10/94 10/24/94 05/25/95

Note: the 5% critical value is ± 1.358 based on Table 1 by Inclán and Tiao (1994)

Table 2. *Results of F-statistics of variance change.*

Coefficients	Market	
	Shanghai Stock Exchange	Shenzhen Stock Exchange
C_k	309.5478	208.8749
C_T	794.7037	581.3895
k	45	45
T	129	129
$F_{54.85}$	0.84	0.96

Note: the 5% critical value is 1.56

Shanghai market and four in the Shenzhen market. The times when structural changes took place are listed in Table 1.

Given the demarkation point of 1 January 1995 when one-day settlement time is extended to the next-day, this study discovers two and three points of variance change before the structural break from the Shanghai and Shenzhen markets respectively. After 1 January 1995, each market experienced one change point. Failing to take these points into consideration can render erroneous results. Consequently, the optimal sample period should start from the beginning of November 1994 to the end of April 1995 in order to remove the impacts of other changes. From fig. 2, there was no noticeable change in return volatility during the optimal sample period.

In order to compare and measure such a volatility change, the F-statistic form the ICSS (Inclán and Tiao, 1994) is employed, versus the modified Levene statistic adopted by Li *et al.* (1997). This study defines the first sample consisting of the observation $a_j, j = k + 1, \ldots, T$ with variance σ_1^2. Then the F-statistic for testing $H_0 : \sigma_0^2 = \sigma_1^2$ against $H_a : \sigma_0^2 < \sigma_1^2$ is

$$F_{T-k,k} = \frac{(C_T - C_k)}{T - k} \cdot \frac{C_k}{k}$$

Given that the new sample period spans from 1 November 1994 to 30 April 1995; this study has 129 data points with the demarcation point at $k = 45$. The estimated F-statistics are reported in Table 2 for the both markets.

From the evidence in Table 2, this study cannot reject the null hypothesis of equal variances for either market. In other words, the settlement time changes does not cause change in volatility. This result is at odds with that by Li

et al. (1997) in which, return volatility was found to have diminished after the institutional change occurred in PRC.

IV. CONCLUDING REMARKS

Applying the ICSS algorithm by Inclán and Tiao (1994), this study has detected multiple variance change points: three for the Shanghai market and four for the Shenzhen market. The impact of these points were not addressed by Li *et al.* (1997) and as such may lead to biased results. These points may well affect return volatility in stock markets. These result indicates that by using Inclán and Tiao's F-statistic the return volatility in both the Shanghai and the Shenzhen markets did not undergo a significant change. The strength of the multiple change-point detection method by Inclán and Tiao lies in isolating such change points before applying the statistical analysis.

ACKNOWLEDGEMENTS

The authors are grateful to the editor and an anonymous referee for valuable suggestions. However, they wish to absorb all the culpability.

REFERENCES

Inclán, C. and Tiao, G. C., (1994) Use of cumulative sums of squares for retrospective detection of changes of variance, *Journal of the American Statistical Association*, **89**, 913-23.

Li, Dong, Lin, S. K. and Li, C., (1997) The impact of settlement time on the volatility of stock markets, *Applied Financial Economics*, **7**, 689-94.

The U.S. and Taiwan Trade Balance Revisited: A Comparison of the Instrument Variable and the VAR Models

Bwo-Nung Huang, Soong-Nark Sohng, and Chin-Wei Yang[*]

The single factor that lies at the core of Taiwan's remarkable economic growth is sustained export growth. During the past twenty-five years, the balance of trade between the U.S. and Taiwan has its share of volatility. Taking variables from different theories, we employ the instrument variable and the VAR model to dissect the problem. It ought to be pointed out that the private sector interest rate instead of the official rate plays a key role in the model. The substantial investment in mainland China from Taiwan has distorted the trade balance picture. (*JEL* Classification: F32, F41)

I. Motivation of the Study

Taiwan, an island in Pacific Ocean with limited natural resources, has no choice but to adopt export-oriented economic policy to sustain her economic growth. The ratio of export to gross

[*]Professor, Department of Economics, National Chung Cheng University, Chia-Yi, 621, Taiwan, (Tel) +886-5-2720411, (Fax) +886-5-2720816, (E-mail) ecdbnh@ccunix.ccu.edu.tw; Professor, Department of Economics, Clarion University of Pennsylvania, Clarion, PA 16214-1232 USA, (Tel) +1-814 -226-2632, (Fax) +1-814-226-1910, (E-mail) sohng@mail.clarion.edu; and Professor, Department of Economics, Clarion University of Pennsylvania, Clarion, PA 16214-1232 USA, (Tel) +1-814-226-2609, (E-mail) yang@mail. clarion.edu, respectively. Financial support from National Science Council of Taiwan (#NSC85-2415-H194-004) is gratefully acknowledged, as is typographical assistance of Ms. C. J. Lin. Usual caveats apply.
[Seoul Journal of Economics 1999, Vol. 12, No. 3]

TABLE 1
TAIWAN'S EXPORTS AND IMPORTS AS THE PERCENTAGE OF THE GNP

Year	Export/GNP	Import/GNP
1965	19.36%	22.36%
1970	30.37%	30.42%
1975	39.86%	43.15%
1980	52.61%	53.80%
1985	53.33%	39.75%
1990	46.54%	41.22%
1991	47.30%	42.37%
1992	43.48%	40.82%

Source: NIAA data base at the Computer Center, Ministry of Education (Taiwan).

TABLE 2
VALUES OF EXPORT AND IMPORT (IN TRILLION TAIWAN DOLLARS) AND ITS SHARE(%) WITH RESPECT TO THE U.S. AND HONG KONG

Year	Total Value of Export	Hong Kong (%)	The U.S. (%)	Total Value of Import	Hong Kong (%)	The U.S. (%)
1980	712	7.87	34.13	711	1.27	23.63
1982	864	7.06	39.47	736	1.63	24.18
1984	1204	6.89	48.84	870	1.72	22.99
1986	1507	7.30	47.71	917	1.53	22.46
1988	1731	9.19	38.71	1423	3.87	26.21
1990	1802	12.76	32.35	1471	2.65	23.05
1992	2047	18.95	28.92	1816	2.48	21.92
1994	2456	22.84	26.14	2261	1.81	21.14

Source: *Industries of Free China* (Executive Yuan of Taiwan)

national product (GNP) had increased from 20% in 1965 to 53.33% in 1985 before it decreased to 43.48% in 1992 (Table 1). Moreover, the U.S. has been the most important market for Taiwan's export; nearly as much as one half of the export went to the U.S. market at its peak. Since 1989 when the Taiwanese government approved investment in mainland China, the amount of export to Hong Kong

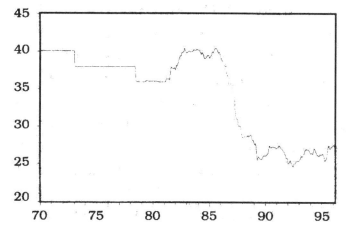

(a) Historical Trend of Taiwan-U.S. Exchange Rate

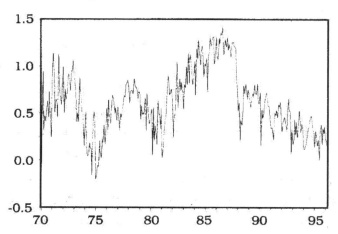

(b) Historical Trade Balances[1] between the U.S. and Taiwan

Note: 1) The relative trade balance graphed here is calculated based on log*EX* − log*IM* in which *EX* and *IM* represent values of exports and imports of Taiwan.

Source: The ARE MOS data bank, computing center of Ministry of Education, Taiwan.

FIGURE 1

has skyrocketed. As direct trade is not allowed, Hong Kong serves as a transshipment center to facilitate the flow of capital and raw materials between Taiwan and mainland China. Consequently, the proportion of export to Hong Kong climbed to 22.84% in 1994, second only to the U.S. (Table 2). The proportion of import from the U.S. has been stabilized around 22% with a peak of 26.21% in 1988 (Table 2) when the Taiwan Dollar appreciated her value drastically. Having had a long-term favorable trade balance against the U.S., the Taiwan government was under pressure to appreciate her currency value from 40 (Taiwan dollars per U.S. dollar) to as much as 25 or a 37% appreciation (Figure 1 (a)). Would such appreciations of the Taiwan Dollar necessarily give license to a more balanced bilateral trade relation? An examination of Figure 1 (b) suggests the existence of a downward level shift in the mean value of trade balance. Could it be attributed to the currency appreciation or the huge emerging Chinese market or increasing amount of export to Hong Kong that serves as the transshipment center?

Although prior studies of this problem are modeled after standard economic theories in selection of variables, they do not consider the variables unique to Taiwan. In addition to the traditional regression analysis, we employ the time series models recently advanced to identify the key factors that explain the bilateral trade balance problem. The results of our study indicate that (1) interest rate of private sector more than that of government is a better predictor for trade balances, (2) the trade between the U.S. and Taiwan has dwindled since investment barriers on China was removed, (3) tariff rates could also be used as a microfoundation variable in explaining trade balances, and (4) the volatility of the exchange rate (New Taiwan Dollars per U.S. Dollar) plays an important role in determining trade balances with the correct sign on the estimated coefficients. The organization of the paper is as follows: The next section presents a comprehensive literature review; Section III includes data and models; Section IV discusses the empirical results; and Section V contains a conclusion.

II. Literature Review

Literature on balance of trade dates back to the elasticity approach known as the Marshall-Lerner condition: Depreciation can lead to an improved trade balance via price and quantity adjustment mechanism. At the other end of spectrum, Alexander (1952) proposed the income-expenditure approach in which income plays an important role in analyzing trade balance. Neither approach, however, takes values of currencies into consideration. According to the Walrasian Law, excess supply of the money market manifests itself in the excess demand of commodity and bond markets. That is, excess demand of the money market spells adverse trade balance. And these three approaches (or their combinations) dominate the empirical models in the choice of variables. For example, Rose and Yellen (1989) propose the following model:

$$TB_t = f(q_t, y_t, y_t^*), \tag{1}$$

where TB_t is the trade balance of time period t; $q_t = p_t^* e_t / p_t$ or real exchange rate in which p_t^* and p_t denote foreign and domestic prices respectively, and e_t is nominal exchange rate; y_t and y_t^* represent domestic and foreign real output respectively. Expanding on this formulation, Bahmani-Oskooee and Pourheydarian (1991) recast the model as

$$TB_t = f(q_t, y_t, y_t^*, m_t, m_t^*), \tag{2}$$

where m_t and m_t^* denote money supply of domestic and foreign economies respectively.

Great majority of studies before Rose and Yellen did not investigate stochastic properties of the time series variables. Examples abound: Miles (1979) regressed changes of trade balance on that of other macroeconomic variables while Bahmani-Oskooee (1985) applied regression technique on trade balance, output, exchange rate and money supply. Such a direct approach has lost its momentum since the seminal work by Nelson and Plosser (1982) in which they found the unit root property in most macroeconomic variables. A direct application of the regression model would very likely lead to the spurious correlation as pointed out by Granger and Newbold (1974). Without doubt, Rose and Yellen (1989) are among earlier pioneers who employ the concept of unit root and cointegration in analyzing the relationship among trade balances and macroeconomic variable. Prior works include the application of distributed lag model to the studies of the J curves (Bahmani-Oskooee 1985; Himarios 1985; Bahmani-Oskooee and Pourheydarian 1991). Other approaches circumscribe a wide variety of econometric techniques, i.e. seemingly unrelated estimates by Miles (1979) and instrument variable estimates by Felmingham (1988) and Rose and Yellen (1989). These reduced-form models suffer from the problem of the correlation due to errors in variables and disturbance terms, and the corresponding inconsistency could render the estimates quite unreliable.

Needless to say, the Granger causality technique is popularized in the trade balance literature as well. One of the most interesting studies is the so-called Feldstein Hypothesis: Would federal budget deficit Granger-cause the current account in the U. S.? Majority of such investigations point to the acceptance of the Feldstein Hypothesis. Darrat (1988) identifies feedback effects between the two variables under the floating exchange rate regime, but fails to locate any causality between the trade balances and other macroeconomic variables (e.g. money stock, inflation rate, and interest rate). In addition, Bachman (1992) finds that financial deficit Granger-causes trade balances, and Huang (1994) investigates the causality between trade balance and real output, money stock and exchange rate.

With the recent advances in the unit root and cointegration techniques, the focus of the research on trade balances has been shifted to the study of long-term equilibrium relationships among trade balance and other macroeconomic variables (e.g. Boucher 1991; Bahmani-Oskooee 1991 and 1992; Bahmani-Oskooee and Payesteh 1993; Arize 1994; Bahmani-Oskooee and Alse 1995). Boucher (1991) investigates the relationship among exchange rates, savings, and trade balances based on the identity of national income account identity:

$$X - M + V \equiv (T - G) + (S - I), \tag{3}$$

where X denotes export; M denotes import; V denotes net income of foreign investment; S denotes domestic private saving; I denotes domestic private investment; and G denotes government expenditure. Using the bivariate model, Boucher finds a significant cointegration between the net national saving and balance of trade. The absence of cointegration relation between the exchange rates and the balance of trade supports the Mundell-Mckinnon Hypothesis. In addition Boucher (1991) employs two sets of equilibrium conditions to explore the cointegration (multivariate) relations with respect to trade balance:

$$TB_t = \alpha + \beta q_t + \delta (y_t - y_t^*) + \theta (p_t - p_t^*) + e_t, \tag{4}$$

$$TB_t = \alpha + \sum_i \beta_i q_{t-i} + u_t, \tag{5}$$

where TB_t, q_t, y_t, y_t^*, p_t, p_t^* are trade balance, real exchange rate, domestic and foreign outputs, domestic and foreign prices respectively. Note the equation (5) describes the J curve phenomenon. Using quarterly data from the first quarter of 1974 to the second quarter of 1988, Boucher does not identify any cointegration relation. Since the Autoregressive Conditional Heteroscedasticity (ARCH) model advocated by Engle (1982) and Bollerslev (1986), much attention has been focused on the speculative time series such as prices of financial assets or exchange rates. In particular, the volatility of exchange rates is found to exhibit the clustering phenomenon. As a result, the clumps of volatilities, not the exchange rates *per se*, play a critical role in trade balance. A plethora of papers have focused their analyses on the exchange rate volatility and its impacts on trade balance.[1] There seems to be a consensus that since the floating exchange system, the volatility of exchange rates has become more profound and profuse (Dornbusch 1989). The economic consequence of the increased risk in foreign exchange rate market manifests itself in the reallocation of resources from exporting sectors to non-exporting sectors, because greater volatility generally leads to decreased export activities as is supported by empirical studies (Edison and Melvin 1990). Majority of the research after 1990 resorts to the techniques of unit root, VAR, ARCH and ECM in order to analyze the impact of the volatility on trade balance (Koray and Lastrapes 1989; Asseery and Peel 1991; Savvides 1992). In particular, Savvides (1992) identifies unanticipated exchange rate volatility as the main cause of decreased real exports.

Another branch of study on trade balance and exchange rate changes is, to a considerable extent, based on microeconomic foundation. Among them are (i) overlapping generation models (Kareken and Wallace 1977; Fried 1980; Buiter 1981; Persson 1983; and Dornbusch 1989), (ii) intertemporal model (Sachs 1981; and Svensson and Razin 1983) and (iii) infinite-horizon overlapping generation model (Obstfeld 1982). As early as 1950, Harberger (1950), and Laursen and Metzler (1950), hypothesized that a deteriorating terms of trade would, in general, lead to a lower savings level which would in turn aggravate the current account. Viewed in this light, domestic savings play a pivotal role in the literature of trade balances. For example, several papers dealt with multiperiod investment behavior (e.g. Razin 1980; Marion and Svensson 1981; Sachs 1981; Bruno 1982; Svensson 1982; and

---

[1]Readers are referred to following papers: Hooper and Kohlhagen (1978); Abrams (1980); Akhtar and Hilton (1984); Gotur (1985); Kenen and Rodrik (1986); Thursby and Thursby (1987); Cushman (1983, 1988); De Grauwe (1988); and Bailey, Tavlas and Hlan (1987).

Helpman and Razin 1984). However, the dynamics of investment were not explicitly considered in these papers. In their overlapping generation model, Persson and Svensson (1985) discuss such dynamics: Temporary or permanent changes of the terms of trade and interest rates can exert different effects. Based on intertemporal models, Sen and Turnovsky (1989) as well as Gardner and Kimbrough (1989) explore the impact of different tariffs (anticipated or unanticipated, temporary or permanent) on the trade balances. The key parameters are found to be intertemporal substitution elasticities, budget deficit (surplus) and preference. In a different vein, Huang (1993) formulates the intertemporal model built upon a small open economy with incomplete capital market. Failing to consider interactive relations among real interest rate, terms of trade and trade balance, Huang does not provide a comprehensive model. Prior studies concentrate largely on trade balances for industrialized countries. Employing advanced time series techniques, the researchers adopt traditional macroeconomic variables to tackle the problem. In this study, we include both microeconomic and macroeconomic variables in the model to analyze the trade balance problem.

III. Model Formulation and Data Description

A. The Model

Variables selection of prior studies depends primarily on the choice of the models, i.e. the macroeconomics-based model (e.g. income, money supply) or intertemporal model (savings, investment, terms of trade). Yet in some cases, exchange rate volatility plays an important role. Our formulation as shown below is a hybrid of these models:

$$TB_t = f(m, m^*, q, r, r^*, y, y^*, Vq, custx, X), \qquad (6)$$
$$\quad\; -\;\; +\;\; +\; -\; +\; -\; +\quad -\quad\; -$$

where $TB_t =$ bilateral trade balance of time period t, and mathematically, $TB_t = \log EX - \log IM$ in which EX and IM are values of exports and imports;

$m, m^* =$ domestic and foreign real money stock respectively;

$q = p_t^* e_t / p_t =$ real exchange rate;

$p_t, p_t^* =$ domestic and foreign prices;

$e =$ bilateral exchange rate (nominal);

$y, y^* =$ domestic and foreign real income;

TABLE 3A

SOURCE OF DATA

Variable Name	Description	Source of Data
EX, IM	U.S.-Taiwan bilateral export and import values (in million Taiwan Dollars)	FSM Data Bank
$e_{\frac{TW}{US}}$	U.S.-Taiwan exchange rate (nominal)	IFS Data Bank
y, y^*	Industrial Production Index (line 66)	IFS Data Bank
m, m^*	currency+quasi-money (line 34 and line 35)	IFS Data Bank
r	Prime rate of First Bank (interest rate of private sector in Taipei)	FSM Data Bank
r^*	Interest rate of the U.S. money market	IFS Data Bank
custx	Tariff revenue	TAX Data Bank
IMP	Aggregate import	TRADE Data Bank
LBP	Labor productivity index (manufacturing sector, starting 1973:01)	WAGE Data Bank
MPIMAT	Price index of imported raw material (starting 1981:01)	PRICE Data Bank
EXTWHK	Value of export from Taiwan to Hong Kong	TRADE Data Bank
i, i^*, i_p	Domestic and foreign real interest rate (net of inflation) and private sector interest rate	Calculated Values
q	Real exchange rate	Calculated Values
EXTWUS	Value of export from Taiwan to U.S.	TRADE Data Bank
IMTWUS	Value of import from U.S. to Taiwan	TRADE Data Bank
TB	logEXTWUS – logIMTWUS	Calculated Values
Vq	Volatility of Taiwan Dollar/U.S. Dollar (real term)	Calculated Values
Vr	Volatility of the interest rate (private sector) of Taiwan	Calculated Values

custx=mean custom tariff=total tariff (monthly) / total import (monthly);[2]

X=other relevant variables such as dummy variables or seasonable dummies taken to explain the structural change; and

Vq=volatility of the exchange rate which is computed based on the definition by Koray and Lastrapes (1990) and Chowdhury (1993)

---

[2]Owing to the availability of custom duty data, we have to compute the general average rather than the weighted average custom duty.

TABLE 3B
CHRONOLOGICAL RECORDS OF MAJOR EVENTS

Time	Description of the Event
December, 1973 to 1974	The First Oil Crisis
October, 1979 to 1980	The Second Oil Crisis
August, 1978	Adoption of Floating Exchange Rate
October 19, 1987	DJIA Plummeted about 500 Points
November, 1987	Visitation of Relatives in Mainland China Approved
October 29, 1987	A Historically Low Yen/US Dollar Ratio of 137.55
August, 1986 and on	Bilateral Trade Negotiation Began
November, 1988	U.S. Negotiation Team Accuse Taiwan of Unfair Trade Practice
August 1990 to March 1991	The Gulf War
June, 1991	A Beginning of the Privatization of Taiwan Banks (15 Banks Were Allowed to Be Privately Owned)

Data Source: *Central Bank Quarterly* (Taiwan)

as shown below:

$$Vq = \left[\left(\frac{1}{m} \right) \sum_{i=1}^{m} (\log Q_{t+i-1} - \log Q_{t+i-2})^2 \right]^{\frac{1}{2}}. \qquad (7)$$

Note that $m = 12$ is the order of moving average, and Q_t is the growth rate of real exchange rate.[3] The signs beneath the variables of equation (6) are expected direction of response from the theories.[4]

B. Data and Sample Period

The sample period extends from January 1973 to March 1996, and the monthly data are obtained from AREMOS Data Bank, Ministry of Education, Taiwan.[5] Source of data, variable description and major events are reported in Table 3A and 3B to facilitate the model presentation.

---

[3] Other values of m are tried (e.g. $m = 6$), but the results remain similar.

[4] Some signs may be ambiguous according to the empirical estimates.

[5] For time series plots of these variables, readers are referred to Huang (1996).

IV. Empirical Model and the Estimation Results of the U.S.-Taiwan Trade Balance

A. Unit Root Tests

Prior studies analyzing trade balances rely on, to a great extent, either single equation techniques or a joint-equation approach. While single equation techniques may render biased and/or inefficient estimate, the joint-equation approach, such as a VAR model does not have theoretical underpinnings and fails to take contemporaneous errors into consideration. In absence of a definitive advantage of either approach, we shall employ both models in hope of reaching a more reliable conclusion. As is well-known in the time series literature, the stationarity of model variables needs to be examined first. We adopt in this paper three different models: The augmented Dickey-Fuller (1979) or ADF, Phillips and Perron (1988) or PP, and Perron's Unit Root Test with structural change (1989). In their seminal paper, Dickey and Fuller (1979) formulated the following model:

$$\triangle y_t = (\rho - 1)y_{t-1} + \sum_{i=1}^{k-1} \theta_i \triangle y_{t-i} + a_t, \tag{8}$$

where a_t = residual that obeys white noise process with $H_0: \rho = 1$. Failure to reject the H_0 implies a unit root for y_t.

A similar model proposed by Phillips and Perron (1988), a nonparametric approach that adjusts for autocorrelation and heteroscedasticity, can be tested based on the following equation:

$$\tau_\mu = \hat{\tau}_\mu \left(\frac{S_a}{S_{Tn}} \right) - \frac{1}{2}(S_{Tn}^2 - S_a^2)\{S_{Tn}^2 T^2 (Y_{t-1} - Y_{-1})^2\}^{\frac{1}{2}}, \tag{9}$$

in which $S_{Tn}^2 = T^{-1} \sum_{t=2}^{T} a_t^2 + 2T^{-1} \sum_{j=1}^{n} w_{jn} \sum_{t=j+1}^{T} a_t a_{t-j}$, with S_a^2 being the sample variance or $S_a^2 = \sum_{t=2}^{T} \frac{a_t^2}{T}$. Note that $w_{jn} = \left\{ 1 - \frac{j}{n+1} \right\}$ is the weight to assure the positivity of the S_{Tn}^2.

It should be pointed out that Fuller's τ_μ table is needed for both ADF and PP tests with a drift term (Fuller 1976). In the absence of the drift term or in the presence of a trend, equations (8) and (9) are revised and compared with critical values of Fuller's τ_τ and τ values (Hamilton 1994). Moreover, the unit root test could lead to erroneous conclusions in the case of the mean shift caused by structural changes (Perron 1990; and Perron and Vogelsang 1992).

TABLE 4

RESULTS OF THE UNIT ROOT TESTS

Variable Name	ADF	PP	I(0) or I(1)	AO-ADF	I(0) or I(1)
EXTWHK	-3.1307	-10.0462*	?	-4.3457**	I(0)
LBP	-2.7430	-3.2195***	?	-3.6186	I(1)
TB	-2.4556	-5.2310*	?	-5.3504*	I(0)
MPIMAT	-1.7757	-1.8129	I(1)	-2.2765	I(1)
custx	-4.3212*	-11.1138*	I(0)		I(0)
q	-0.8838	-0.8448	I(1)	-2.9449	I(1)
m	-1.9117	-2.0758	I(1)	-2.7158	I(1)
m*	-1.8828	-1.7927	I(1)	-2.7787	I(1)
i	-3.2199**	-8.7733*	I(0)		I(0)
i_p	-2.7286***	-6.8337*	?	-7.5038*	I(0)
i*	-2.1439	-2.1093	I(1)	-3.4494	I(1)
Vq	-3.9429**	-3.8647**	I(0)		I(0)
Vr	-3.7932**	-12.8903*	I(0)		I(0)

Note: The critical values ($k=0$) from Table 1 of Perron and Vogelsang (1992) are -5.06, -4.42, -2.16, and -1.64 for 1%, 5%, 95% and 99% significance levels; ADF=Augmented Dickey-Fuller Unit Root Test; PP =Phillips and Perron Unit Root Test; AO-ADF=ADF test with AO-related structural changes. $*=1\%$ significance level, $**=5\%$ significance level, $***=10\%$ significance level.

The level shift is considered a consequence of the existence of additive outliers (AO) and innovation outliers (IO) in the intervention analysis developed by Box and Tiao (1975), Chen and Tiao (1990), and Chen and Liu (1993). According to Perron and Vogelsang (1992), the unit root test in the case of the AO structural change is based on the following equations:

$$y_t = \mu + \delta DU_t + \hat{\varepsilon}_t \quad \text{for} \quad t=1, \cdots, T, \tag{10}$$

$$\hat{\varepsilon}_t = \sum_{t=0}^{k} w_t D(TB)_{t-i} + \rho \hat{\varepsilon}_{t-1} + \sum_{i=1}^{k} L_t \triangle \hat{\varepsilon}_{t-i} + e_t \quad \text{for} \quad t=k+2, \cdots, T, \tag{11}$$

where $DU_t=1$, if $t>T_b$; $DU_t=0$ otherwise. The time when a major event took place is denoted by T_b; $D(TB)_t=1$ for $t=T_b+1$; $D(TB)_t=0$, otherwise. The null hypothesis is $\rho=1$. Similarly, the unit root model in the face of IO-related structural change can be formulated

TABLE 5

UNIT ROOT RESULTS OF THE FIRST-DIFFERENCED VARIABLES

Variable Name	ADF	PP
$\triangle LBP$	-10.54*	-39.88*
$\triangle MPIMAT$	-6.05*	-11.33*
$\triangle q$	-6.06*	-14.06*
$\triangle m$	-6.53*	-13.00*
$\triangle m^*$	-4.71*	-8.48*
$\triangle i^*$	-8.55*	-14.03*

Note: $\triangle$ = first difference

* denotes significant at 1%.

as:

$$y_t = \mu + \delta\, DU_t + \rho\, y_{t-1} + \sum_{i=1}^{k} C_i \triangle y_{t-i} + e_t. \quad (12)$$

The critical values of the two tests can be obtained from Tables 1 and 2 of Perron and Vogelsang (1992) for the hypothesis test.

An examination of Table 4 suggests readily that all the time series variables are found to be stationary except the import price of raw materials (*MPIMAT*), labor productivity of manufacturing sector of Taiwan (*LBP*), U.S.-Taiwan real exchange rate (q), real money supply of the U.S. (m^*), and Taiwan (m), and real interest rate of U.S. money market (i^*). Contrary to the result by Tsung and Hu (1996), the value of export to Hong Kong, interest of Taiwan (private sector), and U.S.-Taiwan trade balance are found to be stationary after taking AO-related or IO-related structural change into consideration. The discrepancy of the result indicates the erroneous conclusion that can be arrived at without considering the outliers in the unit root tests. Furthermore, we take the first-difference of all the variables in Table 4 with the property of I(1), and reapply the ADF and PP tests. The result of Table 5 immediately leads to the conclusion that they are indeed stationary after the first difference.

B. The Single Equation Model

Given that all the variables are stationary, a linear regression

TABLE 6

MODEL ESTIMATES

Instrument Variable Model

$$TB_t = 1.0418 - 0.0149\ i_{pt} + 0.1729\ TB_{t-1} + 0.2346\ TB_{t-2}$$
$$\quad (6.1529)\ (-2.6635)\quad (2.5798)\quad\quad (4.0486)$$
$$\quad - 4.2947\ Vq_t + 12.0567\ Vm2_t - 5.9254\ custx_t$$
$$\quad (-2.2333)\quad\quad (4.5429)\quad\quad\quad (-6.0269)$$
$$\quad - 6.1361\ DSFINL - 0.2648\ DSMLD + \varepsilon_t$$
$$\quad (-3.4651)\quad\quad\quad (-5.4207)$$

$\overline{R}^2 = 0.8121 \quad\quad D\text{-}W = 2.0676$

Ljung-Box Q Statistics:

$Q(4) \quad \chi^2_{(2)} = 0.3915(.983) \quad\quad\quad Q(12) \quad \chi^2_{(2)} = 9.0753(.696)$

Note: Numbers in parentheses are t statistics; TB=U.S.-Taiwan trade balance (adjusted for seasonality); i_p=real interest rate of the private sector (Taipei); Vq_t=volatility of real exchange rate; $Vm2_t$=volatility of $m2$; $custx$=tariff revenue; $DSFINL$=the dummy variable for banding liberalization; $DSMLD$=the dummy variable for investment in mainland China.

model is amenable for the analysis. In addition, dummy variables are employed to capture the structural change as many major events occurred during the sample period. The instrument variable approach is deemed appropriate to avoid the inconsistency arising from the endogeneity of model variables. The result shown in Table 6 reveals that the single equation model explains about 79% of variation in the trade balance with all signs expected. The variables of domestic interest rate, tariff rate, volatility of M2, and volatility of real exchange rate are found to be significant.[6] In addition, two structural dummy variables are also found negatively significant: DSFINL and DSMLD. The privatization of banks in Taiwan (DSFINL) starting 1992 has liberalized the financial market which was under heavy-handed control by its government. As a result, the impact of volatility of the private sectors' interest rate in Taipei can better reflect the change in the U.S.-Taiwan balance of trade. Since the Taiwanese government legalized investment in mainland China

[6]We first estimate these coefficients based on equation (6) and find some are statistically insignificant. Table 6 includes the variables with significance level 10% or less. Due to pronounced seasonality, we deseasonalize the variables using X-11 model before estimation.

(*DSMLD*), there has been a significant structural change regarding the U.S.-Taiwan trade balance: Sizable amount of investment from export sectors in Taiwan has been shifted to China where manufacturing activities take place and finished products are then re-exported to the U.S. This substitution effect accounts for, to a great extent, decreasing trade surplus against the U.S.

We should point out that the interest rate of the private sector rather than the official rate plays a key role in the model.[7] Prior studies invariably employed official interest rates in explaining the variation in the U.S.-Taiwan trade balances, and results are ambiguous and weak.[8] As is well-known in the literature of international trade, the impact of changing interest rate on trade balance is two-fold. On the one hand, demand for domestic currency rises as the interest rate increases. The resulting appreciation of the domestic currency generally has a negative effect on trade balances. On the other hand, rising interest rates discourage current consumption of imports via the so-called intertemporal substitution effect. It would therefore improve the trade balance. The net result depends on the magnitudes of these two conflicting effects. The results by Tsung and Hu (1996) and Lee (1997) have verified that the intertemporal effect is insignificant in the two-tiered financial market. Consequently we expect a negative sign for the variable of domestic interest rate.

Similar to the findings in other studies the volatility of the exchange rate (Taiwan Dollar/U.S. Dollar) had a significant negative impact on the trade balance (5% significance level). As volatility of exchange rate increases, uncertainty would shift the resources away from exporting sectors. Often ignored in such models, the volatility of the domestic money supply (M2) plays an important role in this model: as well as (with the volatility) M2 increases, its impact on economy is positive and transmits itself to export sectors. The effect of a tariff (temporary or not) on trade balances has remained largely unsettled. In the short run, however, it is generally agreed that a tariff would deteriorate the balance of trade, as is witnessed in our analysis.

Without doubt, the financial variables, tariff and labor pro-

---

[7]The real interest rate *i* was found insignificant in the estimation.

[8]The role of private-sector interest rate was first mentioned by Hsu (1980), but was left unaddressed.

ductivity play key roles in the instrument variables model. However, it is the volatility of the exchange rate, not the exchange rate *per se*, that impacted the trade balance with a significance level of 5%. Perhaps, the most important factor in the analysis is the structural dummy variable: The overwhelming investment on mainland China has effectively masked the picture of the U.S.-Taiwan trade relations. In a nutshell, we have found that the appreciation of the New Taiwan Dollar should not be taken as the factor of decreasing the U.S.-Taiwan trade. The determining factors are (i) Taiwan's investment on mainland China, (ii) liberalization of financial institutions and (iii) instable financial policies during the period.

C. VAR Model

The vector autoregression (VAR) model known for its strength in incorporating various combinations of lagged endogenous variables, does not need an *a priori* theoretical foundation. While it is frequently applied in innovation analysis, the VAR model has its limitation: an *ad hoc* decomposition method due to some arbitrary order of variable appearance can give rise to different results. This limitation notwithstanding, VAR models being computationally efficient and theoretically simple are still widely accepted especially during the 1980's. In this section, we employ a five-variable VAR model with variables chosen from the single equation model. Note that all the five variables—tariff rate ($custx$), volatility of the domestic money supply M2 ($Vm2$), volatility of the real exchange rate (Vq), interest rate of private sector in Taipei (i_p), and the U.S.-Taiwan trade balance (TB) are found to be stationary; and hence are amenable to the analysis. With monthly data it is a good idea to include at least twelve lags in the regression. (Hamilton 1994, p. 583). This being the case, the estimated results based on VAR(12) model are employed to calculate impulse response functions (two standard deviations shock). In addition to the five variables, the structural dummies (*DSFINL* and *DSMLD*) are also included as exogenous variables. Shown in Figure 2 (a) are impulse response functions with two standard deviations of average tariff ($custx$), volatility of domestic money stock M2 ($Vm2$), volatility of the real exchange rate (Vq), interest rate of the private sector (i_p), and the U.S.-Taiwan trade balance (TB) respectively. Similarly, the impulse response functions are reported in Figure 2 (b) in the order of $Vm2$, Vq,

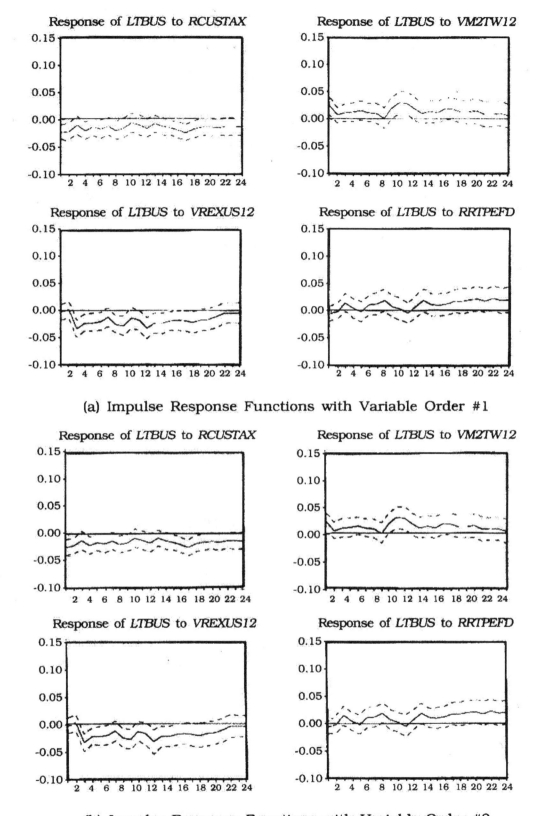

(a) Impulse Response Functions with Variable Order #1

(b) Impulse Response Functions with Variable Order #2

Note: order #1: $custx - Vm2 - Vq - i_p - TB$;
  order #2: $Vm2 - Vq - custx - i_p - TB$.

FIGURE 2

custx, i_p and *TB*. A perusal of Figure 2 (a) and 2 (b) suggests that the order of the variable appearance does not make a noticeable difference. In general, innovation of tariff has some palpable (negative) effects in periods 1, 2, 4, 8, 17 on the U.S.-Taiwan trade balance; the volatility of money supply (M2) exerts its impact (positive) in period 1 and 10; the volatility of the real exchange rate has its negative impact (on the trade balance) felt in periods from 3, through 14 except period 7 and 10. These results are very much in agreement with the single equation estimates. The domestic interest rate, however, does not seem to have a noticeable impact on the trade balance. Barring this, the two models employed in this paper are nearly qualitatively identical.

V. Conclusion

For the past twenty years, Taiwan has witnessed a remarkable economic growth, and in no small part, can it be attributed to sustained export growth. We include in our model a combination of variables from different theories. In general, five key variables — tariff, real interest rate, volatility of the exchange rate, volatility of domestic money supply (M2) and lagged trade balance — explain more than 80% of the variation of the U.S.-Taiwan trade balances. It ought to be noticed that the real interest rate of private sector in Taipei, not the official rate, plays a key role in the model. No less important than these variables, the structural dummy variables are pivotal in the U.S.-Taiwan trade relations. The substantial investment in mainland China from Taiwan has decreased the size of favorable trade balance against the U.S. The "detour" via Hong Kong represents a structural break which is properly addressed in our paper. Consistent with the theory in international trade, a currency appreciation does not necessarily cause improvement in trade balances. Finally, it is interesting to find out that volatility of financial variables plays a major role in explaining U.S.-Taiwan trade balances.

(Received February, 1999; Revised July, 1999)

References

Abrams, Richard K. "International Trade Flows under Flexible Exchange Rates." *Federal Reserve Bank of Kansas City Economic Review* 65 (1980): 3-10.

Akhtar, M. A., and Hilton, R. Spence. "Effects of Exchange Rate Uncertainty on German and U.S. Trade." *Quarterly Review* 9 (1984): 7-16, Federal Reserve Bank of New York.

Alexander, S. "Effect of Devaluation on a Trade Balance." *International Monetary Fund Staff Papers* 2 (1952).

Arize, Augustine C. "Cointegration Test of a Long-Run Relation between the Real Effective Exchange Rate and the Trade Balance." *International Economic Journal* 8 (No. 3 1994): 34-43.

Asseery, A., and Peel, D. A. "The Effects of Exchange Rate Volatility on Exports." *Economics Letters* 37 (1991): 173-7.

Bachman, Daniel D. "Why Is the U.S. Current Account Deficit So Large? Evidence from Vector Autoregressions." *Southern Economic Journal* 59 (No. 2 1992): 232-40.

Bahmani-Oskooee, Mohsen. "Devaluation and the J Curve: Some Evidence from LDCs." *Review of Economics and Statistics* 67 (1985): 500-4.

_____. "Is There a Long-Run Relation between the Trade Balance and the Real Effective Exchange Rate of LDCs?" *Economics Letters* 36 (1991): 403-7.

_____. "What Are the Long-Run Determinants of the U.S. Trade Balance?" *Journal of Post Keynesian Economics* 15 (No. 1 1992): 85-97.

_____, and Alse, Janardhanan. "Is There Any Long-Run Relation between the Terms of Trade and Trade Balance?" *Journal of Policy Modeling* 17 (No. 2 1995): 199-205.

Bahmani-Oskooee, Mohsen, and Payesteh, Sayeed. "Budget Deficits and the Value of the Dollar: An Application of Cointegration and Error-Correction Modeling." *Journal of Macroeconomics* 15 (No. 4 1993): 661-77.

Bahmani-Oskooee, Mohsen, and Pourheydarian, M. "The Australian J-Curve: A Reexamination." *International Economic Journal* 5 (1991): 49-58.

Bailey, Martin J., Tavlas, George S., and Hlan, Michael. "The Impact of Exchange Rate Volatility on Expert Growth: Some Theoretical Considerations and Empirical Results." *Journal of*

Policy Modeling 9 (1987): 225-43.

Blanchard, Olivier J. "Debt, Deficit, and Finite Horizons." *Journal of Political Economy* 93 (1985): 223-47.

Bollerslev, Tim. "Generalized Autoregressive Conditional Heteroscedasticity." *Journal of Econometrics* 31 (April 1986): 307-27.

Boucher, J. L. "The US Current Account: A Long and Short Run Empirical Perspective." *Southern Economic Journal* 58 (1991): 93-111.

Box, G. E. P., and Tiao, G. C. "Intervention Analysis with Application to Economic and Environmental Problems." *Journal of the American Statistical Association* 70 (1975): 70-9.

Bruno, Michael. "Adjustment and Structural Change under Raw Material Price Shocks." *Scandinavian Journal Economics* 84 (No. 2 1982): 199-222.

Buiter, Willem H. "Time Preference and International Lending and Borrowing in an Overlapping-Generations Model." *Journal of Political Economy* 89 (1981): 769-97.

Chen, C., and Liu, L. M. "Joint Estimation of Model Parameters and Outlier Effects in Time Series." *Journal of American Statistical Association* 88 (1993): 284-97.

Chen, C., and Tiao, G. C. "Random Level-Shift Time Series Models, ARIMA Approximations, and Level-Shift Detection." *Journal of Business and Economic Statistics* 8 (1990): 83-97.

Chowdhury, Abdur, R. "Does Exchange Rate Volatility Depress Trade Flows? Evidence from Error-Correction Models." *Review of Economics and Statistics* (1993): 700-6.

Cushman, David. O. "The Effects of Real Exchange Rate Risk on International Trade." *Journal of International Economics* 15 (1983): 45-63.

_____. "U.S. Bilateral Trade Flows and Exchange Risk during the Floating Period." *Journal of International Economics* 24 (1988): 317-30.

Darrat, Ali. "Have Large Budget Deficits Caused Rising Trade Deficits?" *Southern Economic Journal* (1988): 879-87.

De Grauwe, Paul. "Exchange Rate Variability and the Slowdown in Growth of International Trade." *IMF Staff Papers* 35 (1988): 63-84.

Dickey, D. A., and Fuller, W. A. "Distribution of the Estimators for Autoregressive Time Series with a Unit Root." *Journal of the American Statistical Association* 74 (1979): 427-31.

Dornbusch, Rudiger. "Real Exchange Rate and Macroeconomics: A

Selective Survey." *Scandinavian Journal of Economics* 92 (No. 2 1989): 401-32.

Edison, Hali, and Melvin, Michael. "The Determinants and Implications of the Choice of an Exchange Rate System." In William S. Harof and Thomas D. Willett (eds.), *Monetary Policy for a Volatile Global Economy*. Washington, D. C.: AEI Press, 1990.

Engle, Robert F. "Autoregressive Conditional Heteroscedasticity with Estimates of the Variance of U.K. Inflation." *Econometrica* 50 (1982): 987-1007.

_____, and Granger, C. W. J. "Co-Integration and Error Correction: Representation, Estimation, and Testing." *Econometrica* 55 (1987): 251-76.

Felmingham, B. S. "Where is the Australian J-Curve." *Bulletin of Economic Research* 40 (No. 1 1988): 43-56.

Fried, Joel. "The International Distribution of the Gains from Technical Change and from International Trade." *Canadian Journal of Economics* 13 (1980): 65-81.

Fuller, W. A. *Introduction to Statistical Time Series*. Wiley, 1976.

Gardner, G. W., and Kimbrough, W. P. "The Behavior of U.S. Tariff Rate." *American Economic Review* 79 (1989): 211-8.

Gotur, Padma. "Effects of Exchange Rate Volatility on Trade." *IMF Staff Papers* 32 (1985): 475-512.

Granger, C. W. J., and Newbold, P. "Spurious Regressions in Econometrics." *Journal of Econometrics* (1974): 111-20.

Hamilton, James D. *Time Series Analysis*. Princeton: Princeton University Press, 1994.

Harberger, Arnold C. "Currency Depreciation, Income, and the Balance of Trade." *Journal of Political Economy* 58 (No. 1 1950): 47-60.

Helpman, Elhanan, and Razin, Assaf. "The Role of Savings and Investment in Exchange Rate Determination under Alternative Monetary Mechanisms." *Journal of Monetary Economics* 13 (No. 3 1984): 307-25.

Himarios, D. "The Effects of Devaluation on the Trade Balance: A Critical View and Reexamination of Miles's New Results." *Journal of International Money and Finance* 4 (1985): 553-63.

Hooper, Perter, and Kohlhagen, Steven W. "The Effect of Exchange Rate Uncertainty on the Prices and Volume of International Trade." *Journal of International Economics* 8 (1978): 483-511.

Hsu, Jiag-Don. "Official Interest Rate, Black Market Interest Rate, Kickbacks, Income and Economic Growth." *Academia Economic*

Papers (1980): 25-46.

Huang, Chao-Hsi. "An Empirical Study on Taiwan's Current Account: 1961-90." *Applied Economics* 25 (1993): 927-36.

Huang, Bwo-Nung. "A Causal Studies of Trade Balance Exchange Rate and Some Macro Variables: Taiwan vs Japan and Taiwan vs U.S." *Chinese Economic Association Annual Conference Proceedings* (1994): 221-8.

_____. "Trade Balance and Macroeconomic Variables." *Technical Report*, National Science of Council, 1996.

Johansen, S. "Statistical Analysis of Cointegration Vectors." *Journal of Economic Dynamics and Control* 12 (1988): 231-54.

Kareken, John, and Wallace, Neil. "Portfolio Autarky: A Welfare Analysis." *Journal of International Economics* 7 (1977): 19-73.

Kenen, Peter T., and Rodrik, Dani. "Measuring and Analyzing the Effects of Short-Term Volatility in Real Exchange Rates." *Review of Economics and Statistics* 68 (1986): 311-5.

Koray, Faik, and Lastrapes, William D. "Real Exchange Rate Volatility and U.S. Bilateral Trade: A VAR Approach." *Review of Economics and Statistics* 71 (1989): 708-12.

_____. "Exchange Rate Volatility and U.S. Multilateral Trade Flows." *Journal of Macroeconomics* 12 (No. 3 1990): 341-62.

Kroner, K. F., and Lastrapes, W. D. "The Impact of Exchange Rate Volatility on International Trade: Reduced from Estimates Using the GARCH-in Mean Model." *Journal of International Money and Finance* 12 (1993): 298-318.

Laursen, S., and Metzler, L. A. "Flexible Exchange Rates and the Theory of Employment." *Review of Economics and Statistics* 32 (1950): 281-99.

Lee, Jen-Fei "An Over Sensitivity Analysis of Consumption in Taiwan." Unpublished Ph.D. Dissertation, Chia-Yi, Taiwan: National Chung-Cheng University, 1997.

Marion, Nancy P., and Svensson, Lars E. O. "World Equilibrium with Oil Price Increases." Seminar Paper No. 191. Stockholm: University of Stockholm, Institute of International Economic Studies, 1981.

Miles, M. A. "The Effects of Devaluation of the Trade Balance and Balance of Payments: Some New Results." *Journal of Political Economy* (1979): 600-20.

Nelson, C. R., and Plosser, C. I. "Trends and Random Walks in Macroeconomic Time Series." *Journal of Monetary Economics* 10

(1982): 139-62.

Obstfeld, Maurice. "Aggregate Spending and the Terms of Trade: Is There a Laursen-Metzler Effect?" *Quarterly Journal of Economics* 97 (1982): 251-70.

Perron, Pierre. "The Great Crash, the Oil Price Shock, and the Unit Root Hypothesis." *Econometrica* 57 (1989): 1361-401.

_____. "Testing for a Unit Root in a Time Series with a Changing Mean." *Journal of Business and Statistics* 8 (No. 2 1990).

_____, and Vogelsang, Timothy, J. "Nonstationarity and Level Shifts with an Application to Purchasing Power Parity." *Journal of Business and Economic Statistics* 10 (No. 3 1992).

Persson, Torsten. "Deficits and Intergenerational Welfare in Open Economies." NBER Working Paper No. 1083. 1983.

_____, and Svensson, L. E. O. "Current Account Dynamics and the Terms of Trade: Harberger-Laursen-Metzler Model Two Generations Later." *Journal of Political Economy* 93 (No. 1 1985): 73-99.

Phillips, P. C. B., and Perron, Pierre. "Testing for a Unit Root in Time Series Regressions." *Biometrika* 75 (1988): 335-46.

Razin, Assaf. "Capital Movements, Intersectoral Resource Shifts, and the Trade Balance." Seminar Paper No. 159. Stockholm: University of Stockholm, Institute of International Economic Studies, 1980.

Rose, A. K., and Yellen, J. L. "Is There a J Curve?" *Journal of Monetary Economics* 24 (1989): 53-68.

Sachs, Jeffrey D. "The Current Account and Macroeconomic Adjustment in the 1970s." *Brookings Papers Economic Activity* (No. 1 1981): 201-68.

Savvides, Andreas. "Unanticipated Exchange Rate Variability and the Growth of International Trade." *Weltwirtschaftliches Archiv* (1992): 446-63.

Sen, Partha. "Imported Input Price and the Current Account in an Optimizing Model without Capital Mobility." *Journal of Economic Dynamics and Control* 15 (1991): 91-101.

_____, and Turnovsky, S. J. "Tariff, Capital Accumulation and Current Account in a Small Open Economy." *International Economic Review* 30 (1989): 811-3.

Svensson. L. E. O. "Oil Prices, Welfare, and the Trade Balance: An Intertemporal Approach." NBER Working Paper No. 991. 1982.

_____, and Razin, A. "Trade Taxes and the Current Account." *Economics Letters* 32 (1983): 55-7.

Thursby, Jerry G., and Thursby, Marie C. "Bilateral Trade Flows, the Linder Hypothesis and Exchange Rate Risk." *Review of Economics and Statistics* 69 (1987): 488-95.

Tsung, W. L., and Hu, H. T. "A Time Series Analysis of Taiwan's Consumption Function." *Academia Economic Papers* 24 (No. 2 1996): 187-214.

Yaari, Menahem E. "Uncertain Lifetime, Life Insurance, and the Theory of the Consumer." *Review of Economic Studies* 32 (1965): 137-50.

Journal of Financial Studies Vol.5 No.2 October 1997(59-86)

The Random Walk Hypothesis of the Emerging Stock Markets Revisited : A Comparison of Test Power of the Variance Ratio and Rescaled Range Models[*]

Bwo-Nung Huang[a]

National Chung Cheng University

Chin Wei Yang[b]
Clarion University of Pennsylvania

Walter C. Labys[c]
West Virginia University

Abstract

In this paper, we compare the result from the modified rescaled range model and the variance ratio model for the emerging markets. In addition, four different data generating processes are utilized to critically evaluate the robustness of the two models via Monte Carlo simulations. It is found that twelve of the fourteen markets exhibit nonrandom behaviors under the variance ratio model. On the other hand, the null hypothesis is rejected for only 3 emerging markets using the modified rescaled range model. Moreover, when the data follow a MA(1) process even with a small coefficient, the VR model tends to reject the null hypothesis while the R/S model does not.

Keywords: Modified Rescaled Range, Variance Ratio, Random Walk Hypothesis, Monte Carlo Simulation.

---

[*] We are grateful to two anonymous referees for their insightful comments, the remaining errors are ours. The usual caveat applies.

[a] Corresponding author. Professor of Economics, Institute of International Economics, National Chung Cheng University, Chia-Yi, 621 Taiwan, e-mail: ecdbnh@ccunix.ccu.edu.tw.

[b] Professor of Economics, Department of Economics, Clarion University of Pennsylvania, Clarion, Pennsylvania 16214, USA

[c] Benedum Distinguished Scholar and Professor of Resource Economics, Natural Resource Economics Program, West Virginia University, Morgantown, WV 26505-6108, USA

The Random Walk Hypothesis of the Emerging Stock Markets Revisited : A Comparison of Test Power of the Variance Ratio and Rescaled Range Models

1. Introduction

One of the dominant themes in the literature of capital market equilibrium since the 1960s has been the concept of the efficient capital market: a scenario in which security prices fully reflect all available information. The predictability of security prices has become the focal point especially in investment decisions involving multinational stock markets. The results of early studies are in general supportive of the random walk hypothesis in security prices. That is, the returns of securities is often found to exhibit the property of white noise. Consequently, investors cannot predict the stock prices simply based on its previous values.[1]

Several recent studies have shown that some component of the stock returns can be effectively explained by the following models. First, the chaos (Hsieh, 1991; Peters, 1991, 1994) or neural network models began to shed new light on the role of deterministic returns. Second, the models using variance ratio (Lo and MacKinlay, 1988, 1989; Huang, 1995) and rescaled range test (Hurst, 1951; Huang and Yang, 1995) emphasize the importance of long-term dependence in the stock returns. Third, the weekend effect (Keim and Stambaugh, 1984; Huang et al., 1994) and January effect especially for small firms (Reinganum, 1983; and Keim, 1983) have been cited in the literature in different stock markets. Thus, it is worthwhile pointing out that if any of the above three models is supported by some appropriate statistical tests, it signifies that at least a given portion of the stock returns can be explained and predicted. In that case, the efficient market hypothesis may be reexamined in a more rigorous manner. Care must be exercised, however, even if a portion of the stock returns is inconsistent with the random walk hypothesis, it does not necessarily imply a violation of the efficient market hypothesis.

The empirical results are, more often than not, hinged on either the power of the test statistics or the choice of the national stock index. Previous studies largely concentrated on the market indices of industrialized

---

[1] One is referred to the seminal work by Fama (1970, 1991) on the topic of the efficient market hypothesis.

economies, i.e., the U.S., Great Britain, and Japan. Only scant attention has been paid to the emerging markets (e. g., Pacific Rim stock markets). A study on these emerging stock markets is therefore warranted since they are not as well-developed and established, and yet these economies have become more noticeably important. There are signs that the center of gravity in financial market is shifting to Asia (Van Horne, 1990). Besides, a more refined statistical test is needed to overcome some estimation problems in the prior studies. We start by discussing some recent statistical tests on the efficient market hypothesis.

From the late 1980s, the variance ratio (VR) test begins to receive attention for its capability to detect long-term dependence in stock returns (e.g., Fama and French, 1987; Lo and MacKinlay, 1988, 1989; Cochrane, 1988). Lo and MacKinlay (1988) show that a form of the variance ratio $\sqrt{nq}\dfrac{\sigma_b^2}{\sigma_a^2}-1$ approaches normality under some regularity assumptions. Where subscripts a and b denote different holding periods. Besides, the variance ratio test is free from the problems caused by nuisance parameters.[2] Although the VR test possesses the desirable properties, it remains an unsettled issue especially in the case of a nonnormal distribution or with the presence of a short-term autocorrelation. As an alternative to the variance ratio test, Hurst (1951) developed a rescaled range or (R/S) approach to test the existence of long-term persistence in the sequence of water discharge data of river Nile. Unlike the VR test, the R/S approach is robust with respect to the normality assumption: the assumption whose empirical relevance is seriously questioned (e.g., Peters, 1994). Unfortunately, the power of the R/S test is affected by the presence of the short-term autocorrelation especially in the case of high frequency data. To circumvent this problem, Lo (1991) most recently proposed a modified R/S statistic to test the potential existence of long-term memory.

The purpose of this paper is to apply the traditional as well as modified R/S approaches to the emerging stock markets in Asia, America and Europe to test the simple efficiency hypothesis. Well known in the literature, nonnormality and autocorrelation are two main characteristics often found in return data of emerging stock markets. In order to compare the results of the R/S tests with that from the VR procedure, we investigate the responses of both models in the emerging markets. As the deviation from normality and existence of autocorrelation abound in emerging market, the result of the

---

[2] For details, see footnote 6 in Lo and MacKinlay (1988).

413

analysis can be of paramount important for portfolio managers. In addition, we examine the relative power under the R/S and VR tests via Monte Carlo simulations. The strength of the tests under certain condition can be identified. The organization of the paper is as follows. Section 2 offers a brief discussion of the R/S and VR models. Section 3 presents the analysis of the empirical findings of the emerging stock markets. Section 4 contains Monte Carlo simulation and its result. A conclusion is given in section 5.

2. The Variance Ratio Test and the Rescaled Range Test

As an analogy to Einstein's square root of T rule, the VR test makes use of the hypothesis that the variance of the increments of a random walk are linear in the sampling period. In particular, 1/q times the variance of q-difference should approximate a constant, that is, the ratio between 1/q times the variance of q-difference and the 1st-difference should be as close to one as possible. Lo and MacKinlay (1988) derived the asymptotic distribution of this ratio. They found that the ratio followed a normal distribution as shown below:

$$\sqrt{n}\left[\frac{\overline{\sigma}_q^2}{\overline{\sigma}_1^2} - 1\right] \xrightarrow{a} N\left[0, \frac{2(2q-1)(q-1)}{3q}\right] \tag{1}$$

where

$$\overline{\sigma}_1^2 = \frac{1}{n-1}\sum_{t=1}^{n}(Y_t - Y_{t-1} - \hat{\mu})^2 \tag{2}$$

$$\overline{\sigma}_q^2 = \frac{1}{q(n-q+1)(1-q/n)}\sum_{t=1}^{n}(Y_t - Y_{t-q} - q\hat{\mu})^2 \tag{3}$$

$$\mu = \frac{1}{n}(Y_n - Y_0) \tag{4}$$

where n is the total number of observations; and $\xrightarrow{a}$ denotes that the distributional equivalence is asymptotic. The fundamental premise of the random walk hypothesis lies in the following statement: variance of random walk (increments) must be a linear function of time interval. The VR paradigm is hinged upon such a linear dependence. A rejection of the null hypothesis indicates that the random walk hypothesis does not hold, hence the simple hypothesis market hypothesis is very much in doubt.[3]

Well-known in the hydraulics literature, Hurst (1951) developed the R/S statistic to compute reservoir storage required to yield the average flow by computing the cumulative sums of the departures of the annual totals from the mean annual total discharge. The range of the maximum and the minimum of these cumulative totals is taken as the required storage (Hurst, 1951 pp.770). Similarly, the range of partial sums of cumulative deviations of a sequence of security returns from its mean of various lengths, rescaled by its standard deviation constitute the traditional R/S statistic:

$$\widetilde{Q}_n \equiv \frac{1}{S_n}\left[\underset{1\leq k\leq n}{Max} \sum_{j=1}^{k}(r_j - \bar{r}_n) - \underset{1\leq k\leq n}{Max} \sum_{j=1}^{k}(r_j - \bar{r}_n) \right] \tag{5}$$

where $r_1, r_2 \cdots r_n$ represents security returns over n periods; $\bar{r}_n = \frac{1}{n}\Sigma_j r_j$ is the sample mean return; and $S_n \equiv \left[\frac{1}{n}\Sigma_j (r_j - \bar{r}_n)^2\right]^{1/2}$ is an estimator of the standard deviation. The first term in bracket in (5) is the maximum (over k) of the partial sum of the first k deviation of r_j from the sample mean. Since the sum of all n deviations of r_j's from their mean is zero, this maximum is always nonnegative. The second term in (5) is the minimum (over k) of this same sequence of partial sums, and hence it is always nonpositive. The difference of the two quantities, called the range for obvious reasons, is always nonnegative and hence $\widetilde{Q}_n \geq 0$.

For each subperiod of length n>3, an average $\widetilde{Q}_n$ over different starting point is computed for the standard R/S analysis. From some perspective, the R/S statistic is superior to other traditional time series approaches (e.g., autocorrelation, variance ratio, and spectral decomposition models) in detecting long-term dependence. For instance, via Monte Carlo simulations, Mandelbrot and Wallis (1969) were able to show that the R/S statistic is robust in capturing long-range persistence even with the presence of greater magnitudes of skewness and kurtosis in varieties of non-Gaussian distribu-

---

[3] Campbell et al.(1997) give an excellent discussion about the relationship between the random walk and simple efficient market hypothesis (chapter 1 and 2).

tions. Furthermore, the R/S statistic possesses the almost-sure convergence property in the stochastic process with infinite variances: a property that is not found in the tests of the variance ratio or autocorrelation (Mandelbrot, 1972 and 1975). The second advantage of the R/S statistic, according to Mandelbrot (1972), is its capability of testing the existence of nonperiodic cycles. Despite the strength of the R/S statistic, Lo (1991) shows that the test result is extremely sensitive to the presence of short-range dependence such as AR(1).

To account for the potential effect of the short-range dependence in the R/S test, Lo (1991) proposes the modified R/S statistic as follows:

$$Q_n \equiv \frac{1}{\hat{\sigma}_n(q)}\left[Max_{1\le k\le n}\sum_{j=1}^{k}(r_j - \bar{r}_n) - Max_{1\le k\le n}\sum_{j=1}^{k}(r_j - \bar{r}_n) \right] \tag{6}$$

in which

$$\hat{\sigma}(q) \equiv \left\{ \frac{1}{n}\sum_{j=1}^{k}(r_j - \bar{r}_n)^2 + \frac{2}{n}\sum_{j=1}^{q}\omega_j(q)\left\{ \sum_{i=j+1}^{n}(r_i - \bar{r}_n)(r_{i-j} - \bar{r}_n) \right\} \right\}^{1/2} \tag{7}$$

where q is number of lag periods; and the weight is defined as $\omega_j(q) \equiv 1 - \frac{j}{q+1}$ for $q < n$, so that the modified standard deviation estimator in (3) is positive (see Newey and West, 1987). Note that the modified standard deviation estimator also contains weighted autocovariance up to lag q. Andrews (1991) suggests a data-dependent rule to determine the optimum value for q based on some asymptotic property. However, the choice becomes rather arbitrary in the case of a finite sample.

The asymptotic property of the modified R/S or Q_n statistic can be alternatively given by

$$V_n(q) \equiv \frac{1}{\sqrt{n}} Q_n \xrightarrow{a} V \tag{8}$$

in which V, the range of a Brownian bridge, has the following probability distribution (see Kennedy, 1976; and Siddiqui, 1976):

$$F_{\tilde{v}}(v) = 1 + 2\sum_{k=1}^{\infty}(1 - 4k^2v^2)e^{-2(kv)^2} \tag{9}$$

The critical values can therefore be computed from (9),[4] and some frequently-used values were reported by Lo (1991). A rejection of the null hypothesis suggests an existence of the long-memory phenomenon. As a result, the predictable portion of stock returns is noticeable enough to preclude the simple efficient market hypothesis.

3. Data and Empirical Results

Fourteen emerging national stock indices used in this study are from Morgan and Stanley Data Base, including six from Asia (Indonesia, Korea, Malaysia, Philippines, Taiwan, and Thailand), four from south or central American (Argentina, Brazil, Chile and Mexico), three from Europe (Greece, Portugal, and Turkey) and one from middle east area (Jordon).[5] There are a total of 1174 daily prices (excluding dividends) for each index starting from January 1, 1988 to June 30, 1992. The fourteen stock price indices are weighted by market values in U.S. dollars. Table 1 reports some characteristics of these 14 markets. Stock return is defined as the first difference of logarithmic prices or [6]

$$R_t = (\ln P_t - \ln P_{t-1}) \times 100\% \tag{10}$$

where P_t is the current stock price index and R_t is the rate of stock return.

Rising incomes in the emerging economies concomitant with their bulging capital markets are as impressive as they are important. Needless to say, the capital markets of the emerging economies are playing an important role. An examination of Table 1 indicates that all of the 14 countries except Jordan rank within top 40 in terms of both capitalization and traded values

---

[4] Lo(1991) provided detailed computational procedure.

[5] Daily stock return were converted into US dollar via adjusting for exchange rates variations of local currencies. In addition, the Morgan-Stanley data base assigned an initial value of 100 for the first trading day in January of 1988. The Morgan-Stanley data base generally used the previous day's price in the case of holiday(s).

[6] We do not take into consideration the dividends since they are not easily obtainable. In addition, dividends on daily basis do not in general affect the empirical result in a noticeable way (see Poterba and Summers, 1988; Lakonishok and Smidt, 1988).

Table 1   Characteristics of the Emerging Stock Markets

	IND	KOA	MAL	PHI	TWN	THA	ARG
T. R.	40.6	172.2	94.3	25.1	235.5	91.5	33.0
ranking	(25)	(2)	(5)	(37)	(1)	(6)	(33)
Capitalization [a]	32953	139420	220328	40327	195198	130510	43967
as % of World	0.23%	0.99%	1.56%	0.29%	1.38%	0.93%	0.31%
ranking	(31)	(15)	(9)	(29)	(13)	(18)	(26)
Value Traded [b]	9158	211710	153661	6785	346487	86934	10339
as % of World	0.12%	2.88%	2.09%	0.09%	4.72%	1.18%	0.14%
ranking	(29)	(6)	(9)	(33)	(4)	(12)	(28)
listed firms	174	693	410	180	285	347	180
ranking	32	7	15	30	18	17	29
	BRA	CHI	MEX	GRE	JDN	PTG	TUR
T. R.	32.6	7.4	36.8	24.4	33.2	44.0	80.9
ranking	(34)	(49)	(29)	(38)	(32)	(20)	(8)
Capitalization [a]	99430	44622	200671	12319	4891	12417	37496
as % of World	0.71%	0.32%	1.42%	0.87%	0.03%	0.09%	0.27%
ranking	(21)	(25)	(12)	(38)	—	(37)	(30)
Value Traded [b]	57409	2797	62454	2713	1377	4835	23242
as % of World	0.78%	0.04%	0.85%	0.04%	0.02%	0.07%	0.32%
ranking	(18)	(36)	(17)	(37)	—	(35)	(23)
listed firms	550	263	190	143	101	183	152
ranking	(11)	(19)	(26)	(37)	—	(28)	(36)

Source: *Emerging Stock Markets Fact book, 1994*, Published by International Financial Corporation.

Note   that: IND=Indonesia,   KOA=Korea,   MAL=Malaysia,   PHI=Philippines,   TWN=Taiwan, THA=Thailand,   ARG=Argentina,   BRA=Brazil,   CHI=Chile,   MEX=Mexico, GRE=Greece, JDN=Jordan, PTG=Portugal, TUR=Turkey. T.R. = Turnover rate, as % of world = as percentage of world total. a: total world capitalization is 14,100,763 million U. S. dollars for 1993. b: total world traded value is 7,342,265 billion for 1993.

Table 2　Basic Statistics for the Emerging Stock Markets

	IND	KOA	MAL	PHI	TWN	THA	ARG
Mean	0.1210	0.0016	0.0559	0.1026	0.0605	0.0810	0.2180
Std	2.3065	1.6383	1.1822	2.1760	0.7609	1.5601	6.2679
Skewness	7.6690	0.3870	-0.9959	-0.3160	-0.0928	-0.2237	-2.4497
Kurtosis	135.7539	7.9972	12.0018	3.4076	1.3409	6.9731	49.8755
Quintiles							
Min.	-14.8470	-12.6885	-11.5799	-9.4574	-10.6938	-9.2893	-92.2745
Q1	-0.5164	-0.8234	-0.4577	-0.6321	-1.2500	-0.5956	-2.1363
Median	0.0000	-0.0326	0.0723	0.0475	0.0177	0.0239	0.0000
Q3	0.5067	0.7302	0.5669	0.8904	1.5828	0.7912	2.7257
Max.	44.5146	13.5624	5.8428	7.4623	12.6548	9.3333	45.4802
JB	873595	1251	4158	28	136	782	108660
p-value	0.0000	0.0000	0.0000	0.0000	0.0000	0.0000	0.0000

	BRA	CHI	MEX	GRE	JDN	PTG	TUR
Mean	0.0806	0.1497	0.2061	0.0712	-0.0300	-0.0358	0.0122
Std	3.5030	1.4609	1.6440	2.1687	1.4025	1.3028	3.1999
Skewness	-1.0619	-1.3321	0.0695	0.3113	-13.6562	-0.1201	0.1560
Kurtosis	11.2428	17.5563	5.2028	6.9727	315.9142	10.6704	1.3926
Quintiles							
Min.	-31.7917	-16.2308	-9.1579	-11.7485	-33.9164	-10.0707	-12.8195
Q1	-1.6049	-0.5079	-0.5332	-0.9840	-0.1086	-0.6090	-1.7133
Median	0.0297	0.0695	0.1005	0.0000	0.0000	-0.0102	-0.0474
Q3	1.9032	0.8570	1.0250	1.0307	0.1158	0.5000	1.6965
Max.	21.3360	7.8624	8.9895	15.9282	8.9053	8.7989	12.3065
JB	3544	10712	238	791	4826179	2881	131
p-value	0.0000	0.0000	0.0000	0.0000	0.0000	0.0000	0.0000

Note: IND = Indonesia, KOA = Korea, MAL = Malaysia, PHI = Philippines, TWN = Taiwan, THA = Thailand, ARG = Argentina, BRA = Brazil, CHI = Chile, MEX = Mexico, GRE = Greece, JPN = Japan, PTG = Portugal, TUR = Turkey, Std = standard deviation, JB = Jarque-Bera normality statistic. Sample period started from January 2 1988 to June　30 1992 with a total of 1174 daily closing prices (local currency).

in the world. More specifically, the capitalization values of Malaysia, Taiwan and Korea markets are among the top fifteen; and their traded values are ranked in the top ten. Furthermore, in terms of turnover rate, Taiwan and Korea markets have the two highest values. There is no denial that the emerging capital markets are bubbling and there is a pressing need to explore these markets via some appropriate statistical analyses.

Since the statistic characteristics of the emerging stock returns may play an important role in the testing of random walk hypothesis, we report in Table 2 following statistics : mean, standard deviation, skewness, kurtosis, quintiles, range, and Jarque-Bera normality χ^2 statistic. In addition to these statistics, the coefficients of autocorrelation function (ACF) up to twenty lag periods (q=20) with their heteroscedasticity-adjusted standard deviations (in brackets), the Box-Pierce Q statistic with q=20 and the heteroscedasticity-adjusted Q statistics are presented in Table 3.

Generally speaking, stock return distribution of industrialized nations is approximately symmetrical and leptokurtic as is witnessed in the U.S. and U.K. stock markets.[7] In contrast, with the exception of Taiwan, Mexico, Portugal and Turkey, the rest of the stock return distributions are far from being symmetrical. Some of the emerging stock markets exhibit much greater degree of left-skewedness than that of a typical industrialized economy. As for kurtosis the markets of Jordan and Indonesia have two greatest values (315.914 and 135.754) whereas that of Taiwan and Turkey have two smallest values (1.341 and 1.393) which are far below the kurtosis under normality. Likewise the quintile analysis gives the similar results. The three largest values of range are found in the Argentina (approximately 138%), Indonesia (about 59%), and Brazil (53%) markets. The abnormally high values for the markets of Argentina and Brazil can be explained by the presence of hyperinflation in these economies. In sum, all fourteen indices deviate significantly from normality as indicated by the large values of the Jarque-Bera statistics, with the Jordan market manifesting the greatest departure from normality (JB=4826179). The severe departures from normality in these emerging markets suggest that the power of the VR and R/S tests on the simple efficient market needs to   be carefully reexamined.

An inspection of Table 3 reveals that ten of the fourteen markets have the autocorrelation coefficient with q=1 greater than 0.1, the greatest one being from the Indonesia market (0.3894). Some of the coefficients remains

---

[7] The mean, standard deviation, skewness, kurtosis for the U.K. are 0.0329%, 1.0702%, 0.1131, and 3.4424 respectively; they are 0.0435%, 0.9070%, -0.6361, and 3.0305 for the U.S. index.

Table 3-1   Autocorrelation Coefficients of the Emerging Stock Markets

Lags	IND	KOA	MAL	PHI	TWN	THA	ARG
1	0.3894*	-0.0592	0.1134*	0.1154*	0.0866*	0.1161*	-0.0544
	[0.0334]	[0.0345]	[0.0327]	[0.0303]	[0.0310]	[0.0337]	[0.0302]
2	0.1937*	-0.0057	0.1036*	0.0325	0.0645*	-0.0093	-0.1762*
	[0.0303]	[0.0296]	[0.0310]	[0.0301]	[0.0333]	[0.0332]	[0.0318]
3	0.0078	0.0461	-0.0616*	0.0361	0.0330	0.0009	0.1065*
	[0.0307]	[0.0296]	[0.0308]	[0.0305]	[0.0317]	[0.0324]	[0.0294]
4	-0.0025	0.0400	-0.0134	0.0323	0.0133	0.0191	-0.0157
	[0.0307]	[0.0294]	[0.0297]	[0.0303]	[0.0319]	[0.0310]	[0.0299]
5	-0.0260	-0.0221	-0.0424	-0.0227	-0.0176	-0.0104	-0.0769*
	[0.0296]	[0.0295]	[0.0300]	[0.0310]	[0.0314]	[0.0304]	[0.0304]
6	-0.0198	-0.0156	-0.0134	-0.0194	0.0257	-0.0102	-0.0093
	[0.0293]	[0.0290]	[0.0291]	[0.0299]	[0.0322]	[0.0305]	[0.0294]
7	-0.0481	-0.0353	-0.0020	0.0138	-0.0401	-0.0333	-0.0543
	[0.0295]	[0.0293]	[0.0296]	[0.0298]	[0.0316]	[0.0299]	[0.0297]
8	-0.0544	0.0083	0.0175	0.0401	0.0108	0.0085	-0.0059
	[0.0293]	[0.0291]	[0.0295]	[0.0296]	[0.0318]	[0.0302]	[0.0295]
9	0.0020	0.0250	0.0274	0.0337	-0.0050	0.0247	0.0996*
	[0.0292]	[0.0292]	[0.0292]	[0.0301]	[0.0316]	[0.0305]	[0.0303]
10	0.0065	0.0380	0.0298	-0.0012	0.0736*	-0.0061	-0.0108
	[0.0291]	[0.0296]	[0.0294]	[0.0300]	[0.0334]	[0.0301]	[0.0296]
11	0.0356	0.0010	-0.0072	-0.0074	0.0571	-0.0068	0.0188
	[0.0293]	[0.0290]	[0.0291]	[0.0297]	[0.0312]	[0.0311]	[0.0296]
12	0.0459	0.0546	0.0563	0.0578*	0.0597	0.0062	0.0541
	[0.0293]	[0.0290]	[0.0295]	[0.0296]	[0.0313]	[0.0312]	[0.0298]
13	0.0573	0.0532	0.0201	0.0386	0.0066	0.0154	-0.0577
	[0.0292]	[0.0292]	[0.0293]	[0.0301]	[0.0314]	[0.0311]	[0.0300]
14	0.0294	-0.0217	-0.0104	0.0899*	0.0449	-0.0135	-0.0056
	[0.0292]	[0.0293]	[0.0293]	[0.0297]	[0.0313]	[0.0317]	[0.0293]
15	0.0030	-0.0391	-0.0483	-0.0082	0.0262	0.0171	0.0475
	[0.0293]	[0.0293]	[0.0296]	[0.0294]	[0.0321]	[0.0322]	[0.0295]
16	0.0082	0.0580*	-0.0347	0.0318	-0.0272	0.0127	-0.0669*
	[0.0292]	[0.0289]	[0.0292]	[0.0299]	[0.0312]	[0.0304]	[0.0296]

cont.

17	0.0206	0.0190	-0.0273	0.0281	-0.0635	0.0043	-0.0156
	[0.0291]	[0.0289]	[0.0291]	[0.0292]	[0.0315]	[0.0309]	[0.0296]
18	-0.0133	-0.0165	-0.0121	0.0470	0.0459	0.0057	-0.0069
	[0.0292]	[0.0290]	[0.0291]	[0.0302]	[0.0311]	[0.0302]	[0.0293]
19	0.0142	0.0190	-0.0158	0.0358	-0.0302	0.0359	-0.0078
	[0.0292]	[0.0290]	[0.0291]	[0.0303]	[0.0316]	[0.0294]	[0.0300]
20	0.0095	-0.0431	-0.0395	0.0135	0.0142	0.0161	0.0606*
	[0.0291]	[0.0291]	[0.0293]	[0.0300]	[0.0326]	[0.0300]	[0.0298]
Q(20)	239.6559	29.8316	48.6679	45.8378	45.3220	21.7532	96.2528
	(0.0000)	(0.0726)	(0.0003)	(0.0009)	(0.0010)	(0.3541)	(0.0000)
Adjusted	193.9665	28.6196	43.4332	43.2766	38.2145	17.4009	87.8257
Q(20)	(0.0000)	(0.0955)	(0.0018)	(0.0019)	(0.0083)	(0.6268)	(0.0000)

Note: Figure in bracket is the Diebold heteroscedasticity-consistent standard deviation; figure in parenthesis is p value; Adjusted Q(20) denotes the heteroscedasticity-robust Box-Pierce Q statistic with a lag period of 20. Sample period started from January 2 1988 to June 30 1992 with a total of 1174 daily closing prices (US dollar).

Table 3-2   Autocorrelation Coefficients of the Emerging Stock Markets

Lags	BRA	CHI	MEX	GRE	JDN	PTG	TUR
1	0.1866*	0.1955*	0.2449*	0.1350*	-0.0580*	0.1977*	0.2045*
	[0.0330]	[0.0298]	[0.0335]	[0.0320]	[0.0292]	[0.0319]	[0.0328]
2	0.1290*	0.0161	0.0090	-0.0205	-0.0296	0.0655*	-0.0254
	[0.0310]	[0.0293]	[0.0310]	[0.0300]	[0.0292]	[0.0312]	[0.0313]
3	0.0783*	-0.0296	-0.0079	0.0382	0.0282	0.0137	-0.0181
	[0.0316]	[0.0302]	[0.0309]	[0.0301]	[0.0292]	[0.0299]	[0.0308]
4	0.0658*	0.0023	0.0898*	0.0321	0.0272	-0.0140	0.0202
	[0.0297]	[0.0293]	[0.0317]	[0.0297]	[0.0291]	[0.0296]	[0.0311]
5	0.1010*	0.0233	0.0401	-0.0130	0.0468	-0.0304	0.0988*
	[0.0300]	[0.0295]	[0.0329]	[0.0302]	[0.0292]	[0.0295]	[0.0308]
6	0.0718*	0.0098	-0.1141*	-0.0165	-0.0018	-0.0622*	0.0423
	[0.0297]	[0.0292]	[0.0327]	[0.0301]	[0.0292]	[0.0300]	[0.0308]
7	0.0006	-0.0233	-0.0268	-0.0296	-0.0637	0.0161	-0.0107
	[0.0306]	[0.0294]	[0.0311]	[0.0298]	[0.0293]	[0.0294]	[0.0290]

cont.

8	0.0513	-0.0062	0.0012	-0.0341	0.0511	-0.0427	0.0237
	[0.0296]	[0.0293]	[0.0309]	[0.0293]	[0.0292]	[0.0308]	[0.0297]
9	0.0424	0.0079	0.0235	0.0376	-0.0680*	0.0604*	0.0057
	[0.0316]	[0.0293]	[0.0303]	[0.0296]	[0.0292]	[0.0305]	[0.0297]
10	-0.0171	0.0144	-0.0110	0.0769*	0.0069	0.0801*	0.0536
	[0.0320]	[0.0293]	[0.0304]	[0.0309]	[0.0292]	[0.0296]	[0.0303]
11	-0.0048	-0.0217	-0.0179	-0.0137	0.0126	0.0510	0.0375
	[0.0298]	[0.0290]	[0.0307]	[0.0296]	[0.0292]	[0.0293]	[0.0298]
12	-0.0285	0.0098	0.0408	0.0020	-0.0215	0.0415	0.0390
	[0.0300]	[0.0291]	[0.0313]	[0.0303]	[0.0292]	[0.0291]	[0.0297]
13	0.0050	0.0444	-0.0114	0.0629*	-0.0370	-0.0002	0.0204
	[0.0295]	[0.0291]	[0.0305]	[0.0301]	[0.0292]	[0.0292]	[0.0298]
14	-0.0307	0.0274	0.0336	0.0073	0.0103	0.0366	-0.0032
	[0.0297]	[0.0292]	[0.0300]	[0.0304]	[0.0292]	[0.0294]	[0.0301]
15	-0.0145	0.0214	0.0266	0.0122	0.0318	-0.0105	0.0357
	[0.0292]	[0.0291]	[0.0297]	[0.0311]	[0.0292]	[0.0298]	[0.0300]
16	0.0089	-0.0185	0.0459	-0.0132	-0.0431	-0.0290	-0.0099
	[0.0292]	[0.0290]	[0.0295]	[0.0295]	[0.0292]	[0.0290]	[0.0292]
17	-0.0578	-0.0431	0.0316	-0.0103	-0.0282	0.0031	-0.0201
	[0.0294]	[0.0291]	[0.0295]	[0.0301]	[0.0292]	[0.0292]	[0.0288]
18	-0.0058	-0.0229	0.0406	0.0349	0.0057	0.0015	-0.0205
	[0.0294]	[0.0291]	[0.0308]	[0.0295]	[0.0291]	[0.0291]	[0.0305]
19	-0.0471	-0.0046	0.0352	0.0047	-0.1376*	-0.0103	0.0242
	[0.0296]	[0.0290]	[0.0309]	[0.0298]	[0.0301]	[0.0292]	[0.0296]
20	-0.0241	0.0223	-0.0462	0.0414	-0.0057	-0.0372	0.0307
	[0.0295]	[0.0292]	[0.0302]	[0.0303]	[0.0292]	[0.0294]	[0.0297]
Q(20)	105.9887	56.1736	113.0508	45.2284	51.7631	80.7811	76.7535
	(0 0000)	(0.0000)	(0.0000)	(0.0010)	(0.0000)	(0.0000)	(0 0000)
Adjusted	92.1962	54.3997	89.9642	40.0782	50.4135	71.4971	64.2996
Q(20)	(0.0000)	(0.0000)	(0.0000)	(0.0049)	(0.0002)	(0.0000)	(0.0000)

Note: Same as that of Table 3-1.

significant even for q=20 (e.g., Argentina). To facilitate a comprehensive view, we present the signs of the significant coefficients for different q values in Table 4. As shown in the table, the Korea and Chile markets have only one significantly positive coefficient of autocorrelation, and the rest of the markets have at least two. Nine of the fourteen markets have significantly positive coefficients in their ACFs while five have mixed signs on their significant coefficients. The more significant these coefficients of the ACF are, the greater portion of the stock returns can be predicted. In the case when the negative coefficients dominate, one expects a mean-reverting behavior. In comparison with industrialized nations, it seems that there are relatively more significant ACF coefficients in the emerging markets.[8] Does it imply that stock returns in these markets have more predictable component or that some of the significant ACF coefficients are caused by the biasness due to nonnormality assumption ? We answer this question by applying the VR and R/S tests to the fourteen emerging stock markets.

Table 4　Signs and Lags of Autocorrelation

	IND	KOA	MAL	PHl	TWN	THA	ARG	BRA	CHI	MEX	GRE	JDN	PTG	TUR
1	+		+	+	+	+		+	+	+	+	-	+	+
2	+		+		+		-	+					+	
3			-				+	+						
4								+		+				
5							-	+						+
6								+		-			-	
7														
8														
9							+					-	+	
10					+						+		+	
11														
12				+										
13											+			
14				+										
15														
16		+					-							
17														
18														
19												-		
20							+							

Note: + indicates positively significant at 95%; - indicates negatively significant at 95%. Sample period started from January 2 1988 to June 30 1992 with a total of 1174 daily closing prices (local currency).

Note that the VR and the Dickey-Fuller test possess similar power and both tests are more powerful than the Box-Pierce procedure (Lo and Mack-

---

[8] Using the same sample period, we find the autocorrelation coefficients for q=1 to be 0.0179 and 0.0690 for the U.S. and U.K. markets respectively. The existence of autocorrelation is abundantly found in the literature. For example, Campbell et al. (1997), using daily return data of US (1962-1978), estimate the autocorrelation coefficients of first order to be as high as 0.278.

inlay, 1989). Therefore, we focus our attention on the relative performance of the VR and R/S tests. Table 5 and 6 report the results of applying the R/S and VR techniques. The numbers in the first row (q=0) correspond to the traditional R/S method of which short-run autocorrelation is not purged. When the holding period increases (q>0), this corresponds to the revised method proposed by Lo (1991). For example, q=5 and q=20 denote one and four-week holding periods. Via the traditional R/S technique (Table 5), we reject the null hypothesis of no long-term memory for the six emerging markets. However, if short-term dependence is properly accounted for , the test results vary according to different values of q, i.e., there are four rejections (Philippines, Mexico, Greece and Turkey) for the holding period less than three months (q<65). The result of the traditional R/S method may be misleading as shown in Table 5 since the null hypothesis is no longer rejected in the case of Taiwan, Indonesia and Argentina.

Table 5   Empirical Result of the R/S Model

q	IND	KOA	MAL	PHI	TWN	THA	ARG
0	2.13*	1.47	1.31	2.39*	1.81***	1.41	0.82**
5	1.54	1.50	1.18	2.11*	1.61	1.29	0.95
10	1.53	1.50	1.18	2.05**	1.58	1.29	1.02
15	1.51	1.45	1.16	1.97**	1.50	1.29	1.01
20	1.48	1.41	1.17	1.89**	1.47	1.27	1.03
65	1.40	1.38	1.21	1.57	1.30	1.11	1.05

q	BRA	CHI	MEX	GRE	JDN	PTG	TUR
0	1.50	1.58	1.09	2.22*	1.35	1.25	2.10*
5	1.17	1.38	0.89	1.99**	1.40	1.05	1.82***
10	1.07	1.36	0.89	1.98**	1.40	1.05	1.72
15	1.04	1.34	0.88	1.93**	1.42	1.02	1.64
20	1.04	1.33	0.86***	1.89**	1.47	1.00	1.60
65	1.16	1.22	0.91	1.61	1.65	1.00	1.53

Note: 1% critical value are [0.721,2.098],   * = 1% significance level

5% critical value are [0.809,1.862],   ** = 5% significance level

10% critical value are [0.861,1.747],   *** =10% significance level

Quite surprisingly, with the exception of Korea, Jordan, Thailand and Mexico, the null hypothesis of a Gaussian random walk is rejected for all markets and with all different values of q (Table 6) by applying VR model. There are two rejections for the Thailand market and four rejections in the case of Mexico, Evidently, the frequency with which the null hypothesis is rejected under the VR procedure is much greater than that under the R/S methods. Furthermore, the two tests give contradictory results, that is, the conclusion is reversed in eight of the fourteen markets : Indonesia, Malaysia, Thailand, Brazil, Chile, Taiwan, Portugal and Turkey. Only in three markets, (Philippines, Mexico and Greece) is the null hypothesis under both tests is rejected.

Table 6   Empirical Results of the VR Model

q	IND	KOA	MAL	PHI	TWN	THA	ARG
5	1.87**	0.96	1.25**	1.27**	1.26**	1.18**	0.78**
10	1.95**	0.96	1.24**	1.35**	1.32**	1.18	0.66**
15	2.01**	1.02	1.30**	1.47**	1.46**	1.19	0.66**
20	2.11**	1.08	1.30**	1.62**	1.55**	1.22	0.64**
65	2.58**	1.16	1.32	2.43**	2.07**	1.68**	0.65

q	BRA	CHI	MEX	GRE	JDN	PTG	TUR
5	1.55**	1.31**	1.43**	1.24**	0.91	1.39**	1.30**
10	1.93**	1.36**	1.48**	1.26**	0.95	1.31**	1.47**
15	2.08**	1.40**	1.48**	1.34**	0.93	1.43**	1.62**
20	2.11**	1.44**	1.53**	1.40**	0.89	1.51**	1.73**
65	1.68**	1.85**	1.22	2.11**	0.74	1.67**	2.08**

Note: 5% critical values for   q=5, are (0.87,1.13); 5% critical values for q=10, are (0.80,1.20); 5% critical values for q=15, are (0.75,1.25); 5% critical values for q=20, are (0.71,1.29); 5% critical values for q=65, are (0.46,1.54); ** = 5% significance level.

Reliance on daily return data has its undesirable side effect: existence of positive autocorrelation through infrequent or nonsynchronous tradings As previously remarked by Cohen (1983), infrequent and nonsynchronous tradings are considered the culprit of spurious correlation found in the stock returns. In general, such artificial correlations can be attributed to the infrequent trading of small-capitalization stocks. As new information is impounded into the stocks of large capitalization, it trickles down to stocks of small capitalization with lags. Thus a positive serial correlation is generated through the diffusion process. Such a induced positive serial correlation, according to Lo ad Mackinlay (1988), causes more impacts on the equal-weighted index, as it is difficult to ascertain that the positive serial correlation is a direct consequence of infrequent tradings. The presence of positive autocorrelation, be it from infrequent tradings or other unknown causes, indicates the existence of the inevitable problem in using daily returns. An excellent comparison can be found in Table 6 which reports the test results of the VR model. With exactly the same daily return data for the 14 emerging markets, we are able to reject, in majority of cases, the random walk hypothesis. Such a sharp contrast (as compared with the R/S test results) suggests a potential bias manifested via the positive autocorrelation in the daily returns. It is to be mentioned that a slight degree of autocorrelation is tolerated and hence considered to be consistent with the simple efficient market hypothesis according to recent literature.

Lo and Mackinlay (1988) showed that both the VR statistic and the ACF are closely related. As is evidenced in our results, twelve of the fourteen markets have their coefficients (q=1) of the ACF greater than 0.1 and are significant. Consequently, most of them will also have a significant VR statistic. In the case of Mexico and Argentina, the VR statistic is less than one as some of the higher order autocorrelation coefficients are negative. A time series exhibits the phenomenon of a long-term memory if at least one higher order coefficient of the ACF is significant.[9] In this case, the long-term dependence can be detected by either of the two methods. However, as shown in Table 5 and 6, the test results are shown to be abundantly different and contradictory. Given that the modified R/S method is robust with respect to the presence of short-term dependence and nonnormality, coupled with the fact that the VR method requires some form of Gaussian i.i.d., we investigate the impact of these statistical properties on the test results of both the VR and R/S models via Monte Carlo simulations. The purpose for such myriad of simulations is to reveal whether the anomaly is attributed to predictable portion of stock returns or different test powers of the VR and the R/S models.

4. Monte Carlo Simulations and Its Results

Well-known in the literature (Mandelbrot and Wallis, 1969; and Mandelbrot, 1972), the robustness of the R/S model implies that it is insensitive to the result even in the presence of highly skewed distribution and infinite variance. On the other hand, a serious departure from the normality assumption will jeopardize the result of the VR model. In order to examine the power of both models with different underlying distributional assumptions,

---

[9]That is, the autocorrelation decays very slowly.

this paper employs Monte Carlo simulations based on four types of data generating Processes (DGP)

The first DGP is built upon the assumption that the data follows a Gaussian random walk. From it, we derived a set of benchmark values to compare with the results from other DGPs. For instance, the VR and R/S statistics derived from the first DGP should be consistent with the critical values tabulated by Lo and Mackinlay (1988) and Lo (1991).

As is often cited in the literature, infrequent trading could lead to a mild MA(1) phenomenon. However, the MA(1) process does not necessarily imply a violation of the simple efficient market hypothesis. To measure various responses of the VR and the R/S models, while allowing for the MA(1) phenomenon, we design the next DGP. That is, the data of the second DGP are produced based on the first-order moving average model or MA(1) or

$$\Delta Y_t = e_t + \theta_1 e_{t-1} \tag{11}$$

where e_t is a white noise process; and $\theta \in [0.1, -0.1]$, Note that the market efficiency is consistent with MA(1) in the case where the data are characterized by infrequent trading (see Poterba and Summers 1986; and Summers 1988). Indeed, the presence of positive autocorrelation in daily stock returns may or may not be the direct consequence of infrequent tradings. However, it shares some statistical properties of the moving average process. As such, the results from either VR or R/S statistics could be misleading and unsuitable for evaluations without simulating MA(1) process on the return data.

The financial data are in general fraught with lack of symmetry and greater peakedness (leptokurtic). This departure from normality will have an impact on the power of the VR and R/S statistics especially for the former. Evident from Table 2, the great majority of national stock indices are both leptokurtic and asymmetrical. As such, the impact of nonnormality on the test results from the VR and R/S approaches can be of crucial important. In light of it, we follow Fleishman (1978) to present the third DGP in which the lack of normality is described by following process :

$$\Delta Y_t = a + b e_t + c e_t^2 + d e_t^3 \tag{12}$$

A proper choice for the values of a, b, c, and d corresponds to an appropriate distribution for ΔY_t. That is, given the kurtosis and magnitude of skewness, one can obtain a set of values of a, b, c, and d. to obtain ΔY_t. For example given the kurtosis of 93.77 and the skewness of 5.64, we can have a=-0.35, b=-0.91, c=0.35 and d=0.35.[10]

Because the R/S test statistics of four national stock indices exceed the critical values, indicating the potential existence of the long memory, we propose the last DGP in order to test the accuracy of the two model. The last DGP is based on the time series that is characterized by a fractionally-integrated process. Such a process often manifests in itself a slow decay in the ACF (see Granger, 1980, Granger and Joyeux 1980; and Hosking, 1981) as illustrated below :

$$(1-L)^d \Delta Y_t = e_t \tag{13}$$

where d is not an integer and L is the lag operator. Note that equation (13) with a noninteger d suggests the existence of a time series characterized by long-term dependence. Based on the binomial theorem for noninteger powers, we have

$$(1-L)^d = \sum_{k=0}^{\infty} (-1)^k \binom{d}{k} L^k$$

$$\binom{d}{k} \equiv \frac{d(d-1)(d-2)\cdots(d-k+1)}{k!} \tag{14}$$

Substituting $(1-L)^d$ into (13) yields :

$$(1-L)^d \Delta Y_t = \sum_{k=0}^{\infty} (-1)^k \binom{d}{k} L^k \Delta Y_t = \sum_{k=0}^{\infty} A_k \Delta Y_{t-k} = e_t \tag{15}$$

Equation (15) represents one type of autocorrelation model and its autocorrelation coefficient A_k takes the form of the gamma function :

$$A_k = (-1)^k \binom{d}{k} = \frac{\Gamma(k-d)}{\Gamma(-d)\Gamma(k+1)} \tag{16}$$

---

[10]See Fleishman (1978) for details.

In addition, equation (13) can be rearranged into an MA process of infinite order with

$$\Delta Y_t = (1-L)^{-d} e_t = B(L)e_t, \qquad B_k = \frac{\Gamma(k+d)}{\Gamma(d)\Gamma(k+1)} \tag{17}$$

ΔY_t is both stationary and invertible for $d \in \left(-\frac{1}{2}, \frac{1}{2}\right)$ and possesses the long-term memory (Hosking, 1981). The fourth DGP is built on the assumption of $d = \frac{1}{3}$, and within this framework we examine the performance of the VR and R/S tests.

To mitigate the impact by the choice of initial values, we delete the first126 observations with the net sample size 1274. Each DGP consists of 5000 simulations and the results are reported in Table 7.

Table 7   Results of Monte Carlo Simulations under the VR Model

		q=5	q=10	q=15	q=20	q=65
	99%	1.1547	1.248	1.3156	1.3811	1.7342
	95%	1.1097	1.1687	1.2178	1.2518	1.4897
DGP	90%	1.0824	1.1288	1.1606	1.1875	1.3594
	50%	0.9967	0.9954	0.9896	0.9876	0.9629
I	10%	0.9155	0.8719	0.8407	0.818	0.6681
	5%	0.8941	0.8413	0.801	0.7699	0.5952
	1%	0.8585	0.7845	0.7359	0.6902	0.4829
	Mean	0.9988	0.9975	0.9973	0.9969	0.9936
	Std.	0.0647	0.1002	0.1256	0.1466	0.273
	99%	1.3305	1.4671	1.5639	1.6402	2.1042
DGP	95%	1.2803	1.3774	1.4468	1.4985	1.8037
	90%	1.255	1.3302	1.3804	1.4233	1.6423
II	50%	1.1578	1.1743	1.1744	1.1765	1.1593
	10%	1.0677	1.0314	1.0003	0.9712	0.8032
Q=	5%	1.0424	0.9925	0.9562	0.9246	0.7226
0.1	1%	0.9959	0.929	0.8749	0.8259	0.5876
	Mean	1.1596	1.1782	1.1839	1.188	1.1961
	Std.	0.0729	0.1166	0.1489	0.1758	0.3313

cont.

	99%	0.9797	1.0265	1.07	1.1134	1.4077
DGP	95%	0.9343	0.964	0.9932	1.0189	1.2012
	90%	0.913	0.9315	0.9522	0.971	1.0995
II	50%	0.8397	0.8182	0.8113	0.8052	0.7716
	10%	0.7735	0.7198	0.6898	0.6637	0.5377
Q=	5%	0.7543	0.6951	0.6596	0.6301	0.4866
-0.1	1%	0.7239	0.6493	0.6059	0.5718	0.3832
	Mean	0.8418	0.8227	0.8163	0.8129	0.8017
	Std.	0.0551	0.0823	0.102	0.1189	0.222
	99%	1.1688	1.2462	1.3086	1.3739	1.7783
	95%	1.1091	1.175	1.2121	1.2504	1.4959
DGP	90%	1.0804	1.1255	1.1615	1.192	1.3714
	50%	0.9962	0.9933	0.9898	0.9887	0.9656
III	10%	0.9258	0.8845	0.8522	0.824	0.6754
	5%	0.9069	0.8558	0.8149	0.7832	0.6095
	1%	0.8652	0.795	0.7491	0.7122	0.5079
	Mean	1.0004	1.0004	1.0004	1.0002	1.0019
	Std.	0.0626	0.0966	0.1216	0.1435	0.2745
	99%	4.1021	7.2726	9.9810	12.3128	24.9935
	95%	4.0350	7.1053	9.7094	11.9380	22.5031
DGP	90%	3.9878	6.9575	9.4787	11.6235	20.7443
	50%	3.8141	6.4502	8.5561	10.2737	15.9262
	10%	3.6245	5.8977	7.6072	8.9642	11.6163
IV	5%	3.5656	5.7533	7.4177	8.6883	10.4871
	1%	3.4501	5.4376	6.9869	8.0610	8.8451
	Mean	3.8094	6.4475	8.5557	10.2960	16.1554
	Std.	0.1419	0.4038	0.6952	0.9963	3.7009

Note: DGPI through DGPIV are defined as : data generating processes are based on the Gaussian random, MA(1), nonnormal, fractionally-integrated time series. Each simulation consists 5000 replications with a net sample size of 1174.

The first part of Table 7 represents the simulated VR statistics under the assumption of a Gaussian random walk. Critical values of 1%, 5%, 10%, 50%, 90%, 95% and 99% under different lag periods are reported together with their means and standard deviations. These values are comparable to that developed by Lo and Mackinlay (1989). The second part of Table 7 is derived from DGP II with $\theta = 0.1$ and $\theta = -0.1$. An examination of the result indicates that the simulated VR statistics differs noticeably from that under DGPI even with the small value of θ. The mean of the VR statistic is greater (less) than 1 if θ is positive (negative). This implies that the rejection of the null hypothesis of the VR model, to some extent, can be attributed to the fact that the data follow an MA(1) process. It should be pointed out that MA(1) is not inconsistent with weak form efficient market hypothesis. At this point, it is important to highlight this finding. More often than not, one can detect the existence of first-order autocorrelation in daily or intraday stock returns. As such the results from applying the VR statistic would be fairly inaccurate, based on our simulations. It is not surprising that Lo and Mackinlay (1988) circumvent this problem by using the weekly stock returns.[11]

The VR statistics are relatively stable in the case where data are generated from DGPIII. An inspection of the simulated values indicates that only two of the fourteen kurtosis (Indonesia and Jordan) are greater than the kurtosis used in the simulations. Finally, it is rather surprising that the VR statistics are relatively large, indicating a greater tendency to reject the null in the case of DGPIV. In summary, if data are consistent with a simple MA(1) or fractionally-integrated time series, one tends to reject the null hypothesis. While the rejection of the null hypothesis in the former does not necessarily imply an violation of weak form efficient market hypothesis, the rejection in the latter signifies the existence of a long-term dependence; hence, a violation of weak-form efficient market hypothesis. Consequently an effort must be made in the case of a rejected null hypothesis to distinguish whether it comes from MA(1) or fractionally-integrated time series.

In the case of simulated R/S statistics (Table 8) the benchmark values are only slightly greater (smaller) for $\theta > 0$ ($\theta < 0$) if the data are generated from MA(1); but they are all within the range of the tabulated critical values. When the data are generated from the nonnormal population, the simulated R/S statistics are slightly less than that from the normal population. However, the R/S statistics are shown to be greater than the tabulated criti-

---

[11]See Also Huang(1995)

cal values in the case where the data follow the fractionally-integrated time series. But this is exactly what the R/S model is set out to test. Our simulations have illustrated that the R/S statistics are more robust with respect to different distributional properties from which data are generated.

Table 8   Results of Monte Carlo Simulations under the R/S Model

		q=5	q=10	q=15	q=20	q=65
	99%	1.9987	1.9608	1.9473	1.9422	1.9342
	95%	1.7188	1.7205	1.6999	1.7061	1.7063
DGP	90%	1.5908	1.5858	1.5724	1.5898	1.5900
	50%	1.1896	1.1885	1.1840	1.1908	1.1922
I	10%	0.8974	0.8974	0.8935	0.8981	0.8857
	5%	0.8351	0.8354	0.8278	0.8294	0.8230
	1%	0.7302	0.7280	0.7158	0.7231	0.7146
	Mean	1.2256	1.2210	1.2144	1.2198	1.2197
	Std.	0.2751	0.2726	0.2669	0.2700	0.2700
	99%	2.1231	2.1441	2.1619	2.1499	2.1119
DGP	95%	1.8770	1.8665	1.8621	1.8555	1.8573
	90%	1.7332	1.7259	1.7281	1.7119	1.7260
II	50%	1.2919	1.2938	1.2886	1.3027	1.2989
	10%	0.9706	0.9818	0.9670	0.9669	0.9772
Q=	5%	0.9045	0.9101	0.8941	0.8888	0.9034
0.1	1%	0.7962	0.7807	0.7579	0.7811	0.7915
	Mean	1.3263	1.3323	1.3246	1.3267	1.3297
	Std.	0.2997	0.2968	0.3002	0.2955	0.2937
	99%	1.7930	1.7442	1.7546	1.7895	1.7659
DGP	95%	1.5540	1.5166	1.5486	1.5430	1.5499
	90%	1.4384	1.4115	1.4315	1.4276	1.4308
II	50%	1.0753	1.0682	1.0748	1.0714	1.0816
	10%	0.8040	0.8065	0.8064	0.8117	0.8028
Q=	5%	0.7414	0.7508	0.7474	0.7502	0.7405
-0.1	1%	0.6543	0.6545	0.6384	0.6615	0.6472
	Mean	1.1020	1.0929	1.0998	1.1011	1.1050
	Std.	0.2497	0.2371	0.2463	0.2451	0.2463

cont.

	99%	1.8576	1.8430	1.8277	1.8268	1.8268
	95%	1.6222	1.6381	1.6116	1.6287	1.6287
DGP	90%	1.4996	1.5201	1.4962	1.4996	1.4996
	50%	1.1304	1.1375	1.1248	1.1325	1.1325
III	10%	0.8408	0.8571	0.8523	0.8564	0.8564
	5%	0.7826	0.7910	0.7892	0.7923	0.7923
	1%	0.6915	0.6873	0.6952	0.6875	0.6875
	Mean	1.1569	1.1691	1.1519	1.1584	1.1584
	Std.	0.2573	0.2593	0.2501	0.2524	0.2524
	99%	7.4519	7.5136	7.3846	7.5019	7.4217
	95%	6.6219	6.5882	6.5157	6.6478	6.5626
DGP	90%	6.1685	6.1168	6.0823	6.1582	6.0995
	50%	4.5379	4.5556	4.5556	4.5434	4.5598
	10%	3.3506	3.3547	3.3772	3.3196	3.3432
IV	5%	3.0850	3.0871	3.0836	3.0582	3.1060
	1%	2.6778	2.6249	2.6583	2.5871	2.6640
	Mean	4.6565	4.6669	4.6517	4.6544	4.6525
	Std.	1.0832	1.0749	1.0579	1.0924	1.0668

Note: Same as that of Table 7.

Based on the simulations, it seems plausible that the conflicting results between the two models can be explained as follows. The random walk hypothesis of the Indonesia market is rejected in the VR model for all q; but was retained in the modified R/S model. It implies that the Indonesia stock index exhibits a short-term dependence as manifested in the significance of the first two coefficients of the ACF. The same can be said of the markets of Malaysia, Taiwan, Thailand, Argentina, Brazil, Chile, Portugal and Turkey.[12] However, there is strong evidence that the markets of Philippines, Mexico and Greece are inconsistent with weak-form efficient market hypothesis since the null hypothesis is rejected in both models. According to the simulation results, one cannot help arriving at the conclusion: There exist a long memory phenomenon as the null hypotheses are rejected in both

---

[12] When the length of holding period is increased to 130, 195, and 260, the null hypothesis is rejected for the Argentina market indicating a possible existence of a long-term memory.

models.[13]

5. Concluding Remarks

The efficiency of the emerging markets is of primary concerns to multi-national investment firms. Based on the VR test by Lo and Mackinlay (1988), we were able to reject the null hypothesis for twelve of fourteen markets, indicating the existence of predictability of stock returns for most emerging markets. However, we reject the null hypothesis for only three countries (Philippine, Mexico and Greece) using the modified R/S model. How do we resolve these inconsistent results ? From the Monte Carlo simulations, we find that both the VR and R/S models can correctly test the hypothesis if the data are generated from fractionally-integrated time series. However, if the data follow a MA(1) process even with a small coefficient, the VR model tends to reject the null hypothesis while the R/S model does not. Since the presence of MA(1) does not necessarily imply that the stock returns are inconsistent with the weak-form efficient market hypothesis, a rejection of the null hypothesis under the VR model may very well lead to an erroneous conclusion.[14] In summary, care must be exercised in testing the efficient market hypothesis in the emerging markets. Contrary to many findings, we have found that majority of the markets are consistent with weak-form efficient market hypothesis. What is the economic implications of the "long memory" embedded in many finance time series of emerging markets? From the statistical view point, the long memory phenomenon fits comfortably with a set of ACF's that decays slowly. It is important to point out that other advanced statistical techniques can be employed to maximize predictable portion of stock prices in the presence of long memory. It remains an important avenue for future research in equity markets.

Reference

1.   Andrews, D., 1991, "Heteroskedasticity and Autocorrelation Consistent

---

[13]The null hypothesis of the VR model may be rejected due to the presence of MA(1) phenomenon. An examination of Table 3 indicates that the ACF of Greece and Philippines are close to 0.1. A comparison of simulated critical value and the VR statistics reported in Table 6 suggests we fail to reject the null hypothesis at 5% significance level except q=65.

[14]The question of how small the size of the MA(1) coefficient should be in order to be consistent with the simple efficient market hypothesis could be answer by R^2. With the coefficients of about 0.1 ($R^2=0.01$), it is so insignificant that it does not lead to the rejection of the null hypothesis.

Covariance Matrix Estimation," *Econometrica*, 59, 817-858.

2. Campbell, John Y., Andrew W. Lo, and A. Craig MacKinlay, 1997, *The Econometrics of Financial Markets*, Princeton: Princeton University Press.

3. Cohen, K., 1983, "Friction in the Trading Process and the Estimation of Systematic Risk," *Journal of Financial Economics*, 12, 263-278.

4. Cochrane, John H., 1988, "How Big is the Random Walk in GNP?" *Journal of Political Economy*, 96(5), 893-920.

5. Fama, Eugene F., 1970, "Efficient Capital Markets: A Review of Theory and Empirical Work," *Journal of Finance*, 25, 383-417.

6. Fama, Eugene F., and Kenneth R. French, 1988, "Permanent and Temporary Components of Stock Prices," *Journal of Political Economy*, 96(2), 246-273.

7. Fama, Eugene F., 1991, "Efficient Capital Markets: II," *Journal of Finance*, 56(5), 1575-1617.

8. Fleishman, A. I., 1978, "A Method for Simulating Non-Normal Distribution, " *Psychometrika*, 43, 521-32.

9. Granger, C., 1980, "Long Memory Relationships and the Aggregation of Dynamic Models," *Journal of Econometrics*, 14, 227-238.

10. Granger, C., and R. Joyeux, 1980, "An Introduction to Long-Memory Time Series Models and Fractional Differencing," *Journal of Time Series Analysis*, 1, 15-29.

11. Hosking, J. R. M., 1981, "Fractional Differencing," *Biometrika*, 68,165-76.

12. Hsieh, David A., 1991, "Chaos and Nonlinear Dynamics: Application to Financial Markets," *Journal of Finance*, 46(5), 1839-77.

13. Huang, Bwo-Nung, 1995, "Do Asian Stock Market Prices Follow Random Walks? - Evidence from the Variance Ratio Test," *Applied Financial Economics*,5,251-56.

14. Huang, Bwo-Nung, Yu-Jane Liu, and Chin. W. Yang, 1994, "A Comparative Study of the Weekend Effects in the United States, British and The Pacific Rim Stock Markets: An Application of the ARCH model," *Advances in Pacific Basin Business, Economics and Finance*, Vol I., 277-290.

15. Huang, Bwo-Nung, and Chin. W. Yang, 1995, "The Fractal Structure in

Multinational Stock Returns," *Applied Economics Letters*, 2, 67-71.

16. Hurst, H., 1951, "Long Term Storage Capacity of Reservoirs," *Transactions of the American Society of Civil Engineers*, 116,770-799.

17. Keim, D. B., and R. F. Stambaugh, 1984, "A Further Investigation of the Weekend Effect in Stock Returns," *The Journal of Finance*, 39(3), 819-40.

18. Keim, D. B., 1983, "Size Related Anomalies and Stock Return Seasonality: Further Empirical Evidence," *Journal of Financial Economics*, 12(1), 13-32.

19. Kennedy, D., 1976, "The Distribution of the Maximum Brownian Excursion," *Journal of Applied Probability*, 13, 371-376.

20. Lakonishok, Josef and Seymour Smidt, 1988, "Are Seasonal Anomalies Real? A Ninety-Year Perspective," *The Review of Financial Studies*, 1(4), 403-425.

21. Lo, Andrew W., and A. Craig MacKinlay, 1988, "Stock Market Prices Do Not Follow Random Walks: Evidence from a Simple Specification Test," *The Review of Financial Studies*, 1(1), 41-66.

22. Lo, Andrew W., and A. Craig MacKinlay, 1989, "The Size and Power of the Variance Ration Test in Finite Samples — A Monte Carlo Investigation," *Journal of Econometrics*, 40, 203-238.

23. Lo, Andrew W., 1991, "Long-Term Memory In Stock Market Prices," *Econometrica*, 59(5), 1279-1313.

24. Mandelbrot, B., 1972, "Statistical Methodology for Non-periodic Cycles: From the Covariance to R/S Analysis," *Annals of Economic and Social Measurement*, 1, 259-290.

25. Mandelbrot, B., 1975, "Limit Theorems on the Self-Normalized Range for Weakly and Strongly Dependent Processes," *Z. Wahrscheinlichkeitstheorie verw. Gebiete*, 31, 271-285.

26. Mandelbrot, B., and J. Wallis, 1969, "Computer Experiments with Fractional Gaussian Noises. Parts 1, 2, 3, "*Water Resources Research*, 5, 228-267.

27. Newey, W., and K. West, 1987, "A Simple Positive Definite, Heteroscedasticity and Autocorrelation Consistent Covariance Matrix," *Econometrica*, 703-705.

28. Peters, E. E., 1994, *Fractal Market Analysis: Applying Chaos Theory to*

Investment and Economics, (New York, John Wiley & Sons, Inc.)

29. Peters, E. E., 1991, *Chaos and Order in the Capital Markets: A New View of Cycles*, Prices and Market Volatility (New York: John Wiley & Sons, Inc).

30. Poterba, James M., and Lawrence Summers, 1988, "Mean Reversion in Stock Prices: Evidence and Implications." *Journal of Financial Economics*, 22, 27-59.

31. Reinganum, M. R., 1983, "The Anomalous Stock Market Behavior of Small Firms in January: Empirical Tests for Tax-Loss Selling Effect," *Journal of Financial Economics*, 12(1), 89-104.

32. Siddiqui, M., 1976, "The Asymptotic Distribution of the Range and Other Functions of Partial Sums of Stationary Processes," *Water Resources Research*, 12, 1271-1276.

33. Summers, Lawrence H., 1986, "Does the Stock Market Rationally Reflect Fundamental Values?" *Journal of Finance*, 41, 591-600.

34. Van Horne, J. C., 1990, "Changing World and Asian Financial Markets," *Pacific-Basin Capital Markets Research* ed. by S. G. Rhee and R. P. Chang (Amsterdam, North-Holland Publishing Company, 1990): 65-80.

Are Mathematics, Economics, and Accounting Courses Important Determinants in Financial Management: A Rank Order Approach

Rod D. Raehsler[1], Ken Hung[2], Chin W. Yang[3], and Thomas J. Stuhldreher[4]

ABSTRACT

Using an ordered probit model, we examine the determinants of student performance in an introductory, junior level financial management course at a rural state university in the eastern United States. It is discovered that the cumulative grade point average and academic major of a student are marginally significant predictors of that student's final letter grade in the course. There is a much more significant positive relationship between grades received in two prerequisite mathematics courses and the grade in the introductory finance course. The letter grade received in the prerequisite microeconomics course reveals the phenomenon of persistence in that the probability of obtaining a good grade in the finance course is greater if the student received an A or B in the microeconomics course. Final grades earned in prerequisite accounting courses also influence final grades in the introductory finance class. It is also determined that students who are required to take remedial mathematics are at a disadvantage when taking the introductory finance course.

JEL Codes: C25, I22

Keywords: ordered probit, financial management.

Introduction

All educators are interested in determining or verifying factors that influence student academic performance. The identification of new factors that positively influence this performance can often signal a need for changes in an academic curriculum whereas the empirical establishment of linkages between an important sequence of courses and disciplines that are historically expected can support claims that are often made to students and the maintenance of existing curricula. A great deal more attention has been paid to the empirical analysis of possible determinants of academic performance since the work of Spector and Mazzeo (1980). In their seminal work, they employed a logit model to examine the performance of students in intermediate macroeconomics. The subsequent literature is extensive, including work on principles of economics courses by Becker (1983), Borg et al (1989), Park (1990), Watts and Bosshardt (1991), Becker and Watts (1996, 1999, and 2001). Papers on intermediate economics or econometrics are relatively scarce but include Specter and Mazzeo (1980), Ramonda et al (1990), Becker and Greene (2001), and Yang and Raehsler (2005).

While the literature in accounting is mostly limited to gender-related study (Mutchler et al, 1987; Lipe, 1989; Tyson, 1989; Ravenscroft and Buckless, 1992), other studies do exist that focus on income tax

---

[1] Department of Economics, Clarion University of Pennsylvania, Clarion, PA, 16214, (814) 393-2627 (office), rraehsler@clarion.edu.
[2] A. R. Sanchez Jr. School of Business, Texas A&M International University.
[3] Department of Economics, Clarion University of Pennsylvania and National Chung Cheng University, Taiwan.
[4] Department of Finance, Clarion University of Pennsylvania.

439

courses, CPA exams or other related accounting topics (Murphy and Stanga, 1994; Grave et al, 1993; Brahmasrene and Whitten, 2001). In the literature of finance, Berry and Farragher (1987), among others, were the first to survey introductory financial management courses. Since then there have been papers on introductory finance courses (Ely and Hittle, 1990; Paulsen and Gentry, 1995; Chan et al, 1996; Cooley and Heck, 1996; Sen et al, 1997; Chan et al, 1997; Nofsinger and Petry, 1999; VanNess et al, 1999) but no use of more advanced specifications modeling academic performance in any of the existing educational literature. For example, VanNess et al, (1999) employed ordinary least squares and ordered probit models to discover that part-time instructors typically assigned higher grades, cumulative grade point average was a significant explanatory variable for the final grade, and students majoring in either economics or finance had higher average academic performance in the class. In a similar fashion, Sen el al.(1997) identified grade point average, gender, and performance in prerequisite courses as important indicators of final grades in introductory finance. In their work, females tended to outperform males in the sample while grade point average and prerequisite course grades had a positive influence on performance in the finance course. Interestingly, Didia and Hasnat (1998) discovered that female students did not display a grade disadvantage and transfer students may not fare worse than traditional students. A few studies on higher level or graduate finance courses were completed by Rubash (1994), Mark (1998), and Trine and Schellenger, (1999). Trine and Schellenger, (1995) find identify six factors as significant at either ten percent or two percent levels with a coefficient of determination measure of 0.15 in a stepwise multiple regression analysis. Additional work to identify factors in greater detail and with greater precision has not been accomplished in the financial education literature.

Our paper concentrates on the effect of mathematics and economics taken as prerequisites upon the performance in the introductory financial management course (FIN 370), a required course for all business majors at Clarion University. In particular, we are interested in examining a well-known phenomenon of mean reversion in academia (the idea that a higher letter grade in principles of microeconomics leads to a lower grade in FIN 370). Or does an empirical analysis identify "persistence" meaning that a high grade in principles of microeconomics results in a high grade in FIN 370? In addition, we explore the impact of a remedial math course on the letter grades in FIN 370. This is important in determining if remedial math contributes positively to the performance in FIN 370. If so, it is an appropriate prerequisite and confirms the importance of a strong mathematics background in learning introductory concepts in finance. If not, then, in this era of tight budgets and fewer faculty, such remedial coursework could be eliminated. It is important to note at this stage that while the linkage between finance, economics, and mathematics seems inherently obvious, this type of linkage is not always confirmed when actual data is analyzed. Raehsler, Johns, Yang, and Hung (2011), for example, showed that student performance in economics and mathematics courses required for operations management did not provide a statistically strong linkage with final grades. Given the highly mathematical nature of operations management one would anticipate that performance in mathematics, in particular, would be an important variable in determining final grades in the course. As such, it is important to empirically test the relationship of grades in economics and mathematics and how those link to grades in managerial finance rather than assume that the relationship is entirely obvious or unworthy of detailed analysis.

The organization of this paper is as follows: Section II introduces an ordered probit model; Section III presents estimated results of the ordered probit model; and Section IV examines marginal probabilities from changes in mathematics scores and the dummy variables. The conclusion is presented in Section V.

Data Description and the Ordered Probit Model

In this study, students enrolled in managerial finance (FIN 370) from the fall semester of 1999 through the summer session of 2011. As discussed above, FIN 370 is a required core course for all business students in the college and prerequisites consist of two principles of economics courses (macroeconomics and microeconomics) along with an introductory mathematics course, two beginning accounting courses, and business statistics. Academic performance in the principles of microeconomics course is considered here as that course is primarily populated by business college students whereas the principles of macroeconomics course exhibits a student population spread across a wide variety of academic majors. The introductory mathematics courses (MATH 131 and MATH 232) expose students to basic concepts in matrix algebra, linear programming, and financial equations (such as net present value) along with slightly more advanced topics of basic business calculus (simple derivatives). The lengthy time period for this analysis provides a substantial sample of 1452 students enrolled in FIN 370. Variation in grading was minimal throughout this sample as the same instructor taught the course with identical grading methods each term. This and the large sample size helps in providing robust empirical results using the ordered probit model in this study.

It is well known in the econometric literature that when dealing with qualitative measures such as grades or success and failure, the standard ordinary least squares (OLS) regression technique can produce spurious probability estimates (probabilities that are negative or exceed unity) and negative variances (Greene, 2003). To overcome these shortcomings, a binary probit or logit model, which produces consistent probability estimates, provides a better explanation for two outcomes. When more than two outcomes exist, a model of multiple choices such as a multinomial logit or a similar probit model is often used. However, a multinomial logit or probit model suffers from the "independence from irrelevant alternatives" problem. In other words, odds ratios between outcomes i and j are to be independent of all other alternatives; an extremely restrictive condition placed on the model for most types of data analysis. As a result, we opt for the ordered probit model which takes into account the ordinal nature in the dependent variable (Zavoina and McElvey, 1975). The use of the ordered probit model is justified in the multiple-category case (final letter grades in this study), which is considered ordinal in nature. This phenomenon is particularly true in a state-supported university where "curving" the final letter grade and examinations is more common. As a consequence, the distance between various letter grades (the dependent variable Y) is not a fixed interval. In other words, the difference between an A and a B is not the same as that between a B and a C (and so forth). Very often, a greater degree of grading leniency is given to students performing at the lower end of the grading scale and educators find it more difficult to fail students for a wide variety of reasons (budget conditions at academic institutions, student complaints, and even the human nature of instructors). This greater leniency given to low performers compared to high performers in class makes the distribution of different letter grades generally ordinal in nature. Stated more succinctly, a final grade of A is better than a B, which is better than a C, and so on, however, the measured difference between an A and B and a B and C are not necessarily the same. This being the case, consider a latent regression equation in matrix form: (LEFT OFF HERE)

$$y^* = x'\beta + \varepsilon \quad \text{... (1)}$$

Or in linear form:

$$Y = \beta_0 + \beta_1\, GPA_i + \beta_2\, MATH_i + \beta_3\, MAJOR_i + \beta_4\, GENDER_i + \beta_5\, TERM_i +$$
$$\beta_6 D_1 + \beta_7 D_2 + \beta_8 D_3 + \beta_9 D_4 + \beta_{10} D_5 + \beta_{11} D_6 + \beta_{12}\, REM_i + \varepsilon_i \quad \text{... (2)}$$

Where y^* = unobserved latent variable of letter grades

$$y = 0 = D \text{ if } y^* \leq 0 \quad \text{... (3)}$$
$$y = 1 = C \text{ if } 0 < y^* \leq \mu_1 \quad \text{... (4)}$$
$$y = 2 = B \text{ if } \mu_1 < y^* \leq \mu_2 \quad \text{... (5)}$$
$$y = 3 = A \text{ if } y^* \geq \mu_2 \quad \text{.. (6)}$$

and where μ_1 and μ_2 are threshold values by which expected letter grades in FIN 370 are determined. Individual variable definitions are as follows:

GPA	=	grade point average on a 4.0 scale.
MATH	=	average score on two pre-requisite mathematics courses pre- calculus and calculus.
MAJOR	=	'1' for Management or Marketing majors and '0' for Accounting, Economics and Finance majors.
GENDER	=	'1' for male students and '0' for female students.

TERM	=	'1' for the Fall Semester 1999 and 2 for the Spring Semester 2000 and so on. It captures a trend in grading if any.
D_1	=	'1' denotes a student received a D or E in Principles of Microeconomics; $D_1 = 0$ implies he or she received a letter grade of other than a D. It is to be pointed out that the letter grade C is used as the reference group.
D_2	=	'1' denotes a student received a B in Principle of Microeconomics, '0' for other letter grades.
D_3	=	'1' denotes a student received an A in Principle of Microeconomics, '0' for other letter grades.
REM	=	'1' denotes a student was not required to take the remedial math course: '0' if the student was required to take remedial math.
D_4	=	'1' denotes a student received a D or E in Financial Accounting or ACTG 251, '0' for other letter grades.
D_5	=	'1' denotes a student received a B in Financial Accounting or ACTG 251, '0' for other letter grades.
D_6	=	'1' denotes a student received an A in Financial Accounting or ACTG 251, '0' for other letter grades.
ε_i	=	Normally distributed residual with a mean of '0' and variance of '1'.

Empirical Results

Via the use of TSP version 4.5 (2002), we obtain the estimates of the ordered probit model as shown in Table 1.

Table 1: Estimates of Ordered Probit Model

Variable	Coefficient	Standard Error	t ratio	p value
Constant	-1.324	0.228	-5.817*	0.000
GPA	0.636	0.075	8.462*	0.000
MATH	0.470	0.052	8.998*	0.000
MAJOR	-0.143	0.060	-2.400**	0.016
GENDER	0.146	0.060	2.431**	0.015
TERM	-0.010	0.004	-2.924*	0.003
D_1	-0.152	0.107	-1.412	0.158
D_2	0.414	0.071	5.860*	0.000
D_3	0.893	0.111	8.075*	0.000
D_4	-0.143	0.110	-1.296	0.195
D_5	0.110	0.071	1.546	0.122
D_6	0.453	0.095	4.782*	0.000
REM	0.083	0.043	1.931***	0.053
μ_1	1.490	0.058	25.669*	0.000
μ_2	2.871	0.074	38.663*	0.000

Dependent variable = letter grade for introductory finance
Number of observations = 1452
Likelihood ratio = 752.364 (p value = 0.000)
Log likelihood function = -1486.48
Scaled R-squared = 0.439
* = significant at 1%
** = significant at 5%
*** = significant at 10%

An examination of Table 1 indicates that the cumulative GPA is significant in explaining the probability of getting a specific letter grade in FIN 370 (a p-value of 0.00). This is not surprising since a mastery of FIN 370 requires a decent grasp of the material presented in the previously completed pre-requisite coursework and a strong academic background. Therefore, it is not surprising that the GPA as a measure of effort plays an important role in the determination of grades in FIN 370. In addition, the average score in the two pre-requisite mathematics courses plays a key role in grade determination (a p-value of 0.00). It appears that topics such as time value of money, risk management and securities valuation which are covered in FIN 370 are much easier to comprehend when the student has the appropriate preparation in algebra and elementary calculus.

It is interesting to note that students majoring in accounting, economics or finance have greater likelihood of obtaining better grades in FIN 370 with a significantly positive relationship (a p-value of 0.016). This may be attributed to the belief that marketing and management disciplines are not as quantitatively oriented as accounting, economics, and finance and, therefore, students majoring in the marketing or management disciplines might not perform as well in FIN 370 because they are self-selected into majors with less emphasis on quantitative methods. Male students (dummy variable GENDER = 1) tend to score better in FIN 370, and the relationship is statistically significant (a p-value of 0.015). This may confirm the conjecture that male students are better equipped with mathematic skills required for FIN370 or it may display a general difference in interest in finance between genders. The trend variable TERM is included to account for potential grade inflation or deflation over the long sample period and indicates that student average grades in FIN370 has been declining in the sample period (a p-value of 0.003).

The dummy variable D_1 (which indicates that the student received a grade of 'D' or 'E' in principles of microeconomics) is negatively related to the probability of obtaining a better letter grade (in FIN 370) as compared to the reference group of students who received a C in principles of microeconomics. Interestingly, this relationship was not found to be statistically significant (a p-value of 0.158). In contrast, a student was more likely to obtain a better letter grade in FIN 370 if he or she received a B in principles of microeconomics ($D_2 = 1$) and such a relationship was statistically significant (a p-value of 0.000). The same conclusion can be drawn for D_3 ($D_3 = 1$ indicates a student obtaining an A in principles of microeconomics) where the p-value was also 0.000. These findings suggest the attribute of persistence in principles of microeconomics and FIN 370 where persistence identifies that a typical student who does well in ECON 212 will have higher probability of doing well in FIN 370 and doing poorly in ECON 212 indicates a greater probability of doing poorly in FIN 370. This means that, if a student has received an A or B in ECON 212, that grade will not cause the student to reduce the effort needed to do well in FIN 370. Instead, the student is expected to get a good grade in FIN 370 as well. In contrast, even though the statistical relationship is not significant, the student who earned a D in ECON 212 was less likely to do well in FIN 370.

In a similar vein, a student who received an 'A' in ACTG 251 has a greater probability of earning a better letter grade than those who received a 'C' (a p-value of 0.000). The difference for students who earned a B or C is not significant (a p-value of 0.122) as is that between D and C (a p value of 0.195). It appears that the knowledge students obtain in ACTG 251, such as balance sheet and income statement is important in FIN 370 as this relates closely to information presented in capital budgeting. It does not appear that this relation is as significant as the one described in principles of microeconomics throughout the entire grade range.

REM is a dummy variable which indicates whether or not a student is required to take a remedial mathematics course. Remedial mathematics (or MATH 110) is a basic algebra course required of students who come in as freshman and test poorly in mathematics or do not take the mathematics placement test. Since some beginning freshman and transfer students do not take this placement examination despite having reasonable training in mathematics, it is important to include this indicator variable. Using a value of 1 if a student was not required to take remedial mathematics and of 0 if the student was required to , the estimated coefficient for REM is marginally significant (a p-value of 0.053). This reveals that students who are not required to take the remedial math course (REM = 1) have a better chance to do well in FIN 370 as is expected, however, not with the level of significance observed for economics and accounting course grades. Freshman business majors who score unsatisfactorily on the mathematics placement test are required to take the remedial course. Even if students who are required to take remedial math do well in that course, this reveals that those students generally still lack the mathematical competence to master the quantitative topics in FIN 370. This implies that either these students cannot learn higher level mathematics concepts or that the content of MATH 110 needs to be revamped in such a way that the students can actually apply the material they learn in MATH 110 to FIN 370 and other upper level courses which require mathematical sophistication.

Finally, the estimated coefficients of the two threshold variables (μ_1 and μ_2 or 4 categories minus 2) are all statistically significant indicating that the use of the four category ordered probit model is warranted. This type of empirical result is lacking in most papers utilizing the ordered probit specification but of extreme importance in defending resulting estimates from criticism of the model. The scaled r-squared (0.439) is a better measure of goodness of fit than the McFadden r-squared statistic for its consistency in interpretation (Estella, 1998). The value 0.439 is reasonably satisfactory for models of discrete choice which have an unsettled criterion of fit.

Applications of the Ordered Probit Model

For a representative student, average values of GPA, MATH, MAJOR, GENDER, TERM, D_1, D_2, D_3, D_4, D_5, D_6, REM are found to be 2.962, 1.909, 0.517, 0.511, 17.1, 0.092, 0.324, 0.144, 0.094, 0.357, 0.203, 1.093. Substituting these values into Equation (2) yields the expected value of $y = 1.741 = \beta' x$. Coupled with estimated threshold values $\mu_1 = 1.490$ and $\mu_2 = 2.871$, the value of probabilities for an average student to receive a particular letter grade in FIN 370 are shown below:

$$\text{Prob } [y = 0 \text{ or } D] = \varphi(-\beta' x) \quad \dots\dots\dots\dots\dots\dots\dots\dots\dots\dots\dots\dots \quad (7)$$

$$\text{Prob } [y = 1 \text{ or } C] = \varphi[\mu_1 - \beta' x] - \varphi(-\beta' x) \quad \dots\dots\dots\dots\dots\dots\dots\dots\dots \quad (8)$$

$$\text{Prob } [y = 2 \text{ or } B] = \varphi[\mu_2 - \beta' x] - \varphi(\mu_1 - \beta' x) \quad \dots\dots\dots\dots\dots\dots\dots\dots \quad (9)$$

$$\text{Prob } [y = 3 \text{ or } A] = 1 - \varphi(\mu_2 - \beta' x) \quad \dots\dots\dots\dots\dots\dots\dots\dots\dots \quad (10)$$

In each formula φ is the cumulative normal density function. For instance, the probability for a typical student to receive a D or C can be calculated as

$$p(y = 0, D \text{ or } E) = \varphi(-1.741) \approx 0.0409$$

and

$$p(y = 1 \text{ or } C) = \varphi(1.490 - 1.741) - \varphi(-1.741)$$
$$= 0.4013 - 0.0409 = 0.3604$$

The same procedure could be applied to the probability of getting an A or B based on equations (9) and (10). For clarity, we report four such probabilities in Table 2. Since the average score in the two mathematics courses plays a significant role in predicting letter grades in FIN 370, we calculate the marginal probabilities (Greene 2003) as shown below:

$$\partial \Pr ob[Y = 0 \text{ or } D] / \partial Math = -\phi(-\beta' x) * (\hat{\beta}_2) \dots\dots\dots\dots\dots\dots\dots \quad (11)$$
$$= -\phi(-1.741) * (0.47)$$
$$= -0.0878 * 0.47$$
$$= -0.04123$$

$$\partial \Pr ob[Y = 1 \text{ or } C] / \partial Math = [\phi(-\beta' x) - \phi(\mu_1 - \beta' x)] * (\hat{\beta}_2) \dots\dots\dots\dots \quad (12)$$
$$= (0.0878 - 0.3867) * 0.47$$
$$= -0.14035$$

$$\partial \Pr ob[Y = 2 \text{ or } B] / \partial Math = [\phi(\mu_1 - \beta' x) - \phi(\mu_2 - \beta' x)] * (\hat{\beta}_2) \dots\dots\dots (13)$$
$$= (0.3867 - 0.2107) * 0.47$$

$$= 0.08264$$

$$\partial \Pr ob[Y = 3 \; or \; A]/\partial \, Math = \phi \,(\mu_2 - \beta' x)] * (\beta_2) \dots \dots \dots \dots \dots \quad (14)$$

$$= 0.2107 * 0.47$$
$$= 0.098938$$

These probabilities show that students with higher grades in mathematics reduce their probability of receiving a D (-4.123%) or a C (-14.035%) and increase their probability of receiving a B (+8.264%) or an A (+9.8938%). Note that the sum of the marginal probabilities equals zero (see Table 2) and ϕ denotes the normal probability density function, which is required to calculate marginal probabilities (Greene, 2002).

Table 2: Probability and Marginal Probability of Getting Specific Letter Grades in FIN 370 for Typical Student

Letter Grade	Probability p(y)	$\partial \, p(y)/\partial$MATH
y = D	4.090%	-4.123%
y = C	36.040%	-14.035%
y = B	46.950%	8.264%
y = A	15.200%	9.894%

As is evident from Table 2, a great majority of students in FIN 370 received a B (46.95%) or C (36.04%). Students who failed the class would have to repeat the course and pass it or drop out of the program and, therefore, do not show up in this sample. It is interesting to note that some students repeated FIN 370 to get a better grade even if they had passed the course the first time with a C or D. The importance of the average score in the two prerequisite mathematics courses cannot be overstated for these grades distinctly show how students improve their chances of getting an A or B while simultaneously reducing their chances of receiving a C or D in FIN 370. The partial derivatives of the cumulative normal density function are valid for a continuous variable (for example, MATH or GPA in this analysis). In evaluating impacts from a dummy variable, however, we need to substitute the value of the dummy variable (zero and one) into (7), (8), (9), and (10) and calculate the probabilities. Since the procedure is one of the least understood of the econometric models involving discrete variables, we report the results for the dummy variable D_3 (those who earned an A in ECON 212 versus those who received a C) and REM (those who took the remedial mathematics course against those who did not). The results are reported in Tables 3 and 4 below.

As can be inferred from results displayed in Table 3, an average student who received a letter grade of A in principles of microeconomics had a 25.19% greater chance of earning an A in the introductory finance course and a 4.41% greater probability of earning a B in FIN 370. Likewise, a student earning an A in principles of microeconomics would be 24.84% less likely to earn a C and 4.76% less likely to receive a D in FIN 370. Notice that the sum of the marginal probabilities equals zero so that the sum of probabilities of getting specific letter grades in FIN 370 is one. Similarly, an average student who need not take the remedial math course, MATH 110, has 1.8% and 1.32% more chance of getting an A or B respectively while reducing his or her chance to get a C or D by 2.35% and 0.77% respectively in FIN 370. These are small percentages indicating the remedial mathematics course is not as important (albeit with a p-value of 5.3%) as obtaining high grades in principles of microeconomics. The magnitudes are relatively small, but the effects do exist.

TABLE 3: **Impacts of D_3 on the Probability of Getting Specific Grades in FIN 370**

Value	$D_3 = 0$	$D_3 = 1$	Change
$-\beta' x$	-1.612	-2.505	
$\mu_1 - \beta' x$	-0.122	-1.015	
$\mu_2 - \beta' x$	1.259	0.366	
Equation (7) P[y=0 or D]	φ (-1.612) = 0.058	φ (-2.505) = 0.006	-4.76%
Equation (8) P[y=1 or C]	φ (-0.122)- φ (-1.612) = 0.399	φ (-1.015)- φ (-2.505) = 0.150	-24.84%
Equation (9) P[y=2 or B]	φ (1.259)- φ (-0.122) = 0.444	φ (-0.366)- φ (1.015) = 0.488	4.41%
Equation (10) P[y=3 or A]	1 - φ (1.259) = 0.104	1 - φ (-0.366) = 0.356	25.19%

$D_3 = 1$ means received an A in principles of microeconomics. Rounding errors to the third decimal place occurred since we used the cumulative normal table which contains z-values with two decimal places.

Table 4: Impacts of Remedial Math on the Probability of Getting Specific Grades in FIN 370

Value	Remedial = 0	Remedial = 1	Change
$-\beta' x$	-1.650	-1.733	
$\mu_1 - \beta' x$	-0.160	-0.243	
$\mu_2 - \beta' x$	1.221	1.138	
Equation (7) P[y=0 or D]	0.050	0.042	-0.77%
Equation (8) P[y=1 or C]	0.387	0.363	-2.35%
Equation (9) P[y=2 or B]	0.452	0.466	1.32%
Equation (10) P[y=3 or A]	0.111	0.129	1.80%

Remedial = 1 means student was not required to enroll in remedial mathematics. Rounding errors to the third decimal place occurred since we used the cumulative normal table which contains z values with two decimal places.

Performance in the introductory accounting course (ACTG 251) plays an important role on letter grades of FIN370 as well. Knowledge of balance sheets and income statements is necessary to understand capital budgeting and other financial ratios. Receiving an A in ACTG 251 (or D_6=1) has a noticeable effect in the regression (p- value= 0.00). Again, we calculate the relative impacts of a student who received an A in ACTG 251 on letter grades he or she will receive in FIN370. These results are reported in Table 5 and a more detailed examination indicates that a student with "A" in ACTG 251 has 10.94% or 6.01% greater chance of earning an A or B respectively in FIN370 while the same student has a 13.79% and 3.16% smaller chance of receiving a C or D in the finance course.

Table 5: Impacts of Financial Accounting or D_6 on the Probability of Getting Specific Grades in FIN 370

Value	$D_6 = 0$	$D_6 = 1$	Change
$-\beta' x$	-1.649	-2.102	
$\mu_1 - \beta' x$	-0.158	-0.612	
$\mu_2 - \beta' x$	1.222	0.769	
Equation (7) P[y=0 or D]	0.050	0.018	-3.16%
Equation (8) P[y=1 or C]	0.391	0.253	-13.79%
Equation (9) P[y=2 or B]	0.448	0.509	6.01%
Equation (10) P[y=3 or A]	0.111	0.221	10.94%

D_6 = 1 signifies that student received an A in ACTG 251.

Conclusion

This paper focuses on the impact of mathematics, economics and accounting preparation on the grades earned in junior level introductory finance by the typical business student at a rural state university in the eastern United States. The ordered probit model of a large sample (n=1452) indicates that the average score in the two mathematics prerequisite courses is a significant predictor of performance in FIN 370. Grade point average, an effort variable, is found to be significant as is the student's academic major; an accounting, economics or finance major has a better chance to perform better than, on average, students with other majors in the college of business. Gender is found to be an important explanatory variable as male students are found to outperform female students in FIN 370, however, overall average performance in introductory finance has deteriorated over time.

An important finding central to this analysis is that a student with an A or B in principles of microeconomics has greater probability of getting a better grade in FIN 370 and this relationship is statistically significant. The implication is that there appears to be a phenomenon of "persistence" in student performance: those who performed well in prerequisite principles courses will continue to do well in FIN 370. In addition, students who took the remedial mathematics course did not perform as well in FIN

370 compared to those who did not take the course, however, the magnitudes of this relationship are relatively small when compared to the prerequisite courses for FIN 370. This may suggest a need for more coordination regarding topical coverage between the mathematics and finance departments or it could mean that remedial mathematics students lack the ability to grasp the necessary mathematical concepts. Alternatively, it may mean that at least some of these students are less motivated to work hard in math courses. Determining which of these is the case will make for interesting future research. Finally, a student with an 'A' in ACTG 251 has much better chance to earn an 'A' or 'B' and less likely to receive a 'C' or 'D' in FIN370.

The importance of mathematics, economics, and accounting to the expected final grade for students in managerial finance verifies the strong theoretical and quantitative linkage between these disciplines. This overall result has direct implications as to how academic achievement in finance can be improved by promoting a strong business curriculum that encourages a strong foundation in quantitative prerequisites. Future work will focus on differences among genders with regard to academic achievement in introductory finance in addition to the relative decline in academic performance observed over the time period of this study. Gender differences in performance might be linked to variable abilities in mathematics while declining achievement over time could reflect a change in overall academic preparedness among incoming students.

References

Becker, E. William, Jr. 1983. "Economic education research: Part III, statistical estimation methods," *Journal of Economic Education* 14 (3): 4-15.

Berry, T. D. and E. J. Farragher. 1987. "A survey of introductory financial management courses," *Journal of Financial Education* 13(2): 65-72.

Borg, O. Mary, Paul M. Mason, and Stephen L. Shapiro. 1989. "The case of effort variables in student performance," *Journal of Economic Education* 20(3): 308-313.

Chan, Kam C., Connie Shum, and Pikki Lai. 1996. "An Empirical Study of Cooperative Instructional Environment on Student Achievement in Principles of Finance," *Journal of Financial Education* 25(2): 21-28.

Chan, Kam C., Connie Shum, and David J. Wright. 1997. "Class Attendance and Student Performance in Principles of Finance," *Financial Practice and Education* 7(2): 58-65.

Cooley, Philip, and Jean Heck. 1996., "Establishing Benchmark for Teaching the Undergraduate Introductory Course in Financial Management," *Journal of Financial Education* 22 (Fall): 1-10.

Didia, Dal and Baban Hasnat. 1998. "The Determinants of Performance in the University Introductory Finance Course," *Financial Practice and Education* 8(1): 102-107.

Ely, David P. and Linda Hittle. 1990. "The Impact of Math Background on Performance in Managerial Economics and Basic Finance Courses," *Journal of Financial Education* 19(2): 59-61.

Estrella, A. 1998. "A New Measure of Fit for Equations with Dichotomous Dependent Variables," *Journal of Business and Economics Statistics* (April): 198-205.

Graves, O. Finley, Irva Tom Nelson, and Dan S. Dcines. 1993. "Accounting Student Characteristics: Results of the 1992 Federation of Schools in Accountancy (FSA) Survey," *Journal of Accounting Education* 11(21): 221-225.

Greene, W. H. 2003. *Econometric Analysis*, 5th Ed., Upper Saddle River, NJ: Prentice Hall.

Leiber, J. Michael, B. Keith Crew, Mary Ellen Wacker, and Mahesh K. Nalla. 1993. "A Comparison of Transfer and Nontransfer Students Majoring in Criminology and Criminal Justice," *Journal of Criminal Justice Education* 4(1): 133-151.

Liesz, Thomas J. and Mario G. C. Reyes. 1989. "The use of "Piagetian Concepts to Enhance Student Performance in the Introductory Finance Course," *Journal of Financial Education* 18(2): 8-14.

Lipe, Marlys G. 1989. "Further Evidence on the Performance of Female Versus Male Accounting Students," *Issues in Accounting Education* 4(1): 144-152.

Marks, Barry. 1998. "An Examination of the Effectiveness of a Computerized Learning Aid in the Introductory Graduate Finance Course," *Financial Practice and Education* 8(1): 127-132.

Murphy, P. Daniel and Keith G. Stanga. 1994. "The Effects of Frequent Testing in an Income Tax Course: An Experiment," *Journal of Accounting Education* 12(1): 27-41.

Mutchler, Jane F., Joanne H. Turner and David D. Williams. 1987. "The Performance of Female Versus Male Accounting Students," *Issues in Accounting Education* 2(1): 103-111.

Nofsinger, John and Glenn Petry. 1999. "Student Study Behavior and Performance in Principles of Finance," *Journal of Financial Education* 25(Spring): 33-41.

Park, Kang H. and Peter M. Kerr. 1990. "Determinants of Academic Performance: A Multinomial Logit Approach, *Journal of Economic Education* 21(2): 101-111.

Paulsen, Michael B. and James A. Gentry. 1995. "Motivation, Learning Strategies, and Academic Performance: A Study of the College Finance Classroom," *Financial Practice and Education* 5(1): 78-89.

Raehsler, Rod D., Tony R. Johns, Chin W. Yang, and Ken Hung. 2011. "Academic Preparation, Gender, and Student Performance in Operations Management: An Ordered Probit Analysis, *Clarion University Working Paper*.

Raimondo, Henry J., Louis Esposito, and Irving Gershenberg. 1990. "Introductory Class Size and Student Performance in Intermediate Theory Courses," *The Journal of Economic Education* 21(4): 369-381.

Ravenscroft, Susan P. and Frank A. Buckless. 1992. "The Effect of Grading Policies and Student Gender on Academic Performance," *Journal of Accounting Education* 10(1): 163-179.

Rubash, Arlyn R. 1994. "International Finance in an International Environment," *Financial Practice and Education* 4(1): 95-99.

Sen, Swapan, Patrick Joyce, Kathy Farrell, and Jody Toutant. 1997. "Performance in Principles of Finance Courses by Students with Different Specializations," *Finanical Practice and Education* 7(2): 66-73.

Spector, C. Lee and Michael Mazzeo. 1980. "Probit Analysis and Economic Education," *Journal of Economic Education* 11(2): 37-44.

Trine, A. D. and M. H. Schellenger. 1999. "Determinants of Student Performance in an Upper Level Corporate Finance Course," *Academy of Educational Leadership Journal*, 3(2): 42-47.

Tyson, Thomas. 1989. "Grade Performance in Introductory Accounting Courses: Why Female Students Outperform Males," *Issues in Accounting Education* 4(1): 153-160.

VanNess. β . F., VanNess R.S. and R. Kamery. 1999. "The effect of part-time instruction on grades in principles of Finance." *Financial Practice and Education* (Fall/Winter): 105-110

Watts, Michael and William Bossbardt. 1991. "How Instructors Make a Difference: Panel Data Estimates from Principles of Economic Courses," *Review of Economics and Statistics* 73(2): 336-340.

Yang, C. W. and R. D. Raehsler. 2005. "An Economic Analysis on Intermediate Microeconomics: An Ordered Probit Model," *Journal of Economics Educators*, 5(3):1-11.

Zavoina R., and W. McElvey. 1975. "A Statistical Model for the Analysis of Ordinal Levl Dependent Variables," *Journal of Mathematical Sociology* 4(1): 103-120.

About the Book

This book consists of 30 applied econometrics papers authored mainly by Professors Bwo-Nung Huang and/or Chin-Wei Yang or others. The 30 papers are grouped into 4 categories according to different modeling structures: (1) the Granger causality models, (2) Granger causality with panel data, (3) Granger causality with threshold variable and (4) indirect Granger causality, random walk, long term memory and volatility models. These econometric models are then applied to financial, energy markets and world economy in order to capture potential causal relationships. The book can be an excellent text for those who have an interest in applied research in graduate program of Economics or Finance. Dr. Bwo-Nung Huang would like to take this opportunity to thank his dissertation mentor Professor C.K. Liew at University of Oklahoma, Professor C.W.J. Granger at University of California San Diego for his warm encouragement and valuable research guidance, and Professor Ming-jeng Hwang at West Virginia University for the assistance during his Sabbatical Leave year. Professor Chin-Wei Yang would like to thank Professor T. Witt at West Virginia University for two good econometric courses and Dean Hung Jia Jun of National Chung Cheng University for offering him a visiting professor position. The lion's share of this book is on the famed Granger causality models, a major breakthrough in methodology of modern science.

Autobiography of Professor Bwo-Nung Huang

Professor Bwo-Nung Huang (MBA in 1985 and Ph.D. in Economics both from University of Oklahoma in 1989) started his academic career at National Chung Cheng University in Taiwan right after receiving his doctoral degree. Currently he is the Distinguished Professor in Economics of the University. Being a prolific writer and a competent teacher , Professor Huang was granted a position as visiting scholar at University of California at San Diego (07/1997-08/1998) where he had the honor of acquainting Professor C.W.J. Granger (2003 Nobel Laureate in Economics) with the famed causality test. The paper regarding the Asian financial crisis was published (C.W.J. Granger, Bwo-Nung Huang and Chin-Wei Yang) in 2000, which illustrated the Granger causality model predicts outcomes accurately even under tumultuous environments . The paper was visited or downloaded close to 5000 times speaking to its importance in the realm of inductive methodology. In addition, Professor Huang was a visiting scholar during 08/2006~06/2007 at West Virginia University.

During his academic career, he served as Chairperson of Economics Department (08/1998~07/2001), Interim Dean of School of Management (08/1998~01/1999) both at National Chung Cheng University in Taiwan, Dean of School of Management at Providence University in Taiwan (08/2003~07/2004) , Dean of School of Management at National Chia-Yi University (08/2004~07/2006). Beyond that, he served as Dean of Academic Affair (02/2008~07/2009) and Vice President (11/2012~07/2013) both at National Chung Cheng University . Furthermore Dr. Huang is one of the founding members and an evaluator at Higher Education Evaluation & Accreditation Council of Taiwan (HEEACT) for many years with abundant administrative experience .

Professor Huang's research focuses on employing advanced econometrics and its applications to financial markets and macroeconomics including futures market and characteristics of the Taiwan Stock Market. Among other things, he has made important contributions in economic behavior of energy markets, especially in the world oil market: relationship between oil consumption and economic growth. In addition, his papers using sophisticated econometric techniques shed important light on the causal relationship between defense spending and economic growth: arms races for developing countries do not boost economic growth whereas defense spending by developed countries is found to be conducive to economic growth.

Autobiography of Professor Chin-Wei Yang

Professor Yang received his doctoral degree in Economics from West Virginia University in 1979. He served as a post-doctoral research fellow at College of Mineral and Energy Resources of West Virginia University during the 1979-1980 academic year. He was a visiting assistant professor at Indiana University-Purdue University at Fort Wayne for one year. In August 1982, he accepted the position as an assistant professor of Economics at Clarion University of Pennsylvania until his retirement as a full professor in June of 2013. During his 33 years in higher education, he was a visiting professor at National Chung Cheng University of Taiwan in 1992-1993 and 2009-2010 academic years. In May of 2004, Professor Yang was invited to give a talk at Symposium on Elasticity sponsored by Ministry of Education of Taiwan.

As a prolific writer, Professor Yang published numerous papers in different fields in peer-reviewed journals including Operations Research, Journal of Regulatory Economics, Public Finance Quarterly, Energy Journal, Southern Economic Journal, Atlantic Economic Journal, Eastern Economic Journal. He also solved problems in The College Mathematic Journal and Mathematic Magazine. He has the privilege of co-authoring a high-impact paper with Professors C.W.J. Granger (2003 Nobel Prize winner) and B.N. Huang (National Chung Cheng University) in Quarterly Review of Economics and Statistics (Vol.14, pp.337-354, 2000). He also coauthored papers with Professor B.N. Huang in various peer-reviewed journals such as Journal of Comparative Economics, Energy Economics, Peace and Defense Economics, Applied Economics, Agricultural Economics, Manchester School, Economic Modeling, Journal of Competitive Law and Economics, Ecological Economics and Journal of Asian Economics.

Professor Yang is an avid researcher in history and literature (Dream of Red Chambers). He was a columnist on Finance at World Journal Weekly from which he published a book "Money Management from Professor Yang". He published other papers regarding history, literature and commentaries in USA Today, Pittsburgh Post Gazette, China Post, Taipei Review, World Journal, Chiao Bao, Chinese Times, Xaoxing County News Paper and Ke Qiao Daily News.

黃柏農教授自傳

　　黃柏農教授於 1985 年取得美國 University of Oklahoma MBA 學位，1989 於同校獲得了 Ph.D. in Economics 學位，畢業後即回台灣國立中正大學經濟系暨國際經濟研究所服務迄今，目前為該系的特聘教授 (Distinguished Professor)。黃教授曾於 07/1997 ~ 08/1998 赴美國 UCSD 經濟系訪問一年，期間有幸與諾貝爾經濟學獎得主 Professor C. W. J. Granger 共同合作發表文章。黃教授亦曾於 08/2006~06/2007 赴美國西維吉尼亞大學經濟系訪問一年。

　　黃教授在服務學術界的近 30 年當中曾經擔任過國立中正大學經濟系系主任 (08/1998~07/2001)，管理學院代理院長 (08/1998~01/1999)，靜宜大學 (Providence University) 管理學院院長 (08/2003~07/2004)，國立嘉義大學管理學院院長 (08/2004~07/2006)。他在 02/2008~07/2009 期間擔任國立中正大學教務長 (Dean of Academic Affairs)，並於 11/2012~07/2013 兼代國立中正大學副校長，學術行政經驗豐富。此外，黃教授亦擔任財團法人高等教育評鑑中心基金會 (Higher Education Evaluation & Accreditation Council of Taiwan, HEEACT) 系所評鑑及校務行政評鑑委員及召集人多年，學術行政評鑑經驗良多。

　　黃教授學術研究專長在使用計量經濟方法來探討財務變數與總體經濟之間的關係；經濟成長、原油消費與總體經濟變數的因果檢定；經濟成長與國防支出的關係。他在這些專題上做出了傑出的貢獻，尤其是闡明了：發展中國家的大量國防支出不利於經濟成長，但對已發展國家而言，國防支出有利經濟成長。

楊慶偉教授自傳

楊慶偉教授在 1979 年取得西維琴尼亞大學(West Virginia University)經濟學博士學位；並於 08/1979-05/1980 期間,在該校的礦產能源學院擔任博士後研究員.隨後的一年他在印地安那-普渡大學韋恩堡分校(Indiana University-Purdue University at Fort Wayne)擔任訪問助理教授(August 1980~May 1981).從 1982 年 8 月開始,他被賓州州立克萊恩大學（Clarion University of Pennsylvania）聘請為助理教授,直到 2013 年 5 月以正教授退休為止,他在高等教育界服務了三十多年.在這三十多年期間,楊慶偉教授曾經兩度回到國立中正大學客座講學（September 1992~July 1993; September 2009~June 2010).他在 2004 年 5 月被邀請返台灣做由教育部與靜宜大學主辦的彈性理論專題演講.

楊教授著作等身,在不同的領域裡發表了上百篇的學術論文,包括了運籌學(Operations Research),規範經濟學期刊(Journal of Regulatory Economics),財政學季刊(Public Finance Quarterly),能源期刊(Energy Journal),南方經濟學期刊(Southern Economic Journal),東方經濟學期刊(Eastern Economics Journal),大西洋經濟學期刊(Atlantic Economics Journal).他也曾在大學數學期刊(The College Mathematics Journal)和數學雜誌(Mathematics Magazine)成功地提出了証明題的解答.他有幸與 2003 年諾貝爾經濟學獎得主之一的格蘭傑爾(C.W.J. Granger)和黃柏農教授合作,在經濟統計季刊(Quarterly Review of Economics and Statistics Vol.40, pp. 337-354, 2000)上,發表了一篇影響因子很大的文章.此外他與黃柏農教授合著的文章發表在能源經濟學(Energy Economics),比較經濟學(Journal of Comparative Economics),和平與國防(Peace and Defense Economics),應用經濟學期刊(Applied Economics),曼切斯特學院(The Manchester School),農業經濟學(Agricultural Economics),經濟模式(Economic Modelling),競爭法律與經濟學(Journal of Competitive Law and Economics),生態經濟學(Ecological Economics)和亞洲經濟學(Journal of Asian Economics).

楊慶偉教授酷愛歷史研究及「紅樓夢」,曾經是「世界週刊」的專欄作家,著有《楊教授談理財》一書.他的非學術性文章刊登在「世界日報」,「中國時報」,「僑報」,「紹興縣報」,「柯橋日報」,「今日美國（USA Today）」,「匹資堡郵報(Pittsburgh Post Gazette)」「中國郵報（ China Post)」,「台北評論（Taipei Review）」等報紙。

Unit Root, Cointegration, Granger-Causality, Threshold Regression and Other Econometric Modeling with Economic and Financial Data

作　者／黃柏農、楊慶偉（Bwo-Nung Huang & Chin-Wei Yang）

出版者／美商 EHGBooks 微出版公司

發行者／美商漢世紀數位文化公司

臺灣學人出版網：http://www.TaiwanFellowship.org

印　　刷／漢世紀古騰堡®數位出版 POD 雲端科技

出版日期／2018 年 10 月

總經銷／Amazon.com

臺灣銷售網／三民網路書店：http://www.sanmin.com.tw

　　　　　三民書局復北店

　　　　　地址/104 臺北市復興北路 386 號

　　　　　電話/02-2500-6600

　　　　　三民書局重南店

　　　　　地址/100 臺北市重慶南路一段 61 號

　　　　　電話/02-2361-7511

　　　　　　全省金石網路書店：http://www.kingstone.com.tw

中國總代理／廈門外圖集團有限公司

地　　　址／廈門市湖里區悅華路 8 號外圖物流大廈 4 樓

臺灣書店購書專線／0592-5061658、6028707

定　　價／新臺幣 1380 元（美金 46 元／人民幣 322 元）